International Conference on Power Electronics and Energy Engineering 2015

(PEEE 2015)

Advances in Engineering Research Volume 20

Hong Kong, China
19-20 April 2015

Editor:

A. Leung

ISBN: 978-1-5108-0764-8

Printed from e-media with permission by:

Curran Associates, Inc.
57 Morehouse Lane
Red Hook, NY 12571

Some format issues inherent in the e-media version may also appear in this print version.

Copyright© (2015) by Atlantis Press
All rights reserved.
http://www.atlantis-press.com/php/pub.php?publication=PEEE-15

Printed by Curran Associates, Inc. (2015)

For permission requests, please contact Atlantis Press
at the address below.

Atlantis Press
29 avenue Laumière
75019 Paris
France

contact@atlantispress.com

Additional copies of this publication are available from:

Curran Associates, Inc.
57 Morehouse Lane
Red Hook, NY 12571 USA
Phone: 845-758-0400
Fax: 845-758-2634
Email: curran@proceedings.com
Web: www.proceedings.com

TABLE OF CONTENTS

POWER ELECTRONICS TECHNOLOGY AND ENERGY ENGINEERING

SELF-CYCLING MODULARIZED SMART GRID APPLICATION INTEGRATION 1
Y. P. Li, S. Liang, W. Xu, J. B. Xin

DROOP CONTROL OF PARALLEL DUAL-MODE INVERTERS USED IN MICRO GRID 3
C. X. Wen, Z. Y. Liu, Z. X. Li

RESEARCH OF LOSS REDUCTION SCHEME FOR REGIONAL DISTRIBUTION GRIDS
BASED ON G1-GRA 7
W. Zhang, Y. M. Liu, W. Q. Ma

THE ANALYSIS ON THE INFLUENCE FACTORS AND TENDENCY OF POWER
REGULATION IN CHINA 12
Q. Guo, T. Liu, G. W. Gao, Y. Xu

STABILITY CONSTRAINTS OF THE TRANSMISSION CAPACITY ABOUT VSC-HVDC
SYSTEM SUPPLY TO THE PASSIVE NETWORK 14
S. Chen, X. Tang, W. Q. Zhang

DEVELOPMENT OF VLF TEST SYSTEM FOR POWER CABLES 19
C. X. Huang, C. B. Yang, H. J. Sun, H. J. Li, F. Huang, J. M. Guo

SELECTION OF DECOUPLING CAPACITORS FOR POWER DELIVERY NETWORKS WITH
MULTIPLE POWER PORTS 23
Q. D. Wang, Y. C. Wang, X. L. Li, Q. S. Liu

INTELLIGENT MANAGEMENT SYSTEM OF VEHICLE VIOLATION BASED ON
ELECTRONIC LICENSE 27
Y. Z. Cai, Y. R. Li, Z. Y. Cai

RESEARCH OF HTS APPLICATION ON TOKAMAK MAGNETS 30
L. Ren, X. Z. Deng, Y. Xu, H. Liu, J. D. Li, J. Shi, Y. J. Tang

ANALYSIS OF LOSSES IN HIGH SPEED SLOTLESS PM SYNCHRONOUS MOTOR
INTEGRATED THE ADDED LEAKAGE INDUCTANCE 36
B. Q. Kou, H. C. Cao, W. L. Li

THE APPLICATION OF ELECTROCHEMICAL IMPEDANCE IN THE ENVIRONMENTALLY
FRIENDLY SCALE AND CORROSION INHIBITOR POLYEPOXYSUCCINIC ACID 41
H. Y. Tian, B. R. Zhang, F. T. Li

MICRO-PHASE SEPARATION STRUCTURE OF NONIONIC POLYETHER-POLYESTER
POLYURETHANE BASED ON MDI 48
H. Quan, L. Wan, G. F. Wan

A NOVEL HIGH WATER-RESISTANT AQUEOUS ACRYLATE EMULSION, PART 1:
PREPARATION AND CHARACTERIZATION 51
B. K. Ren, K. T. Hu, G. H. Li, A. B. Wu, J. J. Li

TIME-RESOLVED FLUORESCENCE ANISOTROPY STUDY ON INTRAMOLECULAR
INTERACTIONS OF BRANCHED STYRYL DERIVATIVES BASED ON 1,3,5-TRIAZINE 55
Y. C. Wang, G. Q. Wang, D. J. Liu, B. Li

THE DESIGN OF ELECTRODE OF COAL ROADWAY ADVANCED DETECTOR 58
Y. Zhou, Y. M. Lv, J. T. Zhang, Z. M. Liu, X. G. Liu, M. Wu

RESEARCH OF WIND POWER PREDICTION BASED ON THE AUTO-REGRESSIVE MODEL 61
L. Feng, C. H. Liang, H. Huang

MECHANICAL ENGINEERING AND MANUFACTURE

RESEARCH ON FUZZY RELIABILITY PREDICTION METHOD OF CNC MACHINE TOOLS 65
G. B. Zhang, D. M. Luo, Y. Ran, W. Yu

FAILURE ANALYSIS AND DIAGNOSIS OF SPARK EROSION MACHINE TOOLS 69
H. Meng, N. Xin, H. Guo, L. Zhang

DESIGN AND DEVELOPMENT OF MATERIAL MANAGEMENT SYSTEM BASED ON
DIGITAL PULP PIPELINE 71
H. W. Liu, Z. L. Yu, J. D. Wu

TRANSIENT ELASTOHYDRODYNAMIC LUBRICATION ANALYSIS OF SPUR GEAR RUNNING-IN CONSIDERING EFFECTS OF SOLID PARTICLES 74
X. B. Huang, Y. Q. Wang, Q. Liu, N. Dong

ELECTRIC POWER STEERING SYSTEM FOR AGRICULTURAL TRUCK MATCHING DESIGN 79
Z. G. Chen, W. Zhang, Q. Zhang, T. M. Zhou

RESEARCH ON PRECISION FORGING BLADE TENON PROCESSING TECHNOLOGY 83
L. J. Huang, M. Kou, X. T. Tian, Z. Y. Wang

HIGH PRECISION WORK STATION RESEARCH 86
X. M. Cheng, Y. Q. Chen, P. N. Zhang, H. C. Fan, Z. Zhao, C. Z. Li

MATCHING RESEARCH BETWEEN ENGINE AND TRANSMISSION OF VEHICLE BASED ON AVL-CRUISE 89
F. Du, Z. W. Guan, C. H. Liu, Y. S. Wan

A KIND OF RESEARCH ON REGENERATIVE BRAKING ALGORITHM OF HYBRID ELECTRIC VEHICLE 93
Z. L. Liao, Q. Gao, P. Fan, R. F. Hu

STUDY ON MODELING METHOD OF THE PRECISION MACHINED SURFACE GEOMETRY FORM ERROR BASED ON BI-CUBIC B-SPLINE 97
Z. Q. Zhang, X. Jin, Z. J. Zhang

THE SUPPRESSION OF NARROW BAND JAMMERS ALGORITHM AND THE IMPLEMENTATION OF BEIDOU B3 SIGNAL 101
C. M. Li, L. Huang, X. J. Wang

NUMERICAL STUDIES OF ABRASION WEAR ON THE GUIDE VANES IN A SUBMERSIBLE AXIAL FLOW PUMP 106
H. M. Zhang, L. X. Zhang

THE DESIGN OF RICE DIRECT SEEDING MACHINE 109
S. F. Ding, Q. X. Li, Q. Huang, S. F. Ding

DESIGNING A REFRIGERATION AND HEAT-REMOVAL SYSTEM FOR RAPID DETECTOR BASED ON FREEZING POINT 112
X. T. Yu, J. Leng, X. B. Guo

INVESTIGATION OF NEGATIVE INFLUENCES ON RIDE COMFORT PERFORMANCE OF IN-WHEEL MOTOR VEHICLES WITH HIGH UNSPRUNG MASS 115
T. Z. Shi, D. F. Wang, S. M. Chen

ELECTRICAL POWER MANAGEMENT FOR DISTRIBUTED AUTOMOTIVE SYSTEM 118
X. F. Zhang, M. H. Luo, Y. Shen, J. D. Cao

NOVEL 0.1 HZ EXPONENTIAL WAVE GENERATOR BASED ON SEMICONDUCTOR SWITCH 122
Z. Hou, C. B. Yang, H. J. Li, S. C. Ji, J. Li, X. Chen

DYNAMIC ANALYSIS OF AERIAL WORK PLATFORM WORKING DEVICE BASED ON VIRTUAL PROTOTYPE 126
J. Q. Guo, D. W. Liu, J. G. Jia

TFT SUBSTRATE GLASS GEOMETRICAL PARAMETER MEASUREMENT AND DATA PROCESSING 129
H. Guo

ELECTRICAL RETAIL TARIFF MODEL BASED ON THE LOAD CHARACTERISTICS 134
J. J. Wu, H. T. Huang, C. Gao, F. Yu, J. Pan

ULTRASONIC LIVER IMAGE DENOISING BASED ON A HYBRID THRESHOLD METHOD 139
H. J. Zhu, L. Rao

APPLIED MECHANICS

STUDY ON AERODYNAMIC PERFORMANCE OF OFFSHORE WIND TURBINE WITH FLOATING PLATFORM MOTION 143
X. M. Ding, L. Zhang, Y. Ma

A VISUALIZATION EXPERIMENTAL STUDY OF ICING ON BLADE FOR VAWT BY WIND TUNNEL TEST 147
Y. Li, J. Tang, Q. D. Liu, S. L. Wang, F. Feng

KINEMATICS ANALYSIS ON THE CROSS STEP SKILLS OF CHINESE FEMALE JAVELIN THROWER LV HUIHUI 151
M. R. Zhou

MICROSTRUCTURE AND MECHANICAL PROPERTIES OF MICRO-ALLOYING MODIFIED AL-MG ALLOYS 153

J. Zhang, J. J. Zhao, R. L. Zuo

COMPARATIVE STUDY ON MECHANICAL PROPERTIES OF V-SHAPED INSULATOR STRING UNDER FLUCTUATING AND CALM WIND 157

J. J. Huang, X. M. Chen, Y. Xiong, C. L. Liu, Y. Wang

THE INTERNAL STRESS ANALYSIS METHOD OF CEMENT PASTE UNDER CORE RESTRICTED CONDITIONS 160

Y. Li, X. F. Liu, W. L. Bai

RESEARCH AND DEVELOPMENT OF THE TENSION PAY-OFF EQUIPMENT FOR WIRE ROPE MAINTENANCE 163

W. H. Cui, L. Tang, K. Q. Gong, C. Fu

CONVEX-CONCAVE PROPERTY FOR PARABOLA FITTING OF DR. BRIDGE 167

J. Huang, S. X. Ding, H. G. Gan, S. F. Zhang

MATERIAL PERFORMANCE AND OPTIMIZATION

TEM INVESTIGATION ON CERAMIC STRENGTHENING NIAL-BASED COMPOSITE PREPARED BY THERMAL EXPLOSION AND HOT EXTRUSION 170

L. Y. Sheng, C. Lai, T. F. Xi

FLEXURAL BEHAVIOR OF REINFORCED CONCRETE BEAMS STRENGTHENED WITH BFRP BARS 174

G. N. Yang, B. R. Huo, M. X. Zheng

EXPERIMENTAL INVESTIGATING EFFECT OF REPROCESSING ON PROPERTIES OF COMPOSITES BASED ON RECYCLED POLYPROPYLENE 177

F. Gu, P. Hall, N. J. Miles

PROPERTIES OF SNAGCU SOLDERS BEARING AL NANOPARTICLES 185

L. Zhang, L. Sun, Y. H. Guo, Y. Min

A REVIEW: THE WETTABILITY AND OXIDATION RESISTANCE OF SN-ZN-X LEAD-FREE SOLDER JOINTS 188

L. Sun, L. Zhang

TENSILE BEHAVIOR OF HIGH TEMPERATURE CU-CR-ZR ALLOY 191

X. W. Zhang, Q. J. Wang, X. Zhou, B. Liang

POLYPYRROLE/SISAL FIBER COMPOSITES FOR ENERGY STORAGE 195

H. D. Mo, L. M. Zang, C. Yang, C. Wei, F. A. Zhang, S. R. Lu, Z. Q. Wang, X. X. Huang

EFFECT OF ACCUMULATED STRAIN ON THE MICROSTRUCTURE AND PROPERTIES OF TC4 ALLOY PREPARED BY CONTINUOUS VARIOUS CROSS-SECTION RECYCLED EXTRUSION 198

X. L. You, Y. Y. Liu, J. S. Hua, K. Wang, W. Yao

MOLECULAR SIMULATION RESEARCH ON DOMAIN MICRO-STRUCTURE OF TSP-POSS/PU HYBRID COMPOSITES 202

R. Pan, L. L. Wang, Y. Liu

STUDY ON FIRE EXTINGUISHING PERFORMANCE OF SUPERFINE POWDER FIRE EXTINGUISHING AGENT IN A CUP BURNER 206

T. Chen, X. C. Fu, J. J. Xia, L. S. Jing, C. Hu

LAGRANGE MULTIPLYING METHOD BASED CALCULATION OF MATERIAL FLAW THICKNESS EXPLOITING ULTRASONIC MULTIPATH DETECTION 211

X. Z. Shen, S. J. Chen

SIMPLE SYNTHESIS FOR HIERARCHICAL SIO2 TUBES WITH ADJUSTABLE MESOPOROUS 215

Y. H. Zhang, Y. Deng, Y. H. Cai, W. Xiao, L. L. Sun

EFFECT OF SYNTHESIS CONDITIONS ON THE GROWTH OF ZNO NANORODS VIA THE SOLUTION DEPOSITION METHOD 219

C. M. Zhang, T. Meng, S. Y. Yao, S. Huang, D. D. Wang

STUDY OF THE NUMERICAL PARAMETER OPTIMIZATION METHOD BASED ON NANO-INDENTATION PROCESS 222

J. S. Ding, G. Q. Shi, G. F. Shi

GEFITINIB, AS A NEW STENT COATING MATERIAL, SPECIFICALLY INHIBITS SMOOTH MUSCLE CELLS PROLIFERATION THROUGH INHIBITION OF EGFR/AKT PATHWAY PHOSPHORYLATION 225

F. Li, S. Y. Wang, J. Luo, Z. X. Wu, T. Xiao, O. Zeng, J. Yang, C. Chu

A METHOD FOR COMPOSITE WING BOX OPTIMIZATION WITH MANUFACTURING
CONSTRAINTS 228
X. P. Zhong, P. Jin, Q. Han
PREPARATION OF HA/TIO2 BIOLOGICAL COATING ON TITANIUM ALLOY 231
H. P. Shao, S. J. Wu, T. Lin, Z. M. Guo
PREPARATION OF SUB-MICRO SIO2 PARTICLES BY SOL-GEL METHOD 234
T. Lin, H. P. Shao, H. Zheng, L. Zhang

INTELLIGENT NETWORK APPLICATION AND COMPUTER SCIENCE

IMPACT OF COMPUTER MUSIC TECHNOLOGY ON THE EFFECT OF THE INFORMATION
MEMORY OF AUDIENCES 237
Y. Qin, D. J. Li, J. T. Yang
DESIGN AND REALIZATION OF TEMPERATURE MEASUREMENT SYSTEM BASED ON
PROTEUS SOFTWARE 240
C. L. Wang, P. Y. Chen, H. L. Hu
RESEARCH ON THE STRATEGY OF GROUP VEHICLE INTELLIGENT PERCEPTION AND
TRAFFIC ROUTE GUIDANCE OF SEMANTIC CAR ROAD NETWORK 243
L. J. Tai, R. F. Hu, C. W. Chen
FACE RECOGNITION BASED ON GABOR WAVELET TRANSFORM AND MODULAR 2DPCA 245
H. Yan, P. Wang, W. D. Chen, J. Liu
STUDY OF WUHAN METRO VISUAL COMMUNICATION DESIGN UNDER THE
BACKGROUND OF "JIANGCHENG CULTURE" 249
X. Zhang
AN ENERGY CONSUMPTION MODEL OF MOBILE TERMINAL SOFTWARE BASED ON BP
NEURAL NETWORK 252
L. W. Liu, T. F. Zhan, Z. Y. Cai
THE CULTIVATION MODE ABOUT THE ABILITY OF SCIENTIFIC RESEARCH OF
UNDERGRADUATE BASED ON THE DUAL-TUTORIAL SYSTEM 255
X. D. Yuan, J. Li, G. L. Yuan
TRAINING MODE ON GRADUATION DESIGN OF ENGINEERING STUDENTS BASED ON
THE DUAL-TUTOR SYSTEM 258
X. D. Yuan, X. Liu, G. L. Yuan
THE CONNECTIVITY OF FAULTY FOLDED HYPERCUBE NETWORKS 261
D. Yuan, H. M. Liu, M. Z. Tang
RESEARCH AND APPLICATION OF FUZZY CONTROL WITH MULTIPLE WEIGHTED
FACTORS BY GENETIC ALGORITHM 266
L. J. Dong
ROUTING PROBLEMS IN EMERGENCY LOGISTICS BASED ON IMPROVED DATA
ENVELOPMENT ANALYSIS 270
X. X. Zhu, S. Y. Wang
THE RESEARCH OF NUMERICAL SIMULATION ABOUT THE IMPROVED DENSE MEDIUM
CYCLONE 273
X. B. Li, Q. Q. Huang, X. Li
A STUDY ON SOLVING THE NONLINEAR SEEPAGE FLOW MODEL OF THREE-REGION
COMPOSITE RESERVOIR 277
Y. Wang, X. X. Dong, S. C. Li, H. E. Li, D. D. Gui
A STUDY ON SOLVING THE LINEAR SEEPAGE FLOW MODEL WITH RWE OF COMPOSITE
RESERVOIR 281
Y. Wang, X. X. Dong, S. C. Li, H. E. Li, D. D. Gui
A FUZZY TRUST EVALUATION MODEL IN MOBILE COMMERCE 285
J. Chen, Z. Y. Cai, J. Peng
A NOVEL METHOD FOR 3D MORPHING BY DEFORMATION MATRIX WITH TRIANGLE
MESHES 289
C. L. Peng, T. W. Xing, Y. Yu, Y. Zhou, S. D. Du
A METHOD OF PUBLIC POLICY REFINEMENT BASED ON OWL AND LINEAR TEMPORAL
LOGIC 294
D. P. Lang, S. B. Huang, L. S. Shen, T. Zhang, H. Chen
STOCHASTIC ECONOMIC DISPATCH USING BACTERIAL SWARM ALGORITHM 298
M. S. Li, Y. Hu, X. Zhang

STUDY ON THE COMPREHENSION DIFFERENCE BETWEEN MANAGERS AND FRONT-LINE EMPLOYEES 302
W. Jiang, Z. M. Zhu, L. N. Li

ANALYSIS ON TECTONIC STYLE AND FORMING MECHANISM OF THE CAMBRIAN SYSTEM IN THE CENTRAL UPLIFT BELT, TARIM BASIN 306
W. Yin, T. L. Fan

EFFECT OF THE PLAN FOR EDUCATING AND TRAINING EXCELLENT ENGINEERS ON THE EXPERIMENTAL TECHNICAL ABILITY OF UNDERGRADUATE 310
X. D. Yuan, X. J. Yang, G. L. Yuan

STUDY ON THE EXPERIMENTAL TEACHING FOR UNDERGRADUATES IN REQUIREMENTS OF THE PLAN FOR EDUCATING AND TRAINING EXCELLENT ENGINEERS 313
X. D. Yuan, X. J. Yang, G. L. Yuan

STUDY ON THE OPENING EXPERIMENT TEACHING FOR UNDERGRADUATES IN REQUIREMENTS OF THE CREDIT MANAGEMENT SYSTEM 316
X. D. Yuan, X. J. Yang, G. L. Yuan

STUDY ON THE TALENT CULTIVATION MODE IN REQUIREMENTS OF THE CREDIT MANAGEMENT SYSTEM 319
X. D. Yuan, X. J. Yang, G. L. Yuan

ANALYSIS ON THE CHARACTERISTICS OF PARTICLES SPECTRAL OF ICE CAUSED BY FREEZING RAIN IN NAN-YUE MOUNTAIN OF HUNAN PROVINCE 322
L. C. Zhang, J. L. Yang, B. Liu, K. J. Zhu, X. Y. Li

CHANGES IN CONTENT AND COMPONENT OF PURPLE CORN (ZEA MAYS L.) ANTHOCYANIN DURING THE EXTRACTION AND PREPARATION 325
D. Wang, Y. Ma, P. P. Liu, C. Zhang, X. Y. Zhao

STUDY AND PRACTICE ON TEACHING SYSTEM REFORM OF ENGINEERING GRAPHICS BASED ON TRAINING ENGINEERING CONSCIOUSNESS 328
D. T. Xu, Y. J. Feng, J. L. Shi

Author Index

International Conference on Power Electronics and Energy Engineering (PEEE 2015)

Self-Cycling Modularized Smart Grid Application Integration

Y.P. Li, J.B. Xin

State Grid Jiangxi Electric Power Research Institute
Nanchang, 330096, China

S. Liang, W. Xu

Nanrui Technology Co., Ltd
Nanjing, 211106, China

Abstract—The essence of the construction of smart grid demonstration project is an integration process of various types of technology, system and device. With the integration of various kinds of adjustable resources including distributed power supply, reactive power compensation devices, and energy storage devices etc, the traditional passive distribution network is evolving into active distribution network, and low carbon power grid comprehensive demonstration project is a composition of distributed generation, energy storage, active load and the associated control equipment. On the basis of establishing unified information platform including various sub modules, it is significantly important to achieve low carbon optimized dispatch through the coordination of the modules.

Keywords-modularity; smart grid; low carbon

I. INTRODUCTION

With the continuous increase of the global resource environment pressure, social requirements on environmental protection, energy saving and emission reduction and sustainable development is being required greatly; at the same time, the continuous development of electricity market and users requirement for electricity quality and reliability decides that the electric networks of the future must be able to provide more safe, reliable, clean and high quality power supply, to adapt to the needs of various types of energy power generation mode, and to adapt to the need of highly market-oriented power transactions.

A smart grid is the use of sensors, communications, computational ability and control to enhance the overall functionality of the electric power delivery system. This permits several functions which allow optimization in combination of the use of bulk generation and storage, transmission, distribution, distributed resources and consumer end uses toward goals which ensure reliability and optimize the use of energy, mitigate environmental impact, manage assets, and contain cost.

Low carbon power grid comprehensive demonstration project is a composition of distributed generation, energy storage, active load and the associated control equipment [1-4]. On the basis of establishing an unified information platform including various sub module, it is significantly important to achieve low carbon optimized dispatch through the coordination of the modules [5-6].

II. UNIFIED INFORMATION PLATFORM

Gongqing city is located in the north of Jiangxi province,

China, the middle of the Nanchang-Jiujiang industrial corridor. The existing area of Gongqing is 170 km2, with the population of 100 thousands. In June 2011, the Japanese NEDO launched the Gongqing "smart community" demonstration project with Jiangxi Province officially in Tokyo. The demonstration project has been constructed by Toshiba, Itochu, and State Grid of China Corporation etc, and is expected to be complete in 2014. The core idea of the project is to construct an intelligent technology demonstration area by considering both economic and low carbon development.

Based on the already constructed integrated visualization platform for the Jiangxi Gongqing city smart grid, we researched and developed an unified information support platform including hardware, software, historical database and middle ware data bus etc, to provide support for the six professional application modules of distribution network dispatch, operation and maintenance, marketing, planning, distributed power supply and low carbon evaluation and validation.

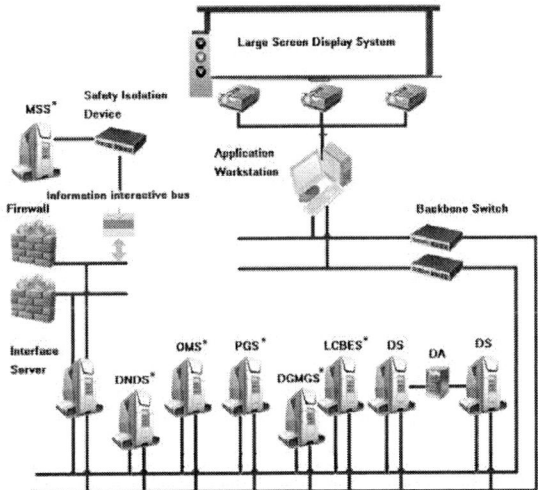

Note: DNDS: Distribution network dispatch server; OMS: Operation and maintenance server; PGS: Planning programming server; DGMGS: Distributed generation and micro grid server; LCBES: Low carbon benefits evaluation server; MSS: Marketing service server; DS: Data server; DA: Disk array   *: Application module devices

FIGURE I. THE UNIFIED INFORMATION PLATFORM: HARDWARE STRUCTURE.

The illustration of the unified information platform

© 2015. The authors - Published by Atlantis Press

hardware structure is shown in Figure 1: the two data servers and the disk array to form a local data center, the six application servers to demonstrate the visualization data of six professional application modules and finish the WEB page login. And the software structure of the platform is illustrated in Figure 2.

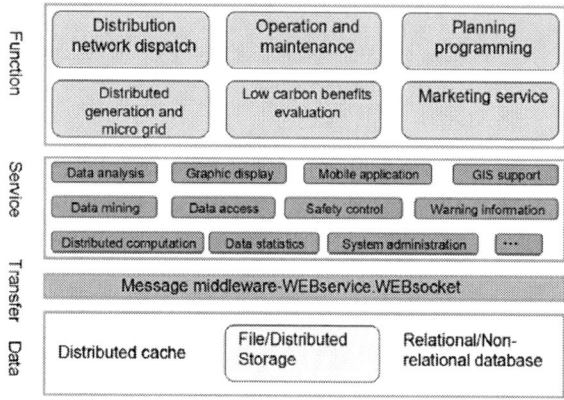

FIGURE II. THE UNIFIED INFORMATION PLATFORM: SOFTWARE STRUCTURE.

By utilizing the IEB information interactive bus, the platform can realize the data exchange, data sharing and application integration with the currently existed information systems including production management system, GIS system, marketing management system, 95598 system, the electric energy data acquisition system, dispatching automation system, power quality monitoring system, power voltage monitoring system, WSN device status monitoring, electric vehicle charging/exchange station, and intelligent residential community master station system etc.

III. SELF-CYCLING APPLICATION INTEGRATION

To abbreviate, we use DNDM for distribution network dispatch module, OMM for operation and maintenance module, PGM for planning programming module, DGMGM: distributed generation and micro grid module, LCBEM: low carbon benefits evaluation module, and MSM for marketing service module.

Based on the current power grid operation state, in order to reduce the network loss as the constraint conditions and improve the efficiency and safety of power grid operation as the objective, the DNDM module utilize the power grid real-time data to establish the data analysis model, make optimization scheduling strategy of regional power grid operation, photovoltaic power generation system, the electric vehicle charging and discharging system and storage system. On this basis, online simulation is performed to generate simulation data that meets operating requirements, including three phase power flow, network loss and power. Then the simulation data was provided to both the OMM and LCBEM modules.

After that, the OMM module performs the power grid operation risk assessment, including the comparison of the

operation status of switch, bus and load with the ideal model, and submit the evaluation results to the DNDM module. At the same time, the OMM module performs the online risk evaluation of all the transformers, lines and load etc in the whole area, then the evaluation results is provided to the PGM module.

When the DGMGM module received scheduling orders from the DNDM module, it makes the corresponding adjustment of photovoltaic power generation, energy storage devices and micro grid operation state, and then the operation condition information is passed back to the DNDM module.

After the MSM module received scheduling orders from the DNDM module, it makes the corresponding adjustment of the charging and discharging strategy for electric vehicles, orderly charging control of electric vehicles, and then the operation condition information is passed back to the DNDM module.

Based on the current grid operation state, the LCBEM module computes both carbon emission index and real-time line loss, also, this computation was performed on the simulation data received from the DNDM module, then the results of both carbon and loss reduction was formed. The above computation was done again after receiving the power grid operation data with optimal adjustment. In this way, the effectiveness of both the multi-energy complementary low carbon optimization scheduling and the carbon reduction was verified.

IV. CONCLUSION

We developed an unified information support platform including hardware, software, historical database and middle ware data bus etc, to provide support for the six professional application modules of distribution network dispatch, operation and maintenance, marketing, planning, distributed power supply and low carbon evaluation and validation. On the basis of the platform, both the multi-energy complementary low carbon optimization scheduling and the carbon reduction can be realized through the coordination of six professional application modules.

REFERENCES

[1] Kang Chongqing, Chen Qixin, et al. Prospects of low-carbon electricity[J]. Power System Technology, 2009, 33(2): 1-7.

[2] Lu Siyu, Lou Suhua, et al. A model for generation expansion planning of power system based on carbon emission trajectory model under low-carbon economy[J]. Transactions of China Electrotechnical Society, 2011, 26(11): 175-181.

[3] Hondo H. Life cycle GHG emission analysis of power generation systems: Japanese case[J]. Energy, 2005(30): 2042-2056.

[4] Zhou Tianrui, Kang Chongqing, et al. Analysis on distribution characteristics and mechanisms of carbon emission flow in electric power network[J]. Automation of Electric Power Systems, 2012, 36(15): 39-44.

[5] Xiao Xiangning, Chen Zheng, et al. Integrated mode and key issues of renewable energy sources and electric vehicles' charging and discharging facilities in microgrid[J]. Transactions of China Electrotechnical Society, 2013, 28(2): 1-14.

[6] Li Da-yong, Ma Dongxue, et al. Application of network information visualization[J]. Power System Protection and Control, 2009, 37(23): 156-158.

Droop Control of Parallel Dual-Mode Inverters Used in Micro Grid

C.X. Wen, Z.Y. Liu, Z.X. Li

Power Electronic and Motor Drive Engineering Research Center
North China University of Technology
Beijing, China

Abstract—**Grid-connected and island control of parallel inverters used in micro grid based on a variety of micro-source were introduced in this paper. Micro-grid in the connected mode should be able to operate automatically with the grid frequency and output high quality electricity in PQ control, and in island mode it can realize load power sharing of the parallel DGs in Droop control [1]. Simultaneously, it should also ensure the stability of the load voltage and frequency in island mode with the droop-based controller which can make the micro-grid smoothly switching between the two operation modes. Finally, simulation of two inverters in the connected mode and island mode was introduced in MATLAB / SIMULINK; the simulation results show the effect of droop control and the load-sharing function. The results also indicate the feasibility and correctness of the control strategy.**

Keywords-micro grid; parallel dual-mode inverters; PQ control; droop control

I. INTRODUCTION

With the rapid development of power electronic technology, in order to ensure the reliability and stability of the power output, more and more distributed generation (DG) systems are used in many cases. The large capacity inverters are gradually replaced by parallel system composed of many small inverters which greatly increase the flexibility and reliability of the system.

The power generation systems of large-scale distributed energy which usually connect to connect the grid by inverters compose of micro-grid. Micro-grid is low-voltage electrical distribution networks, heterogeneously composed of distributed generation, storage, load, and management system with the large primary network. Micro-grid can connect to the wide area electric power system through PCC (Point of Common Coupling), and can operate independently.

Mostly, micro-grid works in grid-connected mode. When the main grid has faults or the power quality can't meet the load requirements, micro-grid should transfer seamlessly from grid-connected mode to island mode and should be reconnected after the main grid returns to normal status. In order to eliminate the transient current and voltage rush associated with the micro grid operation mode transitions, seamless switch control is investigated in this paper.

II. MICRO-GRID STRUCTURE AND CONTROL ANALYSIS

The micro-grid structure used in this paper is shown [2] in Figure 1. This micro-grid includes two DGs. Every DG connected to the micro-grid AC bus through a static switch. The micro-grid connected to the main grid via a smart switch (SS).

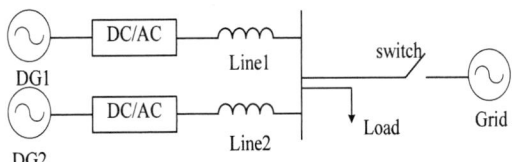

FIGURE I. MICRO-GRID STRUCTURE DIAGRAM.

A. Single Inverter Grid-Connected PQ Control

The main purpose of single inverter grid-connected PQ control is to ensure PQ control of distributed power output to maintain active and reactive power in the range of the reference power.

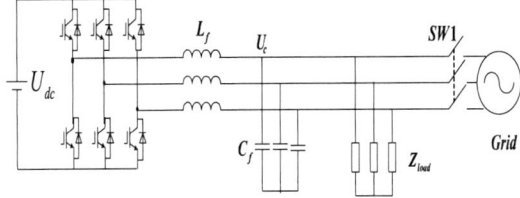

FIGURE II. THE GRID-CONNECTED STRUCTURE DIAGRAM.

Under the DQ coordinate system, the inverter output voltage equation as well as active and reactive power of expression is

$$P_{ref} = u_d i_d + u_q i_q \quad Q_{ref} = u_q i_d - u_d i_q$$

$$\begin{cases} i_{dref} = \dfrac{P_{ref}}{u_d} \\ i_{qref} = -\dfrac{Q_{ref}}{u_d} \end{cases} \tag{1}$$

$$\begin{cases} v_{Fd} = \left(k_{dP} + \dfrac{k_{dI}}{s}\right)\left(i_{dref} - i_d\right) - \omega L i_q + u_d \\ v_{Fq} = \left(k_{qP} + \dfrac{k_{qI}}{s}\right)\left(i_{qref} - i_q\right) + \omega L i_d + u_q \end{cases} \tag{2}$$

© 2015. The authors - Published by Atlantis Press

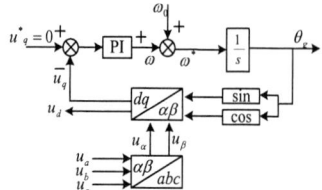

FIGURE III. PRINCIPLE OF SPLL.

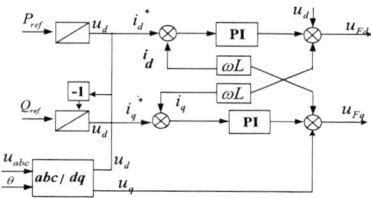

FIGURE IV. DG1 INVERTER PQ CONTROL ON GRID-CONNECTED.

When the micro-grid is in the connected-grid mode, the main inverter operates at a constant current control mode. The reference voltage and frequency are provided and guaranteed by the synchronous rotating coordinate system of d-axis orientation on the grid voltage vector direction, then in the same grid voltage, the inverter output active and reactive power coordinated to its current output respectively Therefore, the inverter output active and reactive power can be controlled to the grid.

B. Droop Control in Island Mode

According to the analyses above, when the output impendence is inductive, the active power can be controlled by power angle and the reactive power can be controlled by the amplitude of the output voltage. The P-f and Q-V droop control can realize power sharing between parallel inverters. The frequency and the amplitude of the inverter output voltage reference can be expressed as below

$$f = f^* - m(P - P^*)$$
$$V = V^* - n(Q - Q^*) \qquad (3)$$

Where f*and V* are frequency and amplitude of the output voltage respectively, P* is the active power reference and Q* is the reactive power reference. M and n are the droop frequency and amplitude coefficients.

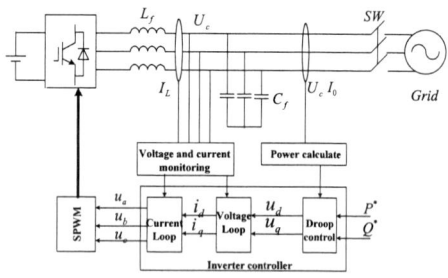

FIGURE V. DROOP CONTROL ON ISLAND MODE.

Droop control measures the value of distributed power output of active and reactive power, and uses the relevant Droop characteristics to determine the frequency and amplitude of the voltage reference value, that is P / f and Q / V control. The main power control strategy includes outer and inner voltage/current loop to achieve the two components which the inverter output voltage is regulated. By DQ coordinate system, the instantaneous power:

$$P_{grid} = u_d i_d + u_q i_q \quad Q_{grid} = u_q i_d - u_d i_q$$

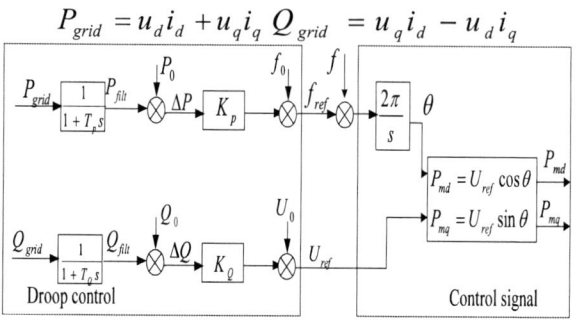

FIGURE VI. DROOP CONTROL SIGNAL FORMING.

III. SIMULATION AND ANALYSIS

Based on the above analysis, two inverter-based control and off-grid method in which a single inverter adopts PQ control and two inverters adopt net P/f & Q/V droop control strategies are proposed in MATLAB/SIMULINK simulation platform [3]. These methods and strategies can efficiently convert the energy and maintain the load stability and balance. Figure 7 illustrates the diagram of the micro-grid test system. Two three-phase inverters are used to research inverters in parallel operation which can eliminate the effect of one inverter suddenly putting into the system or suddenly removing from the system.

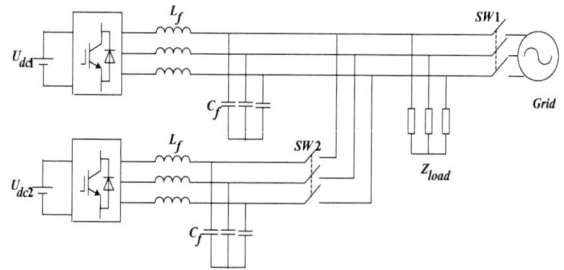

FIGURE VII. SIMULATION OF MICRO-GRID STRUCTURE DIAGRAM.

C. The Simulation of Dual Mode Inverters

When the micro-grid is in grid-connected mode, the parameters of each part of the system are shown in TABLE I.

TABLE I.   SYSTEM PARAMETERS.

Parameter	INVERTER1	INVERTER2
Input Dc voltage	800V	800V
Filter Inductance	6.76mH	3mH
Filter Capacitor	15uF	500uF
DG output power	2kW	13kW
SW1,SW2 time		0.2s

TABLE II. SYSTEM PARAMETERS.

GRID PARAMETERS	
Grid Voltage	380V
Grid Frequency	50Hz
LOAD PARAMETERS	
Load Voltage	380V
Load Frequency	50Hz
Active power	15kW

Mode 1: Inverter1 operating in the grid-connected mode. Switch1 is close and switch2 is cut off at t=0.2s. The voltage of Inverter1 meets the demands of load and main grid. The current of load is supported by inverter1 and main grid, it can make the connected grid and the simulation results are shown in Figure 8 (a. the voltage of Inverter1, b. current of Inverter1).

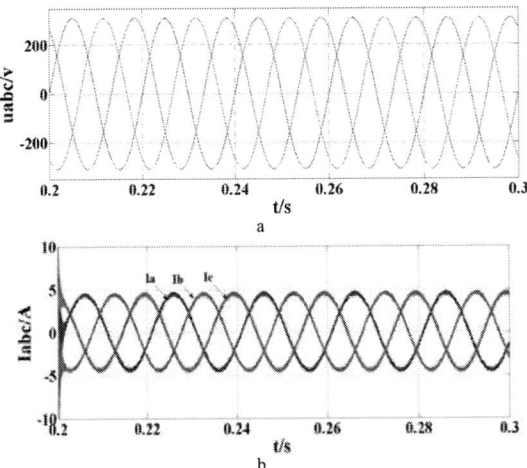

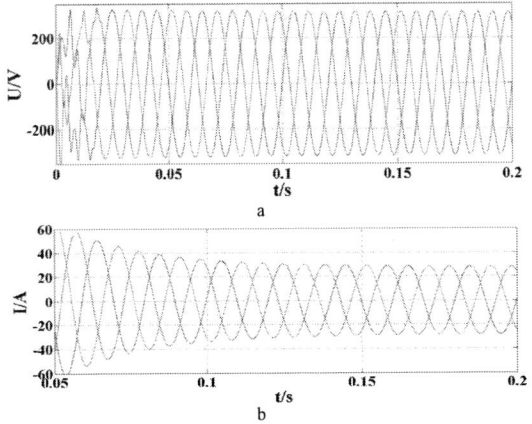

FIGURE VIII. THE SIMULATION OF GRID-CONNECTED MODE.

Mode 2: Inverter2 operating in the island mode. Switch1 is cut off at t=0.2s. The current of load is supported by inverter1 and inverter2; it can meet the demand of load and make it stably operating. The simulation results are shown in Figure 9 (a. the voltage of Inverter 2, b. current of Inverter 2).

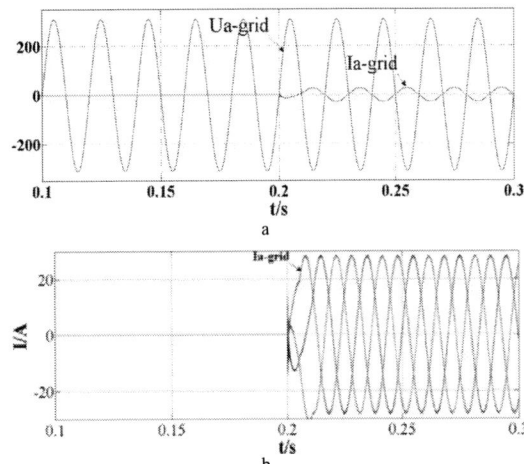

FIGURE IX. THE SIMULATION OF ISLAND MODE.

Transition from island mode to grid-connected mode: The

voltage is unchanged from island mode to grid-connected mode and meets the demand of load. Firstly, inverter1 and inverter2 operating in island mode and the current is zero in grid-connected mode before t=0.2s, at t=0.2s, switch1 is cut in and switch2 is cut off, the grid current can achieve a good transition between two modes [4]. The simulation results are shown in Figure 10 (a. voltage and current of grid, b. current of grid).

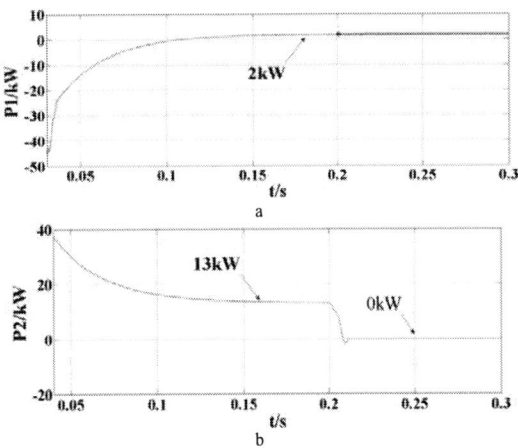

FIGURE X. THE VOLTAGE AND CURRENT FROM ISLAND TO GRID-CONNECTED MODE.

Active power transition from island mode to grid-connected mode: The active power of load is unchanged from island mode to grid-connected mode and meets the demand of load. Firstly, inverter1 and inverter2 operating in island mode and the total active power is 15kW before t=0.2s, at t=0.2s, inverter1 and main grid operating in grid-connected mode, the active power of load can meet demands and operate stably. The simulation results are shown in Figure 11 (a. active power of inverter1, b. active power of inverter2), Figure 12.

FIGURE XI. INVERTER1 AND INVERTER2 OUTPUT ACTIVE POWER.

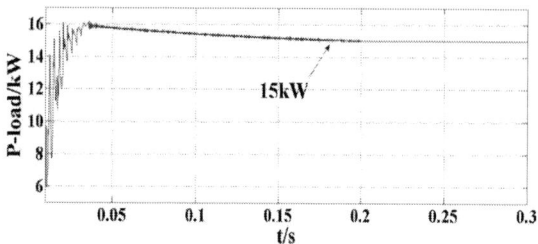

FIGURE XII. ACTIVE POWER ABSORBED BY THE LOAD.

IV. Conclusion

Micro-grid often operates under two typical modes. Micro-grid in the grid-connected mode can make it run automatically with the grid frequency and output high quality electricity. Furthermore, micro-grid in island mode can realize load power sharing of the parallel distributed power of micro-grid. Meanwhile, it can also ensure the stability of the load voltage and frequency. Moreover, the droop controller [5] can make the micro-grid smooth switching between the two kinds of operation modes. Finally, the simulation results indicated that the phase and magnitude of the load voltage are successfully matched to the grid voltage at the point of transfer from island mode to grid-connected operation without any distortions. Similarly, when the mode returns back, the load voltage can quickly approach its desired voltage without voltage and current rush by using the proposed method and strategy.

References

[1] Hu, Shang-Hung., Kuo, Chun-Yi., Lee, Tzung-Lin & Josep M.Guerrero, "Droop-Controlled Inverters with Seamless Transition between Islanding and Grid-Connected Operations," ECCE, Phoenix, AZ, Sept 2011, pp.2196-2201.

[2] Yang Zhan gang, Wang Cheng Shan, Che Yan Bo, "A Small-scale Micro-grid System with Flexible Modes of Operation," Automation of Electric Power Systems, vol.33, No.14, July 2009, pp.89-92.

[3] Kanellos,F.D., Tsouchnikas,A.I., & Hatziargyriou, N.D., "Micro-grid simulation during grid-connected and islanded modes of operation," presented at the Int. Conf. Power Systems Transients (IPST), Montreal, QC, Canada, 2005, Paper IPST05-113.

[4] Chen, C.L. Wang, Y.B. & Lai, J.S., "Design of parallel inverters for smooth mode transfer micro-grid applications," in Applied Power Electronics Conference and Exposition, 2009 IEEE.APEC 2009, pp. 1288-1294.

[5] Lee, C.T. Jiang, R.P. & Cheng, P.T., "A grid synchronization method for droop controlled distributed energy resources converters," in Energy Conversion Congress and Exposition, 2011 IEEE. ECCE 2011, pp. 743-749.

Research of Loss Reduction Scheme for Regional Distribution Grids Based On G_1-GRA

W. Zhang
Taiyuan University of Technology
Taiyuan, China

Y.M. Liu, W.Q. Ma
Yangquan Power Supply Company
Yangquan, China

Abstract—this paper proposed a comprehensive loss reduction evaluation method for regional distribution grids based on G_1 method and grey relational analysis (GRA). The method is good at avoiding the prominent disadvantages of analytic hierarchy process (AHP) by determining the weights of evaluation indices. Otherwise, the method of grey relational analysis was used for dimensionless processing. Therefore, the authors put forward an implementation scheme about the preferred area for loss reducing and the optimal index, which is proved to be practical, significant, and applicable.

Keywords-distribution network; g1 method; grey relational analysis; combination evaluation

I. INTRODUCTION

Line loss rate is the main economic and technical index of power network for planning, operating, and managing. Therefore, according to the need of the mechanism of power market operation, it is significant to reduce the line loss rate to a reasonable level. So the operation status of economic power grid as well as the economic benefits of power enterprises can be improved.

Among all the segments of power system, the distribution network is of significant importance for its complex network structure, serious line loss and poor reliability. Based on the above consideration, the distribution network has a large space in terms of efficiency. In order to improve the level of energy efficiency of the distribution network, it is necessary to expand the related theory. Therefore, energy efficiency index system needs to be established to evaluate the efficiency level of distribution network comprehensively. On this account, it is significant to obtain the actual conditions of distribution network and diagnose the weak line in energy efficiency, thus providing direction and suggestions for enhancing the efficiency level of distribution network.

Many domestic and foreign literature studies have involved the assessment method aiming at the reduction of distribution network loss. The model in [1] put forward a comprehensive renovation investment plan for regional distribution network based on grey connectedness weighting, which is about the combination of a variety of energy saving measures. Analytic hierarchy process is adopted in [2] to determine the weight of individual index. Based on the research about energy-saving index of distribution network, a energy efficiency index system of distribution network is established. The correlation between network loss factors and line loss rate is analyzed in [3] via the method of AHP-gray

relational analysis. moreover, GM(1,N) model has been developed through strongly correlated loss factors, for the purpose of predicting ideal line loss rate and calculating respective and comprehensive loss reduction potential. In the process of determining index weight by AHP method, a matrix needs to be constructed. And then conduct the consistency check. However, the matrix constructed is often hard to meet the requirement of consistency. So the judgment matrix usually needs to modify. The G1-GRA method is adopted in [4] to evaluate the comprehensive performance of coal-fired power plant. And the scheme is of practical significance in the comprehensive evaluation of distribution network.

Comprehensive evaluation method of loss reduction scheme concentrates on two aspects. On the one hand, the index consistent dimensionless method needs to be too proposed. On the other hand, the index weight needs to determine. Furthermore, the index characteristics of distribution network and the difference between the quantitative indicators and qualitative indicators are not considered in classical mathematics method of indicators dimensionless solutions. Subordinating degree function is attached to the fuzzy mathematics method. And it is often difficult to determine suitable function for the process of selecting the subordinating degree function is optional. Besides, in order to find out the subordinating degree function of each index, the index must be substituted into the corresponding subordinating degree function, which adding unnecessary errors. Grey relational analysis made improvement on the above mentioned method. On the process of partial treatment for single index data, the overall scheme is considered, which reduced the impact of incomplete information or asymmetry in the very great degree. In addition, the method of determining the weight generally includes the expert evaluation method, AHP method, G1 method, principal component analysis method and so on.

II. CONTENT AND STEPS OF G1 METHOD

G1 method is a subjective weighting method, which is presented by professor Y.J.Guo of Northeastern University. This method to determine index weight makes improvements based on the analytic hierarchy process and avoids the shortcoming of AHP method effectively. In the process of determining index weight, there is no need to construct judgment matrix as well as check the consistency. Compared to the analytic hierarchy process, the computation of G1 method has significantly reduced [5-7]. Specific content is as follows:

A. Determination of Order Relationship

Definition: if the important degree of evaluation index X_i, relative to a certain evaluation criteria (or target), is greater than (or less than) X_j, denoted by $X_i > X_j$.

Definition: if the indices of X_1, $X_2 \ldots X_m$, relative to a certain evaluation criteria (or target) has the relation $X_i > X_j > \ldots > X_k$, where i, $j \ldots k = 1, 2 \ldots M$, then the order relation is determined according to the token ">".

B. The Relative Importance of the Ratio Between x_k-1 and x_k

According to steps as follows to build the order relation for the evaluation index set $\{x_1, x2, \ldots, x_m\}$.

1) Find out the most important evaluation index in the index set $\{x_1, x_2, \ldots, x_m\}$, denoted by x_1^*;

2) Find out the most important evaluation index among the rest of the m-1 indicators, denoted by x_2^*; thus determining the only order relation $x_1^* > x_2^* > \ldots > x_m^*$, where x_i^* means the i'th evaluation index according to the order relation of ">" sorted.

3) Express the relatively important degree between the indices with the ratio between adjacent index important degrees of x_{k-1} and x_k, denoted by r_k. Namely, $r_k = \omega_{k-1}^* / \omega_k^*$. The value of r_k is shown in Table 1.

TABLE I. 2013 STATIC DATA Of EXPEDITION FACTORS.

r_k	The degree of importance(compare x_{k-1} with x_k)
1.0	Equal importance
1.1	Intermediate level
1.2	Moderate importance
1.3	Intermediate level
1.4	Strong importance
1.5	Intermediate level
1.6	Very strong
1.7	Intermediate level
1.8	Extreme importance

4) Calculate the weight coefficient. If the rational assignment of r_k is presented to meet the inequality: $r_{k-1} > 1/r_k$, then, ω_m^* can be expressed by:

$$\omega_m^* = \left[1 + \sum_{k=2}^{m} \prod_{i=k}^{m} r_i \right]^{-1} \tag{1}$$

$$\omega_{k-1}^* = r_k \omega_k^* \tag{2}$$

$k=m, m-1, m-2, \cdots 3, 2$. where, ω_k^* is the weight of the k'th index.

III. CONTENT AND STEPS OF GREY RELATIONAL ANALYSIS

Grey relational analysis (GRA) is a kind of multiple factors analysis method. Identify the superiority of proposed scheme through calculating the degree of association between the proposed scheme and ideal scheme[8,9]. The higher the correlation, the closer for the proposed scheme to the ideal one. Moreover, the comprehensive performance is superior, and vice versa.

The solving steps of grey relational analysis are as follows:

1) According to the purpose of evaluation to determine evaluation index system and collect evaluation data. The formation of the following matrix based on the n datum sequence:

$$(X_1, \ X_2, \ldots X_n) = \begin{pmatrix} x_1(1) & x_2(1) & \cdots & x_n(1) \\ x_1(2) & x_2(2) & \cdots & x_n(2) \\ \vdots & \vdots & \ddots & \vdots \\ x_1(m) & x_2(m) & \cdots & x_n(m) \end{pmatrix} \tag{3}$$

2) Determine the reference data columns

Reference data series is an ideal standard of comparison. The optimal value (or worst) of each index can be used to constitute the reference data column. It is also allowed to choose other reference value according to the evaluation purpose. Denoted by:

$$X_0 = (x_0(1), x_0(2), \cdots, x_0(m)) \tag{4}$$

3) Method of initial value is used to quantified the criterions. Thereinto, the initial value equation of the index data, which has positive correlation relative to the reference factor X0 is:

$$x_j'(k) = x_j(k) / x_1(k) \tag{5}$$

For the negative correlation factors, using the convert method. Convert them into positive correlation factor at first. Then conduct the initial value transformation. Calculation equation is:

$$x_j'(k) = (1 / x_j(k)) / (1 / x_1(k)) = x_1(k) / x_j(k) \tag{6}$$

$$(X_0', \ X_1', \ldots X_n') = \begin{pmatrix} x_0'(1) & x_1'(1) & \cdots & x_n'(1) \\ x_0'(2) & x_1'(2) & \cdots & x_n'(2) \\ \vdots & \vdots & \ddots & \vdots \\ x_0'(m) & x_1'(m) & \cdots & x_n'(m) \end{pmatrix} \tag{7}$$

4) Figure out the absolute difference between each appraisal object index sequence (the comparative sequence) and the reference sequence corresponding element. Get the difference sequence and find out the maximum as well as minimum value.

5) Determine the resolution coefficient ρ. resolution coefficientρcan be any value between zero to one. Its value will bring influence on determining the correlation value. Here's limitation of the coefficient values range: $0 < \rho < 1$.

If $\Delta_{max} > 3\Delta_v$, then $\varepsilon \leq \rho \leq 1.5\varepsilon$;

If $\Delta_{max} \leq 3\Delta_v$, then $1.5\varepsilon \leq \rho \leq 2\varepsilon$;

Where $\varepsilon = \Delta_v / \Delta_{max}$. Δ_v is the mean value of the Absolute Difference.

$$\Delta_v = \frac{1}{mn} \sum_{j=1}^{n} \left[\sum_{k=1}^{m} \left| x_0(k) - x_j(k) \right| \right] \quad (8)$$

$$\Delta_{max} = \max_{1 \le k \le m} \left[\max_{1 \le j \le n} \left| x_0(k) - x_j(k) \right| \right] \quad (9)$$

6) Calculate the correlation coefficient. Each reference sequence and comparative sequence are calculated separately and figure out the corresponding element of correlation coefficient via Eq.8, Eq.9.

$$\xi_j(k) = \frac{\Delta(\min) + \rho \times \Delta(\max)}{\Delta_j(k) + \rho \times \Delta(\max)} \quad (10)$$

The specific expression is:

$$\xi_j(k) = \frac{\min_j \min_k \left| x_o(k) - x_j(k) \right| + \rho \times \max_j \max_k \left| x_o(k) - x_j(k) \right|}{\left| x_o(k) - x_j(k) \right| + \rho \times \max_j \max_k \left| x_o(k) - x_j(k) \right|} \quad (11)$$

7) Calculate the correlation degree. the correlation degree between comparative sequence and reference sequence is reflected by M correlation coefficients. The equation is:

$$r_j = \frac{1}{n} \sum_{k=1}^{m} \xi_j(k), \; j = 1, 2, \cdots n \quad (12)$$

Determine the weighted arithmetic average according to the G1 method and get the correlation between them.

$$r_j = \sum_{k=1}^{m} \xi_j(k) \omega_j, \; j = 1, 2, \cdots, n \quad (13)$$

IV. EMPIRICAL STUDY

In order to realize the optimization evaluation of line loss scheme, six factors are selected to be investigated, namely, feeder branch length, average load rate of distribution transformer, average load rate of feeder, insulation rate of feeder, cable rate of feeder and ring rate of feeder, denoted by $X_1, X_2, X_3, X_4, X_5, X_6$ respectively.

C. The Loss Reduction Region Optimization

Table 2 shows the static result of four regions in a given area in 2013.

TABLE II. 2013 DATA STATICS OF EXPEDITION FACTORS.

region	X1 (km)	X2(%)	X3(%)	X4(%)	X5(%)	X6(%)
A	9.6	62.7	59.6	82.7	81.3	67.6
B	11.7	60.9	62.5	84.1	80.2	86.2
C	10.5	63.4	64.9	85.7	81.7	83.4
D	10.3	61.2	62.6	86.2	82.6	57.3

According to expert evaluation, index importance ranks as follows:

$X_2 > X_3 > X_1 > X_5 > X_4 > X_6$

Comparison results of the importance of adjacent indices are listed as follows:

TABLE III. ADJACENT INDEX COMPARISON RESULTS.

No.	2	3	4	5	6
r_k	1.1	1.2	1.4	1.2	1.4

Use the G_1 method to determine the index weight coefficients:

$r_2 r_3 r_4 r_5 r_6 = 3.105$; $r_3 r_4 r_5 r_6 = 2.822$; $r_4 r_5 r_6 = 2.352$; $r_5 r_6 = 1.680$; $r_6 = 1.200$.

TABLE IV. THE WEIGHT VALUES OF G1 METHOD.

index	X_2	X_3	X_1	X_5	X_4	X_6
weight	0.219	0.199	0.166	0.118	0.099	0.082

According to grey correlation analysis, feeder branch length and average load rate of feeder are reverse indices, the rest are positive indices. So the selected reference sequence is:

TABLE V. THE REFERENCE SEQUENCE VALUE.

region	X_1 (km)	X2(%)	X3(%)	X4(%)	X5(%)	X6(%)
Reference sequence	9.6	63.4	60.3	86.2	82.6	86.2

Results after the initial treatment are as follows:

TABLE VI. INITIAL RESULTS.

region	X_1	X_2	X_3	X_4	X_5	X_6
A	1.000	0.989	0.988	0.959	0.984	0.784
B	1.219	0.961	1.036	0.976	0.971	1.000
C	1.094	1.000	1.076	0.994	0.989	0.968
D	1.073	0.965	1.038	1.000	1.000	0.665
Reference sequence	1.000	1.000	1.000	1.000	1.000	1.000

Denote A*, B*, C*, D* as the absolute difference between the reference sequence and comparative sequence.

TABLE VII. ABSOLUTE DIFFERENCE.

A^*	0.000	0.011	0.012	0.041	0.016	0.216
B^*	0.219	0.039	0.036	0.024	0.029	0.000
C^*	0.094	0.000	0.076	0.006	0.011	0.032
D^*	0.073	0.035	0.038	0.000	0.000	0.335

Then, $\Delta(\max)=0.335$, $\Delta(\min)=0$, $\Delta v=0.056$, $\varepsilon = \Delta v / \Delta \max = 0.167$. It is known from Eq.8, Eq.9, if $\Delta \max > 3 \Delta v$, then $0.167 \le \rho \le 0.2505$, take the median, then $\rho = 0.21$.

Grey relational coefficient list:

TABLE VIII. GREY RELATIONAL COEFFICIENT

$\xi_1(k)$	1.000	0.864	0.858	0.634	0.817	0.246
$\xi_2(k)$	0.243	0.641	0.658	0.743	0.708	1.000
$\xi_3(k)$	0.429	1.000	0.480	0.924	0.866	0.684
$\xi_4(k)$	0.491	0.670	0.648	1.000	1.000	0.173

Obtain the grey correlation degree via the G1-GRA method.

TABLE IX. GREY COEFFICIENT DEGREE OF EACH REGION.

region	A	B	C	D
Grey coefficient degree	0.705	0.551	0.635	0.588

It is can been seen that the network loss reduction condition of region A is the best of all and region B is the worst. So region B if preferred to implement the loss reduction scheme.

D. The Loss Reduction Index Optimization

Table 10 shows the static results of corresponding indices of region B in 2009-2013.

TABLE X. STATISTIC DATA OF REGION B IN 2009-2013.

year	X_0(%)	X_1 (km)	X_2(%)	X_3(%)	X_4(%)	X_5(%)	X_6(%)
2009	6.48	12.3	59.7	65.7	75.3	71.7	59.4
2010	6.35	12.1	61.3	64.3	77.2	74.6	65.7
2011	6.17	11.9	61.3	63.1	81.3	75.3	81.2
2012	6.02	11.9	60.5	63.1	82.3	79.2	84.6
2013	5.99	11.7	60.9	62.5	84.1	80.2	86.2

Determine the weight of each index via G1 method. The results are shown in Table 4. The grey correlation degree calculation results are as follows:

$$(X_0, X_1, X_2, X_3, X_4, X_5, X_6) = \begin{pmatrix} 6.48 & 12.3 & 59.7 & 65.7 & 75.3 & 71.7 & 59.4 \\ 6.35 & 12.1 & 61.3 & 64.3 & 77.2 & 74.6 & 65.7 \\ 6.17 & 11.9 & 61.3 & 63.1 & 81.3 & 75.3 & 81.2 \\ 6.02 & 11.9 & 60.5 & 63.1 & 82.3 & 79.2 & 84.6 \\ 5.99 & 11.7 & 60.9 & 62.5 & 84.1 & 80.2 & 86.2 \end{pmatrix}$$

Through initialization operators, the original data can be transformed into dimensionless data. The shorter the length of feeder X_1 and the smaller the average load rate of feeder X_3, the smaller the rate of the comprehensive line loss. So X_1 and X_3 are positive correlation factors, compared with X_0. The rest of the indices are negative correlation factors. Therefore, after initialization and the reciprocal treatment, the results are shown as follows:

TABLE XI. INITIALIZATION RESULTS.

year	X_0	X_1	X_2	X_3	X_4	X_5	X_6
2010	1.000	1.000	1.000	1.000	1.000	1.000	1.000
2011	0.980	0.984	0.974	0.979	0.975	0.961	0.904
2012	0.952	0.967	0.974	0.960	0.926	0.952	0.732
2013	0.929	0.967	0.987	0.960	0.915	0.905	0.702
2014	0.924	0.951	0.980	0.951	0.895	0.894	0.689

Compute the absolute difference value of the comparative sequence and the reference sequence. And then list the difference sequence below.

TABLE XII. DIFFERENCE SEQUENCE.

X^*_1	X^*_2	X^*_3	X^*_4	X^*_5	X^*_6
0.000	0.000	0.000	0.000	0.000	0.000
0.004	0.006	0.001	0.005	0.019	0.076
0.015	0.022	0.008	0.026	0.000	0.221
0.038	0.058	0.031	0.014	0.024	0.227
0.027	0.056	0.027	0.029	0.030	0.235

Then, Δ(max) =0.235, Δ(min)=0, Δv=0.040, ε=Δv /Δmax=0.170. It is known from Eq.8, Eq.9, if Δmax>3Δv, then 0.170≤ρ≤0.255, take the median, then ρ=0.21.

The equation and calculation results of the correlation coefficient show as follows:

$$\xi_j(k) = \frac{\Delta(\min) + \rho \times \Delta(\max)}{\Delta_j(k) + \rho \times \Delta(\max)}$$

TABLE XIII. GREY RELATIONAL COEFFICIENT OF EACH INDEX.

$\xi_1(k)$	$\xi_2(k)$	$\xi_3(k)$	$\xi_4(k)$	$\xi_5(k)$	$\xi_6(k)$
1.000	1.000	1.000	1.000	1.000	1.000
0.928	0.891	0.975	0.916	0.724	0.394
0.763	0.694	0.857	0.655	0.999	0.183
0.562	0.461	0.611	0.778	0.675	0.179
0.648	0.469	0.647	0.630	0.619	0.173

The correlation of each index with the method of grey relational analysis only is:

TABLE XIV. GREY COEFFICIENT DEGREE RESULTS VIA GRA ONLY.

X_1	X_2	X_3	X_4	X_5	X_6
0.780	0.703	0.818	0.796	0.804	0.386

Obtain the grey correlation degree via the G1-GRA method.

TABLE XV. GREY COEFFICIENT DEGREE RESULT VIA G1-GRA.

X_1	X_2	X_3	X_4	X_5	X_6
0.647	0.769	0.814	0.393	0.476	0.159

From the table, it can be seen that when implementing loss reduction scheme for region B, priority should be given to X1, X2 and X3. Taking measures to reduce the length of feeder, improve the average rate of distribution transformer and reduce the average load rate of the feeder. Thus realizing the purpose of power loss reduction and the network management optimization.

V. CONCLUSION

In this pater, the G1-GRA combination model is adopted to put forward a comprehensive loss evaluation scheme for regional distribution network. Through analyzing the statistic data of four region in a certain area, the optimal region to implement the loss reduction scheme is found out. What's more, this method figure out the first–rank indices from the selected six ones via G1-GRA. The combination evaluation model proved to be simple and operable. Moreover, the evaluation model avoids the problem of inconsistency of judgment matrix when using the AHP method. It helps the power supply enterprise master the operation efficiency of the distribution network and provide guidance for reformation and schematization of distribution network.

ACKNOWLEDGEMENTS

The authors would like to thank the project sponsor. The project is supported by science and technology projects Fund of Yangquan Power Supply Company of China. (No. SGSXYQOOXTJS (2014)263)

REFERENCES

[1] Zhang Yongjun, Shi Hui. Distribution network energy-saving investment compact planning based on grey connectedness weighting [J]. Automation of Electric Power Systems, 2010, 34(22):46-50.

[2] Yang Xiaobin, Li Heming. Finite set optimal predictive control for neutral point clamped three-level grid-connected inverter [J]. Automation of electric power systems, 2013, 37(21): 146-150.

[3] Qiu Zejing, Xiang Tieyuan, Chen Hongkun, Yin Jie. Evaluation of Distribution Network Loss Reduction Potential Based on Improved Gray Relation [J]. China Rural Water and Hydropower, 2013: 106: 108.

[4] Huang Yuansheng, Song He, Tang Xinfa, Tian Lixia, Wan Youwei. Research on G_1-GRA based comprehensive performance evaluation model for thermal power plant [J]. Journal of North China Electric Power University, 2014, 41 (2):99-102.

[5] Chen Shu, Yu Di, Wu Liming. Fatigue risk fuzzy evaluation for high-risk operations based on G_1 method [J]. Journal of Safety Science and Technology, 2014, 10(4)90-95.

[6] Guo Yajun. Comprehensive evaluation theory and method [M]. Beijing: Science Press, 2002.

[7] Wang Xuejun, Guo Yajun. Based on the G_1 method of the consistency of the judgment matrix analysis [J]. Chinese Management Science, 2006, 6(3):65-70.

[8] Liu, S.F, et al. Gray System Theory and its Application [M].Beijing: China Science Press, 2000.

[9] Zhang Daotian, Yan Zheng, Han Dong, Zhang Nana, Chen Hui, Yu Nanhua. A Grey Clustering Based Evaluation on Technical Advancement of Smart Substation [J]. Power System Technology, 2014, 38(7): 1724-1730.

The Analysis on the Influence Factors and Tendency of Power Regulation in China

Q. Guo, T. Liu, G.W. Gao, Y. Xu
State Grid Energy Research Institute
China

Abstract—**Regulation is very useful in remedy the market failure effect caused by monopoly, especially in natural monopoly industry such as electricity. This paper first reviews the regulation situation in China's power industry, and then we analyze the influence factors that have influence on the regulation in power industry. At last we suggest a development path for China's power regulation.**

Keywords-power regulation; influence factors; China power industry

I. REGULATION SITUATION IN CHINA'S POWER INDUSTRY

Regulation refers to the restrictions that the government implements to the enterprises within an industry, including market access, pricing and service quality, health, safety and environmental regulations. Electricity Grid as a typical natural monopoly industry is needed to design effective mechanism government regulation to make up market failure caused by monopoly.

A. Regulatory Pattern

In 2002, after the electricity reform, the former state power company was separated into five power generation group and two power grid companies. The industry structure was transformed from the traditional vertical integration to the competition in power generation and government regulation in power transmission and distribution. Meanwhile, since the feed-in tariff was still approved by the national development and reform commission, the real competitive feed-in tariff has not been formed. In power grid, due to natural monopoly characteristics of grid, distribution electricity price was strictly regulated by the government. The transmission price is computed by the difference between the buying prices and selling, actually it is the cost-plus pricing method.

B. Regulatory System

In March 2003, the state electricity regulatory commission (hereinafter referred to as SERC) was established was to fulfill the national electricity regulatory responsibilities. The SERC gradually established a power regulatory organization system in national, regional and provincial level. In February 2005, the electricity regulatory ordinance was carried out, which became the SERC's regulation basis in legal. Electricity regulator subsequently carried out a series of regulation rules, which covered power safety and quality of power supply, market construction and supervision work. In March

2008, the national energy administration was established, which was responsible for the formulation and organize the implementation of the energy industry planning, industrial policy and standards, the development of new energy, promote energy conservation, etc. In March 2013, the SERC was canceled, and was combined with the national energy administration.

So far, China has established the basic framework of power regulation, with the gradual reform of electricity market, power regulation will dynamically change to meet the requirements of market-oriented reform.

II. THE MAIN FACTORS INFLUENCING THE CHINESE POWER REGULATION

With the new government came to effect, and gradually shifted the way of economic development, the government paid more attention to the quality of economic development. The year 2014 was regarded as the new "first year of reform", the political system and economic system reform has entered a new historical period. Power industry as the pillar industry of national economy, the development of this industry became mature, and the industry structure became more reasonable day by day. The main factors influencing the power regulation include economic development environment, government reform, electric power industry development, and electricity market reform.

C. The Economic Development Environment

The new government gradually abandoned the GDP standard development pattern, and government gradually controlled the development of energy-intensive industries. With economic growth gradually slowed down, economic development will face the more tightly environmental constraint. China's urbanization will get into the fast track, the development of urbanization rate will increase at a rate of 1% per year, and the rapid growth of urbanization will further enhance the requirement of power supply quality and safety. However, the growth rate of GDP in China would still in a mid-high track, the network regulation should be to achieve the sustainable development of national economy as the ultimate goal, to ensure the power industry's long-term capacity adequacy, and get enough allowed revenues for power grid companies to ensure that the power grid enterprises have the ability to pay the necessary operation cost and investment cost, and have the incentives to conduct electricity infrastructure construction as well as the transmission network investment enthusiasm in the long term.

D. The Government Reform

The transformation of government function, reduce the decentralization is the core of government reform. In 2013 the government institutional reform combined SERC with the national energy bureau. At the same time, the state council canceled administrative approval several times.

After decentralization, the regulation on power grid would have some changes in the content and key point. One is that in the market regulation, the decentralization has greatly simplified the advance program, and focuses more on strengthening the regulation afterwards. The second change is regulation content. Before and decentralization, power regulation mainly depends on investment approval, market access regulation, which is not good to stimulate the market and the vitality of enterprises. After the decentralization, regulation focuses more on the fair access, order of the electricity market and the power quality.

E. The Development of the Power Industry

With the economic development slow down, the overall demand of Chinese power industry will grow steady, and the structure of the generation become cleaner. The exploitation of power generation mainly concentrated and scattered supplemented. The ability of wide range resource allocation on power will markedly be enhanced. The power consumption will be more on green resource and be more efficient and intelligent interactive.

One is the development of new energy will have a great influence on power grid interconnection access, and the higher request for the power grid dispatching. Second, with the development of the UHV technology, the power regulation in the future will focus more on the safe and stable operation of power grid. Third, with the development of smart grid, power system security and stability is the primary task of electricity network regulation.

F. The Power Market-Oriented Reform

Along with the advancement of electric power market reform, power regulation will guide the power grid enterprises to increase renewable energy transmission, key technology such as optimization of electric power dispatching and operation control in the field of research and development, and guide enterprises to carry out more long-term planning, investment and construction, strengthen the whole assets life cycle management. All these will provide the necessary economic incentives the power grid enterprises to reduce costs and improve efficiency, and provide a solid foundation to the direct trading between the large industry users and generation firms.

III. THE ANALYSIS ON THE REGULATORY DEVELOPMENT PATH OF CHINA'S POWER INDUSTRY

According the regulatory status and influencing factors of China's electricity industry, we put forward the regulation development path of China's electricity industry, including the near future, medium-term and long-term three stages.

In the near future, the government should focus on the improvement of the function, change the focus of management. Enhance comprehensive energy management departments coordination function, make the regulation function of regulator be more perfect, and strengthen the construction of the supervisory ability of the regulator. Management focus should turn from the energy production and supply to focus more on demand, from economic regulation to social regulation.

In the medium-term, the government should focus on the reform of government institutions, straighten out the relationship between the different regulatory bodies, and straighten out the relationship between the central regulation and local regulation. In order to enhance the government's administrative execution ability as the main line, the energy management institutions should be reformed. From the organization guarantee, system guarantees two respects, to realize the goal of the central and local consistency.

In the long run, the regulation should be formed in accordance with the law, effectively promote the sustainable development of the new management system and a long-term mechanism. Forming perfect energy management and supervision system of laws and regulations, and focus on energy saving and improving energy efficiency, ensuring the safety and renewable energy development and so on.

REFERENCES

[1] Estache. A, Rossi. M. A. Do Regulation and Ownership Drive the Efficiency of Electricity Distribution, Evidence from Latin America, Economics Letters, 86(2), 253-257, 2005.

[2] Richard A. Posner, "Theory of Economic Regulation," Bell Journal of Economics and Management Science, pp.335-338, 1974.

[3] Sappington, D.E.M. and Sibley, D.S. "Regulating without Cost Information: The Incremental Surplus Subsidy Scheme", International Economic Review, pp.297-306, 1988.

[4] Sappington, D.E.M. and Sibley, D.S. "Strategic Nonlinear Pricing under Price-Cap Regulation", Rand Journal of Economics, pp.1-19, 1992.

[5] Coen, D. and Thatcher, M., "Network governance and multilevel delegation: European networks of regulatory agencies", Journal of Public Policy, 49–71, 2008.

[6] Finger, M. and Varone, F., "Bringing technical systems back in: towards a new European model of regulating the network industries", Competition and Regulation in Network Industries, 87–106, 2006.

Stability Constraints of the Transmission Capacity about VSC-HVDC System Supply to the Passive Network

S. Chen, X. Tang, W.Q. Zhang

College of Electrical and Informaton Engineering
Changsha University of Science and Technology
Changsha 410000, Hunan Province, China

Abstract—the external volt-ampere characteristic of the rectifier and inverter station of the VSC-HVDC (Voltage Sourced Converter-High Voltage Direct Current) system is simplified and the dc network equivalent circuit of VSC-HVDC system is built. The small signal model of the VSC-HVDC system is then derived. The transmission capacity is calculated under small-signal stability constraints and three influence factors are also obtained. Simulation is conducted by using PSCAD/EMTDC software. The simulation results demonstrate that the transmission capacity of VSC-HVDC system supply a passive network is be quite affected by the DC voltage, DC-side capacitance, and the resistance and reactance of the DC transmission lines.

Keywords-VSC-HVDC; equivalent circuit; small signal stability; transmission capacity

I. INTRODUCTION

With the electric field and areas expanding, the scale of the grid is expanding rapidly. HVDC is characterized by a number of technical merits: relatively lower cost in the long-distance and large-capacity power transmission, without the phenomena of system operational stability constraint, and smaller charging capacitance of DC cable. With these advantages, the HVDC has a broad application in asynchronous ac-grid interconnection, long-distance and large-capacity power transmission, and power transmission with cables, etc [1]-[3]. The recently developed HVDC technology based on voltage-sourced converters (VSC), namely VSC-HVDC, have some clear advantages. It can control the active and reactive power independently, and there is no need for the commutation-capacity. Based on these, the new developed technology is being applied in the connection of wind power [4]-[6], and feeding of weak grid [7]-[8] and the passive loads [9]-[12].

The main factors that affecting the maximum transmission power on long-distance AC transmission can be known as follows: thermal limits, constraints of voltage loss and the stability constraints of maintaining the synchronous running on the both ends of connected AC system.

The grid frequency of two regional can be decoupled through HVDC interconnection. It is no longer restricted by the stability constraints of synchronous operation. However, large enough commutation-capacity need to be provided by receiving grid in the traditional high voltage dc transmission system, so the strength of receiving grid (or the size of the short-circuit

ratio) is an important impacting factor on the power/voltage stability [13]-[16].

For the VSC-HVDC is concerned, commutation-capacity is no longer needed. That is to say, the strength of receiving grid has a little effect on maximum transmission capacity. In order to ensure the power quality, VSC-HVDC system supplying to passive network can control the voltage of inverter connect to AC bus with fast and indifference. The responding speed of AC bus voltage to the disturbances of DC side is under milliseconds [17]. So the load of the DC system (as viewed from the DC side of the inverter station) presents constant power characteristics, which weakened DC voltage stability of the network, thereby restricting the maximum transmission power of the VSC-HVDC system feeding of the passive network.

Due to the responding speed of the supply-side disturbances controlled by inverter station of VSC-HVDC systems supplying to passive network bellows milliseconds and line electromagnetic transient response belong to the same time scale. Therefore, in this paper, through the establishment of the VSC-HVDC system dynamic model that consider the line of electromagnetic transient and feeding of passive network, the voltage stability of DC network is analyzed. And then the quantitative calculation formula for the transmission capacity of the VSC-HVDC systems is deduced.

II. MODEL OF SYSTEM

A. Description of the System

Fig.1 shows the structure of the VSC-HVDC system supply to the passive network. As shown, converter station is the voltage-source cnverter with sinusoidal pulse width modulation; the inverter station and the rectifier station are connected by dc transmission lines, while rectifier station connects with the large power grid. The series combination of AC side inductance L and resistance R indicate converter transformer and converter reactors between the PCC (Point of Common Coupling) and converter station; The series combination of inductance LS and resistance RS represents the equivalent impedance of the AC system; AC filter and converter reactors are used to filter out the high-frequency components generated by the switching converter station; DC capacitor is used to reduce the DC voltage fluctuation; Where Req and Leq are equivalent resistance and reactance of the DC transmission line, Ceq represents equivalent capacitance in parallel by the distributed capacitance and DC capacitor.

© 2015. The authors - Published by Atlantis Press

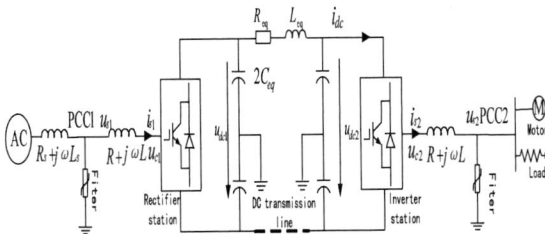

FIGURE I. THE STRUCTURE OF THE VSC-HVDC SYSTEM.

B. The Constant Power Load Characteristics of DC Network

When the VSC-HVDC system supply power to the passive network, its inverter station usually adopted AC voltage and fixed frequency control strategy, in order to ensure power quality, the response time of disturbance affected by voltage control is under milliseconds, therefore, DC network presents a constant power load characteristics (the view of the inverter station from the DC side). Fig. 2 shows the constant power load characteristic curve. The volt-ampere characteristics are obtained in (1).

$$i = \frac{P_{CPL}}{v} \tag{1}$$

Where i is the current flowing into the load, v is the voltage at the end of the load, P_{CPL} is the load power. Formula ($I=P_{CPL}/V$) can be turned into a linear in the equilibrium point, as given by (2).

$$i = 2\frac{P_{CPL}}{V} - \frac{P_{CPL}}{V^2}v \tag{2}$$

As can be seen from the above equation, the impedance characteristics of the constant power load are nonlinear, and the input impedance of small-signal is negative.

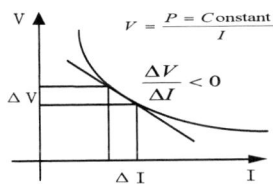

FIGURE II. THE VOLT AMPERE CHARACTERISTICS OF THE CONSTANT POWER LOAD.

C. Simplified Model of VSC-HVDC System

In this paper, rectifier station of the VSC-HVDC system supply power to passive network using fixed-voltage control, to maintain the level of DC voltage and active power balance of the entire network, its external characteristics performs a constant voltage source, thereby, as shown in fig.3, the simplifying circuit system can be obtained, and thus the equilibriuem quations of the system is given by (3) and (4).

$$\frac{di_l}{dt} = \frac{1}{L}(v_{c1} - v_{c2}) - i_l R_{eq} = \frac{1}{L}(v_{c1} - v_{c2}) - i_l R_{eq} \tag{3}$$

$$\frac{dv_{c2}}{dt} = \frac{1}{c}(i_l - i_{CPL}) = \frac{1}{c}(i_l - f(v_{c2})) \tag{4}$$

Where f(v_{c2}) is the volt-ampere characteristics of constant power load.

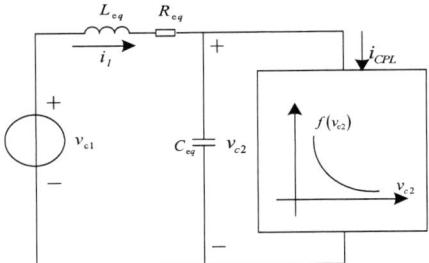

FIGURE III. SIMPLIFIED CIRCUIT DIAGRAM OF THE SYSTEM.

III. SMALL SIGNAL-STABILITY ANALYSIS OF SYSTEMS

Due to the volt-ampere characteristics of constant power load is nonlinear, In the point of the power balance (IL=P_{CPL}/V_{C2}), according to equation (2) to linearize the volt-ampere characteristics, (5) is obtained.

$$f(v_{c2}) \approx f(V_{c2}) + \frac{1}{r_L}\Delta v_{c2} \tag{5}$$

In the formula, $r_L = -V_{c2}^2 / P_{CPL}$.

Appling small perturbations with $i_l=I_L+\triangle i_l$ and $v_{c2}=V_{C2}+\Delta v_{c2}$. $\triangle i_l$ and $\triangle v_c$ were small interfering increments of currents and voltages. According to fomula (5), after the load was linearized, substituting the preturbed states into the fomula (3) and (4), then (6) is obtained.

$$\begin{cases} \dfrac{d\Delta i_l}{dt} = -\dfrac{R_{eq}}{L_{eq}}\Delta i_l - \dfrac{1}{L_{eq}}\Delta v_{c2} \\ \dfrac{d\Delta v_{c2}}{dt} = \dfrac{1}{C}\Delta i_l - f(v_{c2})\Delta v_{c2} \end{cases} \tag{6}$$

$$\lambda^2 + (\frac{R_{eq}}{L_{eq}} + \frac{1}{C_{eq}r_L})\lambda + \frac{R_{eq}}{L_{eq}C_{eq}}\frac{1}{r_L} + \frac{1}{L_{eq}C_{eq}} = 0 \tag{7}$$

The eigenvalues formula of the system are then expressed as

$$\lambda^2 + (\frac{R_{eq}}{L_{eq}} + \frac{1}{C_{eq}r_L})\lambda + \frac{R_{eq}}{L_{eq}C_{eq}}\frac{1}{r_L} + \frac{1}{L_{eq}C_{eq}} = 0 \tag{8}$$

For pure resistive load, $r_L > 0$, the eigenvalues of the system is in the left half plane, and when the load is constant power, $r_L < 0$, the eigenvalues of the system is turned into the right half plane, the small signal stability conditions of the constant power load system can be calculated as follows:

$$\begin{cases} \dfrac{R_{eq}}{L_{eq}} + \dfrac{1}{C_{eq} r_L} > 0 \\[2mm] \dfrac{R_{eq}}{L_{eq} C_{eq}} \dfrac{1}{r_L} + \dfrac{1}{C_{eq} L_{eq}} > 0 \end{cases} \tag{9}$$

Corresponding to the above analysis, the stability sonstraints of the transmission power can be obtained, as given by (10) and (11).

$$P_{CPL} < \frac{V_C^2 R_{eq} C_{eq}}{L_{eq}} \tag{10}$$

$$P_{CPL} < \frac{V_{C2}^2}{R_{eq}} \tag{11}$$

Because Req is small, the condition of formula (11) is easily meet. According to equation (10), there are three main factors that limit the maximum transmission power of the system, there are voltage level, the type of line and the value of DC side capacitance, the higher the voltage level, the greater the value of the DC side capacitor, the larger the transmission powe, the transmission power ability of submarine cable is larger than overhead transmission. Fig.4 shows the relationship between maximum transmission power and the value of dc voltage and the capacitor of DC side. In order to further study the transmission limit power affected by the transmission distance, ignore the loss of the line, (12) is obtained.

$$V_{C2} \approx V_{C1} - D r_0 \frac{P_{CPL}}{V_{C1}} \tag{12}$$

Where D is the distance of transmission, r0 is the equivalent resistance of line per kilometer length. Using (12) in (10) and solving the resultant, (13) is obtained.

$$P_{CPL} < \frac{1}{2} \frac{V_{C1}^2}{r_0^3 D^2} \frac{l_0}{C_{eq}} \left(\frac{2 D r_0^2 C_{eq}}{l_0} + 1 \right) - \frac{V_{C1}^2}{D r_0} \sqrt{\frac{1}{4} \frac{(2 D r_0^2 C_{eq} + l_0)^2}{r_0^4 C_{eq}^2 D^2} - 1} \tag{13}$$

Where l_0 is the equivalent inductance of line per kilometer length.Fig.5 shows the relationship between the transmission distance and the maximum transmission power and the linetype.

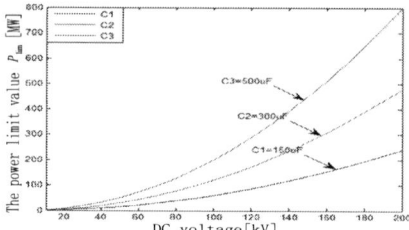

FIGURE IV. POWER LIMIT VALUES UNDER DIFFERENT DC VOLTAGE.

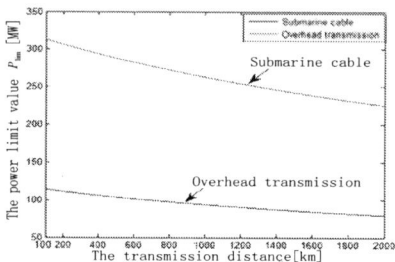

FIGURE V. POWER LIMIT VALUES UNDER DIFFERENT TRANSMISSION DISTANCE.

IV. SIMULATION

In order to verify the correctness and validity of the analytical model proposed in this paper, the simulation model of the passive network powered by VSC-HVDC system in PSCAD/EMTDC is established, Fig.1 showns the structure of simulation model. Table 1 showns the AC grid parameters connected with VSC-HVDC system.

TABLE I. THE PARAMETERS OF AC GRID.

Us	AC rated voltage of rectifier	110 KV
L	The inductance of the Commutation reactance	0.0006 H
R	The resistance of the commutation reactance	0.005 Ω
f	The frequency of the system	50 Hz

A. Instability Phenomenon

The parameters of DC side in the simulation system are given by the following: V_{C1}=118kV, C_{eq}=500μF, L_{eq}=0.18H, R_{eq}=3.14Ω. The ultimate power of VSC-HVDC system, P_{lim}=121MW can be obtained by formula. Fig.6 shows the power of system increased to 121MW and 135MW respectively when strating 100MW in five seconds, it can be seen from the figure the system remains stable when the power increased to 121MW, but the power appeared concussion when the power of the system has increased to 135MW. Because of the rectifier and inverter station adopted simplified model, 121 MW from the calculation of maximum transmitted power value is smaller than 135 MW from the simulation to get the maximum transmission power value.

16

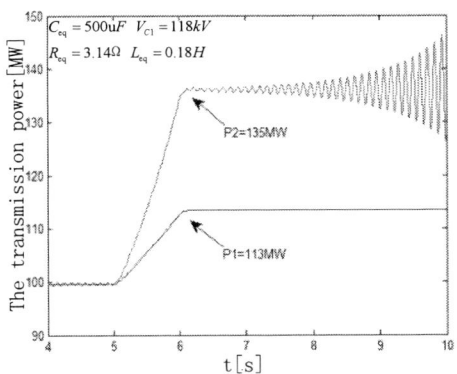

FIGURE VI. THE TRANSMISSION POWER OF THE SYSTEM.

B. The Impact of Maximum Transmission Power on DC Capacitorer

To substitute the parameters of the DC side capacitance (C_{eq}=600μF) into the simulaton 3.1. According to the formula (9), the maximum transmit power of the system is 145MW. Fig.6 shows the power of system increased to 135MW and 155MW respectively when strating 100MW in five seconds, as can be seen from the Figure 7, with value of the DC side capacitor increasing, the value of the system transmission system power increased to 155MW. The system appears a power oscillations, that is mean increase the value of DC side capacitor can increase the maximum transmission power of the system.

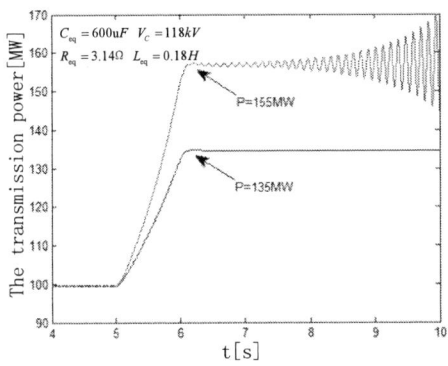

FIGURE VII. THE TRANSMISSION POWER OF THE SYSTEM WHEN DIRECT CURRENT CAPACITY IS 600MF.

C. The Impact of Maximum Transmission Power on DC Voltage

To substitute the parameters of the dc votltage (V_{C1}=130kV) into the simulaton 3.1. According to the formula (9), the maximum transmit power of the system is 147MW. Fig.8 shows the power of system increased to 135MW and 158MW respectively when strating 100MW in five seconds, as can be seen from the figure, with value of the DC voltage increasing, the value of the system transmission system power increased to 158MW. The system appears a power oscillations, that is mean increase the value of DC voltage can increase the maximum transmission power of the system.

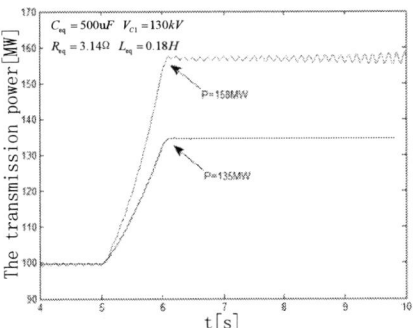

FIGURE VIII. THE TRANSMISSION POWER OF THE SYSTEM WHEN DC-SIDE VOLTAGE IS 130KV.

D. The Impact of Maximum Transmission Power on the Type of DC Line

To substitute the cable parameters of the over-head transmission into the simulaton 3.1 (the equivalent para-meter of cable: reference value R0=5m Ω /kM, L0=0.1085mH/kM [18-19]). According to the formula (12), the transmission capacity of the system is 310MW. Fig.9 shows the curve when transmission power of the system reach 135MW with two kinds of line parameters, the use of overhead line transmission make the system appear a power oscillation, and the use of cable can operation stability. That is to say that the ratio of line parameters R0 and L0 affect the maximum transmission power of the system.

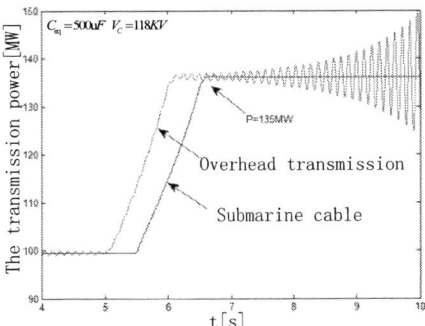

FIGURE IX. THE TRANSMISSION POWER OF THE SYSTEM UNDER DIFFERENT TRANSMISSION LINES.

The formula (12) shows that the length of the HVDC transmission line will also affect the maximum transmission power of the system. Fig. 10 shows the waveform of transmission power, when the transmission power of the over-head transmission line reached 135MW at 100kM and 200kM .Graphic shows the longer the transmission line, the lower the transport capacity of system.

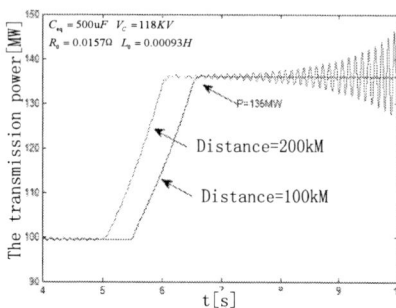

FIGURE X. THE TRANSMISSION POWER OF THE SYSTEM UNDER DIFFERENT TRANSMISSION DISTANCE.

V. CONCLUSION

In this paper, a small-signal stability of the VSC-HVDC system was analysed. The important affecting factors on transmission capacity of passive network powered by VSC-HVDC, such as the parameters of DC voltage, DC capacitor and transmission distance, are quantitatively analyzed and simulated. From the simulation results, the following conclusion can be drawn.

1. The higher the DC voltage and DC capacitor, the greater transmission capacity of system;

2. The farther the transmission distance, the smaller transmission capacity, the capacity by the cable is larger than the overhead line transmission;

3. The Control parameters of rectifier and inverter stations have a certain influence on the transmission capacity of the system, the rectifier take a DC voltage source description, the maximum transmission capacity calculation tend to be conservative.

REFERENCES

[1] Li Xingyuan,Zhao Rui,Liu Tianqi,Wang Yuhong,Wang Xi.Research of Conventional High Voltage Direct Current Transmission System Stability Analysis and Control[J].Transactions of China Electro Technical Society,2013,10:288-300.

[2] TANG Guang fu,HE Zhi yuan,PANG Hui. Research, Application and Development of VSC-HVDC Engineering Technology[J]. Automation of Electric Power Systems ,2013,15:3-14.

[3] Long Willis,Nilsson Stig.HVDC transmission:yesterday and today[J].IEEE Power and Energy Magazine,2007,5(2):22-31.

[4] ZHAO Jing1,ZHAO Chengyong1,SUN Yiying1,ZHANG Jianpo1,JING Huabing.Low Voltage Ride-Through Technology for Wind FarmsConnected to Power Grid via MMC-Based HVDC Transmission[J]. Power System Technology. 2013,03:726-732.

[5] Bresesti P,Kling W L,Hendriks R L,et al.HVDC connection of offshore wind farms to the transmission system[J].Energy Conversion,IEEE Transactions on, 2007,22(1):37-43.

[6] Wei Xiaoguang,Tang Guangfu.Effect of VSC-HVDC applied on improving wind farm voltage stability[J].Power System Technology,2007,31(8):27-31(in Chinese).

[7] Xu Zheng.Characteristics of HVD Connected To Weak AC Systems Part1,HVDC Transmission Capability[J].Power System Technology,1997, 01:12-16.

[8] Beccuti G, Papafotiou G,Harnefors L. Multivariable Optimal Control of HVDC Transmission Links With Network Parameter Estimation for Weak Grids[J].

[9] GUO Xiaojiang1, GUO Qiang2, MA Shiying2,XU Zhengxiong2,ZHANG Yantao2,BU Guangquan2,WANG Chengshan1. Research on System Interconnection Requirements of DC Island Sending Systems[J], Proceedings of the CSEE, 2012, 34:42-49+8.

[10] Zhang L, Harnefors L,Nee H P.Modelling and control of VSC-HVDC connected to island systems[C].Power and Energy Society General Meeting,2010 IEEE. 2010:1-8.

[11] CHEN Hai-rong, XUZheng.Control Design for VSC- HVDC Supplying Passive Ntwork[J]. proceedings of the CSEE,2006,23:42-48.

[12] LIANG Hai-feng,LI Geng-yin,LI Guang-kai,ZHANG Kai,ZHOU Ming.Simulation Study of VSC-HVDC Sytem Connected to Passive Network[J].Power System Technology,2005,08:45-50.

[13] ZHANG Yingmin1,HE Yang2,LI Xingyuan1, ZHAO Rui1,WANG Pengfei1,CHEN Hu1.Analysis on Power Stability of Multi-Infeed HVDC Power Transmission System[J] .Power System Technology,2011,35(6):50-54.(in Chinese).

[14] Yu J,Karady G G,Gu L.Applications of embedded HVDC in power system transmission[C]//Power Engineering and Automation Conference (PEAM), 2012 IEEE.IEE.2012:1-6.

[15] CHEN Hu1,ZHANG Yingmin1,HE Yang2,LI Xingyuan1,ZHAO Rui1, WANG Pengfei1. Analysis on Power Stability of Multi-Infeed HVDC Power Transmission System[J].Power System Technology,2011,06:50-54.

[16] LIN Wei-fang, TANG Yong, BU Guang- quan. Definition and Application of Short Circuit Ratio for Multi-infeed AC/DC Power Systems [J]. Proceedings of the CSEE, 2008,3:1-8.

[17] Tang Xin,Li Jianlin,Teng Benke.Enhancement of Voltage Quality in a Passive Network Supplied by a VSC-HVDC System Under Disturbances [J].Transactions of China Electro Technical Society,2013,09:112-119.

[18] Li SHaskew T A,Xu L.Control of HVDC light system using conventional and direct current vector control approaches[J].Power Electronics,IEEE Transactions on,2010,25(12):3106-3118.

[19] UNITECH Power Systems,Shell.''Long distance electrical power supply for subsea installation'' accessed in the Internet on January 2008.

International Conference on Power Electronics and Energy Engineering (PEEE 2015)

Development of VLF Test System for Power Cables

C.X. Huang, C.B. Yang, H.J. Sun, H.J. Li
Xi'an Jiaotong University
28 West Xianning Road, Xi'an, 710049, P.R. China

F. Huang, J.M. Guo
Guangxi Power Grid Electric Power Research Institute Co Ltd
6-2 Democratic Road, Nanning, 530012, P.R. China

Abstract—**with the rapid growth of the using of power cables in distribution network, a large demand for mobile high-voltage cable test systems is expected in the near future. VLF test system has a great equivalence of AC test, so it is wildly used in power cable test. In this paper, a novel VLF high voltage generator, based on LCC resonant converter, is introduced detailed. The Extended Describing Function (EDF) is used to select parameter of LCC resonant converter. In order to generate a true sinusoidal test voltage, the control strategy, concluding changing duty cycles of LCC resonant converter and load resistors, is proposed. The generator can generate a sinusoidal test voltage of 20 kV at 0.1 Hz and the result shows that the total harmonic distortion can meet the requirement of IEEE 400.2 standard.**

Keywords-VLF test system; power cables; extended describing function (EDF)

I. Introduction

As the bridge of power transmission, cable is an important part of power system. There are certain amounts of water in cable when it is manufactured, installed and used, which will contribute to water trees in cable. Water trees won't produce partial discharge, so it is difficult to discover. AC test has a good test result, but it is not easy to use in on-site test. Because with the increasing of length of the test cable, the devices of AC test become larger and heavier. Very low frequency (VLF) test system developed rapidly in recent years, which has a very good equivalence of AC test [1]. Not only withstand voltage test but also measuring of the dielectric loss Angle and partial discharge of the cable can be carried out by using VLF test system, so that comprehensive analysis on the status of the cable can be made easily.

In this paper, a novel topology of VLF high voltage generator is proposed. This generator is supplied by 220V, 50Hz Alternating Current. Firstly, alternating current becomes direct current after a rectifier. And then, the high frequency resonance current is produced in resonant cavity. Meanwhile, there is a high voltage on the power cables, which can be

adjusted via changing the load resistance and the duty cycle of inverter. LCC series-parallel resonant converter, combining the advantages of series resonant converter (SRC) and parallel resonant converter (PRC) can reduce the EMI noise and switching loss.

II. Basic Principle of Test System

The test system consists of two subsystems responsible for the positive and negative half waves. Except for the direction of high voltage silicon stack, the two subsystems are all the same, so that only one subsystem is analyzed in this paper. As illustrated in Figure 1, there are Rectifier Bridge, LCC resonant converters, high-frequency transformers, double voltage circuit and adjustable load resistor in a subsystem. In order to simplify the analysis, all the components in this circuit are considered as ideal components.

Once the topology of the circuit is known, a mathematical formulation for its behavior can be done. Based on the circuit illustrated in Figure 1, three nonlinear states can be obtained:

$$\frac{di_L(t)}{dt} = \frac{u_{AB}(t) - R_s * i_L(t) - u_s(t) - u_p(t)}{L_s}$$

$$\frac{du_s(t)}{dt} = \frac{1}{C_s} * i_L(t)$$

$$\frac{du_{out}(t)}{dt} = \frac{1}{C_o} * (i_D(t) - \frac{u_{out}(t)}{R_{out}}) \quad (1)$$

The equation 1 cannot be solved because the number of unknown variables is much more than the number of equations. Through analyzing the circuit, Extended Describing Function (EDF) can be used for solving this task [2]. The variables in equation (1) can be approximated by its first-order coefficients or its zero-order coefficients of their respective Fourier series. So, they can be written as follows:

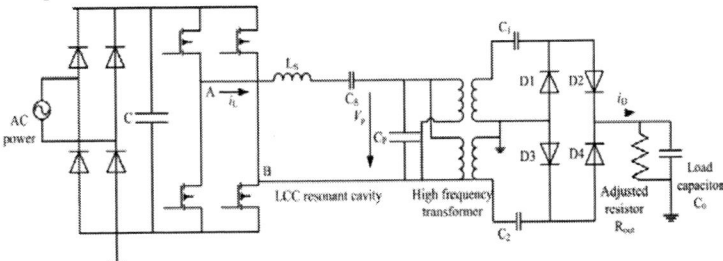

FIGURE I. POWER CIRCUIT OF THE SUBSYSTEM.

© 2015. The authors - Published by Atlantis Press

$$i_L\left(t\right) \approx i_{LS}\left(t\right)*\sin(\omega\,t) + i_{LC}\left(t\right)*\cos(\omega\,t)$$

$$u_s\left(t\right) \approx u_{SS}\left(t\right)*\sin(\omega\,t) + u_{SC}\left(t\right)*\cos(\omega\,t)$$

$$u_p\left(t\right) \approx u_{PS}\left(t\right)*\sin(\omega\,t) + u_{PC}\left(t\right)*\cos(\omega\,t)$$

$$u_{AB}\left(t\right) \approx \frac{\pi}{4}u_d*\sin(\frac{d*\pi}{2})*\sin(\omega\,t)$$

$$u_{out}(t) \approx u_{oavg}$$

$$i_D\left(t\right) \approx i_D\left(i_L, u_{oavg}\right) \tag{2}$$

Through complex mathematical analysis [5], the equation (1) can be rewritten as equation (3):

$$\frac{di_{LS}\left(t\right)}{dt} = \frac{1}{L_s}[\frac{\pi}{4}u_d*\sin(\frac{d*\pi}{2})*\sin(\omega\,t)$$

$$-R_S*i_{LS}-u_{SS}-\frac{i_{LS}*\sin^2\psi+i_{LC}*\mu}{\pi*Cp*\omega}+L_S*\omega*i_{LC}]$$

$$\frac{di_{LC}\left(t\right)}{dt} = \frac{1}{L_s}[-R_s*i_{LC}-u_{SC}-$$

$$\frac{i_{LC}*\sin^2\psi-i_{LS}*\mu}{\pi*Cp*\omega}+L_S*\omega*i_{LS}]$$

$$\frac{du_{SS}}{dt} = \frac{1}{C_s}*i_{LS}+\omega*u_{SC}$$

$$\frac{du_{SC}}{dt} = \frac{1}{C_s}*i_{LC}-\omega*u_{SS}$$

$$\frac{du_{out}(t)}{dt} = \frac{1}{C_o}*(\frac{\sqrt{i^2_{LS}+^2_{LC}}}{\pi}*[1+\cos(\psi)]-\frac{u_{out}(t)}{R_{out}}) \tag{3}$$

III. SYSTEM IMPLEMENTATION

A. Resonant Parameter Selection

From the equation (3), it can be seen that once the value of C_P, C_S, L_S, f_S and the turns ratio are determined, both the dynamic response and the steady state response of the circuit can be deduced. Based on the dynamic response and the steady state response of the circuit, the optimal parameter can be decided. The parameter α (equal to C_P/C_S) has a great influence to the circuit performance. From the Figure (2), the higher the α, the higher the voltage gain and the resonant current. The smaller the α, the smoother the carve of normalized voltage conversion. According to the paper [3], the optimum range of α is 0.5 to 0.8. So α=0.5 is chosen.

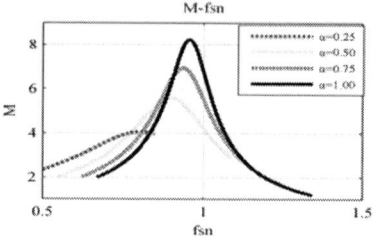

FIGURE II. CALCULATED NORMALIZED VOLTAGE CONVERSION VS. NORMALIZED FREQUENCY.

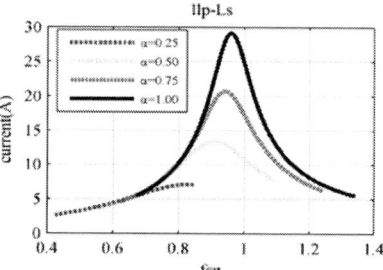

FIGURE III. RESONANT CURRENT VS. NORMALIZED FREQUENCY.

The peak of resonant current is an important parameter, which not only determines the maximum voltage of C_P、C_S and mosfet, but also reflects the power loss of the resonant circuit. Therefore, the peak of resonant current should be minimized. When the output voltage reaches the limit, the operation state is as follow: U_{in}=300V; f_S=25 kHz; n=30; I_L=5A; the series resonant frequency $f_0 = f_S / 1.2$; standardized coefficient of voltage and current are respectively 1.2 and 1.5[4]. The resonance inductor and capacitor can be calculated by the following formula.

$$\frac{I_L}{I_{L\_N}} = \frac{U_{IN}}{U_{IN\_N}}*\sqrt{\frac{C_S}{L_S}} = \frac{5}{1.5}$$

$$f_0 = \frac{1}{2\pi\sqrt{L_S C_S}} = 25000/1.2 \tag{4}$$

After calculated, L_S=560μH; C_P=50nF; C_S=100nF. Taking the above parameters into the equation 3, the output voltage versus switching frequency f_S and duty cycle d are shown in Figure 3 and Figure 4. In order to expand the range of output voltage from 0 to 27kV, the best way is to adjust duty cycles.

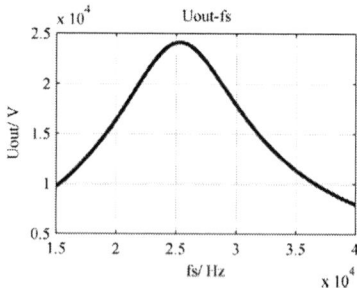

FIGURE IV. THE OUTPUT VOLTAGE VS. SWITCHING FREQUENCY.

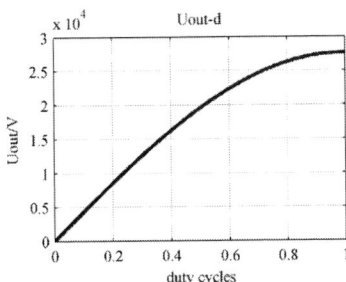

FIGURE V. THE OUTPUT VOLTAGE VS. DUTY CYCLES.

B. Control Strategy

From Figure 4, the output voltage will change from 0 to 27kV when duty cycles changes from 0 to 1, so we can regulate the output voltage by changing the duty cycles. Through analyzing the relationship between output voltage and duty cycles, output voltage is a linear function of sin ($\pi/2*d$) (Figure 5). Through the simulation, significant distortion can be found at zero-crossing point (Figure 6), so only changing duty cycles cannot get the true sine wave. This is because in the vicinity of the zero crossing point, the sinusoidal voltage has large variation rate, which is much larger than the voltage drop rate on load capacitance. Even though d=0, the output voltage doesn't drop rapidly. And then, significant distortion will happen.

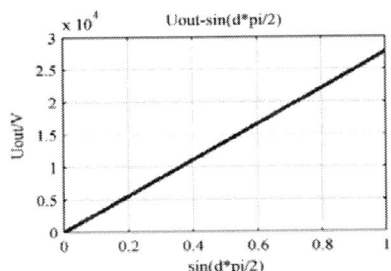

FIGURE VI. THE OUTPUT VOLTAGE VS. SIN(D*Π/2)/.

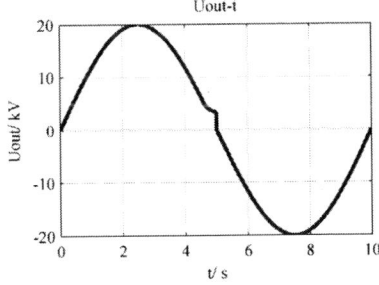

FIGURE VII. THE OUTPUT VOLTAGE WITHOUT CHANGE LOAD RE-SISTOR.

In order to improve the waveform, the large load resistor can be replaced by some small resistors in series, and each small resistor parallels with high voltage IGBT, shown as Figure 7. And then, sequentially close IGBT at the appropriate time to reducing output resistance and the time constant of the RC discharge. Coupled with the duty cycles change, the output voltage waveform distortion is small. Improved waveform is shown in Figure 8 and the block diagram of the control system is shown in Figure 9.

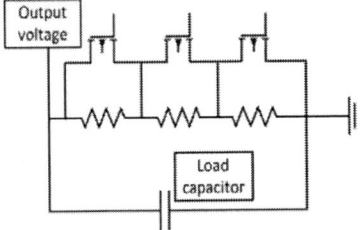

FIGURE VIII. IMPROVED LOAD RESISTOR.

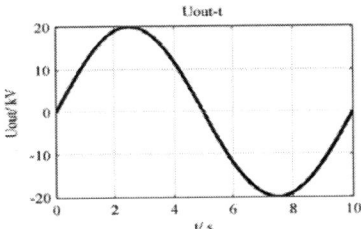

FIGURE IX. IMPROVED OUTPUT VOLTAGE.

At the beginning, duty cycle is 0, then the output voltage is 0 and the resonant current IL is approximately 0; when the output voltage doesn't equal to pre-set voltage which changes as sine, the Microprocessor calculate the $\triangle d$ and required output resistance. At the same time, the resonant current should be detected, if the resonant current exceeds warning value, duty cycles will be set as 0; if not, the program turns back to the next circulation.

The whole system is shown in Figure 10; the control system is implemented on a microcontroller which is integrated in the resonant circuit. The test cable is instead of a capacitor. Output voltage is shown in Figure 11, total harmonic distortion (THD) levels does not exceed 5%, so it can meet the requirement of IEEE 400.2 standard.

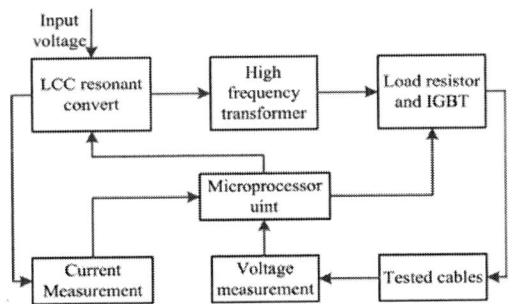

FIGURE X. THE BLOCK OF CONTROL UNITS.

IV. CONCLUSION

In this paper, a VLF high voltage test system based on LCC resonant converter is introduced. A design method of the LCC resonant converter is provided according to Extended Describing Function. The feedback control is used in this test system in order to achieve sinusoidal output voltage and protect the circuit. Experimental results show the correctness and feasibility of the theory.

FIGURE XI. OVERVIEW OF THE TEST SYSTEM.

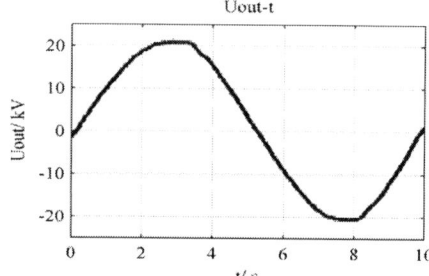

FIGURE XII. EXPERIMENTAL RESULTS.

REFERENCES

[1] Seesanga S, Seesanga S, Kongnun W, Kongnun W, Chotigo S, Chotigo S. Microcontroller modulation for VLF high voltage generator rate 3 kV peak using DSPIC-30F2010.2008 IEEE 2nd International Power and Energy Conference. IEEE; 2008:62-66.

[2] Martin-Ramos, J., & Diaz, J., and so on. Dynamic and steady-state models for the PRC-LCC resonant topology with a capacitor as output filter. IEEE Transactions on Industrial Electronics, 54(4), 2262-2275.

[3] Martin-Ramos, J., & Diaz, J., and so on. Power Supply for a High-Voltage Application. IEEE Transactions on power electronics.23(4) 1608-1619.

[4] Xiangdong Sun, Long Duan and so on. Analysis and Design of HVDC LCC resonant converter. China Electrotechnical Society.2002.17(5) 60-64.

[5] Hu M, Froehleke N, Boecker J. Small-signal model and control design of LCC resonant converter with a capacitive load applied in very low frequency high voltage test system. ; 2009.

Selection of Decoupling Capacitors for Power Delivery Networks with Multiple Power Ports

Q.D. Wang, Y.C. Wang, X.L. Li
State Key Laboratory of Power Transmission Equipment and System Security and New Technology
Chongqing University
Chongqing, 400044, China

Q.S. Liu
Chongqing Vehicle Test & Research Institute, National Coach Quality Supervision and Test Center
Chongqing EMC Engineering Technology Research Center
Chongqing, 401122, China

Abstract—in high-speed printed circuit board (PCB), decoupling capacitors are usually used to reduce noise of power ports. The number and location of capacitors is related to design cost and quality. This paper introduces a method to determine the suitable value and placement of decoupling capacitors for multiple power ports on the same board. Linear network theory is applied to characterize power delivery networks (PDN) with a high accuracy and then a PSO is customized for choosing decoupling capacitors, so that multiple-impedance of all power ports can meet the requirement of the target impedance after placing these capacitors. Finally, a practical board with specific capacitors was simulated and measured. The simulation and experimental results all meet the requirement of target impedance, which confirms the proposed method is valid.

Keywords-decoupling capacitors; linear network theory; PSO; multiple-input impedance

I. INTRODUCTION

Generally, the connection of the chip to the power and ground planes (PGP) is represented by the power port. To maintain an acceptable power noise level, decoupling capacitors are put onto the PCB to decrease the impedance of power ports, while it will increase the power consumption and component costs. Therefore it is significant to select capacitors as effectively as possible. Some formulas derived from experience can help determine the suitable value of capacitor parameters [1]. Choosing capacitors including their types and locations can be automated by genetic algorithms [2-4].

Methods discussed above can help to lower PDN impedance successfully and they all focus on the PDN with only one power port. However, there are usually multiple power ports on the PCB. Power noise of these ports may interconnect with each other. In this paper, linear network theory is employed to characterize PDN as a group. Then this PDN is integrated with circuit models of capacitors to find the optimum values and locations of decoupling capacitors, which is processed by PSO technique.

II. IMPEDANCE OF POWER PORTS

When the voltage ripple tolerance for one power port is ripple, the target impedance is given by

$$Z_{target} = \frac{V_{dd} \times ripple}{0.5 I_{peak}} \tag{1}$$

Where V_{dd} is the supply voltage, I_{peak} is the maximum current drawn by the port over an entire clock period. Assuming that there exist n power ports totally, as shown in Figure 1, the target impedance of each power port can be calculated according to Eq. 1. In addition each port takes the input impedance as its PDN impedance. According to the linear network theory, the electric potential on port j is

$$\dot{U}_j = Z_{j1}\dot{I}_1 + Z_{j2}\dot{I}_2 + \cdots + Z_{jn}\dot{I}_n = \sum_{l=1}^{n} Z_{jl}\dot{I}_l \tag{2}$$

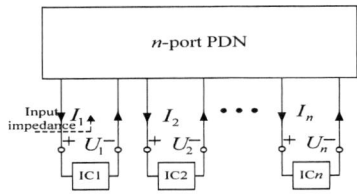

FIGURE I. DIAGRAM OF *N*-PORT PDN.

Hence, the input impedance of port j is then

$$Z_j = \frac{\dot{U}_j}{\dot{I}_j} = \frac{1}{\dot{I}_j} \sum_{l=1}^{n} Z_{jl}\dot{I}_l \tag{3}$$

Where Z_j is defined as multiple-input impedance of port j[5]. In the above calculation all effects of other power ports are reflected by the transfer impedance. Therefore the multiple-input impedance can characterize the power port with a high accuracy. On the contrary, if there is only one power port, self-impedance is often used as PDN impedance in traditional decoupling design.

III. SELECTION OF DECOUPLING CAPACITORS BASED ON PSO

A. Decoupling Design

In order to reveal the effect of a capacitor on the power port, the connection of a capacitor to the PGP is replaced by a new port, which is defined as capacitor port here. Firstly Z parameter of all ports can be extracted by numerical methods or measurement. When one capacitor is mounted or removed during iterations, Z matrix will be changed. Assuming that there exist p power ports and q capacitor ports in PDN, the power ports and capacitor ports are indexed as 1 to p and p+1 to p+q, respectively.

$$\left[\dot{U}_1 \quad \cdots \quad \dot{U}_p \mid \dot{U}_{p+1} \quad \cdots \quad \dot{U}_{p+q}\right]^T = \left[\boldsymbol{U}_{ic} \mid \boldsymbol{U}_{cap}\right]$$
$$\left[\dot{I}_1 \quad \cdots \quad \dot{I}_p \mid \dot{I}_{p+1} \quad \cdots \quad \dot{I}_{p+q}\right]^T = \left[\boldsymbol{I}_{ic} \mid \boldsymbol{I}_{cap}\right]$$
$$\begin{bmatrix} Z_{11} & \cdots & Z_{1p} & \mid & Z_{1(p+1)} & \cdots & Z_{1(p+q)} \\ \cdots & \cdots & \cdots & \mid & \cdots & \cdots & \cdots \\ Z_{p1} & \cdots & Z_{pp} & \mid & Z_{p(p+1)} & \cdots & Z_{p(p+q)} \\ \hline Z_{(p+1)1} & \cdots & Z_{(p+1)p} & \mid & Z_{(p+1)(p+1)} & \cdots & Z_{(p+1)(p+q)} \\ \cdots & \cdots & \cdots & \mid & \cdots & \cdots & \cdots \\ Z_{(p+q)1} & \cdots & Z_{(p+q)p} & \mid & Z_{(p+q)(p+1)} & \cdots & Z_{(p+q)(p+q)} \end{bmatrix} = \begin{bmatrix} \boldsymbol{A} & \mid & \boldsymbol{B} \\ \hline \boldsymbol{C} & \mid & \boldsymbol{D} \end{bmatrix} \tag{4}$$

$$\begin{bmatrix} \boldsymbol{U}_{ic} \\ \boldsymbol{U}_{cap} \end{bmatrix} = \begin{bmatrix} \boldsymbol{A} & \boldsymbol{B} \\ \boldsymbol{C} & \boldsymbol{D} \end{bmatrix} \begin{bmatrix} \boldsymbol{I}_{ic} \\ \boldsymbol{I}_{cap} \end{bmatrix} \tag{5}$$

Considering that capacitor port k(k=p+1,p+2,...,p+q) has capacitors in parallel in one capacitor port, as shown in Figure 3, current drawn by port k is

$$\dot{I}_k = -Y_k \dot{U}_k \tag{6}$$

Where Y_k is sum of the admittance of these capacitors. If there is no capacitor on port k, then Y_k is 0.

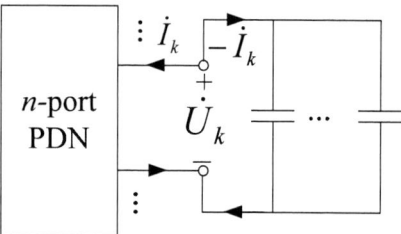

FIGURE II. DIAGRAM OF PORT WITH CAPACITOR IN PARALLEL.

Thus potential and current of these q ports can be written in a matrix form,

$$\boldsymbol{I}_{cap} = \boldsymbol{Y}_C \, \boldsymbol{U}_{cap}$$
$$\boldsymbol{Y}_C = -diag\left(Y_{p+1}, Y_{p+2}, \cdots, Y_{p+q}\right) \tag{7}$$

When all capacitors and their ports are determined, Eq. 5 together with Eq. 7 can eliminate q capacitor ports. Then the Z matrix of p power ports after placing capacitor is derived by

$$\boldsymbol{Z} = \boldsymbol{A} + \boldsymbol{B}(\boldsymbol{I} - \boldsymbol{Y}_C \boldsymbol{D})^{-1} \boldsymbol{Y}_C \boldsymbol{C} \tag{8}$$

After that, the multiple-input impedance of the power ports is calculated by Eq. 3 and compared with their target impedance to determine whether decoupling design is acceptable or not. If not, PSO will change capacitors or their ports by iterations and YC will be changed at same time. After several iterations the Z matrix is optimized and finally the type of capacitors and their ports will be derived.

B. Optimal Technique

Particle Swarm Optimization (PSO) is inspired by the movement of birds while flocking. Each particle claims to be a solution of the problem within the search space, intending to attain optimal value of fitness, which is the objective function of optimization problem. In next iterations, particles keep the track of their best positions obtained so far and the best position among all the particles. Then their velocities and positions are updated by the following equations

$$v_i(t + \Delta t) = \omega(t) v_i(t) + p_1 r_1 (x_l - x_i) + p_2 r_2 (x_g - x_i)$$
$$x_i(t + \Delta t) = x_i(t) + v_i(t + \Delta t)\Delta t$$
$$\omega(t) = (\omega_i - \omega_f)\frac{t_{max} - t}{t_{max}} + \omega_f \tag{9}$$

Where x_i is the position of a particle and v_i is the velocity, t is the current iteration number, r_1 and r_2 are random numbers uniformly distributed in the range [0, 1]. The parameter ω is inertia who's initial and final values are respectively ω_i and ω_f, p_1 and p_2 are the acceleration coefficients. The x_g represents the best position attained by globally best particle and x_l is the best position gained by particle i so far. t_{max} is the maximum number of iterations. Since there are thousands of kinds of decoupling capacitors, PSO has advantage over some manual methods in searching the best composition of capacitors due to its global searching ability.

As mentioned in last section, some capacitor ports are set and marked in advance here. Then all capacitors are sorted according to their resonance points because the frequency range throughout which one capacitor has low impedance is associated with the resonance point. In this case study, the particle has two dimensions and can have discrete values only, one dimension represents capacitor number, and the other represents port number. When capacitor number and port number are initialized, they are corresponding with each other. Then fitness is determined by Eq. 10.

$$\min \; fit = N_C + \sum_{i=1}^{m} (\max(Z_i(f)) - Z_{it\arg et}) + P \qquad (10)$$

$$P = \begin{cases} 0, \max(Z_i(f)) \le Z_{it\arg et} \; (i=1,2,\cdots m) \\ k \sum_{i=1}^{m} \max(Z_i(f)), else \end{cases} \qquad (11)$$

Where N_C is the number of capacitors chosen by PSO, $Z_i(f)$ is the magnitude of the multiple input impedance of power port i, $Z_{itarget}$ is the target impedance of power port i, m is the number of power ports, P is a penalty function and k is a constant such as 50. Eq. 10 and Eq. 11 are integrated into PSO to search the capacitor number and port number which contribute to the minimum value of fitness.

C. Practical Analysis

A practical board with two layers and a pair of PGP, was designed and manufactured, as shown in Figure.4. The board has 3 power ports (P1~P3) on the top layer and 16 capacitor ports (P4~P19) on the bottom layer. It also has a dielectric thickness of 0.4866 mm, and a metal thickness of 0.036 mm. The dielectric has a permittivity of 4.8 and a loss tangent of 0.02. There is one SMA connector attached to every power port. The SMA connectors are used to connect the port with the measuring instrument, Agilent E5061B Network Analyzer. The board including VRM, packages and bulk capacitors was modeled in Ansoft SIwave. The VRM is modeled by an inductor of 5nH in series with a resistor of 50 mΩ. There is a 10uF bulk capacitor mounted besides the VRM.

FIGURE III. A PRACTICAL BOARD FOR CASE STUDY.

The supply voltage of power ports is 3.3V. The maximum voltage ripple allowed is all 5% and maximum current is respectively 0.12A, 0.10A and 0.8A. According to Eq. 1, the target impedance is 2.750Ω, 3.300Ω and 4.125Ω, respectively. The frequency range of decoupling design is about 1MHz to 200MHz. When there are no decoupling capacitors on board, the calculated and experimental multiple-input impedance of the power ports is shown in Figure 5.

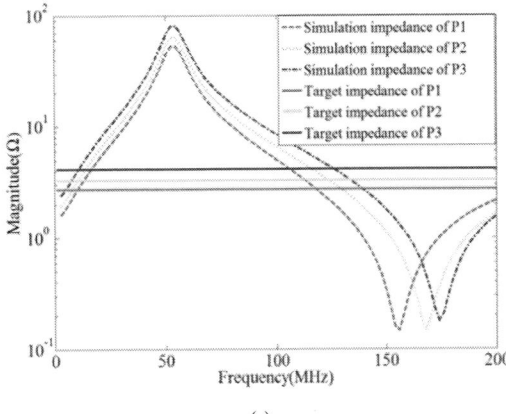

(a)

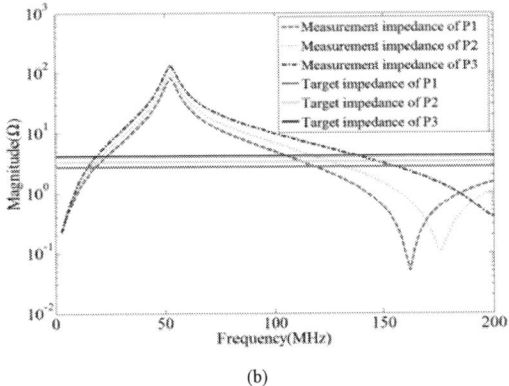

(b)

FIGURE IV. PDN IMPEDANCE WITHOUT CAPACITORS: (A) SIMULATION,(B) MEASUREMENT.

The simulation results agree with experimental tests. Traces in Figure 5 are not low enough in a very large frequency range, thus decoupling capacitors are picked out by PSO in this case and they are listed in Table 1.

TABLE I. STYLES AND LOCATIONS OF CAPACITORS.

Capacitor Name	Manufacturers	Ports
CL21C471JBANNNE	Samsung	P5
CL21F103ZBANNNE	Samsung	P7
CL21C102JBANNNE	Samsung	P10
CL21F105ZAFNNNE	Samsung	P8 P9 P11 P14

Decoupling capacitors were mounted both in theoretical model and practical board. The magnitude of simulated and experimental impedance of the three ports after decoupling design is shown in Figure 6.

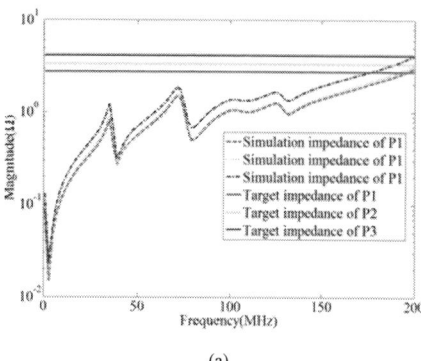

(a)

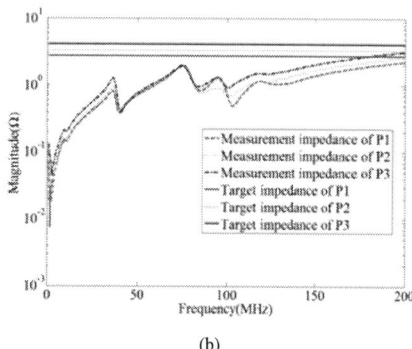

(b)

FIGURE V. PDN IMPEDANCE WITH CAPACITORS: (A) SIMULATION,(B) MEASUREMENT.

The behavior between simulated and experimental multiple-input impedance keeps the similar trend. The resonances of simulated results have little difference from those of experiment because the parasitic parameters caused by soldering capacitors in experiment may be not the same as those used in simulation and they are difficult to be controlled when frequency is high. In short, the results with selected capacitors by PSO are much better than those without capacitors and they meet the requirement of the target impedance.

IV. CONCLUSION

Linear network theory was used to describe the complex connection among multiple power ports of chips and it is proven that decoupling design based on multiple-input impedance has a high accuracy. Then optimal technique PSO was employed to select decoupling capacitors including their types, numbers and locations for multiple power ports at the same time. The selected capacitors are mounted on a practical board and consequently the decoupling results derived by experiments are acceptable.

ACKNOWLEDGMENT

This work is supported by Natural Science Foundation Project of China (No. 51177183), Natural Science Foundation Project of Chongqing (No.CSTC 2009BB3033) and Application Development Project of Chongqing (No.cstc2013yykfA60001).

REFERENCES

[1] Pan Siming, Achkir Brice. Optimization of power delivery network design for multiple supply voltages, 2013 IEEE International Symposium on Electromagnetic Compatibility, pp. 333-337.

[2] Swaminathan M, Kim J, Novak I, et al. Power distribution networks for system-on-Package status and challenges, IEEE Transactions on Advanced Packaging, 2004, 27(2): 286-300.

[3] Choi Jae Young, Swaminathan Madhavan. Decoupling capacitor placement in power delivery networks using MFEM, IEEE Transactions on Components Packaging and Manufacturing Technology, 2011, 1(10): 1651-1661.

[4] Tripathi Jai Narayan, Mukherjee Jayanta, Apte Prakash R., et al. Selection and placement of decoupling capacitors in high speed systems, IEEE Transactions on Electromagnetic Compatibility Magazine, 2013, 2(4): 72-78.

[5] Mu-Shui Zhang, Yu-Shan Li, Li-Ping Li. Analyze and design high-speed power delivery networks using new multiinput impedances in printed circuit boards, IEEE Transactions on Microwave Theory and Techniques, 2009, 57(7): 1818-1831.

International Conference on Power Electronics and Energy Engineering (PEEE 2015)

Intelligent Management System of Vehicle Violation Based on Electronic License

Y.Z. Cai, Y.R. Li, Z.Y. Cai

College of Computer and Information Technology
China Three Gorges University
Yichang, China

Abstract—**Traditional vehicle management system is based on metal plate mainly identified by the video and image processing techniques which are low accuracy and less efficiency. However based on RFID technology, electronic license has the unique factory curing ID number and the vehicle physical binding, making it incomparable anti-counterfeiting. Here the design scheme of the whole system including database, detecting processing, and collision algorithm are discussed. Then the prototype of the proposed system is presented, and effectively solves the violation management and identification problem.**

Keywords-RFID; electronic license; mobile internet electronic; intelligent management system

I. INTRODUCTION

Radio Frequency Identification (RFID) technology originated from the 1990s, went through three stages, electromagnetic induction or electromagnetic propagation mode, non-contact identification of the target tracking and two-way data communication of the new automatic identification technology[1].

With the increasing application of radio frequency identification technology, electronic identification is appeared in automatic vehicle identification [2]. It can be set for two-way communication path between the inlet passage of the reader device in order to achieve the purpose of the electric vehicle target identification and data exchange[3]. Electronic license system includes a database server, the card issuing terminals, terminal management, and display terminals, and is helpful for automatic plate detection by electronic vehicle license, identification, access control and information management, and other related functions[4]. Electronic license is stored in the vehicle identification database, which can only be accessed and operated by an authorized RFID readers[5]. The transport corridors set up monitoring station connected with a central server via WLAN and Police linked with PDA by network security professionals [6].

Toll and remote real-time monitoring, vehicle safety inspection and intelligent traffic management can help solve the car monitoring, and other functions to improve the traffic situation[7]. In this paper, this problem is primarily based.

II. DESIGN SYSTEM ANALYSES

Because vehicle has their own unique electronic identification, corresponding label system and database should be developed with vehicle electronic information management

to record basic information of all vehicles. RFID readers can be placed in each city traffic junctions to read the traffic data and transmit them to the central processor via ZigBee network. Processors can determine whether the vehicle is illegal or not by query the database of vehicle electronic information management. If a vehicle is illegal, the alarm equipment bounded together with readers would alarm the law enforcement officers nearby.

By corresponding algorithm design, this system monitor whether a vehicle entering the monitored area and whether it is violated. In the case of a car, it is gonging to generate the illegal recording, uploading them to the central control subsystem, gathering other illegal information, and providing a basis for illegal punishment.

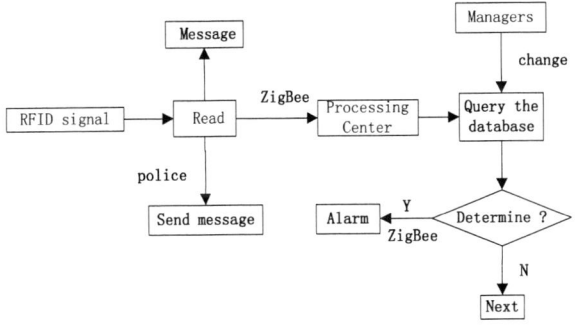

FIGURE I. SYSTEM SCHEME.

III. SYSTEM ANALYSIS AND DESIGN MODULES

A. Database Design

The system needs business process analysis tool to establish a database table for the system to save various data, as shown in Table 1.

© 2015. The authors - Published by Atlantis Press

TABLE I. BASIC SYSTEM TABLE.

Name	Explanation
U_Panel	User Control Panel
U_AddPanel	User additional control panel
U_Groups	User Groups
U_Users	User
U_UserAttr	User Properties
U_System	The basic properties of the system
U_LoginSet	User login settings
U_UseLogin	User login status table
U_UpFileSet	Upload file settings User Group
C_Unit	Vehicle personnel affiliations table
C_Chauffeur	Vehicles person table
C_CarInfo	Vehicle Information Sheet
C_Violation	Vehicle illegal information table
C_Accident	Vehicle Accident Information Sheet
C_YAuditing	Vehicles examined table

B. Detection Processing System

The handheld system terminals is used for transmitting the data from UHF RFID reader device by 51 processors, then the processors connected with the microcontroller and ZigBee wireless communication module can communicate with the processing center, as shown in Fig 2.

C. Red light Violation Monitoring

In this unit when the red light on, the readers installed in road intersection are used to judge whether a car in the monitoring area is illegal. If there is an illegal car, then it determine whether it is a normal driving, such as turn right; if it is not normal driving, the vehicle is recorded as an electronic tag ID number, and other relevant information is also collected, such as time, place, names and other road intersections. Finally, recorded electronic tag's ID number, and other illegal information will be precluding to the data collection center.

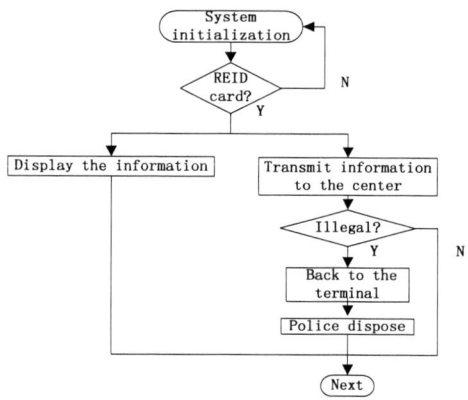

FIGURE II. FLOWCHART OF HANDHELD TERMINAL.

D. Monitoring Illegal Parking

Illegal parking phenomenon often occurs in the city, which is one of the important causes of urban traffic congestion. The basic principle is to monitor illegal parking is that the reader is reasonably set to monitor in the prohibit parking area defined by traffic management departments.

E. Collision Algorithm Descriptions

The system uses a binary searching algorithm, the tag reader in the work area continues to be divided into P subsets (P> l), and then a subset of the same division is continuously divided into a subset of the more or fewer within the tag number to achieve a successful identification tag reader, until the number of tags within a subset of 1. When the tag is read being completed within a certain subset, the reader will search back using other waiting to read the label. This can be seen as a process of tag grouping by all labels according to the grouping scheme from the root to leaf nodes and processing diversion layer by layer. Only all leaf node labels are searched, can the process be successfully read out.

Performance analysis of algorithms:

To find a separated tab is required to repeat from relatively large number of labels. The average search depends on the number of readers within scope of the total number of tags which is identified as n:

$$I = \log_2 n + 1 \tag{1}$$

In the N pending identification tags, average recognition algorithm requires a label search. Obviously, the identification tag within the read range will be reduced to number of completed tag, and total required search cable time BS of identification N is:

$$I_{bs} = (\log_2 n + 1) + [\log_2 (n-1) + 1] + ... + (\log_2 2 + 1) + (\log_2 1 + 1) = n + \log_2(N)$$
$$B_{bs} = I_{bs} * k = (n + \log_2(n!)) * k \tag{2}$$

Because each request is passed to the tag reader by instruction, its argument is the length of the entire sequence number, so the binary bits to be transmitted to the reader is kth in the total number of searching algorithm and the product serial number length is labeled as:

$$L_{bs} = I_{bs} * k = (n + \log_2(n!)) * k \tag{3}$$

IV. SYSTEM TEST

Firstly, each modules of the system are all tested respectively, and it turned out that all modules were normal. The exchanging information test between the different modules is showed as normal data transfer speed, with rapid response to electronic license. The login interface is shown in Fig 3.

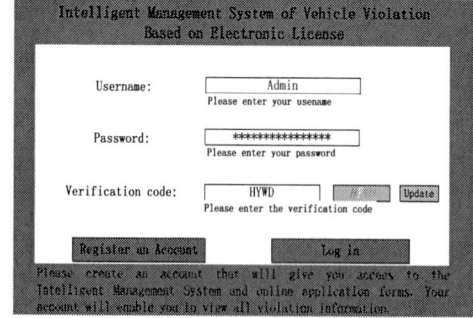

FIGURE III. SYSTEM LOGIN INTERFACE.

The Table 2 shows the test results of system database, where primarily record information of vehicle owners includes vehicle violation information.

TABLE II. DATABASE TEST TABLE.

Test Module	Test function	Test Results
User Login	User login, authentication code updates, user duplicate login	Good
User Control Panel Management	Add module control node, modify the module node, set the module node, the mobile module nodes, remove the module node	Good
User Group Management	Adding user groups, modify user	Good
User group permission settings	Add, modify, save the user module operating authority	Good
Manage Users	Add users, change user, delete user, disable and enable the user to lock and unlock the user	Good
Unit Management	Modify the unit, delete unit, disable and enable the unit to lock and unlock the unit	Good
Personnel management	Add personnel, modify personnel, personnel delete, disable and enable staff to lock and unlock the staff	Good

By vehicle registration database, the system can display basic information of the owners to identify the fake cards, deck, theft, illegal operations. Therefore achieved automatic vehicle identification can provide an illegal vehicle blacklist and the vehicle identification authenticity, preventing the vehicle from counterfeiting, scrapping tax evasion and other phenomena, and combating with car thefts, car snatching, decks and other criminal activities. The figure 4 shows the integrated test of owner information.

In the system test, the entire system works well, demonstrating the system design stability and the usability of intelligent management system of vehicle violation.

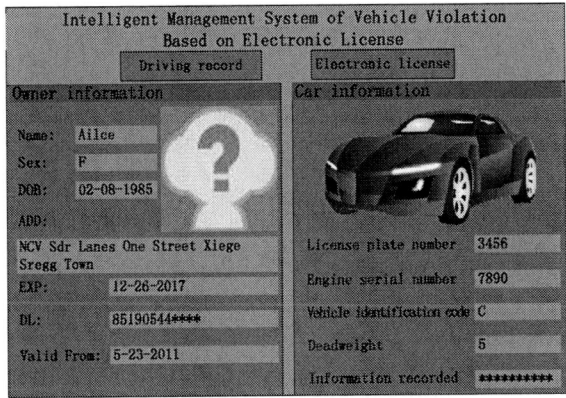

FIGURE IV. INTEGRATED TEST.

V. SUMMARY

The electronic license management system identify illegal vehicles based a variety of sensors and a given unique ID number of vehicle, and can analyze current vehicle in real-time collection and display the information at any place. Different from traditional metallic plate, the embedded terminal platform facilitate the transport sector to grasp and analyze vehicle information, integrating the intelligent control and human control into an intelligent vehicle management system.

ACKNOWLEDGEMENTS

This research was supported by the National Natural Science Foundation of China (No. 71471102), Key projects of science and technology research program of Hubei Provincial Department of Education in China (Grant No. D20101203).

REFERENCES

[1] Slapsinskas P and Zilys M.Efficiency of the Electronic License Plate Recognition System[J].ELEKTRONIKA IR ELEKTROTECHNIKA, 2013, v19,n8:59-64.

[2] Chen Wei, Yu Jianding, Liu, Xiangjun. The Design of Electronic License Plate Recognition Terminal System based on nRF24LE1[C]. 5th International Symposium on Computational Intelligence and Design (ISCID).OCT 28-29, 2012:127-129.

[3] Liu Qiang, Yang Shuchen and Wang Jun ects. Based on wireless sensor networking and video-aware navigation and find parking system[C]. 3rd International Conference on Energy, Environment and Sustainable Development.NOV 12-13, 2013, n860-863:2876-2879.

[4] Ruttenberg, Clara. Finding the tool that fits best: Cloud based task management for electronic resources[J]. (McKeldin Library, University of Maryland, College Park, College Park, MD, United States) Source: OCLC Systems and Services, 2013.v 29, n 3: 151-160.

[5] Negny, Stéphane; Dupros, Fabrice; Vautrin, Benoît .Collaborative Simulation and Scientific Big Data Analysis: Illustration for Sustainability in Natural Hazards Management and Chemical Process Engineering[J]. Computers in Industry, April 2014 .v 65, n 3:521-535.

[6] Brandwein, Dennis; Uckelmann, Dieter; Beenken, Bjoern. Using RFID in License Plates and Vignettes for Electronic Vehicle Identification Structured Testing of Passive UHF Systems for Vehicle Identification[C]. 1st International Conference on the Impact of Virtual, Remote and Real Labs in Logistics.FEB 28-MAR 01, 2012,n282:148-155.

[7] Tiiro, Samuli. Spectrum Sharing in MIMO Cognitive Radio Systems With Imperfect Channel State[J] .IEICE Transactions on Communications. 2014,v E97-B, n 4: 867-874.

International Conference on Power Electronics and Energy Engineering (PEEE 2015)

Research of HTS Application on Tokamak Magnets

L. Ren, X.Z. Deng, Y. Xu, H. Liu, J.D. Li, J. Shi, Y.J. Tang

State Key Laboratory of Advanced Electromagnetic Engineering and Technology
R&D Center of Applied Superconductivity
Huazhong University of Science and Technology
Wuhan, 430074, P. R. China

Abstract—in this paper, the application feasibility of HTS tapes on tokamak magnets is discussed and the existing four kinds of HTS high current conductors are summarized. Furthermore, an YBCO conductor structure suitable for a 10 GJ tokamak TF magnet is designed and the current carrying capacity is evaluated. The simulation result shows that the critical current of the YBCO conductor is about 75 kA @ 10 T, 30 K.

Keywords-HTS Application; Tokamak Magnets; YBCO conductor

I. INTRODUCTION

It is essential to use superconducting coils for future fusion power generation devices to decrease the electric energy consumption and thus ensure the magnet systems operating safely and stably. From the mid-seventies to the late eighties, the applications of superconductivity on tokamak entered a golden age, with plenty of large devices and prototype coils being developed in Russia, US, Japan and Europe. The increase of dimensions and stored energy drove an interesting variety of conductor and coil design. So far, these superconducting coils are all made of low temperature superconducting cable-in-conduit conductor (LTS CICC), such as NbTi, NbZr and Nb3Sn. Because of the transverse load degradation and self-field induced quench, it is difficult for LTS CICC to be larger dimensions. Therefore, it is urgent to design next generation tokamak conductors.

With the development of HTS materials, it is possible for HTS conductors to be applied in tokamak magnets. Compared to LTS magnets, HTS magnets can greatly save cooling cost and have better thermal stability. In ITER, first generation (1G) HTS conductors have been used for some components such as current leads and bus bars. Moreover, Bi-2212 wires have been previously bundled into Rutherford cables for magnet applications. But, as some relevant problems, such as high current conductor and coil development, and high-cost obstacle to the use of HTS materials, have not yet been resolved, HTS conductors have not been introduced into tokamak magnets on a large scale. At present, YBCO commercial products have been available. The mechanical properties are better than that of BSCCO [1]. In the near future, if single YBCO tape can be made long enough to meet the requirements for winding tokamak coils and the cost fall down greatly, YBCO tapes will go into practical application on tokamak magnets.

In this paper, the application feasibility of HTS tapes on tokamak magnets is discussed and the existing four kinds of HTS high current conductors are introduced. Furthermore, an YBCO high current conductor structure suitable for tokamak magnets is designed and the critical current is calculated.

II. HTS MATERIALS MAY BE APPLIED IN TOKAMAK MAGNETS

There are two possible options for HTS materials used in tokamak magnets: 1G BSCCO and 2G YBCO conductors. Bi-2212 is the only HTS conductor which can be produced as a round multi-filamentary wire. Therefore, it is isotropic and conventional winding methods could be used. But it is brittle and strain sensitive. The Bi-2223 multi-filamentary tape has a strong anisotropy. The engineering critical current densities of Bi-type conductors are in the range of 70-125 A/mm2 (77 K, self-field). The minimum bending radius is 25 mm. A typical maximum axial tensile stress value is in the range of 200-250 MPa (77 K). The critical currents of Bi2223 decrease rapidly with the increasing of the magnetic field at temperatures above 50 K. In order to obtain higher critical current densities and lower field sensitivity, all present application of BSCCO have operating temperatures lower than 30 K [2]. At 77 K, the perpendicular magnetic field that Bi2223 can withstand is below 1T [3].

The engineering critical current densities of YBCO coated conductors (CCs) are in the range of 200-400 A/mm2 (77 K, self-field). The tensile stress of up to 700 MPa can be tolerated with no irreversible degradation of critical currents (about 0.6% strain). The minimum bending radius is about 10 mm. Compared to Bi-type conductors, the in-field performance and mechanical properties have been improved greatly. Even at fields well above 10 T, the operation temperatures are high enough to make a profit. They have high critical currents at relatively high temperature (less than 50 K) and that makes it feasible to be cooled by cryocoolers instead of by liquid helium. YBCO had already been proposed to be applied in tokamak magnets in 2004 [4].

Figure 1 shows the electromagnetic characteristics of Bi2223 (made by Innost, China) and YBCO (made by SuperPower, USA) obtained by experiments. The tests have been finished in liquid nitrogen. When a HTS conductor is applied in magnetic field perpendicular to the tape surface, the critical current of Bi2223 falls faster than that of YBCO does. When the magnetic field angle θ (the angle of the tape surface

© 2015. The authors - Published by Atlantis Press

and the magnetic field) changes, it has a less effect on the critical current of YBCO than it does on that of Bi2223.

Briefly, YBCO CCs have more potential to be used for tokamak magnets because of its impressive performance.

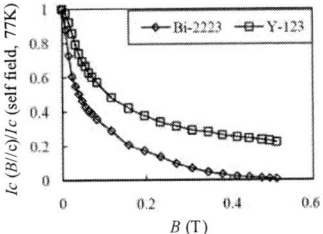

(a) Normalized current dependent on external field

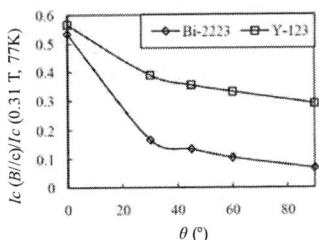

(b) Normalized current dependent on angle θ

FIGURE I. ELECTROMAGNETIC CHARACTERISTICS OF BI2223 AND YBCO.

III. DEMONSTRATED CONDUCTOR DESIGN

A. Concept Demonstration of High Current Conductors

There are four kinds of high current HTS conductors that have been proposed. They are the rectangular conductor type, Rutherford cable type [5], helical winding cable type [6] and helical twisted stacking-tape cable type [7].

In 2001, a Bi2212 conductor was designed for the TF coil of A-SSTR2. It is composed of a rectangular HTS cable and a rectangular copper cable (see Figure 2). The conductor is cooled by a rectangular cooling channel which contacted its one side surface. The operating temperature is 20 K. The critical current density is 1000 A/mm2 at 20 K and 23T.

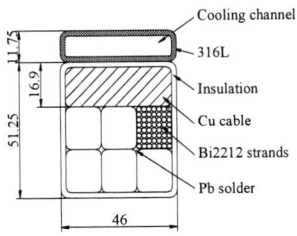

FIGURE II. STRUCTURE OF RECTANGULAR CONDUCTOR TYPE.

In 2005, Roebel technique was applied for YBCO CCs to achieve high current carrying capabilities for low ac-loss application. In 2011, S. I. Schlachter, et al, developed Roebel cables consisting of up to 50 tapes with current carrying capabilities up to 2.6 kA (77 K, self-field) and presented a concept for CC Rutherford cable using Roebel subcable as strands. Figure 3 shows the structure.

Another approach is a REBCO CC helically winding cable which is constructed by spiral-winding tapes around an insulated round copper cable former (see Figure 4). Each superconducting layer is wound in the direction opposite to the winding direction of its neighboring layer. The cable consists of 24 tapes in 8 layers (3 tapes per layer). The individual CC has Ic of about 125 A at 76 K. The critical current of the cable is 2.8 kA at 76 K, self field.

The forth approach is to twist stacked-tape cable in a shaped former. Figure 5 shows three-helical-groove CICC of 32-YBCO- tapes in each groove. A stacked and twisted conductor of 96 YBCO tapes (Ic = 90 A) in three grooves may carry approximately up to 1.53 kA current at 77 K, self field.

The four demonstrated designs indicate that HTS materials, especially YBCO could be made into high current conductors for large coils, such as tokamak magnets. The first design is similar to LTS CICC conductor. It is a good choice only for Bi2212 round conductor. The other three designs can be applied to 2G HTS tapes. The Rutherford cable type needs to cut out a part of the conductors so as to strand them into a rectangular cable. Hence, the cost is relatively expensive. The third approach is not adequate for much higher current cables because there is much difference between the tape lengths of each layer. For the fourth approach, it may be not easy for the cable to be cooled by cryogenic liquid because of lack of cryogenic liquid channels in the cable interior.

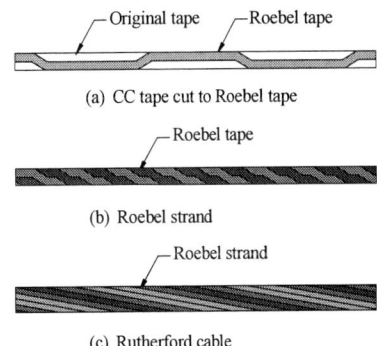

FIGURE III. STRUCTURE OF RUTHERFORD CABLE TYPE.

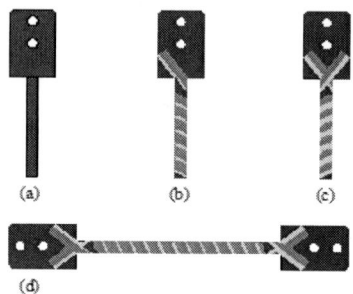

FIGURE IV. STRUCTURE OF HELICALLY WINDING CABLE TYPE.

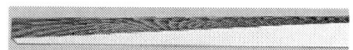

(a) A stacked-tape cable twisted with a twist pitch

(b) Three helical grooves in a rod

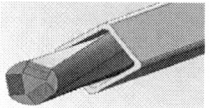

(c) Three-helical-channel CICC cable

FIGURE V. STRUCTURE OF TWISTED STACKING-TAPE CABLE.

B. Design Requirements

High current conductors are usually made by conductor strands in parallel. In order to minimize nonuniform current distribution at different location, the strands have to transpose each other along the length of the conductors. The thin and flat shape is a barrier for HTS tapes to be twisted repeatedly.

In general, HTS conductors for tokamak should meet these conditions: they should have high-current and high in-field engineering current density windings; be flexible and mechanically robust; and have low conductor anisotropy, low magnetization losses. In addition, it is necessary to keep a certain ratio of copper to superconductor for stabilizing conductor and quench protection although the ratio could be reduced greatly considering that YBCO conductors usually conclude copper stability layers. Above all things, the conductor design must consider the operating condition of the HTS coils.

In the design of HTS coils, the operating temperature of the coil is a key issue. When operating temperature is 30 K instead of 4.5 K, the capacity of cryogenic system is expected to be reduced by about 40% [8]. Furthermore, the mechanical performance of structural materials are decreased if YBCO tapes operating at more than 30 K. Therefore, 30 K is chosen as the operating temperature. In this paper, a high current conductor is designed for a future tokamak TF coil with 50 kA nominal current and 10 T maximal magnetic field. Considering the design margin, the critical current of the HTS conductor will be designed as 75 kA @10 T, 30 K. The temperature rise

of the conductor due to Joule heat generation during quench is less than 200 K. The terminal voltage of the TF coil is limited to 20 kV. What follows is a design of an YBCO high current conductor. The main parameter requirements are listed in Table 1.

TABLE I. PARAMETERS OF THE HIGH CURRENT CONDUCTOR.

Critical current (10 T, 30 K)	75 kA
Maximum field	10 T
Operating temperature	30 K
Ratio of copper to superconductor	>1: 4
Jc in YBCO at 30 K and 10 T	>100 A/mm^2
YBCO: Cu: Ag of a CC tape	1: 40: 2
Copper cross section	300 mm^2

IV. PHOTOGRAPHS AND FIGURES

A. Conductor Configuration

According to the above principles and parameter requirements, an YBCO conductor design is given in Figure 6. YBCO tapes from SuperPower Inc. are used. The width and thickness of a single tape are 4 mm (some of them are 2 mm) and 0.095 mm, respectively. The corresponding critical currents are 120 A for 4 mm width tape and 60 A for 2 mm width tape.

The conductor is mainly composed of three parts: a former, YBCO tape stack groups and a stainless tube. The former, which is made of a copper rod of 30 mm diameter, has four helical grooves which are orthogonal.

In each groove, there are 5 twisted rectangular YBCO stacks. They constitute nearly 1/4 of the cross section. In order to fix the YBCO tapes, they are sandwiched with two 0.1 mm copper strips at the bottom and top of each stack. The stacks are twisted with a twist pitch of 180 mm. Four stack groups and the former form a 30-mm-diameter circle and are embedded in a tube of 32 mm outer diameter. The center hole and the gap outside the circle can serve as cooling channels. The detailed parameters are shown in Figure 7.

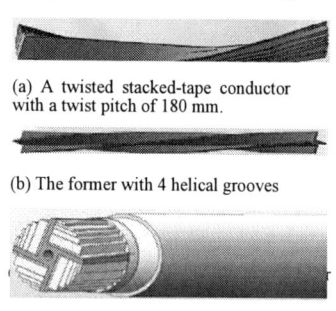

(a) A twisted stacked-tape conductor with a twist pitch of 180 mm.

(b) The former with 4 helical grooves

(c) The four-helical-channel CICC conductor

FIGURE VI. THE STRUCTURE OF THE CICC CONDUCTOR.

32

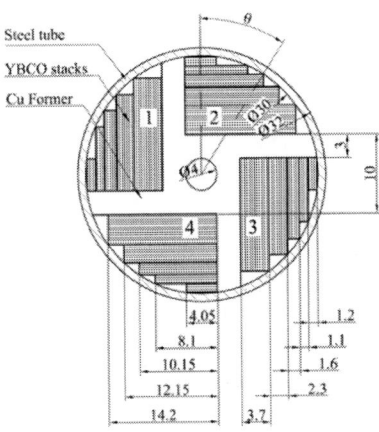

FIGURE VII. THE CROSS SECTION OF THE CONDUCTOR.

B. Conductor Configuration

The in-field critical current of a HTS tape should be determined by the crossing point of the load line and the critical current characteristic curve of the YBCO tape. When the critical current of a single tape in self-field is determined, the critical current of the CICC conductor could be calculated. Therefore, in the critical current calculation, we must measure current density dependence on magnetic field characteristic curve (Ic-B curve) and find out the load line of the HTS tape. Figure 8 is the experimentally obtained Ic-B curve of SCS 4050 YBCO tape at different magnetic field angles. The normalized current is shown in Figure 9.

From Figure 9, it can be seen that the critical current degradation is greater in the normal magnetic field, especially in the perpendicular field, than in the parallel one. Furthermore, the current degradation becomes more severe with the increase of the magnetic field B. When B is 0.05 T, the critical current in perpendicular field ($Ic_{\perp B}$) is 70.5% of that in parallel field, 77 K, self field ($Ic_{//B, \text{77 K, self field}}$). When B is 0.55 T, $Ic_{\perp B}$ drops down to 22% of $Ic_{//B, \text{77 K, self field}}$.

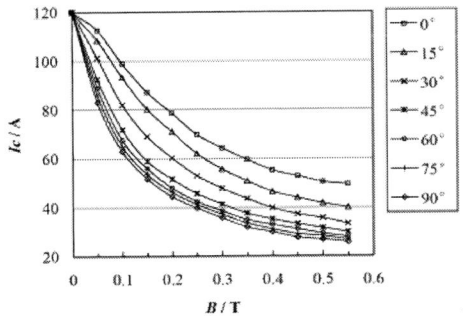

FIGURE VIII. CRITICAL CURRENT OF YBCO TAPE, 77 K.

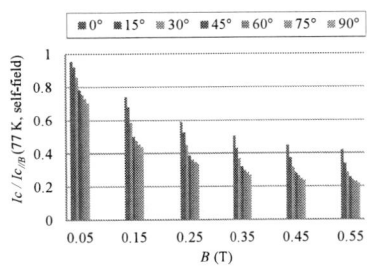

FIGURE IX. NORMALIZED CURRENT OF THE YBCO TAPE, 77 K.

C. Conductor Configuration

For the CICC conductor, when the magnetic field is perpendicular to the surfaces of YBCO stacks 2 and 4 ($\theta = 90°$), it is parallel to those of stacks 1 and 3 ($\theta = 0°$). Thus, current redistribution occurs, and YBCO stacks 1 and 3 will share more current than stacks 2 and 4. Therefore, compared to the case of magnetic field perpendicular to all 4 tape surfaces, the whole current carrying capacity has a promotion. Figure 10 shows the normalized current of the YBCO CICC conductor at different field angles (θ =0°/ 90°, 15°/75°, 30°/60° and 45°/45°), which is the mean value at two different magnetic field angles.

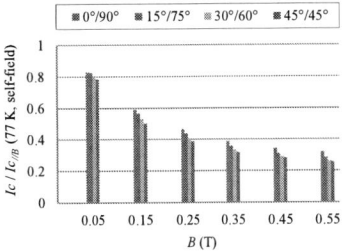

FIGURE X. MEAN NORMALIZED CURRENT OF THE YBCO CICC CONDUCTOR, 77 K.

When B is 0.05 T, the mean critical current in 45° field ($Ic_{\angle 45°B}$) is the least among those in the normal magnetic field, which is 78.5% of $Ic_{//B, \text{77 K, self field}}$, and $Ic_{\angle 0°B}$ is about 83% of $Ic_{//B, \text{77 K, self field}}$. When B is 0.55 T, $Ic_{\angle 45°B}$ is 32% of $Ic_{//B, \text{77 K, self field}}$ and $Ic_{\angle 0°B}$ is 25.5% of $Ic_{//B, \text{77 K, self field}}$. Compared Figure 10 with Figure 9, it can be seen that anisotropy of the YBCO CICC conductor becomes weaker than that of the single YBCO tape.

D. Conductor Configuration

To get the load characteristic, we assume that each tape of the conductor is fed with 120 A for a 4 mm width tape or 60 A for a 2 mm width tape. Then, the magnetic field distribution of the conductor could be found by simulation. As the critical current of the tape in perpendicular magnetic field is the lowest, the perpendicular magnetic field components applied on the tapes in two quarters of CICC conductor and then parallel magnetic field components in the other two are calculated as shown in Figure 11 (a). Furthermore, for the CICC conductor, as the critical current at 45° field is the lowest, Figure 11 (b) shows 45° magnetic field distribution of the tapes in two quarters of CICC conductor.

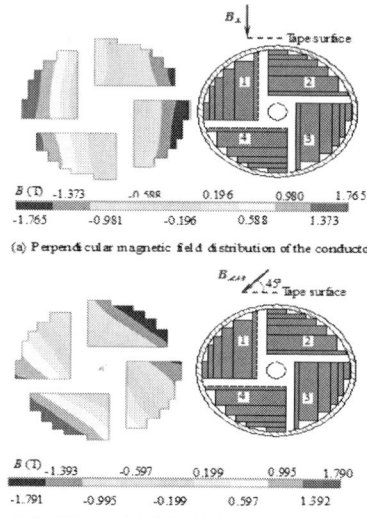

(a) Perpendicular magnetic field distribution of the conductor

(b) The 45° magnetic field distribution of the conductor

FIGURE XI. MAGNETIC FIELD DISTRIBUTION OF THE CONDUCTOR AT 77 K, SELF FIELD.

From Figure 11, it can be seen that the maximum vertical and 45° magnetic fields are 1.765 T and 1.791 T, respectively. The intersecting points of the magnetic field characteristic and the corresponding load characteristic curves are the critical operating points. The lower of two values is the critical current of single tape in the CICC conductor at 77 K, self-field. Figure 12 shows the critical current of the CICC conductor at 77 K. For convenience, the current values are all expressed as that of a single tape in the CICC conductor.

Because the amount of the tapes is certain, the critical current of the CICC conductor can be calculated if the critical operating point of a single YBCO tape is found out. The critical current of the CICC conductor is the sum of the critical current of each single YBCO tape at magnetic field.

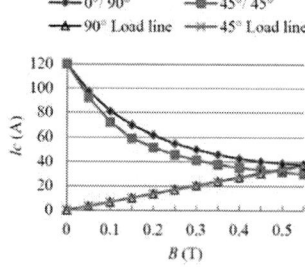

FIGURE XII. MAGNETIC FIELD AND LOAD CHARACTERISTIC AT 77 K.

From Figure 12, it can be seen that: When θ is 0° or 90°, the critical current of a single 4 mm width tape is 37.59 A at 0.55 T, 77 K, and it is the maximal value. When θ is 45°, the critical current is minimal, which is 32.16 A at 0.48 T, 77 K. The CICC conductor consists of 940 YBCO tapes of 4 mm width and 208 YBCO tapes of 2 mm width. Hence, the critical current of the YBCO CICC conductor is conservatively estimated at 33.5 kA at 0.48 T, 77 K.

E. Conductor Configuration

Similarly, we can obtain the critical current of CICC conductor at 30 K. Assuming each tape of the conductor is fed with 120 A for a 4 mm width tape or 60 A for a 2 mm width tape, the corresponding magnetic field components applied on the tapes are simulated as shown in Figure 13.

The maximum vertical and 45° magnetic fields are 10.696 T and 10.854 T, respectively. Then, the load lines and magnetic field characteristic curves of a single YBCO tape in the CICC conductor dependent on perpendicular and 45° magnetic field can be depicted in Figure 14. The current values are expressed by the lift factor [Ic (30 K)/ Ic (self field, 77 K)] from SuperPower Inc.

As shown in Figure 14, when θ is 0° or 90°, the lift factor of a single 4 mm width tape is 3.34 at 6 T, 30 K. When θ is 45°, the lift factor is 2.78 at 5 T, 30 K. When the conductor is applied in 10 T, 30 K, the lift factor is 2.24. Considering the sum of YBCO tapes and Ic (self field, 77 K), the critical current of the YBCO CICC conductor is about 75 kA @ 10 T, 30 K.

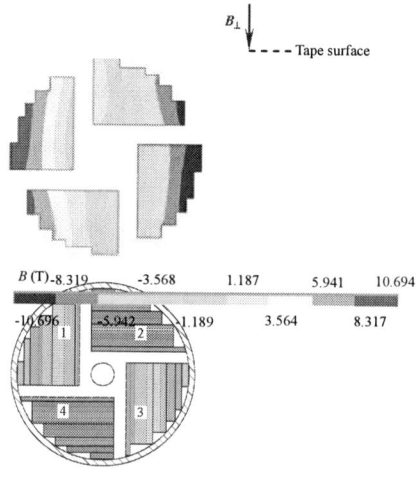

(a) Perpendicular magnetic field distribution of the conductor

FIGURE XIII. THE MAGNETIC FIELD DISTRIBUTION OF THE CONDUCTOR AT 30 K.

V. Conclusion

Compared to 1G HTS conductors, 2G HTS tapes show higher in-field critical current and better mechanical properties. The four demonstrated designs indicate that HTS materials, especially YBCO materials could be used for high operation current tokamak magnets.

An YBCO CICC conductor for a 10 GJ tokamak concept magnet is designed. It includes a copper former with 4 helical grooves, a stainless tube and 4 YBCO tape stack groups which consist of 940 YBCO tapes of 4 mm width and 208 YBCO tapes of 2 mm width. It is easy to be cooled by cryogenic liquid or gas. The calculation shows that the critical current values can get to 33.5 kA @ 0.48 T, 77 K and 75 kA @ 10 T, 30 K.

Acknowledgement

This work was supported in part by Specialized Research Fund for ITER under Grant 2011GB113004 and HUST Foundation for Independent Innovation under Grant 2014TS1145.

References

[1] R.M. Scanlan, D. R. Dietderich, H. C. Higley, et al, Fabrication and test results for Rutherford-type cables made from BSCCO strands, *IEEE Trans. Sppl. Superconduct.*, vol. 9, no. 2, pp. 130-133, June 1999.

[2] D.C. van der Laan, J.W. Ekin, H.J.N. et al. Davidson and J. Schwartz, Effect of tensile strain on grain connectivity and flux pinning in $Bi_2Sr_2Ca_2Cu_3O_x$ tapes, *Applied Physics Letters*, vol. 88, pp. 022511, 2006.

[3] T. Ando, T. Kato, K. Ushigusa, et al, Design of the toroidal field coik for A-SSTR2 using high Tc superconductor, *Fusion Engineering and Design*, vol. 58-59, pp. 13-16, 2001.

[4] W.H. Fietz, S. Fink, R. Heller, et al, High temperature superconductors for the ITER magnet system and beyond, *Fusion Engineering and Design*, vol. 75-79, pp. 105-109, November 2005.

[5] W. Goldacker, R. Nast, G. Kotzyba, et al, High current DyBCO-ROEBEL assembled coated conductor (RACC), EUCAS2005, 11-15 Sept. 2005, Vienna Austria.

[6] D.C. van der Laan & X. F. Lu, $RE-Ba_2Cu_3O_{7-\delta}$ coated conductor helical cables for electric power transmission and SMES, EPRI Superconductivity Conference Oct. 12th, 2011, Tallahassee, FL.

[7] M. Takayasu, Joseph V. Minervini & L. Bromberg, HTS twisted stacked-tape cable development, *Supercond. Sci. Technol.*, vol. 25, pp. 014011, 2012.

International Conference on Power Electronics and Energy Engineering (PEEE 2015)

Analysis of Losses in High Speed Slotless PM Synchronous Motor Integrated the Added Leakage Inductance

B.Q. Kou, H.C. Cao
Harbin Institute of Technology
Harbin, Heilongjiang, China

W.L. Li
Beijing Jiaotong University
Beijing, China

Abstract—this paper researched the loss of high speed slotless permanent magnet synchronous motor (PMSM) with appended leakage inductance structure. This PMSM adopted novel stator structure to add the leakage inductance for solving the problem that high speed slotless PMSM inductance value is small. Compared with traditional motor, this motor adopted outer slot-tooth structure in stator iron; meanwhile it would add the iron loss. Therefore, this paper adopts the Bertotti separating iron model to calculate the iron loss of the stator with outer slot-tooth, then the iron loss in the yoke of stator was compared with the outer slot-tooth. The result proved that, the outer slot-tooth structure has less influence to stator iron loss. Secondly, the influence law between the appended leakage inductance and rotor eddy loss was analyzed. The paper analyzed rotor eddy loss under the condition of three kinds of winding that has different space distribution, and obtain the influence law between the winding space distribution and rotor eddy loss.

Keywords-PMSM; leakage inductance; influence law

I. INTRODUCTION

Due to advantages of high speed, high power density, small moment of inertia, etc., research of high-speed permanent magnet synchronous motor (HS-PMSM) has been more and more extensive application in fields of distributed power generation, flywheel energy storage, high speed centrifugal compressor, electric spindle and other. Now it has become the current research focus in the field of international electrotechnical (Paulides, Jewell & Howe 2004, Wang 2006, Cho et al. 2011, Kong, Wang & Xing 2012).

With the advantages of HS-PMSM, such as no cogging effect, high speed slotless permanent magnet synchronous motor (HSS-PMSM) could significantly reduce the torque ripple of the motor, so it is suitable for the occasion for low loss, low vibration and high accuracy. However, as less number of pole pairs and winding turns and large air gap than a slot motor, winding inductance value is usually very small and it would lead to produce a large number of current harmonics in the stator windings and a lot of eddy current losses in the rotor.

In response to these problems, the authors proposed a novel type of HSS-PMSM. The motor stator is composed of two parts of slotless stator core with outer slot and back around winding. It can produce additional leakage inductance use the outer slot to restrain stator current harmonic. The value of additional leakage inductance could adjust in large range by ad-

justment of parameters of outer slot. Compared to traditional method that inductance is series connection in the external of motor in (Zwyssig, Round & Kolar 2008, Koshio et al. 2009), the volume of motor itself is increased, but the integration of the system is greatly improved and the reliability of the system is increased.

The structure of outer tooth has greater diameter of novel core than the ordinary slotless motor, with increasing iron loss. But the greater value additional leakage inductance integrated in motor could significantly reduce the high-order harmonic current in the winding, and then reduce the eddy current loss in rotor. Therefore, a detailed calculation of iron loss and eddy current loss in the new structure motor is needed to compare.

II. STRUCTURE OF NOVEL HSS-PMSM

The stator structure of the integrated additional leakage inductance slotless motor is shown in Figure 1. The stator core is a slotless core with outside slot, the stator winding is a back wound winding and the inner side of the windings is fixed by sheet winding frame.

The wound back winding has been used as the motor's winding and the winding arrangement in a 1/4 region is shown as Figure 2. After placing the first coil pieces (A_1) in inner slot, the coil should be directly around into the corresponding outside slot(A'), and back from the stator end of the other side for continuing to place next coil side A_2 in the same pole group.

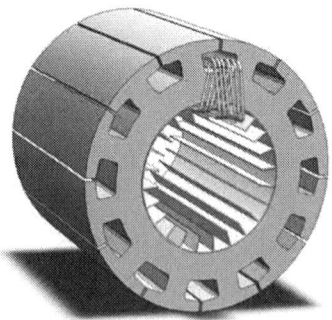

FIGURE I. STATOR OF THE INTEGRATED ADDITIONAL LEAKAGE INDUCTANCE SLOTLESS MOTOR.

© 2015. The authors - Published by Atlantis Press

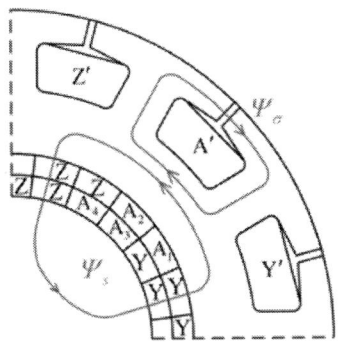

FIGURE II. WINDING ARRANGEMENT AND DISTRIBUTION OF THE FLUX.

When one phase winding is energized, the path of the winding flux is shown as Figure 2. The flux linkages Ψs surround the inside coil through the stator inner space, interacts with the magnetic field of the rotor to realization of electromechanical energy conversion. The flux linkages Ψ_σ surround the outer coils only through the outside slot gap; do not enter inside of the stator. It can be equivalent as the leakage inductance series outside the motor to reduce the harmonic current. So, it can be called as additional leakage inductance.

A HSS-PMSM has been designed and the main parameters are shown in Table 1.

The winding self-inductance L_{s1}=0.152mH and the additional leakage inductance L_σ =0.152mH, about 13 times of the winding self-inductance L_{s1}.

III. IRON LOSS ANALYSIS

Due to the high frequency of magnetic field in the stator of the HS-PMSM, stator iron loss is great.

TABLE I. THE MAIN PARAMETERS OF PROTOTYPE MOTOR.

Parameter	Value	Parameter	Value
Phase	3	Air gap (mm)	1
Power (kW)	5	Pole pairs	2
Rated speed (rpm)	15000	Inside slots	24
Rated voltage (V)	380	Outside slots	12
Core length (mm)	114	Conductors per phase	192
Inner diameter of stator (mm)	70	Conductor per inside slot	12
Outer diameter of stator (mm)	126	Width of outside teeth (mm)	11
Inner diameter of PM (mm)	26	Width of outside slot (mm)	1.4
Outer diameter of PM (mm)	50	High of outside slot (mm)	5
Thickness of sleeve (mm)	1	Width of outside slot (mm)	18

In order to calculate the iron loss, the iron loss calculation model should be established. At present, calculation method based on Bertotti separation iron loss model is broad applications. This method considers the loss caused by the rotation of the magnetization and alternating magnetic in the core simultaneously. According to the reasons of loss generated, iron loss caused by arbitrary magnetic flux density waveform is decomposed into hysteresis loss, eddy current loss and additional loss.

A. Iron Loss Calculation Model

In the alternating magnetic field, the loss due to ferromagnetic materials is constantly repeated magnetization and magnetic domain constantly mutual friction is called hysteresis loss, its expression is:

$$P_h = k_h B_m^\alpha f \tag{1}$$

Where, k_h is coefficient of hysteresis loss, f is frequency of magnetic field, B_m is magnitude of the flux density, α is calculation parameters of hysteresis loss.

When the alternating magnetic field through the core, the eddy currents induced in the iron will cause eddy current losses, and its expression is:

$$P_c = k_c B_m^2 f^2 \tag{2}$$

Where, k_c is coefficient of eddy current.

In addition, there is a part of loss neither belongs to hysteresis loss, also don't belong to eddy current loss, known as the additional loss, its expression is:

$$P_e = k_e B_m^{1.5} f^{1.5} \tag{3}$$

Where, k_e is coefficient of additional loss.

Thus, according to the Bertotti separation iron loss model, iron loss power density per unit weight of iron core at a fixed magnetic field frequency can be expressed as follows:

$$dP_{Fe} = k_h B_m^\alpha f + k_c B_m^2 f^2 + k_e B_m^{1.5} f^{1.5} \tag{4}$$

B. Analysis of Iron Loss in the Novel Type Slotless Stator

A simulation model of the motor is built in Flux. And the core flux density distribution of the motor under no-load and load conditions can be obtained, as shown in Figure 3.

It can be seen, that the flux density in inner side stator yoke have little difference in no-load and load condition. This is because the smaller winding inductance of slotless motor makes armature reaction flux in load is small, less effect on the flux density of the inner stator yoke. And the magnetic flux density in stator outer teeth large changes in no-load and load conditions. This is due to the outside teeth is the main magnetic circuit of leakage flux produced by outside winding. At no-load, magnetic flux leakage produced by little current winding is also small, so flux density in the outer tooth of stator is small. At load, magnetic flux leakage produced by large current winding is also large, so flux density in the outer tooth of stator becomes larger.

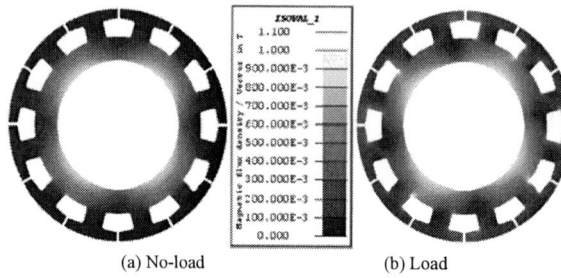

| (a) No-load | (b) Load |

FIGURE III. IRON CORE FLUX DENSITY DISTRIBUTION.

In order to analyse the iron loss caused by the outside cogging structure of stator, the stator core area is divided into two parts, the inner ring part and the outer teeth part, as shown in Figure 4. Area of the inner ring is 2512mm2, and the area of the outside tooth is 4164mm2, about 1.66 times as much as the area of inside ring.

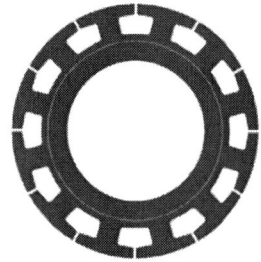

FIGURE IV. THE TWO SECTION OF THE STATOR CORE.

At different current frequencies, the two parts of the stator core loss is calculated by the analytical method and FEM simulation method, respectively. The results are shown in Figure 5. It can be seen, the iron loss of the outside tooth is very small, although its area is larger than the inner part. At no-load, iron loss in the outside teeth is only about 18% of that in the inner ring; and at load, iron loss in outside teeth is about 40% of that in the inner ring. Iron loss Power density in outer teeth is far lower than that in the inner ring.

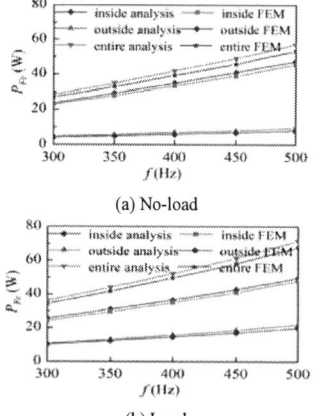

(a) No-load

(b) Load

FIGURE V. IRON LOSS WITH FREQUENCY.

Therefore, in the actual design process, flux density in the outside tooth can be reduced as much as possible through rational design to reduce iron loss in outside teeth.

IV. EDDY CURRENT LOSS ANALYSIS

Both permanent magnet and the rotor sleeve of HSS-PMSM are conductors and will produce eddy current loss in the alternating magnetic field. In HS-PMSM, factors such as non-sinusoidal distribution winding magneto motive force, winding harmonic current induced by PWM chopper, et al, will produce a large number of harmonics in the air gap magnetic field. They will produce not to be neglected eddy current loss in the rotor. Meanwhile, most of the HS-PMSM is a sealing structure; condition of rotor cooling is poor. If the rotor's eddy current loss is too large, it will cause the temperature of permanent magnet increased significantly, and even cause demagnetization. In this paper, the relationship between the structure of stator and rotor eddy current loss is analysed.

A. Influence on Eddy Current Loss by Additional Leakage Inductance

In order to compare the HS-PMSM system without the additional leakage inductance and the HS-PMSM system with the additional leakage inductance, a motor that have the same inside winding parameters of the prototype but without the outside slot and additional leakage inductors is designed. Following, the motor with additional leakage inductance is named as motor A, and the motor without additional leakage inductance is named as motor B. The system simulation of the two motors is studied using the Simulink. Modulation mode of the driver is SVPWM. The parameters of the two motors are shown in Table 2.

TABLE II. THE COIL PARAMETERS OF TWO MOTORS.

Prameters	Motor A	Motor B
R_s (Ω)	0.288	0.183
L_d (mH)	2.282	0.248
L_q (mH)	2.282	0.248
Ψ_f (Wb)	0.0907	0.0907
p	2	2

Winding currents of the two motors are obtained by Simulink, as shown in Figure 6. It can be seen, that the harmonic current of the motor with additional leakage inductance is well suppressed.

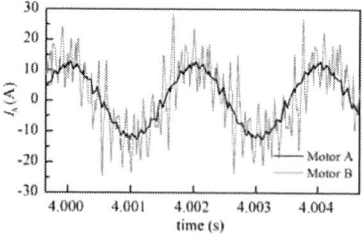

FIGURE VI. COMPARISON OF WINDING CURRENT OF TWO PROTOTYPES.

38

FEM analysis of the two motors is carried out by using the current in Figure 6, the eddy current density of the rotors are obtained, as shown in Figure 7.

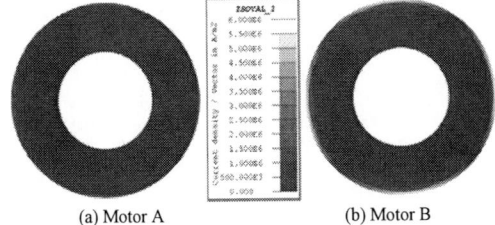

(a) Motor A   (b) Motor B

FIGURE VII. COMPARISON OF EDDY CURRENT DENSITY OF TWO PROTOTYPES.

As less harmonic current in the motor A, eddy current density of the induction that generated in the rotor is relatively small, so the rotor eddy current loss is very small. The eddy current loss of the two motors was calculated, which in motor A is 2.35W, in motor B is 86.73W. Consequently, structure of additional leakage inductance can obviously reduce the rotor eddy current loss.

B. Influence on Eddy Current Loss by Spatial Distribution of Winding

In the slotless motor, the distribution of armature winding is uniform in space. In (Zhu et al. 2004), the spatial distribution of winding equivalent current sheet is directly decided by the spatial distribution of the stator winding. So, the spatial distribution of slotless motor stator winding has a direct relationship with rotor eddy current loss.

In order to analysis the relationship between the spatial distribution of winding and the eddy current loss, three kinds of armature winding are discussed in the following text. Figure 8 shows the diagram of space structure of the three kinds of windings. Among them, Figure 8a is a 12 slot single layer winding. This structure is relatively simple, but the harmonic of back EMF and harmonic of spatial distribution is large. Figure 8b is a 24 slot double layer short pitch winding adopted in the prototype, which is used in motor frequently. And in this paper, a newly quasi-sinusoidal winding based on the 24 slot double layers winding is proposed, as shown in Figure 8c. The distribution of winding under each pole phase group is similar to the sine along with the changes of space angles.

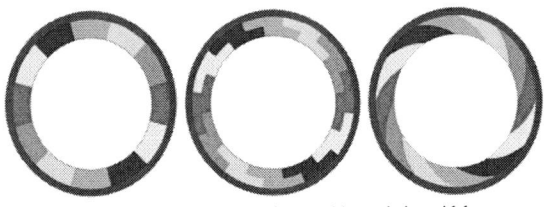

(a) 12 slot   (b) 24 slot   (c) quasi-sinusoidal

FIGURE VIII. SPATIAL DISTRIBUTION OF THREE KINDS OF WINDINGS.

At time t =0, the equivalent current sheet waveform of the three kinds of winding are shown in Figure 9. It can be seen,

the equivalent current sheet waveform of the 12 slot single layer windings is very rough because its winding distribution is more concentrated. The equivalent current sheet waveform of 24 slot double layers winding has greatly improved. And the equivalent current sheet waveform of quasi sinusoidal winding is very close to the sine wave.

The total harmonic distortion (THD) is used to evaluate the properties of the three kinds of winding structure, the contrast results are shown in Table 3. It can be seen from Table 3, the difference of fundamental amplitude between the three kinds of winding is very small, but the difference of THD is very.

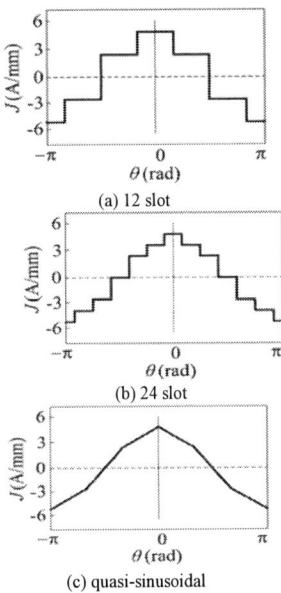

(a) 12 slot

(b) 24 slot

(c) quasi-sinusoidal

FIGURE IX. SPACE DISTRIBUTION OF WINDING'S EQUIVALENT CURRENT SHEET.

Obvious. Among them, the THD of the quasi sinusoidal winding is minimum, only 4.67%. Therefore, from the perspective of space harmonic, the quasi sinusoidal winding which has fundamental amplitude close to the full pitch winding and also has a very small THD, it's a very excellent winding structure.

TABLE III. COMPARISON OF THREE KIND OF COIL STRUCTURE.

	12 slot	24 slot	Quasi-sinusoidal
Fundamental amplitude (A/mm)	5.290	5.135	5.076
THD (%)	31.35	16.78	4.67

Figure 10 is the distribution of eddy current density in sleeve and permanent magnet of the rotor of the motors with three kinds of winding with the same current. From the sleeve region in Figure 10, it can be seen, that the eddy current density produced by 12 slot winding is highest, and the eddy current density produced by quasi sinusoidal winding is the lowest.

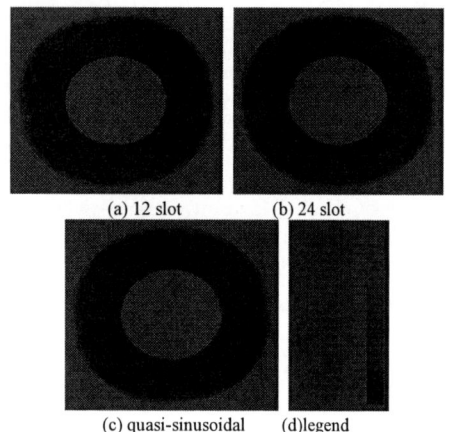

(a) 12 slot   (b) 24 slot

(c) quasi-sinusoidal   (d)legend

FIGURE X. DISTRIBUTION OF EDDY CURRENT DENSITY IN ROTOR.

The rotor eddy current loss of motors with three kinds of winding is shown in Table 4. It can be seen, that the rotor eddy current loss of 12 slot winding is maximum, and the eddy current loss of quasi sinusoidal winding is minimum. Moreover, substantial of eddy current loss in sleeve is more than in the permanent magnet.

TABLE IV. EDDY CURRENT LOSS OF THREE KIND OF WINDING.

Eddy current loss	12 slot	24 slot	Quasi-sinusoidal
In sleeve (W)	1.60	1.49	1.45
In PM (W)	0.93	0.86	0.84
Total (W)	2.53	2.35	2.29

Therefore, during the design of the motor, to minimize the rotor eddy current losses, we should try to make the spatial distribution of winding under each pole phase group nearly sinusoidal.

V. CONCLUSION

(1) This paper introduces a new kind of HSS-PMSM. This motor integrates additional leakage inductance inside, that can reduce the winding current harmonics and meanwhile will increase the area of the stator core.

(2) Iron loss of motor which is analyzed by Bertotti separation iron loss model is compared with the simulation results. The compare result describes that additional leakage structure leads that the stator core area increased by 160%, but the stator iron loss increase only 19.79W, not to the iron loss of inner part 40%.

(3) The rotor eddy current loss with or without appended inductance are analyzed by Simulink tool. The Simulink result could be proved that additional leakage inductance structure could restrain the winding current harmonics well, and can reduce the eddy current loss more than 95%.

(4) Three kinds of winding space distribution structure are analyzed and compared. The results show that, with the winding distribution close to the sinusoidal distribution in space, eddy current losses are also reduced.

ACKNOWLEDGEMENT

This paper was supported by International Cooperation (2013DFR60510), Ministry of Science and Technology of China. Their powerful supports are grateful.

REFERENCES

[1] Cho, H.W., Ko, K.J., Choi, J.Y. & et al. 2011. Rotor natural frequency in high-speed permanent magnet synchronous motor for turbo-compressor application. *IEEE Transactions on Magnetics* 47(10): 4258-4261.

[2] Kong, X.G., Wang, F.X. & Xing, J.Q. 2012. Losses calculation and temperature field analysis of high speed permanent magnet machines. *Transaction of China Electrotechnical Society* 27(9): 166-173.

[3] Koshio, N., Kubota, K., Miki, I. & et al. 2009. Improvement of current waveforms of position sensor-less vector controlled permanent magnet synchronous motor at high frequency region. *International Conference on Electrical Machines and Systems, Tokyo, Japan*: 1-5.

[4] Paulides, J.J.H., Jewell, G.W. & Howe, D. 2004. An evaluation of alternative stator lamination materials for a high-speed, 1.5 MW, permanent-magnet generator. *IEEE Transactions on Magnetics*, 40(4): 2041-2043.

[5] Wang, F.X. 2006. Study on design feature and related technology of high speed electrical machines. *Journal of Shenyang University of Technology* 28(3): 258-264.

[6] Zhu, Z.Q., Ng, K., Schofield, N. & Howe, D. 2004. Improved analytical modeling of rotor eddy current loss in brushless machines equipped with surface-mounted permanent magnets. *IEE Proc. Electric Power Applications*, 151(6): 641-650.

[7] Zwyssig, C., Round, S. & Kolar, J. 2008. An ultrahigh-speed, low power electrical drive system. *IEEE Transactions on Industrial Electronics*, 55(2): 577-585.

The Application of Electrochemical Impedance in the Environmentally Friendly Scale and Corrosion Inhibitor Polyepoxysuccinic ACID

B.R. Zhang
College of Environmental Science & Engineering
State Key Laboratory of Pollution Control and Resource Reuse
Tonji University
Shanghai, 200092, PR China
Key Laboratory of Yangtze River Water Environment Ministry of Education
College of Environmental Science and Engineering
Tongji University
1239 Siping Road, Shanghai 200092, PR China

H.Y. Tian
College of Environmental Science & Engineering
State Key Laboratory of Pollution Control and Resource Reuse
Tonji University
Shanghai, 200092, PR China

F.T. Li
College of Environmental Science & Engineering
State Key Laboratory of Pollution Control and Resource Reuse
Tonji University
Shanghai, 200092, PR China

Abstract—**Polyepoxysuccinic acid (PESA) is a new, non-phosphorous, biodegradable, environmentally friendly scale and corrosion inhibitor implemented in industrial water treatment. The corrosion and scale inhibition properties of PESA were investigated in the test water by using static beaker test, weight-loss method, polarization curves and electrochemical impedance spectroscopy (EIS). It is found that, PESA has better scale inhibition efficiency on CaCO3, BaSO4 as well as is a phosphate-free environment-friendly inhibitor with multi-scale properties. There is a limited corrosion inhibition effect on carbon steel provided by PESA used alone at low concentration, and with the increase of PESA concentration, corrosion inhibition is gradually strengthened. The pH value has significant effect on corrosion inhibition property of PESA. There is low corrosion inhibition efficiency in acidic condition, as well as better corrosion inhibition efficiency of PESA in alkaline condition. PESA is mainly as the function of anodic corrosion inhibitor. PESA should be a non-phosphorous, biodegradable, environmentally friendly scale and corrosion inhibitor implemented in industrial water treatment.**

Keywords-polyepoxysuccinic acid; scale and corrosion inhibition; polarization curve; electrochemical impedance spectroscopy

I. INTRODUCTION

The formation of mineral scales can cause a series of problems in many processes ranging from desalination to oil production, such as blockage of pipeline, damage of production system, increase of energy consumption, decrease of productivity and unscheduled equipment shutdown[1-10], and eventually result in unwanted increase of operating costs[11–13]. The most practical and economical method to combat this problem is the use of chemical scale inhibitors. Recently, the impact of chemicals on the environment is an issue of increasing global importance.

From the middle 1970s to early 1990s, phosphonates had been developed and are extensively used as scale inhibitors. Phosphonates such as HEDP, ATMP, EDTMP, DTPMP, PBTCA, which contain one or more $R_3C-P(O)(OH)_2$ groups rather than a $R_3C-O-P(O)(OH)_2$, possess a number of superior qualities: high chemical stability under extremes of pH and temperature, the ability to complex metal, the ability to adsorb strongly onto metal coatings, and some dispersancy activity towards suspending matter. However, it is well known that even at moderate conditions of calcium hardness, pH, and temperature, phosphonates can react with calcium ion and precipitate as calcium-phosphonate. Furthermore, phosphonates can also cause eutrophication.

In the past three decades, many water soluble anionic polymers compositions and molecular weights have been developed as scale inhibitors. These polymers are mainly homo-, co-, or ter-polymers having acrylic or maleic as the initial monomer in conjunction with other monomers containing acrylamide, ester, sulfonic acid, phosphonic acid, etc. They possess the excellent ability as a scale inhibitor and metal ion stabilizer. Acrylic acid-based scale inhibitors such as polya-

© 2015. The authors - Published by Atlantis Press

crylic acid (PAA), acrylic acid/2-acrylamido-2-methypropane sulfonic acid copolymer (AA/AMPS), acrylic acid/2-acrylamido-2-methypropane sulfonic acid/2-hydroxypropyl-acrylate copolymer (AA/AMPS/HPA), although generally of low toxicity, are also unfortunately non-biodegradable.

In the 21st century, biodegradability, the capability of being broken down into simple, non-toxic materials by the action of microorganisms and fungi, becomes an important mechanism for limiting the build-up of chemicals in the environment. Polyepoxysuccinic acid (PESA) is a novel scale inhibitor first developed by Betzdearbon (U.S.). It is a representative of green scale inhibitors because of its non-nitrogenous, non-phosphorus and biodegradable features. PESA with good scale-inhibiting performances was also synthesized successfully in China [1, 2]. Versatile scale inhibition performances of PESA were analyzed as well [3-5]. According to the advantages of less dosage, high scale inhibition efficiency in the water of high alkalinity and high solid content, as well as the performance of corrosion inhibition, PESA has become a research hotspot now [6]. The aims of this paper are to present the corrosion and scale inhibition properties of PESA investigated by using static beaker test, weight-loss method, polarization curves and electrochemical impedance spectroscopy (EIS) in the test water.

II. EXPERIMENT SECTION

A. Main Materials

Polyepoxysuccinic acid (PESA) used in this study was obtained from Wuxi Ecolom Chemical Co. Ltd., China. The PESA was as an aqueous solution (40 wt %) in the presence of sodium polyepoxysuccinic acid. Molecular weight M_w of the PESA was approximately 2100, as shown in Figure 1.

Carbon steel specimens used in experiments were composed of the following (wt %): $0.17 \sim 0.24\%C$, $0.17 \sim 0.37\%Si$, $0.35 \sim 0.65\%Mn$, $\leq 0.035P$, $\leq 0.035S$, $\leq 0.025Ni$, $\leq 0.025Cr$, $\leq 0.025Cu$, and balance Fe.

B. Static Beaker Test for Calcium Carbonate Scale Inhibition Efficiency

Static beaker calcium carbonate test was conducted by using deionized water and reagent grade chemicals. The static beaker test involved the adding of an amount of the water treatment agent to a solution containing an amount of Ca^{2+} and an amount of HCO_3^- at the required pH .The beakers were incubated in a water bath for 18 hour at 80℃. After cooling, an aliquot was filtered through 0.22 μm filter paper, the calcium concentration in the filtrate was measured by using the standard ethylenediaminetetraacetic acid (EDTA) titration method. The static scale inhibition efficiency was calculated by

$$\eta(\%) = \frac{V_1 - V_0}{V_2 - V_0} \times 100$$

where V_1 is the amount of consumed EDTA of the sample with the addition of the water treatment agent after incubation, V_2 is the amount of consumed EDTA of the sample with

the addition of the water treatment agent before incubation, V_0 is the amount of consumed EDTA of the sample without the addition of the water treatment agent after incubation.

C. Static Beaker Test for Barium Sulfate Scale Inhibition Efficiency

Static beaker barium sulfate test was conducted by adding of an amount of the water treatment agent to a solution containing an amount of Ba^{2+} and an amount of SO_4^{2-} at the required pH .The beakers were incubated in a water bath for 18 hour at 80℃. After cooling, an aliquot was filtered through 0.22μm filter paper; the barium concentration in the filtrate was analyzed by inductively coupled plasma atomic emission spectroscopy (ICP). Barium sulfate inhibition efficiency was calculated by

$$\eta(\%) = \frac{[Ba^{2+}]_1 - [Ba^{2+}]_0}{[Ba^{2+}]_2 - [Ba^{2+}]_0} \times 100$$

where $[Ba^{2+}]_1$ is the concentration of barium ion of the sample with the addition of the water treatment agent after incubation, $[Ba^{2+}]_2$ is the concentration of barium ion of the sample with the addition of the water treatment agent before incubation, $[Ba^{2+}]_0$ is the concentration of barium ion of the sample without the addition of the water treatment agent after incubation.

D. Static Beaker Test for Corrosion Inhibition Efficiency

Static beaker corrosion test was tested by weight loss experiment, which was conducted in a 2 liter beakers equipped with an air/CO2 sparge. The beaker was immersed in a water bath at 45℃. The carbon steel sheets were polished with different grades of emery paper, degreased with acetone, and rinsed with distilled water. Having been dried and accurately weighed, the carbon steel specimens were immersed in the beaker with the test water in the absence and presence of inhibitors for a period of 72 hours. A typical analysis of the test water is given in Table 1. After test, the carbon steel specimens were taken out, rinsed with water thoroughly, dried and accurately weighted. Each set of experiments was repeated two times to ensure reproducibility. The corrosion inhibition efficiency was calculated by

$$R(\%) = \frac{W_0 - W_1}{W_0} \times 100$$

where R is the corrosion inhibition efficiency, W0 and W are the value of the weight loss of carbon steel immersed in test solution without and with inhibitor, respectively.

TABLE I. ANALYSIS OF THE TEST WATER.

Parameter	Value
pH	7.5
Total hardness (mg·L^{-1}, as CaCO$_3$)	142
Ca^{2+} (mg·L^{-1}, as CaCO$_3$)	129
Alkalinity (mg·L^{-1}, as CaCO$_3$)	80
Chloride (mg·L^{-1})	62
Sulfate (mg·L^{-1})	59
Turbidity (NTU)	0.22
Total dissolved solids (mg·L^{-1})	324

E. Test Method of Tafel Polarization Curve

The Tafel polarization curve measurements were carried out using a computer controlled potentiostat (Autolab PGSTAT30) in a three electrode cell assembly at a scan rate of 1 mV/s. A platinum electrode and a saturated calomel electrode (SCE) were used as auxiliary and reference electrode respectively. The work electrode (WE) was a carbon steel specimen of area 1 cm2. All the experiments were carried out at 30±2℃ with the test water (shown in table 1) as an electrolyte under static and naturally aerated condition. The Tafel polarization curve studies were conducted in the test water containing various concentrations of inhibitor. The experiments were repeated to ensure reproducibility. The inhibition efficiency of the inhibitor for the carbon steel corrosion was calculated by

$$R(\%) = \frac{W_0 - W_1}{W_0} \times 100$$

F. Electrochemical Impedance Spectroscopy (EIS)

The same instrument of the Tafel polarization curve measurements was used for electrochemical impedance measurement (Autolab PGSTAT30) at 30±2℃. Electrochemical impedance measurement was carried out in the region of 0.01 Hz to 100 KHz with the perturbation amplitude of 5 mV.

III. RESULTS AND DISCUSSION

A. Calcium Carbonate Scale Inhibition Performance

In accordance with the experiment method 2.2, the additional experimental conditions are as follows: the mass concentration of Ca^{2+} and HCO$_3^-$ are 500 mg·L^{-1} and 750mg·L^{-1} respectively, pH value of the solution is 9.0. The comparison of static scale inhibition efficiency between PESA and other typical phosphonates is shown in Figure 1. It can be seen that although the CaCO$_3$ scale inhibition efficiency of PESA is a little less than PBTCA, but more than polycarboxylic acid inhibitors PAA and PMAAA, sulfonate inhibitors AA/AMPS, and polyphosphoric acid inhibitor SHMP. Therefore, PESA has good scale inhibition efficiency on CaCO$_3$.

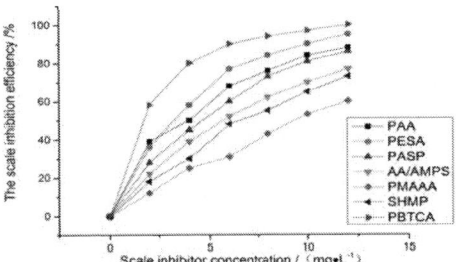

FIGURE I. INFLUENCE OF SCALE INHIBITOR CONCENTRATION ON CaCO$_3$ SCALE INHIBITION EFFICIENCY.

B. Barium Sulfate Scale Inhibition Performance

According to the experiment method 2.3, the additional experimental conditions are as follows: the mass concentration of Ba^{2+} and SO$_4^{2-}$ are 20 mg·L^{-1} and 100 mg·L^{-1} respectively, pH value of the solution is 7.0. As demonstrated, PESA and SHMP have good scale inhibition effect on BaSO$_4$, while the PAA, PMAAA, AA/HPA/AMPS and PBTCA still have no BaSO$_4$ scale inhibition effect even at the concentration of 15 mg·L^{-1}. The impacts of PESA, SHMP and PASP concentration on BaSO$_4$ scale inhibition efficiency are shown in Figure 2. As demonstrated in Figure 2, the BaSO$_4$ scale inhibition efficiency of PESA is superior to the traditional BaSO$_4$ scale inhibitor SHMP; as a result, PESA has a very strong scale inhibition effect on BaSO$_4$.

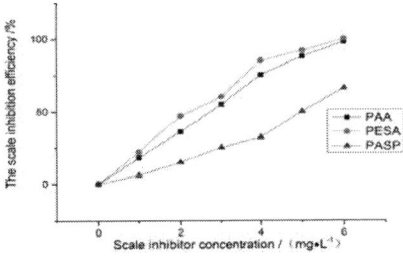

FIGURE II. INFLUENCE OF SCALE INHIBITOR CONCENTRATION ON BaSO4 SCALE INHIBITION EFFICIENCY.

C. Static Beaker Test for Corrosion Inhibition

The static corrosion inhibition efficiency of PESA to carbon steel at different concentration is shown in Table 2 and Figure 3, which was tested by weight loss experiment.

Table 2 and Figure 3 illustrate that, under the condition of this water quality, the corrosion inhibition effects of PESA used alone to carbon steel at low concentrations is weak, and with the increase of PESA concentration, corrosion inhibition is gradually strengthened. Therefore, to achieve a better inhibition affect, the required quality concentration of PESA should be higher.

43

TABLE II. THE WEIGHT LOSS EXPERIMENT RESULTS AT DIFFERENT PESA CONCENTRATION.

PESA(mg·L⁻¹)	Corrosion velocity(mm·a⁻¹)	Corrosion inhibition efficiency (%)
0	1.3855	-
50	1.3524	2.39
100	1.2654	8.67
150	0.4799	65.36
200	0.3482	74.87
800	0.2454	82.29

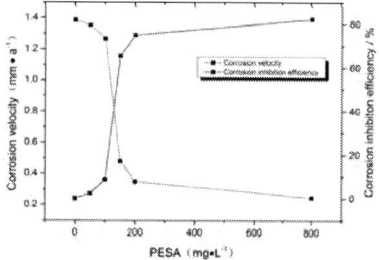

FIGURE III. PESA CONCENTRATION—CORROSION VELOCITY—CORROSION INHIBITION EFFICIENCY.

Table 3 illustrates the effect on the static corrosion inhibition efficiency of PESA to carbon steel at different pH values according to the weight loss experiment. The pH values of the test solution were adjusted to 2, 4, 9 respectively, and then dosing 200 mg·L⁻¹ of PESA to carry out weight loss experiment.

As shown in Table 2, it demonstrates that pH value has significant effect on corrosion inhibition property of PESA. There is low corrosion inhibition efficiency in acidic condition; as the pH value increases, the corrosion velocity of carbon steel obviously decreases as well as corrosion inhibition efficiency increases, namely corrosion inhibition enhances. Moreover, in alkaline condition, the corrosion inhibition efficiency of PESA is better and can reach about 80%. It is assumed that, due to the alkaline environment, PESA may dissociate, which contributes to its adsorption on the carbon steel surface.

TABLE III. THE WEIGHT LOSS EXPERIMENT RESULTS OF PESA AT DIFFERENT PH VALUES OF TEST SOLUTION.

pH	PESA(mg·L⁻¹)	Corrosion velocity(mm·a⁻¹)	Corrosion inhibition efficiency (%)
2	0	4.5357	-
2	200	3.9118	13.76
4	0	1.6462	-
4	200	0.6587	60.00
12	0	0.1378	-
12	200	0.0227	83.53

D. The Tafel Polarization Curve Measurement

The polarization curve experiments of carbon steel at different pH values of test solution (pH=2, 4, 9) were carried out and the corresponding results are given in Figure 4-6. As they illustrated, it is found that there are all greater impact on the anodic process of electrodes by adding PESA at different pH values of test solution; this effect amplified as the pH value increases, and the self-corrosion potential turns to the positive shift. In other words, PESA belongs to anodic corrosion inhibitor.

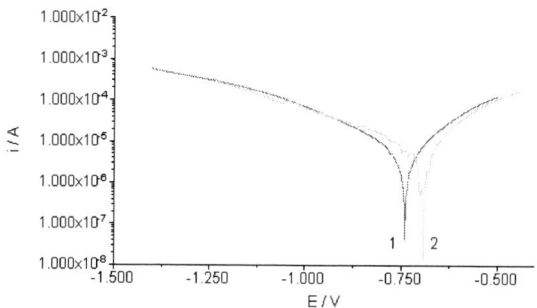

FIGURE IV. THE POLARIZATION CURVE OF PESA AT PH=2. (1-PESA, 2-BLANK)

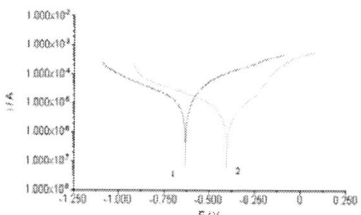

FIGURE V. THE POLARIZATION CURVE OF PESA AT PH=4 (1-PESA, 2-BLANK).

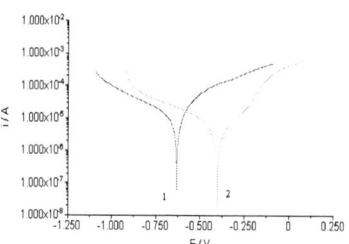

FIGURE VI. THE POLARIZATION CURVE OF PESA AT PH=9 (1-PESA; 2-BLANK.)

Figure 7 shows the results of polarization curve tests to carbon steel at different concentrations of PESA (50 mg·L⁻¹, 100 mg·L⁻¹, 150 mg·L⁻¹, 200 mg·L⁻¹ and 800 mg·L⁻¹, respectively). The parameters, such as polarization resistance (Rp), self-corrosion potential (Ecorr), anodic Tafel Slope, cathodic Tafel slope, etc., as well as corrosion velocity, can be obtained by using the software of PGATAT30 electrochemical workstation to carry out the Tafel extrapolation to the active part of the polarization curves in Figure 4[7, 8], and then corrosion inhibition efficiency can be calculated according to the corrosion velocity of the blank specimen. The data is shown in Table 4.

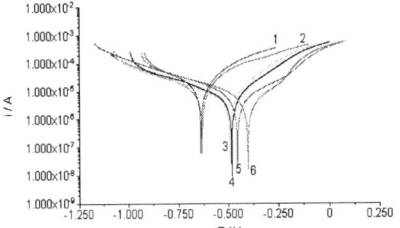

FIGURE VII. THE POLARIZATION CURVES OF PESA AT DIFFERENT CONCENTRATIONS.

(1-800mg·L^{-1}; 2-Blank; 3-50mg·L^{-1}; 4-100mg·L^{-1}; 5-200mg·L^{-1}; 6-150 mg·L^{-1})

As it is observed in Table 4, the corrosion inhibition efficiency is better by adding more PESA in the neutral solution medium. When PESA is used alone, aiming to achieve better corrosion inhibition efficiency, the concentration of PESA should reach 200 mg · L^{-1} or more. The OCP of PESA at different concentrations increase obviously compared to the blank, which demonstrates that PESA is mainly as the function of anodic inhibitor [9, 10]. The reason contributes to this

is that PESA adsorbs on the carbon steel surface and reacts with Fe^{3+}, but also forms the membrane which hampers the proliferation of Fe^{2+} in the anodic area, so that it comes anodic polarization and anodic potential rise, and eventually leading Ecorr to be higher.

From Figure 7, it is shown that there is no significant inhibition efficiency for the anode reaction of carbon steel by adding 100 mg · L^{-1} of PESA, indicating that the PESA concentration is not high enough, the formed membrane has poor density and may even only covers part of the surface of carbon steel. The formed membrane by PESA, which is incomplete, makes effect on its inhibition efficiency, and thus the anode reaction cannot be effectively curbed.

Compared with the blank, the anodic Tafel Slope rises markedly with the PESA concentration increases, which illustrating that the presence of PESA controls the anode in the reaction, so as to inhibit the anodic reaction of corrosion process. Accompanying with the concentration of PESA reaching above 200 mg · L^{-1}, the corrosion potential Ecorr decreases as well as the Tafel Slop of cathodic polarization curves rises, in accordance with which it implies that the process of cathodic reaction begins to be inhibited as the concentration of PESA increases. PESA can chelate Ca^{2+}, Mg^{2+} and other ions in the test solution and then disperse in it; with the increased concentration of PESA, part of the chelate deposit in the cathode region and thus the cathodic process is suppressed [11].

TABLE IV. THE POLARIZATION PARAMETERS OF DIFFERENT CONCENTRATIONS OF PESA AT 30℃.

Mass concentration (mg·L^{-1})	OCP (mV)	Ecorr (mV)	Rp ×10^{2}ohm	Ba (m·dec^{-1})	Bc (m·dec^{-1})	Corrosion velocity (mm·a^{-1})	Corrosion inhibition efficiency (%)
0	-490	-635	4.00	170	368	0.1126	-
50	-475	-498	6.91	128	395	0.0702	37.66
100	-391	-480	7.81	204	406	0.0681	39.52
150	-320	-390	1.06	206	492	0.0439	61.01
200	-335	-450	8.90	271	435	0.0312	72.29
800	-571	-630	3.01	212	397	0.0209	81.44

As the concentration of PESA further up to 800 mg · L^{-1}, the cathode Tafel Slop of PESA polarization curve increases, it is inferred there may be a reaction between Mg^{2+} and OH$^-$ generated in the cathode reaction, and the product Mg(OH)$_2$ deposits in the cathode region as to prevent the spread of oxygen and thus inhibit the cathode reaction [12-16].

E. Electrochemical Impedance Measurement

The EIS Nyquist plots of carbon steel electrode after the tests of dosing different concentrations of PESA in the solution have been shown in Figure 8. Since the impedance is plural, the horizontal axis is for the real part while the vertical axis is for the imaginary part. As can be seen from Figure 8, impedance spectra suffers different degrees of large semicircular features; under the condition that the dosing concentration of PE-

SA is less than 100 mg · L^{-1}, the increasing range of Rp value (charge transfer resistance) with the rising of the concentration. It demonstrates that corrosion inhibition efficiency is not very good in terms of the low dosing concentration of PESA; when the concentration to be more than 100 mg · L^{-1}, Rp value increases by a large margin. The greater Rp value indicates that the adsorbed membrane on the surface of carbon steel is thicker by the action of corrosion inhibitor, and it also comes out greater absorption area as well as the better inhibition efficiency, due to which the increase of the Rp value illuminates that PESA has significant corrosion inhibition effect to carbon steel.

The high-frequency capacitance arc appearing in the high-frequency zone of the impedance spectroscopy shows that there is a more complete adsorbed membrane forming on the surface of electrode, and the metallic corrosion reaction is inhibited, moreover, the charge transfer reaction plays a leading role in metal corrosion process. The chord length from semi-circle to the real axis (Z' axis) corresponds to the charge transfer resistance Rp. In terms of the corrosion inhibitor system, Rp reflects the resistance for the process of metal ionization by the covering layer of corrosion inhibitor, or reflects the rate of corrosion reaction. The greater Rp value is, indicating the greater resistance for the process of metal ionization while the smaller corrosion rate of metals.

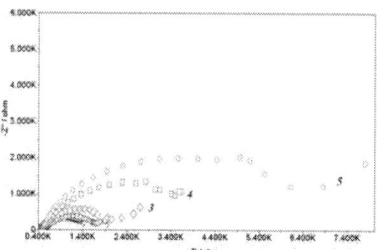

FIGURE VIII. THE EIS NYQUIST PLOTS OF CARBON STEEL ELECTRODE IN THE SOLUTIONS ADDED DIFFERENT CONCENTRATIONS OF PESA.

(1-Blank; 2-50 mg·L^{-1}; 3-100 mg·L^{-1}; 4-150 mg·L^{-1}; 5-200 mg·L^{-1})

Table 5 shows the corrosion inhibition efficiency at different concentrations of PESA calculated by

$$\eta=（1-Rp_0/Rp）\times100\%$$

where Rp and Rp$_0$ are the charge transfer resistance in test solution with and without corrosion inhibitor, respectively. The greater the concentration of PESA, the stronger the corrosion inhibition effect within the scope of the experiment, which is indicated in Figure 9.

TABLE V. AC IMPEDANCE AT DIFFERENT CONCENTRATIONS OF PESA.

PESA(mg·L^{-1})	Rp(ohm)	η(100%)
0	4.99971×10^2	-
50	1.19138×10^3	58.03
100	1.75014×10^3	71.43
150	3.86054×10^3	87.05
200	6.62408×10^3	92.45

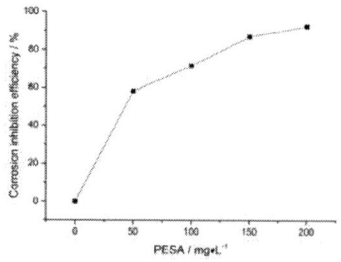

FIGURE IX. PESA CONCENTRATION—CORROSION INHIBITION EFFICIENCY.

IV. CONCLUSION

Under the experimental conditions in this paper, comparing with the scale inhibitors, such as PBTCA, PAA, PMAAA and AA/AMPS, PESA and SHMP has scale inhibition effect on CaCO$_3$ (12 mg·L^{-1} of the scale inhibitor's concentration) and BaSO$_4$ (6 mg·L^{-1} of the scale inhibitor's concentration), furthermore, the scale inhibition efficiency on CaCO$_3$, BaSO$_4$ of PESA is much better than SHMP. Meanwhile, SHMP is easy to decompose into the precipitation of calcium phosphate, reducing its scale inhibition efficiency, but also has high phosphorus content, while the PESA is a phosphate-free environment-friendly inhibitor with versatile scale inhibition properties, and thus it has a very broad application prospects.

There is limited corrosion inhibition effect on carbon steel provided by PESA used alone at low concentration (<120 mg·L^{-1}), and with the increase of PESA concentration, corrosion inhibition is gradually strengthened.

The pH value has significant effect on corrosion inhibition property of PESA. And PESA has corrosion inhibition effects to carbon steel by 120mg·L^{-1} of the concentration to some degree. However, there is low corrosion inhibition efficiency in acidic condition; as the pH value increases, the corrosion velocity of carbon steel obviously decreases as well as corrosion inhibition efficiency increases, namely corrosion inhibition enhances. Moreover, in alkaline condition, the corrosion inhibition efficiency of PESA is better and can reach about 80%. PESA is mainly as the function of anodic corrosion inhibitor.

PESA curbs metallic corrosion process by forming the absorbed membrane on the surface of carbon steels, as well as the charge transfer reaction plays a leading role in metal corrosion process.

PESA should be a non-phosphorous, biodegradable, environmentally friendly scale and corrosion inhibitor implemented in industrial water treatment.

REFERENCE

[1] SUN Yonghong, XIANG Wenhua, WANG Ying, Study on polyepoxysuccinic acid reverse osmosis scale inhibitor.Journal of Environmental Sciences, Volume 21, Supplement 1, 2009, Pages S73-S75

[2] Xiong R C, Zhou Q, Wei G, 2003. Corrosion inhibition and synergistic effect of green scale inhibitor polyepoxysuccinic acid. Journal of Chemical Industry and Engineering, 54(9): 1323–1325.

[3] Lei W, Wang F Y, Xia M Z, Wang F H, 2006. Synthesis and its scale inhibition effect of green scale inhibitor polyepoxysuccinic acid. Journal of Chemical Industry and Engineering, 57(9): 2207–2213.

[4] Zhang R B, Li F T, 2002. Versatile scale inhibition of polyepoxysuccinic acid. Industrial Water Treatment, 22(9): 21–24.

[5] Zhou W S, Du Q Y, Yu R X, Zhang L, Wei W Y, 2006. Evaluation of the scale inhibition effect of PESA on RO system. Industrial Water Treatment, 26(10): 58–60.

[6] Carter, C. G., Fan, L. G., Fan, J. C., Kreh, R. P. and Jovancicevic, V. 1994. Method of inhibiting corrosion of metals using polytartaric acids. EP Patent No. 0609590.

[7] NIE Lijun, TAN Chengyu, 2005. Corrosion behavior of A3 steel in sulfuric acid solutions. Corrosion & Protection, 26(10):439-432.

[8] S. Keritit, B. Hammouti. Corrosion inhibition of iron in 1M HCl by 1-phenyl-5-mercapto-1, 2, 3, 4-tetrazole. Applied Surface Science, 1996,(93):59-66.

[9] K.F. Khaled, N. Hackerman, Mater. Chem. Phys, 82(2003)949.

[10] K.F. Khaled, N. Hackerman, Electrochimica Acta, 82(2003)949.

[11] Yu Hui, Wu Jianhua, Wang Hongren, et al, 2002. Studies on The Inhibition of YKI-05 Corrosion Inhibitor to 907A Steel in Different Immersed Environments. Electrochemistry, 8(4):439-444.

[12] TAN Weigang, LU Zhu, LI Yan, 2001. A study of tungstate-based water treatment agents without phosphate for inhibiting corrosion and scaling. Corrosion & Protection, 22(6):237-239.

[13] Li Yan, Lu Zhu, 2000. Application of surface analysis technology in study on corrosion inhibition mechanism of tungstate. Corrosion & Protection, 21(10):447-450.

[14] WANG Chao, CHEN Xinping, LIANG Limin. Green Chemistry and Corrosion and Scale Inhibitors. Chemistry and Adhesion, 4 (2001):171-173.

[15] ZHANG Jianqiang, YAN Lianhe ,WANG Ying,2002. Research Advancement of Green Revolution in High Polymeric Chemicals for Water Treatment. Jiangsu Chemical Industry, 30(4):27-30.

[16] Harvey J. Flitt, D. Paul Schweinsberg. Evaluation of corrosion rate from polarisation curves not exhibiting a Tafel region. Corrosion Science 47 (2005) 3034–3052.

Micro-Phase Separation Structure of Nonionic Polyether-Polyester Polyurethane Based On MDI

H. Quan

Shaanxi University of Science and Technology
Xian 710021, PR China

L. Wan, G.F. Wan

Global Chemicals International Ltd
523556 Dongguan, PR China

Abstract—Segmented polyether-polyester polyurethanes with an amorphous hydrophilic soft segment phase were prepared from 4,4'-diphenylmethane diisocyanate (MDI), polybutylene adipate glycol 2000 (PBA2000) and polyethylene glycol 1000 (PEG1000) with 1,4-butanediol (BDO) as the chain extender. Furthermore, the micro phase separation structure of the polyurethanes was studied. The studies show, the micro-structure of nonionic polyurethane has been remarkable influenced by the structure, molecule and concentration of its soft segments.

Keywords-polyurethane; micro-structure, polyether; polyester

I. INTRODUCTION

Polyurethanes usually have frozen hard domains and reversible soft domains (microphase structure) due to the incompatibility of the constituent segments[1], and almost all of the application properties of polyurethane materials have consanguineous correlativity with its microstructure[2,3].

Generally, the polyurethane with different microphase separation structure will show an obvious difference on their macroscopical properties even if they had a same chemical structure. The microstructure of polyurethane macromolecule is strongly dependent on the chemical structure, components, relative content, molecular weight, crystallinity of polymeric monomers and the cross-linking degree of soft and hard segment phases etc [4].

II. EXPERIMENTAL

A. Materials and Preparation of Polyurethane

Both PBA2000 and PEG1000 are employed as the mixed soft segment of polyurethanes and MDI as the hard segment, BDO is used as chain extender, dimethyl formamide (DMF) as solvent.

The two-step procedure is followed to synthesize polyurethanes, which gives less random and more block polymer [5].

B. Preparation of Specimens and Analytical Methods

Polyurethanes are casted onto a glass plate. After dried at 80oC, these polyurethanes are further heat treated for 10min at 150oC. Finally, these polyurethanes were broken to pieces for thermal analysis.

Phase inversion temperature (PIT) of these polyurethanes is detected with modulated differential scanning calorimeter, DSC (Du Pont DSC-2000). The specimens (about 10mg) are heated from –50oC to 150oC at the heating rate of 20oC/min. Wide angle X-ray diffractometer is adopted to investigate the crystallinity (D/max-3C, Japan) of samples at the conditions of 2θ=0-50o, 3o/min, Cu Kα, 40KV, 30mA and room temperature.

III. THE STRUCTURE OF POLYURETHANE AND ITS PIT

TABLE I. POLYESTER-POLYETHER POLYURETHANES MATERIAL AND THEIR PROPERTIES.

Item / Serial number	Group A					Group B					Group C				
	1	2	3	4	5	1	2	3	4	5	1	2	3	4	5
Hard segment con. %	15.8	19.8	23.5	*26.8*	30.9	17.5	21.9	25.8	*29.4*	33.6	19.6	24.4	28.6	*32.4*	36.9
Hard segment mol. wt	319	420	522	*622*	760	318	420	521	*624*	759	316	419	520	*623*	760
Ether bond content %	5.5	5.1	4.9	*4.7*	4.4	10.0	9.5	9.0	*8.6*	8.0	15.7	14.8	14.0	*13.2*	12.4
Soft segment mol. wt			*1700*					*1500*					*1300*		
Soft charge ratio[①]			*3:7*					*1:1*					*7:3*		
PIT ºC	23.0	23.0	30.0	*37.0*	45.0	28.0	28.0	32.0	*39.0*	39.0	36.0	35.0	38.0	*44.0*	44.0

Notes: Soft charge ratio means PEG1000:PBA2000.

© 2015. The authors - Published by Atlantis Press

IV. The Microstructure of Samples

With regard to the glass transition range (Figure 1), all the endothermic peaks of samples are neither a typical glass transition graph of amorphous soft segments nor a typical watery fusion graph of crystalline soft segment; they have both features of amorphous and crystalline polymer simultaneously.

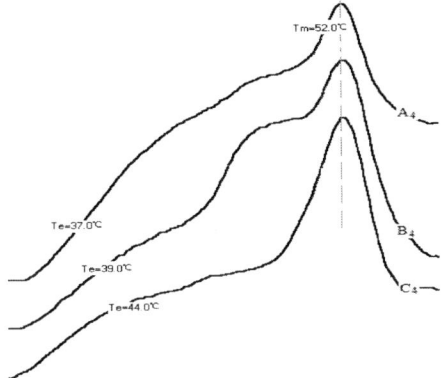

FIGURE I. THE DSC SPECTRUMS OF A4/B4/C4.

In fact, the phenyl existed in hard monomer MDI and the polar ester groups existed in soft monomer PBA2000, which can form steady hydrogen bonding with carbamate groups (hard segment), which results in a poor micro phase separation degree of samples. Also, the crystallization ability of soft segments of polyester-polyether polyurethanes has been impaired markedly because the soft monomer PEG1000 has a shorter chain segment and a interchain attraction between PEG and PBA[4].

We consider that the most of soft segment existed as amorphous structure, but quite a number of soft segment also existed as crystalline particle and crystalline orientation. These crystalline particles have different crystalline modification, crystal perfection and crystalline dimension. The consecutive change of the slope of DSC spectrums implies that the crystalline particles have different melt temperature.

Actually, as the amount of PEG is increased, a co-continuous PEG and PBA morphology forms in the soft matrix. Absorption of water vapor by PEG segments will further enhance the phase separation in the soft matrix since the polarity difference will be greater between PEG and PBA phases. In the absence of water, PEG and PBA show partial miscibility when mixed. This also suggests that in copolymers containing low levels of PEG, it will be dissolved / distributed within continuous PBA matrix. In contrast, as PEG content is increased above a critical level and the system is brought in contact with water vapor, permeability of moisture will be improved markedly. This will dramatically change the polarity of PEG domains and force them to phase separate from PBA, resulting in a co-continuous PEG / PBA soft matrix.

The foregoing analysis for the supramolecular structure of samples can be strongly verified by their wide angle X diffraction (WAXD) datum (Table 2)

TABLE II. THE WAXD IMAGE DATUM OF POLYESTER-POLYETHER POLYURETHANES MATERIAL A4, B4 AND C4.

PU	Items	1	2	3	4	5	6	7	8	9	10
	2θ	10.10	13.78	22.68	23.54	26.32	31.26	37.74	40.60	41.78	43.76
	d value	8.75	6.42	3.92	3.78	3.38	2.86	2.38	2.22	2.16	2.07
	Count	031	031	036	036	033	033	025	024	024	023
A4	Relative intensity	86	86	99	98	90	90	68	66	65	64
	Half-high width	0.02	0.02	0.02	0.02	0.02	0.02	0.02	0.02	0.02	0.02
	Integra intensity	33	33	38	38	35	35	26	25	25	25
	2θ	10.38	15.38	17.72	21.74	24.24	28.64	37.00	44.36	—	—
	d value	8.52	5.76	5.00	4.08	3.67	3.11	2.43	2.04	—	—
	Count	028	025	031	032	034	032	025	026	—	—
B4	Relative intensity	83	73	91	94	99	94	73	75	—	—
	Half-high width	0.02	0.02	3.32	11.16	16.12	24.96	0.02	0.02	—	—
	Integra intensity	30	27	5502	19080	29106	42834	27	27	—	—
	2θ	7.68	8.76	21.30	21.94	24.98	29.26	31.42	34.06	36.00	41.68
	d value	11.50	10.09	4.17	4.05	3.56	3.05	2.84	2.63	2.49	2.16
	Count	023	027	031	032	033	033	031	028	029	029
C4	Relative intensity	71	80	94	96	98	99	94	84	88	86
	Half-high width	0.02	0.02	0.02	0.02	0.02	45.60	0.02	0.02	0.02	0.02
	Integra intensity	25	28	33	34	35	79717	33	30	31	30

Several wider diffraction peaks of samples imply that there are a mass of crystalline particle whose crystal perfection is very poor existing in polyurethanes.

The datum of relative intensity and integra intensity show that there are a large deal of amorphous macromolecule and a small quantity of crystallized macromolecule existing in samples simultaneously. Further more, the distribution of half-high width of WAXD spectrums becomes wider and wider ($0.020 \sim 0.020$, $0.020 \sim 24.960$ and $0.020 \sim 45.600$ respectively) with the decline of soft segment average molecular weight (from A4 to C4) of polyurethanes. This implies that the crystallization ability of mixed soft segment has been impaired from Group A to C, and the reason has been analyzed in the foregoing discussion. However, from Group A to C, the crystallinity of PEG1000 in soft phase has been improved notwithstanding the average molecular weight of soft segments is decreasing (Figure1 and Table 2). Just as we analyzed, it is because that the microphase separation of PEG in soft domain from PBA has been improved with the increment of PEG in the mixed soft phase.

According to Table 2, all the WAXD spectrums have $8 \sim 10$ bragg diffraction peaks with different half-high width, every diffraction peaks is the reflection of many crystalline particles with uniform crystallization modality. Without doubt, there are some of crystallized macromolecules with different crystallization structure existing in samples.

V. CONCLUSIONS

For the mixed soft segments of PEG1000 and PBA2000, the crystallinity of PEG1000 in soft phase has been improved with the increasing dosage of PEG1000 in the mixed soft monomers even if the average molecular weight of soft segments is decreasing.

Within a group, the phase inversion temperature of polyurethanes moves to higher temperature with the increase of their hard segment content. It is because of the worse micro phase separation.

The incensement of the soft segment molecular weight results in a decrease of the soft phase PIT. However, when the hard segment content is higher, this tendency becomes not so much clear anymore because of a higher compatibility of the two phases.

REFERENCES

[1] H. M. Jeong, S. Y. Lee, B. K. Kim. Shape memory polyurethane containing amorphous reversible phase [J]. Journal of materials science, 2000, 35:1579-1583.

[2] J. H. Yang, B. C. Chun, Y. C. Chung et al. Comparison of thermal/mechanical properties and shape memory effect of polyurethane block-copolymers with planar or bent shape of hard segment[J]. Polymer, 2003, 44:3251-3258.

[3] Ashish Aneja, Garth L. Wilkes. Hard segment connectivity in low molecular weight model 'trisegment' polyurethanes based on monols[J]. Polymer, 2004, 45:927-935.

[4] D. J. HOURSTON, G. WILLIAMS, R. SATGURU et al. A Structure-Property Study of IPDI-Based Polyurethane Anionomers[J]. Journal of Applied Polymer Science, 1998, 67: 1437- 1448.

[5] C. G. Mothe, C. R. de Araujo. Properties of polyurethane elastomers and composites by thermal analysis [J]. Thermochimica acta, 2000, 357-358:321-325.

International Conference on Power Electronics and Energy Engineering (PEEE 2015)

A Novel High Water-Resistant Aqueous Acrylate Emulsion, Part 1: Preparation and Characterization

B.K. Ren

School of Biological and Chemical Engineering
Zhejiang University of Science and Technology
College of Chemical Engineering
Zhejiang University of Technology
China

K.T. Hu

School of Biological and Chemical Engineering
Zhejiang University of Science and Technology
China

G.H. Li

College of Chemical Engineering
Zhejiang University of Technology
China

A.B. Wu, J.J. Li

Hangzhou Special Paper Industry Co.
Ltd, China

Abstract—The acrylate copolymer emulsion with core-shell structure was successfully prepared via pre-emulsified semi-continuous seeded polymerization using SDS and OP-10 as compound emulsifier, and methyl methylacrylate (MMA), butyl acrylate (BA) and dodecafluoroheptyl methacrylate (DFMA) as main monomers. SEM and OCA were used to characterize the structure and the product, meanwhile, the influence of core-shell monomer ratio, the dosage of emulsifier, initiator and fluoride monomers on the reacting product properties were studied. Obvious ball structure could be seen clearly by the scanning electron microscopes when the hard monomer MMA matched with a moderate amount of DFMA as the shell in an optimized reaction condition. Detected with the help of particulate size description analyser, the latex particle which had uniformly size was found mono-dispersed.

Keywords-fluorinated acrylate; aqueous emulsion; core-shell structure; water-resistance; uniform particle size

I. INTRODUCTION

Aqueous acryl ate emulsions have become one of the major research and development fields due to its environmental friendliness, high weather ability, excellent film-processing ability and constructability[1-4].However the obvious disadvantages including poor resistance to heat temperature and little water-resistance restrict its application range[2,5]. In addition, when they were used as architectural coatings and encountered low temperature, they would become fragilely. Thus, to avoid these disadvantages, attempts have been made by some researchers to synthesize a series of novel latexes using sorts of functional monomers to modify[5-8].

Among numerous functional monomers, fluorine-containing monomers have attracted increasing attention of many investigators[7,9,10], because the acryl ate copolymer films modified by them not noly keep the original properties of polyacrylate, such as good adhesion to matrics, but also have their unique and excellent properties, which include high thermal, chemical, aging and weather resistance. Furthermore, they have low levels in dielectric constants,

refractive index, surface energy and flammbility, execellent inertness to solvents, hydrocarbons, acids, alkalis and moisture adsorption as well as intriguing oil and water repellency due to the low polarizability and the strong electro-negativity of the fluorine atom[11-12]. Therefore, they have been used progressively in a wide-range of applications[13-15]. However, the price of fluorinated monomers is rather which leads to a limitation of large-scale use for fluorinated polyacrylate(PFA) products.

To control the amount of DFMA, we used a new polymerization technology based on seed emulsion polymerization which was called core-shell emulsion polymerization. We arranged them in the shell part which was coated on of the core part. Herein, the fluorinated polyacrylate emulsion with core-shell structure was prepared via pre-emulsified semi-continuous seeded polymerization using SDS and OP-10 as compound emulsifier, methyl methylacrylate (MMA) and butyl acrylate (BA) as main monomers and dodecafluoroheptyl methacrylate (DFMA) as functional monomers. The effects of polymerization conditions on the conversion and poly-merization stability were studied, meanwhile, FT-IR, SEM, TEM and PSDA (particulate size description analyser) were used to characterize the structure and the product. The results showed that DMFA effectively invovled in the emulsion copolymerization, in addition, uniform particle size and a clear core-shell structure were attained.

II. EXPERIMENTAL

A. Material

Methyl methacrylate (MMA, 99+%), polyoxyethylated alkyphenol (OP-10, 99+%), sodium dodecysulfate (SDS, 99+%) were purchased from Shanghai Lingfeng Chemical Reagent Co. Ltd., China. Butyl acrylate (BA, 99+%) and Dodecafluoroheptyl methacrylate (Actyflon-G04, 96+%) were obtained from Sinopharm Chemical Reagent Co. Ltd., China and XEOGIA Fluorine-Silicon Chemical Co. Ltd., China

© 2015. The authors - Published by Atlantis Press

respectively. Ammonium persulphate (APS, 99+%) was purchased from Chinasun Specailty Products Co. Ltd., China. Ammonia Solution (25%) and filtrating base paper were from Hangzhou Liren Pharmaceutical Co. Ltd., China and a certain Hangzhou Special Paper Co. Ltd., China.

1) Synthetic technology roadmap: Synthetic technology roadmap is shown in Figure 1.

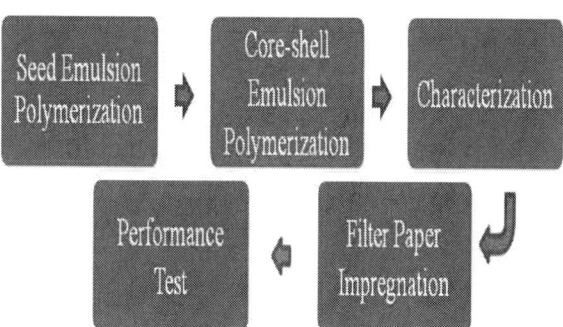

FIGURE I. SYNTHETIC TECHNOLOGY ROADMAP.

2) Core polymerization:

The reaction was conducted in a 250mL four-neck flask equipped with a reflux condenser, a mechanical stirrer, a thermometer, and a nitrogen gas inlet. Core part pre-emulsification: OP-10, SDS and H2O were added to the reactor as follows. When all the emulsifier solute and reached 40oC, the mixtures added with MMA and BA were pre-emulsified with a high stirring speed under nitrogen atmosphere for about 20 mins.

Seed emulsion polymerization: The mixture was heated to 72oC in a water bath and a part of APS solution (1.6g APS dissolved in 40.0g water) was added to trigger reaction. When the mixture in the reactor appeared blue fluorescence, the reaction was kept for another 30 mins and then cooled to room temperature.

3) Shell polymerization:

Shell part pre-emulsification: OP-10, SDS and H2O were added to a four-neck jacketed glass reactor as above. After the emulsifier stirred and dissolved, certain amount of MMA and DFMA were added.

Core-shell emulsion polymerization: All the shell emulsion were replaced by core emulsion, then the shell emulsion and APS solution were added using the monomer-starved method at 72oC in about 2h. When the mixture and APS solution was added completely, the reaction was kept for another 1h. At last, ammonia solution was dropped into the synthesized emulsion to adjust the pH value to the range of 6.0-7.0.

4) Impregnation of industrial filter paper:

Filtrating base paper was impregnated into the above dilute fluorinated acrylate emulsion. The glue amount was controlled at 20±2% by press out the redundant impregnation agent from the bottom of filter paper. After dried by electricity

heat drum wind drying oven at 110oC, the products were obtained.

5) Water resistance and paper stiffness test: Water resistance test referred to JIS P8122-2004. Paper stiffness test referred to GB/T 22364-2008.

6) Characterization and measurements:

Conversion rate and gel fraction were tested by weighing method. The emulsion particle size and its distribution were determined by a granularity measurement instrument (Zetasizer Nano ZEN3600) from Malvern Co. (England). Scanning electrons microscope (SEM, Hitachi S-4700II, Japan) was used to investigate the surface morphology of emulsion particles.

B. Results and Discussion

1) Effect of emulsifier:

The effect of m (SDS)/m (OP-10) ratio on the steady-time of the pre-emulsion was shown in Figure 2. As could be seen in the figure, with the increase in the m (SDS)/m (OP-10) ratio, the emulsion stability increased firstly and then declined. When SDS mass-fraction reached 33.3%, in other words m (SDS)/m (OP-10) ratio equaled 1/2, the steady time stayed the longest. This is because when using the nonionic emulsifier OP-10 and ionic emulsifier SDS as compound emulsifier, they were adsorbed on the surface of latex particles alternatively and SDS molecules were inserted into OP-10 molecules. This resulted that the static tension of the latex particle surface was greatly reduced and the adsorption force with latex particles increased. Simultaneously the thick hydro-layer formed on the surface by OP-10 guarantee the stability of polymer emulsion with the help of synergistic effect [16]. Above all, when SDS/OP-10 equaled 1/2, the combination between OP-10 and SDS seemed best and thus led to a best emulsifying effect.

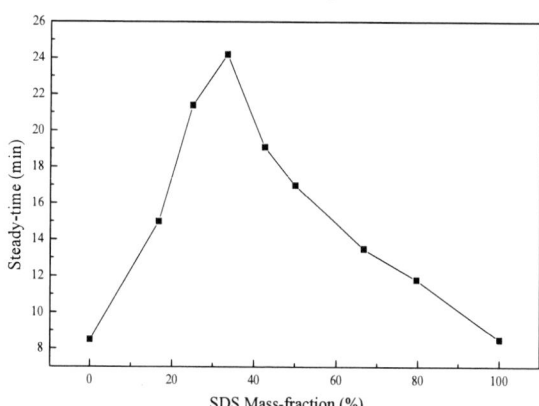

FIGURE II. EFFECT OF M (SDS)/M (OP-10) RATIO ON THE STEADY-TIME OF THE PRE-EMULSION.

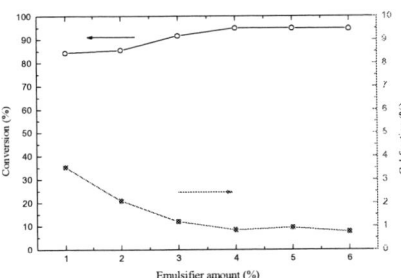

FIGURE III. EFFECT OF EMULSIFIER AMOUNT ON THE
CONVERSION AND GEL FRACTION.

As shown in Figure 3 was the effect of emulsifier amount on the conversion and gel fraction. As we see, with the increase of emulsifier dosage, the final conversion of polymer emulsion increased gradually and when the emulsifier dosage arrived at 4.0%, it became balanced. However, the gel fraction was trending downward and varied little after 3.0%. This is partly because the total gel fraction was only a little that the collection of the gel had great chance to bring error. On the other hand, this precisely illustrated when about 3.0%, the emulsifier had met the requirement of the system stability and the increase of particle size. Overall considered, the compound emulsifier amount was controlled at 3.5(wt) %-4.0(wt) % of the monomer amount.

2) Effect of initiator amount:

Figure 4 showed the effect of APS initiator amount on the final conversion and polymerization stability. As shown here, with the increase of dosage of initiator, both conversion and gel fraction were on the rise as a whole. Gel fraction maintained upward trend, while conversion rate was first increased and then decreased a little.

When the initiator amount arose from 0.15% to 0.62%, the concentration of free radicals increased in the water phase, so the ones adsorbed onto the particle surface increased. The increasing free radicals ultimately resulted in the progressive reaction and connection between monomers and particles. With the further reaction, the conversion rate improved much higher. However when the dosage of initiator continued to be arise, the polymerization speed was accelerated and thus resulted that the Brownian motion[17] of latex particles was more active which could enlarge the collision rate. But the more active system brought lower polymerization stability, a little lower conversion rate and higher gel fraction.

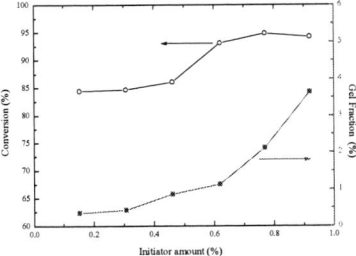

FIGURE IV. EFFECT OF INITIATOR AMOUNT ON THE CONVERSION
AND GEL FRACTION.

3) SEM Analysis

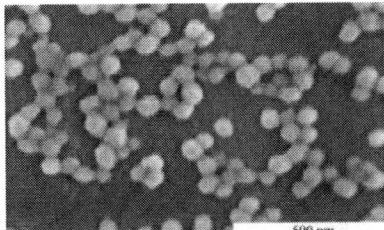

FIGURE V. SEM PHOTOGRAPHS OF FLUORINE-CONTAINING
ACRYLATE LATEX PARTICLES.

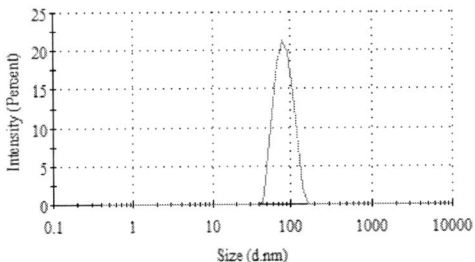

FIGURE VI. PARTICLE SIZE DISTRIBUTION OF
FLUORINE-CONTAINING ACRYLATE LATEX PARTICLES.

Figure 5 and Figure 6 were SEM photographs and particle size distribution of fluorine-containing acrylate latex particles. As Figure 5 showed, the latex particle which had uniformly size was mono-dispersed and nearly spherical. As seen in Figure 6, the particle size was about 84 nm, dispersity index PdI was 0.027 (<0.06) and they also confirmed the above result made from SEM figure. The core-shell polymerization was employed with Semi-continuous Hunger State Changing Mode, the low concentration monomer of core layer on the surface were soon reacted to keep the latex particles' uniformity -- that is why the uniform particles were obtained.

III. CONCLUSION

A type of fluorine containing acrylate emulsion with core-shell structure were successfully synthesized using a mixture emulsifier system via an emulsion polymerization approach. While the dosage of the mixture emulsifier, the initiator APS and the monomer DFMA were employed as 4%, 0.6% and 6% (on basis of the total monomer weight), m(SDS)/m(OP-10) equaled 1/2, the latex particles with distinct core-shell structure prepared could be acquired with narrower size distribution, smaller particle size and stronger water resistance.

ACKNOWLEDGEMENT

This work was supported by the National Natural Science Fund of China (No. 21173193, No. 21301154) and the Commonweal Project from Science Technology Department of Zhejiang Province (No. 2014C33038).

REFERENCES

[1] A. Bal, G. Guclu, T.B. Iyim, S. Ozgumus, Effects of Nanoparticles on Film Properties of Waterborne Acrylic Emulsions, Polym.-Plast.

Technol. Eng. 50 (2011) 990–995.

[2] X. Qu, N. Wang, P.A. Lovell, Preparation and Characterization of the Latexes with Different Particle Sizes by Semibatch Emulsion Polymerization and the Influence on Properties of Waterborne Pressure-Sensitive Adhesives, J. Appl. Polym. Sci. 112 (2009) 3030–3040.

[3] K. Zhang, H. Shen, X. Zhang, R. Lan, H. Chen, Preparation and Properties of a Waterborne Contact Adhesive Based on Polychloroprene Latex and Styrene-Acrylate Emulsion Blend, J. Adhes. Sci. Technol. 23 (2009) 163–175.

[4] D. Wu, F. Qiu, H. Xu, J. Zhang, D. Yang, Preparation, Characterization, and Properties of Environmentally Friendly Waterborne Poly(urethane acrylate)/Silica Hybrids, J. Appl. Polym. Sci. 119 (2011) 1683–1695.

[5] J. Xu, H. Hu, Preparation and Characterization of Styrene Acrylate Emulsion Surface Sizing Agent Modified with Rosin, J. Appl. Polym. Sci. 123 (2012) 611–616.

[6] J.Y. Seo, M. Han, Multi-functional hybrid coatings containing silica nanoparticles and anti-corrosive acrylate monomer for scratch and corrosion resistance, Nanotechnology. 22 (2011) 025601.

[7] C. Ai, Y. Ke, Y. Yi, J. Guan, F. Pan, Preparation and Properties of Poly(fluorated-acrylate)/Montmorillonite Composite Emulsion, Integr. Ferroelectr. 136 (2012) 156–168.

[8] H. Li, F. Lei, P. Li, W. Duan, J. Zhou, X. Tan, Selective Adsorption of Berberine Hydrochloride Using Molecularly Imprinted Polymers with Modified Rosin as Cross-linker, Asian J. Chem. 25 (2013) 7421–7426.

[9] L. Zhang, L. Zang, H. Zhang, J. Guo, Synthesis of fluorine-containing latexes with core-shell structure by UV-initiated microemulsion polymerization, Iran. Polym. J. 22 (2013) 93–100.

[10] W. Xu, Q. An, L. Hao, L. Huang, Synthesis, film morphology, and performance of cationic fluorinated polyacrylate emulsion with core-shell structure, J. Appl. Polym. Sci. 125 (2012) 2376–2383.

[11] J. Wang, X.-R. Zeng, H.-Q. Li, Preparation and characterization of soap-free fluorine-containing acrylate latex, J. Coat. Technol. Res. 7 (2010) 469–476.

[12] L. Junyan, H. Ling, Z. Yuansuo, Synthesis and Property Investigation of Three Core-Shell Fluoroacrylate Copolymer Latexes, J. Appl. Polym. Sci. 112 (2009) 1615–1621.

[13] X. Huang, X. Wen, J. Cheng, Z. Yang, Sticky superhydrophobic filter paper developed by dip-coating of fluorinated waterborne epoxy emulsion, Appl. Surf. Sci. 258 (2012) 8739–8746.

[14] T. Kubota, Recent Development of Fluorine-containing Transparent Polymers, J. Synth. Org. Chem. Jpn. 71 (2013) 196–206.

[15] Z.-Y. Wang, J.-C. Ho, W.-J. Shu, Studies of fluorine-containing bismaleimide resins part I: Synthesis and characteristics of model compounds, J. Appl. Polym. Sci. 123 (2012) 2977–2984.

[16] T. Cao, Q. Liu, J. Hu, Synthesis Principle, Property and Application of Polymer Emulsion, second ed., Chemical Industry Press, Beijing, 2007, pp. 172-256.

[17] T. Poggio, V. Kapeliouchko, V. Arcella, E. Marchese, Multimodal fluoropolymer dispersions, Prog. Org. Coat. 48 (2003) 310–315.

Time-Resolved Fluorescence Anisotropy Study on Intramolecular Interactions of Branched Styryl Derivatives Based On 1, 3, 5-Triazine

Y.C. Wang
Department of Physics
Dalian Maritime University
Dalian, China
*Corresponding author

G.Q. Wang
Department of Physics
Dalian Maritime University
Dalian, China

D.J. Liu
Department of Physics
Dalian Maritime University
Dalian, China
*Corresponding author

B. Li
Department of Electronic Engineering
East China Normal University
Shanghai, China
*Corresponding author

Abstract—A series of branched styryl derivatives based on 1, 3, 5-triazine have been studied by time-resolved fluorescence anisotropy method to study the intramolecules interferaction between branches. The obtained results further confirmed the TPA enhancement mechanism, the anisotropy of trimer shows faster decay and small residual value indicates there are strong intramolecules interactions among branches, this maybe the enhancement mechanism of TPA properties for the trimer.

Keywords-two photon absorption; fluorescence; time-resolved fluorescence anisotropy

I. Introduction

Materials with large two-photon absorption (TPA) properties are of great interest in many fields [1-3], such as optical limiting, three-dimensional microfabrication, optical storage, and so on. Design and synthesis of the organic and/or polymeric functional materials with excellent TPA performance have thus stimulated extensive research activities across the world in recent years [4].

Although there already are a large number of papers and patents concerning preparation of TPA materials, including one-dimensional dipolar [5], quadrupolar [6], and multibranched chromophores [7], development of the molecular materials with large TPA cross section still draws much attention and presents an ongoing challenge.

To design and synthesize more excellent molecules with outstanding TPA properties, a proper understanding of the dynamics of two-photon excitation process in the materials is of great importance. Femtosecond (fs) pump-probe experiment and time-resolved fluorescence and time-resolved fluorescence anisotropy are very useful technique to obtain the information of the relaxation processes of the excited states. Theodore Goodson,

III, Varnavski and co-workers et al. employed pump-probe, time-resolved photoluminescence, and three-pulse photon echo measurements to measure dynamics of some molecules including multibranched chromophores and organic conjugated dendrimers [8-9]. Anisotropy study is a powerful tool to give additional information about the energy redistribution and the dynamics of electronic coupling in multi-branched molecules. It has been widely used in experiments on photosynthetic reaction center. Goodson et al. reported plenty of significant results based on time-resolved fluorescence up-conversion measurement on optical dendrimers[10-11]. Dynamics of molecules with two-photon absorption properties are, however, still inadequate especially for the conjugated molecules. In our previous work, we reported the TPA character and excited state dynamics of several molecules and polymers with both linear and tri-branched structure [12-14]. Recently, we investigated a tri-branched materials, by using time-resolved fluorescence anisotropy methods, The tri-branched materials displays very good two photon absorption ability and nonlinear enhancement according to monomer case. The time-resolved photoluminescence anisotropy results indicate the enhancement mechanism of TPA properties.

II. Materials and Experimental Methods

The structures of T01, T02 and T03 are shown in Figure 1. The synthesis method, UV-visible absorption spectra, fluorescence as well as TPF/TPA properties have been reported elsewhere in detail [15]. Two-photon absorption (2PA) cross-sections measured by the open aperture Z-scan technique were determined to be 77, 90 and 410 GM for T01, T02 and T03, respectively.

© 2015. The authors - Published by Atlantis Press

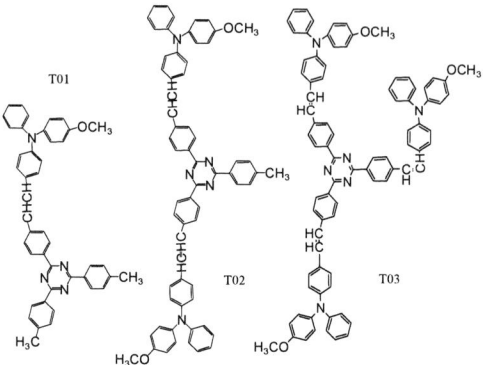

FIGURE I. THE STRUCTURE OF T01, T02 AND T03.

The ultrafast responses of the polymer were investigated by fs time-resolved photoluminescence (TRPL) experiments. The setup is shown in Figure 2. Briefly, the pump beam at 800 nm after passing an optical delay line was used as a gate beam to open "Kerr gate" through photo-induced birefringence of Kerr Material (CS2), while the second part of 800 nm beam was frequency doubled by using a BBO crystal to act as the probe beam. The second harmonic pulses with vertical polarization were used to pump samples efficiently and the collected fluorescence was set either parallel or perpendicular to that of incident beam by a polarizer (P1). Another polarizer with orthogonal polarization to P1 was placed after sample to study the polarization effect. Dispersed by a monochromator, the signal was detected by a photomultiplier (Hamamatsu R1104) connected to a lock-in amplifier (SR830, Stanford Research Systems). The polarization of the gate beam was set at 45°with respect to that of SHG. In doing so, we can get TRFL signals under different configuration for anisotropy study.

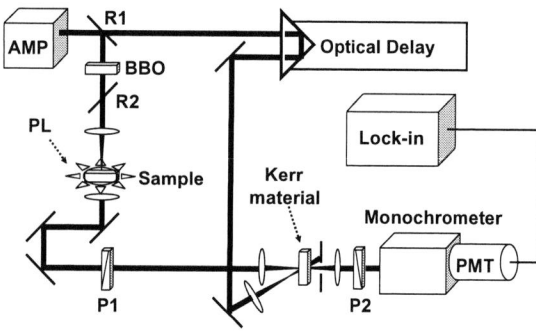

FIGURE II. EXPERIMENTAL SETUP FOR THE OKG METHOD. R1: BEAM SPLITTER; R2: HIGH REFLECTIVE MIRROR AT 800 NM; P1 AND P2: POLARIZERS (CROSS-POLARIZATION); KERR MATERIAL: CS2 IN 5 MM CELL; PMT: PHOTOMULTIPLIER TUBE.

The fs pulses employed in ultrafast dynamics measurements were generated by amplification stage of the used fs laser system (Spitfire, Spectra-Physics). The average output power from the Spitfire was about 300 mW. The pulse duration was 140 fs, the wavelength was 800 nm and the repetition rate was 1kHz. In the experimental investigation, the solvent CHCl3 has been used

without further distillation. All the experiments were carried out at room temperature.

III. RESULTS AND DISCUSSIONS

As we all know, anisotropy can be decided by

$$r(\tau) = \frac{I_{//} - I_{\perp}}{I_{//} + 2I_{\perp}} \tag{1}$$

Where $I_{//}$ ($I_{\perp}$) denote the fluorescence intensity whose polarization is parallel (perpendicular) to that of excited beam. However, analyzing TRPL experimental data based on OKG technique is a great challenge because it is more complicated than the analysis of transient absorption and fluorescence up-conversion experimental data. In the latter, the system response can be simply considered as Gaussian which is convenient for deconvolution treatment. In our experiment, the system response should be the response profile of CS2, which is nearly exponential type. Thus we used the following two formulas to calculate:

$$I_{//} = \int \sigma(t-\tau)[A_1 e^{-t/\tau_1} + A_2 e^{-t/\tau_2} + A_3 e^{-t/\tau_3}][1 + 2\{(r_0 - r_1)e^{-t/\tau_r} + r_1\}]dt \tag{2}$$

$$I_{\perp} = \int \sigma(t-\tau)[A_1 e^{-t/\tau_1} + A_2 e^{-t/\tau_2} + A_3 e^{-t/\tau_3}][1 - \{(r_0 - r_1)e^{-t/\tau_r} + r_1\}]dt \tag{3}$$

Where τ_r is the time constant of anisotropy decay, r_0 and r_1 represent initial value and residual value, respectively.

Figure 3 shows the TRFL anisotropy experimental results of T01, T02, and T03 in THF solution at 510 nm and 550 nm, respectively. The fitting results are summarized in Table I. It can be seen from the results that, there is a fast anisotropy decay process at wavelength of 510 nm for all the three compounds. The lifetimes of the process are ~500fs, 415fs, 350fs for monomer, dimmer and trimer, respectively. The initial anisotropy value and resistant value is 0.65 and 0.46 for monomer T01. For dimmer T02, the initial anisotropy value and residual value is 0.57 and 0.43. Trimer T03 hold the smallest initial anisotropy value and residual value (0.52/0.36), the more branches, the smaller two values. When the probe wavelength tuned to 550 nm, no obvious anisotropy decay process was observed for all three compounds. The two values remain at some value. The value is also decrease with increase of number of branches. Usually, large initial anisotropy value as well as fast decay indicates strong intramolecular interactions which directly affect energy redistributions and bring large nonlinear optical effect [10]. This may be a reason why TPA cross section of trimer T03 is is about 5.32-fold larger than that of monomer T01, the anisotropy decay time is shorter than that of monomer T01.

56

TABLE I. THE TIME-RESOLVED FLUORESCENCE ANISOTROPY
RESULTS OF MONOMER, DIMER AND TRIMER T01, T02 AND T03.

	T(fs)	R1	R0	R 550nm
T01	≈500	0.65	0.46	0.4
T02	≈415	0.57	0.43	0.38
T03	≈350	0.52	0.36	0.2

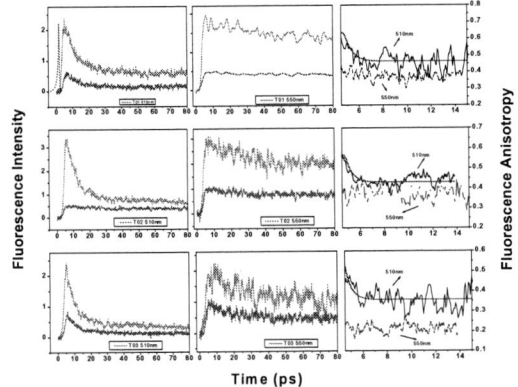

FIGURE III. EXPERIMENTAL SETUP FOR THE OKG METHOD. R1:
BEAM SPLITTER; R2: HIGH REFLECTIVE MIRROR AT 800 NM; P1 AND
P2: POLARIZERS (CROSS-POLARIZATION); KERR MATERIAL: CS2 IN
5 MM CELL; PMT: PHOTOMULTIPLIER TUBE.

IV. SUMMARY

In this study, the anisotropy of monomer T01, dimer T02 and trimer T03, are investigated by time-resolved fluorescence anisotropy technique. Obvious different between monomer, dimer and trimer are observed. It was found that, dimer and trimer show shorter depolarization time and less residual values in comparison with monomer. The anisotropy of trimer shows faster decay and small residual value indicates there are strong intramolecules interactions among branches, this maybe the enhancement mechanism of TPA properties for the trimer.

ACKNOWLEDGEMENT

We sincerely thank the financial support from National Natural Science Foundation of China (11404048, 61205154, 11375034) and the Fundamental Research Funds for the Central Universities (3132013104, 3132015233, 3132015152, 3132013106, 3132014231, 3132014337) and the program for Liaoning Excellent Talents in University (Grant No. LJQ2014051)

REFERENCES

[1] J. D. Bhawaolar, G. S. He and P. N. Prasad, Nonlinear multiphoton process in organic polymeric materials, Rep. Prog. Phys. 59(1996) 1041-1070.

[2] Q. D. Zheng, S. K. Gupta, G. S. He, L-S. Tan, P. N. Prasad, Synthesis, Characterization, Two-Photon Absorption, and Optical Limiting Properties of Ladder-Type Oligo-p-phenylene-Cored Chromophores, Adv. Funct. Mater. 18(2008)2770-2779.

[3] W. Denk, J. H. Strickler, and W. W. Webb, Two-photon Laser Scanning Fluorescence Microscopy, Science. 248(1990) 73-76.

[4] M. Albota, D. Beljonne, J. L. Bredas, J. E. Ehrlich, J. Y. Fu, A. A. Heikal, S. E. Hess, T. Kogej, M. D. Levin, S. R. Marder, D. McCord-Maughon, J. W. Peny, H. R6ckel, M. Rumi, G. Subramaniam, W. W. Webb, X. L. Wu and C. Xu, Design of Organic Molecules with Large Two-Photon Absorption Cross Section, Science 281(1998) 1653-1656.

[5] B. A. Reinhardt, L. L. Brott, J. J. Clarson, Highly active two-photon dyes: design, synthesis and characterization toward applicatioin, Chem. Mater. 10(1998)1863-1874.

[6] M. Albota, D. Beljonne, J. L. Bredas, J. E. Ehrlich, J. Y. Fu, A. A. Heikal, S. E. Hess, T. Kogej, M. D. Levin, S. R. Marder, D. McCord-Maughon, J. W. Peny, H. R6ckel, M. Rumi, G. Subramaniam, W. W. Webb, X. L. Wu and C. Xu, Design of Organic Molecules with Large Two-Photon Absorption Cross Section, Science 281(1998) 1653-1656.

[7] T.-C. Lin, Y.-F. Chen, C.-L. Hu, C.-S. Hsu, Two-photon absorption and optical power limiting properties in femtosecond regime of novel multi-branched chromophores based on tri-substituted olefinic scaffolds, J. Mater. Chem. 19(2009)7075-7080.

[8] M. I. Ranasinghe, Y. Wang, T. III. Goodson, Excitation energy transfer in branched dendritic macromolecules at low (4K) temperatures, J. Am. Chem. Soc. 125(2003) 5258.

[9] Y. Wang, G. S. He, P. N. Prasad, T. Goodson III, Ultrafast Dynamics in Multibranched Structures with Enhanced Two-Photon Absorption, J. Am. Chem. Soc. 127(2005) 10128-10129.

[10] O. P. Varnavski, J. C. Ostrowski, L. Sukhomlinova, R. J. Twieg, G. C. Bazan, T. Goodson III, Coherent Effects in Energy Transport in Model Dendritic Structures Investigated by Ultrafast Fluorescence Anisotropy Spectroscopy, J. Am. Chem. Soc. 124(2002) 1736-1743.

[11] T. III. Goodson, Optical excitations in organic dendrimers investigated by time-resolved and nonlinear optical spectroscopy, Acc. Chem. Res. 38(2005) 99.

[12] Y. C. Wang, Y. L. Yan, B. Li and S. X. Qian, The study on the two-photon absorption/fluorescence properties and the ultrafast dynamics of the organic materials, Progress In Physics. 32(2012) 1-30.

[13] Y. C. Wang, Y. H. Jiang, J. L. Hua, H. Tian, and S. X. Qian, Optical limiting properties and ultrafast dynamics of six-branched styryl derivatives based on 1,3,5-triazine, J. Appl. Phys. 110(2011) 033518.

[14] Y. C. Wang, D. K. Zhang, H. Zhou, J. L. Ding, Q. Chen, Y. Xiao, S. X. Qian, Nonlinear optical properties and ultrafast dynamics of three novel boradiazaindacene derivatives, J. Appl. Phys. 108(2010) 033520.

[15] B. Li, R. Tong, R. Y. Zhu, F. S. Meng, H. Tian, S. X. Qian, The Ultrafast Dynamics and Nonlinear Optical Properties of Tribranched Styryl Derivatives Based on 1,3,5-Triazine, J. Phys. Chem. B 109(2005) 10705-10710.

The Design of Electrode of Coal Roadway Advanced Detector

Y. Zhou
School of Mechanical Electronic & Information Engineering
China University of Mining and Technology (Beijing)
Beijing, China

Y.M. Lv
School of Mechanical Electronic & Information Engineering
China University of Mining and Technology (Beijing)
Beijing, China

J.T. Zhang
School of Mechanical Electronic & Information Engineering
China University of Mining and Technology (Beijing)
Beijing, China

Z.M. Liu
School of Mechanical Electronic & Information Engineering
China University of Mining and Technology (Beijing)
Beijing, China

X.G. Liu
School of Mechanical Electronic & Information Engineering
China University of Mining and Technology (Beijing)
Beijing, China

M. Wu
School of Mechanical Electronic & Information Engineering
China University of Mining and Technology (Beijing)
Beijing, China

Abstract—**In order to realize ahead forecast of geological structure of coal roadway and reduce coal mine accidents, an advanced detector based on Dynamic and Directional Electric Field Excitation Method was designed by my research group. As an important part of the detector, the electrode has a great influence on the formation of the electric field and detection accuracy. According to the requirements of the instrument, this paper presents the requirements and original parameters of the electrode, the main consideration is to reduce electrode grounding resistance. And by using of Matlab software simulation to get the change law of the grounding resistance with the depth into the rock, we got the basic size of electrode.**

Keywords-advanced detector; electrode; grounding resistance; simulation study

I. INTRODUCTION

"There must be a search before digging and excavation after probing" is the principle of the coal mine production process which has to be adhering to. According to statistics, there are about 65% of water inrush accident and more than 80%of coal and gas outburst occurred in heading stage [1]. Therefore, the effective prediction of the concealed geological structure in front of the mine heading face is of great significance for the safe production of coal mine.

Based on the research status of coal mine roadway excavation and advanced detection methods, to solve the existing forecast problems of poor real-time, poor operability, poor anti-interference ability, my research group proposed Dynamic and Directional Electric Field Excitation Method [2]. The advanced detection method is based on dual frequency induced polarization method and focus deflection effect of electric field. We have got a principled sample machine of coal roadway advanced detector which mainly comprises of a transmitting device and a receiving device.

As an important part of the detector, the electrode has a great influence on the formation of the electric field and detection accuracy. Common electrode currently used underground can't satisfy the requirements of our instrument. Therefore, a kind of electrode is designed in this paper. The electrode is consistent with the principle of Dynamic and Directional Electric Field Excitation Method. It can also meet the special environment of coal mine and form good electric field.

II. ELECTRODE REQUIREMENTS TO THE GROUNDING RESISTANCE

Analyzing the instrument requirements to electrode and the special environment of the coal mine, we got the original parameters of the electrode design and established an original parameters table as design reference.

According to the transmitting device requirements to electrode, the maximum voltage value through the emitter electrode, constraint electrode and grounding electrode is100V, the maximum voltage value through the receiving electrode is 10V.According to the receiving device requirements to electrode, the maximum current value through the emitter electrode, constraint electrode and grounding electrode is 100mA, the maximum current value through the receiving electrode is very weak and its magnitude is between10-3~10-4mA.

In order to form a good electric field of emission current, the partial pressure of electrode must be small. The electrode is good conductor, its resistivity is at a level of $10-8\Omega\cdot m$. For the electrode dimensions are in cm level, we can get the magnitude

© 2015. The authors - Published by Atlantis Press

of electrode resistance should be between10-3~10-4Ω, the magnitude of electrode pressure drop is between 10-5~10-4V at the maximum current value of 100mA.So the influence on the measurement results could be neglectful. According to the requirements of transmitting device, the magnitude of grounding resistance is at a level of10K through the theoretical calculation, so the real value of grounding resistance is less than it.

Based on the above analysis, establish the original parameter table of the electrode design in the following Table 1.

TABLE I. THEORIGINAL PARAMETER TABLE OF THE ELECTRODE DESIGN.

	Emitterelectrode	Constraintelectrode	Grounding electrode	Receiving electrode
Maximum voltage（V）	100	100	100	10
Maximum current(mV)	100	100	100	10^{-3}~10^{-4}
Electrode resistance(Ω)	10^{-3}~10^{-4}	10^{-3}~10^{-4}	10^{-3}~10^{-4}	10^{-3}~10^{-4}
Grounding resistance(KΩ)	<10	<10	<10	<10

III. ELECTRODE DESIGN

In the measurements of electrical prospecting, leakage detection, lightning detection, soil resistivity test, all need grounding the electrode, electrode grounding will inevitably produce grounding resistance [3, 4, 5]. The value of grounding resistance will have effects on the measurement results which can't be ignored [6, 7]. If the grounding resistance is small, it can save power, improve the accuracy, enhance anti-interference ability of the system, work effectively and ensure the security of system. Therefore, reducing electrode grounding resistance is a major consideration in design of electrode. Electrodes used at present are usually of cylindrical stick shape and made of steel, it's the basis of the design.

Access to information, calculate the grounding resistance according to the definition of conductor resistance. In the practical application, in order to operate conveniently and guarantee the electrode have a good contact with the surrounding rock, cylindrical stick electrodes are always used, as shown in Figure I. In the figure, ρ represents the resistivity of rock and assume the rock is a kind of infinite homogeneous isotropic medium here, the stick electrode has a radius of r0 and l represents the depth into rock.

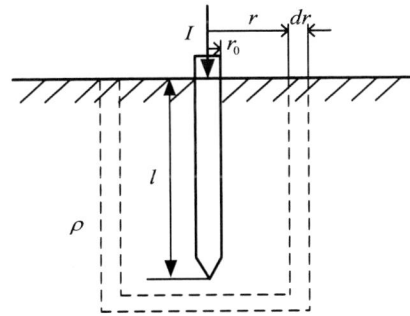

FIGURE I. THE PRACTICAL APPLICATION MODEL OF ELECTRODE.

$R_l$represents the resistance from superficial layer $r_0$of the electrode to a cylindrical layer with a radius of r, calculated as following:

$$R_1 = \int_{r_0}^{r} \frac{\rho dr}{2\pi r(l+r-r_0)+\pi r^2} dR = \frac{\rho}{2\pi(l-r_0)} \ln \frac{r(r_0+2l)}{r_0(3r+2l-2r_0)} \quad (1)$$

When r→∞, the theory value of grounding resistance as following:

$$R = \frac{\rho}{2\pi(l-r_0)} \ln \frac{r_0+2l}{3r_0} \quad (2)$$

(1) Determination of the electrode radius

Analyze the function relationship, the relationship between the grounding resistance R and the radius$r_0$of stick electrode is as following: the grounding resistance R markedly decreases with the increase of stick electrode radius r_0. Therefore, in the actual measurement, the radius r_0 of electrode should be large as far as possible if condition permission. After full consideration of the factors such as handling, installation and demolition, choose the length of 0.5cm as the radius of electrode.

(2) Determination the length of the electrode and the depth into rock

The surrounding rocks of coal mine are complex and majority of them are shale, argillaceous shale and sandstone shale. Assume the main component is shale, according to the information provided, the resistivity of shale is5×103Ω·m.

By the resistivity of shale is5×103Ω·m, the radius of electrode is 0.5cm, according to the formula (2), the relationship between grounding resistance R with the depth into rockl is shown in Figure II. From the figure we can see the grounding resistance increases with the decrease of the depth into rock, when the depth is more than 50cm, the curve flats gradually. Considering the larger the depth into rock, the more difficult wedging and pulling out the electrode, so the electrode length ranges of 50~70cm and the depth into rock 35~50cm, select the depth into rock 35cm here.

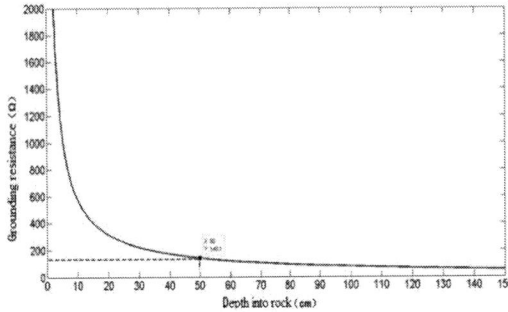

FIGURE II. THE RELATIONSHIP BETWEEN GROUNDING RESISTANCE WITH THEDEPTH INTO ROCK.

Put determined parameter values into the formula (2), calculate the value of grounding resistance R=8.881×103Ω. Compared to the original parameter table, this value can satisfy the requirements of instrument, namely both the length of electrode and the depth into rock can meet the design requirements.

According to the depth into rock, combined with the length of exposed external electrode, the total length can be designed as 50cm, namely the length of depth into rock is 35cm and the exposed part 15cm.

We test whether the resistance value can meet the requirements of the original parameter table after determining the basic sizes of electrode. Steel resistivity is variable due to the internal components, but all are good conductors. In approximate computation, the value can be replaced by the resistivity of iron which is the main component of it, namely 9.78 × 10-8Ω·m.The electrode resistance equals 1.56 × 10-4Ωaccording to the property of resistance, which meets the resistance requirements of electrode in original parameter table. So the radius and total length of the electrode are designed reasonable.

(3) Determination the taper of electrode head

In order to install easily, we designed a taper of 1:3 on the electrode head. Because 3cm is relatively small compared to total size, there is no significant impact to physical and chemical properties of electrode. Therefore, we assume the electrode is of ideal cylindrical in the above calculation.

(4) Integral structure of the electrode

Through the above design, schematic diagram of the integral electrode structure is shown in Figure III:

FIGURE III. SCHEMATIC DIAGRAM OF THE INTEGRAL ELECTRODE STRUCTURE.

IV. CONCLUSIONS

In this paper, through the analysis of concrete requirements of the advanced detector, we design the electrode mainly take account of reducing the grounding resistance. With the help of simulation analysis by Matlab software, we get better theoretical value of basic parameters of electrode size and meet the requirements of the instrument.

ACKNOWLEDGEMENTS

This work was supported by the National High Technology Research and Development Program of China (863 Program) (No. 2012AA06A405) and the Fundamental Research Funds for the Central Universities (No.2010YJ01).

REFERENCES

[1] Liu Qingwen. The application of seismic prospecting technology during the development of gypsum mine [J]. Coal Geology & Exploration, 2001, 29 (5): 60-62.

[2] Zhang Weijie. Research of advanced detection technology based on dynamic and directional electric field excitation method of coal roadway driving [D].China University of Mining andTechnology(Beijing), 2012.

[3] Fu Liangkui. Electrical exploration tutorial [M]. Beijing: Geological Publishing House, 1983.

[4] Guan Shaopeng, Pan Junfeng, NengChangxin, etc. Grounding resistance of electrode in landfill leakage detection using electrical method [J].Research of Environmental Sciences, 2008, 21 (6): 39-42.

[5] Mei Ji, Wang Yingbo, Sun Yanbing. Analysis of familiar questions about measuring of earth resistivity [J].Environmental Science & Technology, 2011, 34 (6G): 246-248.

[6] Zhang Lingyun, Liu Hongfu, Li Chengyou. Experiment on ground resistance in high-density electrical measurement [J].Progress in Exploration Geophysics,2010, 33 (3): 179-183.

[7] Li-Hsiung Chen, Jiann-Fuh Chen, Tsorng-Juu Liang, and Wen-I Wang.A Study of Grounding Resistance Reduction Agent Using Granulated Blast Furnace Slag[J]. IEEE Transactionson Power Delivery, 2004, 19(3):973-978.

International Conference on Power Electronics and Energy Engineering (PEEE 2015)

Research of Wind Power Prediction Based on the Auto-Regressive Model

L. Feng

School of Electrical and Information
Changchun Institute of Technology Jilin
Changchun, China

C.H. Liang

School of Electrical and Information
Changchun Institute of Technology Jilin
Changchun, China

H. Huang

Logistics Services Center Jilin Province Electric Power Limited Company
Changchun, Jilin, China

Abstract—**This paper discusses in detail the reason for the inaccurate result from the present system for wind farm output power prediction. Time series analysis method was applied for improving existing problem in prediction model and treatment method of basic data. Improving self-regressive mathematical model was established and taken the model identification. Using SPSS software simulates and further assists model identification and using given wind farm historical output power data to forecast one and multi-wind power unit output power in odd-number days and a week. Finally, this paper compares and analyses the getting prediction power and expound the next step work that improves the wind power prediction accuracy.**

Keywords-wind power prediction; time series analysis; self-regressive mathematical model; simulation

I. INTRODUCTION

Wind power technology is current renewable energy utilization of the most mature technology and the most large-scale development and commercialization development prospects. With the development of wind power technology development, its malpractice appears gradually. Intermittency and fluctuation of wind power itself unique determines the power of intermittent and fluctuating. In order to satisfy the power supply demand, ensuring the reliability of the stable operation of the power grid and power supply system, so, we need to effectively plan and schedule for the power supply system. While the wind power intermittent itself unique and uncertainty, increase the difficulty of grid scheduling and the reserve power capacity. In order to solve the distribution problem of instability and reduce the power grid reserve capacity cost, we must predict the output power of large-scale wind farms. Through real-time accurate prediction for wind power generation, power dispatching department can advance the scheduling plan according to the wind power change in order to ensure the power balance and the safety operation of the power grid, and adjust accordingly to effectively mitigate the adverse impact of wind power on power grid and effectively reduce the operation cost and improve the efficiency of wind power generation. Therefore, how to forecast the wind power as accurately as possible is an urgent problem to be solved.

II. AR MATHEMATICAL MODELING

We apply the stochastic process theory and mathematical statistical methods to study statistical regularities followed by random data sequence. Random data is arranged based on chronological sequence. It possesses good predictability using time series analysis to analyze smooth time series data. The ARMA model consists of two models of the AR (p) model and the MA (q).

A. ARMA Model Identification Criteria

If stationary series x (t) is tailing the autocorrelation function, partial autocorrelation function for the censored sequence, x (t) is the AR sequence. If stationary series x (t) is the autocorrelation function of the truncation, partial autocorrelation function is tailing sequence, then, x (t) is the MA sequence. If the autocorrelation function and partial autocorrelation function of stationary series x (t) is all tailing, x (t) is the ARMA series. Where the self-correlation function is $\rho_k = r_k / r_0$. The self-covariance function is $r(k) = \frac{1}{n} \sum_{t}^{n-k} (x_{t+k} - \bar{x})(x_t - \bar{x})$, among, $\bar{x} = \frac{1}{n} \sum_{t=1}^{n} x_t$ is the sample mean. Based on model identification criteria, we analyze data and complete the establishment of the mathematical model.

B. Model Identification

For example, wind power data of certain wind power plant; we adopt historical data to model from May 10 to May 30, 2006 and for the test data on May 31. We use the historical data to draw a line chart from May 10 to May 30, 2006, Figure. I show:

© 2015. The authors - Published by Atlantis Press

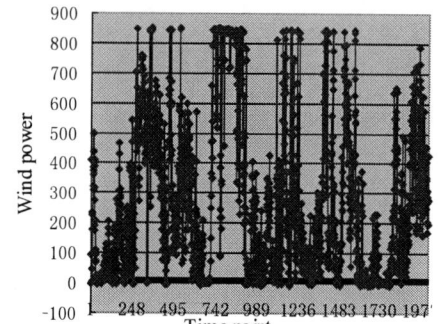

FIGURE I. A WIND FARM WIND POWER HISTORY DATA.

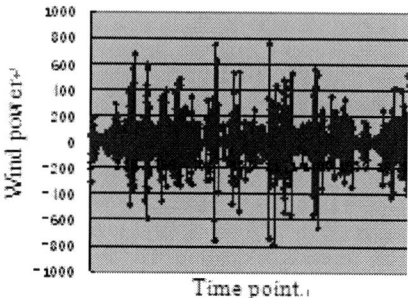

FIGURE II. A FIRST-ORDER DIFF.

As can be seen from the Figure I, the data has large fluctuations and is not stable related time series data, and therefore, we need to make odd transform for data and difference calculation. We take first-order differential operation for the data, the point data shown in figure 2.

By a first-order differential diagram, we can know, the value of wind turbine power is always in the vicinity of a constant random fluctuation, and fluctuations are of the range bounded, no clear trend and cycle characteristics, we regard it as smooth sequence. Therefore, we can further analyze and model based on this data.

III. AR SIMULATION MODELING

We apply SPSS software to simulate self-correlogram and assist further to make model identification, autocorrelation shown in Figure 3 and Figure 4:

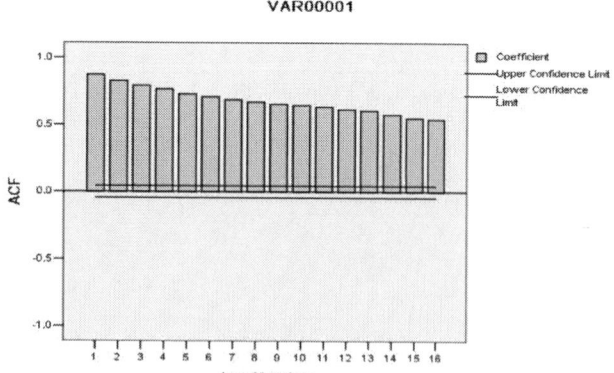

FIGURE III. AR CORRELATION.

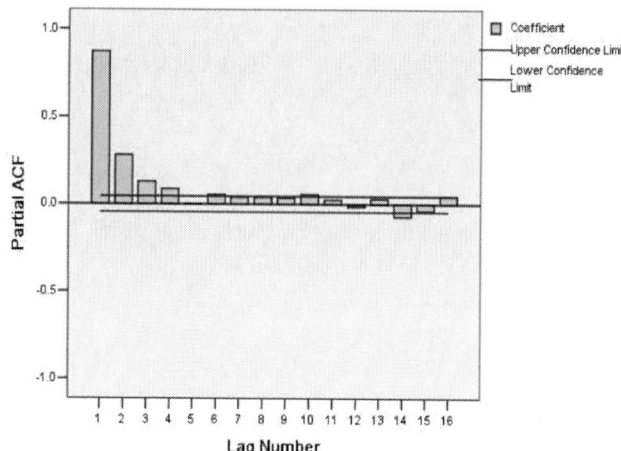

FIGURE IV. MA CORRELATION.

As can be seen from Figure 3, model AR (P) analyzed for auto-correlation function of stationary sequence x (t) which is derived from data analysis is tailing. The correlation of partial autocorrelation function correlation established by the MA model is censored, namely, AR (P) model accord with time series analysis and establish AR (p) model, take a P=2.

A. Establishing Autocorrelation Function

We establish the stationary model

$x_t = \phi_1 x_{t-1} + \phi_2 x_{t-2} + \ldots\ldots + \phi_p x_{t-p} + \varepsilon_t$, and multiply at the equal sign by $x_{t-k}, \forall k \geq 1$, and then obtain its expectations, gain the equation:

$$E(x_t x_{t-k}) = \phi_1 E(x_{t-1} x_{t-k}) + \ldots\ldots + \phi_p E(x_{t-p} x_{t-k}) + E(\varepsilon_t x_{t-k}), \forall k \geq 1$$

According to the conditions of the AR (p) model, we can get $E(\varepsilon_t x_{t-k}) = 0, \forall k \geq 1$, so we can get the recurrence formula of self-covariance function:

$$r_k = \phi_1 r_{k-1} + \phi_2 r_{k-2} + \ldots\ldots \phi_p r_{k-P}$$

Because of $\rho_k = r_k / r_0$, r_k is divided by the variance function r_0, and gain recurrence formula of self-correlation function $\rho_k = \phi_1 \rho_{k-1} + \phi_2 \rho_{k-2} + \ldots\ldots + \phi_p \rho_{k-p}$.

We can calculate on the data basis of the above formula obtained and obtain $\overline{x}$, r_k, ρ_k, and then $\phi_1, \phi_2, \ldots\ldots \phi_p$.

Using SPSS software analysis, we get simulation results shown in Figure 5:

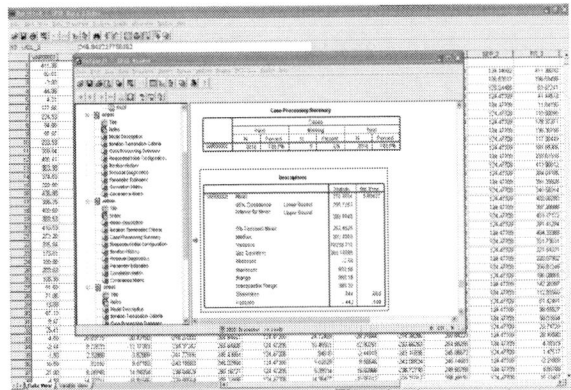

FIGURE V. SPSS SIMULATION RESULT.

From the Figure, we obtain the value of the upper bound 289.8845 and lower bound values 266.7263 at the 95% confidence interval.

B. Wind Power Prediction of Single-day and One Week

We use AR (p) to predict 96 point-in-time sequence wind power data for the 1st wind turbines of certain wind farm from 0:00 to 23:45 on May 31, shown in Table 1:

We use AR (p) to gain wind power forecast value and the actual value of 58 wind turbine for certain wind farm, line graph shown in Figure 6:

TABLE I. AR(P) MODEL FOR FORECAST OF WIND POWER DATA.

Time Point	Power Prediction	Time Point	Power Prediction	Time Point	Power Prediction
1	214.46	33	175.2	65	144.12
2	172.13	34	333.33	66	171.19
3	213.76	35	168.81	67	184.11
4	263.7	36	333.33	68	193.96
5	208.98	37	175.98	69	333.33
6	163.43	38	333.33	70	333.33
7	206.88	39	218.65	71	175.27
8	237.07	40	148.37	72	333.33
9	229.88	41	165.56	73	333.33
10	204.65	42	173.9	74	333.33
11	202.94	43	146.02	75	333.33
12	197.02	44	161.55	76	172.79
13	204.06	45	185.88	77	213.2
14	212.89	46	174.36	78	182.47
15	226.14	47	333.33	79	183.25
16	196.69	48	160.34	80	188.58
17	156.59	49	333.33	81	208.69
18	184.32	50	199.06	82	219.55
19	185.86	51	333.33	83	151.29
20	181.57	52	333.33	84	182.79
21	207.74	53	333.33	85	333.33
22	197.95	54	240.67	86	333.33
23	190.11	55	226.57	87	198.43
24	198.56	56	333.33	88	333.33
25	139.15	57	333.33	89	204.39
26	83.15	58	263.94	90	207.71
27	128.06	59	185.89	91	333.33
28	92.29	60	333.33	92	333.33
29	144.13	61	138.16	93	333.33
30	151	62	216	94	220.93
31	125.9	63	152.84	95	218.81
32	160.9	64	191.18	96	181.14

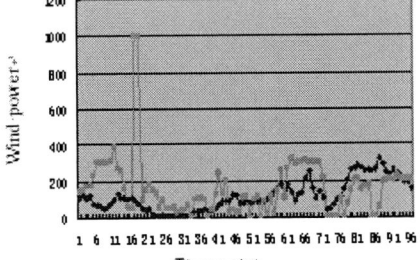

FIGURE VI. COMPARISON OF 58 WIND TURBINE WIND POWER FORECAST VALUE AND THE ACTUAL VALUE.

IV. SIMULATION MODELING RESULTS

According to accuracy rate formula of the wind farm power forecasting

$$r_1 = (1 - \sqrt{\frac{1}{N}\sum_{k=1}^{N}(\frac{P_{Mk} - P_{Pk}}{Cap})^2}) \times 100\%$$

And the pass rate is calculated

$$r_2 = \frac{1}{N}\sum_{k=1}^{N}B_k \times 100\%$$

Based on prediction results, we analyze comparatively for prediction data and the actual data from the two aspects of accuracy and pass rate, as shown in Table 2 and Table 3:

TABLE II. MAY 31ST 00:00 TO 23:45 FORECAST ACCURACY.

Wind Turbine	Accuracy	Pass rate
No1	77.29%	86.32%
No2	76.34%	87.49%
No3	75.64%	84.53%
No4	79.47%	82.48%
4 Wind Turbine	80.42%	79.53%
58 Wind Turbine	80.94%	76.48%

TABLE III. MAY 31ST 00:00 TO 23:45 IN JUNE 6TH FORECAST ACCURACY.

Wind Turbine	Accuracy	Pass rate
No1	77.05%	81.86%
No2	76.55%	87.57%
No3	74.92%	83.82%
No4	79.79%	81.89%
4 Wind Turbine	80.32%	79.52%
58 Wind Turbine	80.14%	76.54%

We analyze the forecast data obtained through simulation modeling, the modeling and analysis of time series method is suitable for linear problems. When the prediction step is relatively larger, it reflects good predictability. With the increase of the predicted time, the superiority of the time-series analysis model is increasingly apparent. Moreover, with the increase of the predicted time, the advantage of the ARMA model is gradually increasing and improve more obvious. Therefore, time series method is to be more effective for forecast the output power in the short-term real-time forecast.

ACKNOWLEDGEMENTS

This work was supported by the Science and Technology development plan project of Department of Science and Technology of Jilin Province (120130169), China.

This work was supported by the Science and Technology Research Project in 12th Five-Year Periods of Department of Education of Jilin Province (2013, 304), China.

REFERENCES

[1] Xiuyuan Yang, Yang Xiao, and Shuying Chen, "Research of Wind Speed and Wind Power Prediction," Proceedings of the CSEE, 2005, vol. 25 (11): 3, pp.1–5.

[2] Wei Xiong, Operational Research, Beijing: Machinery Industry Press, 2005.

[3] Xingjia Yao, Wind Power Generation Test Technology, Beijing: Electronic Industry Press, 2011.

[4] Guixing Yang, Xiqiang Chang, Weiqing Wang, and Xiuping Yao, "Discussion of the Forecast Precision for Wind Power Forecasting System," Power System and Clean Energy, 2011, vol. 27 (1), pp67-71.

International Conference on Power Electronics and Energy Engineering (PEEE 2015)

Research on Fuzzy Reliability Prediction Method of CNC Machine Tools

G.B. Zhang

College of Mechanical Engineering
Chongqing University
Chongqing, china

D.M. Luo

College of Mechanical Engineering
Chongqing University
Chongqing, china

Y. Ran

College of Mechanical Engineering
Chongqing University
Chongqing, china

W. Yu

College of Mechanical Engineering
Chongqing University
Chongqing, china

Abstract-**Traditionally the reliability data is very insufficient in the early designing stage of CNC machine tools and the factors that affect reliability prediction are complicated and fuzzy. So, in this paper, a fuzzy comprehensive evaluation model based on vague set algorithm (VSA) is proposed for reliability prediction. Firstly, the GO methodology is used to establish the system reliability model of CNC machine tools. Secondly, the reliability prediction model is built based on vague set. Finally, an example of rotary table in machining center is taken to prove the availability of the model. Considering the four chief influence factors (design, test, manufacture and maintenance), the VSA model is adopted to make quantitative analysis and prediction, by which the feasibility and applicability of the proposed method are verified, and simultaneously the references are provided for the reliability design of next stage.**

Keywords-CNC machine tools; reliability prediction; vague set; GO methodology

I. INTRODUCTION

Reliability prediction is the main component of system reliability design, which is generally used to pre-estimate whether the designed system, can achieve the specified reliability requirements under the given operating condition. It is of great influence to product planning, maintenance decision, reliability assessment, the fault detection and risk evaluation of the manufacturing process [1]. However, the reliability data is seriously insufficient during the initial design stage, and the factors that affect the reliability prediction are complicated and fuzzy. Therefore it is difficult to deal with such item with fuzzy uncertainty.

According to the property that the fuzzy numbers are suitable for quantifying fuzzy information, Zhao et al. proposed an approach of multi-stage fuzzy synthetic assessment based on fuzzy numbers [2]. Hao et al. presented a reliability prediction method introducing interval analytic hierarchy (AHP), which

exhaustively analyzed the difference between the evaluation object and the similar product, and realized the mutual complement of certain and fuzzy information [3]. Chen et al. proposed an electronic equipment reliability prediction method (EERPM) combined with functional analysis and physics of failure [4]. Ren et al. did a research on the CNC reliability prediction based on grey system theory [5]. Zhao et al. put forward a mechanical reliability prediction based on fuzzy comprehensive and fuzzy inference method [6]. The methods above extraordinarily depend on the basic sample data. But the data insufficient is serious in the initial design stage of machine tools, coupled with many uncertain factors, which made all the methods above inapplicable. Vague set [7], the generalization of the fuzzy set, is more intuitive and operable, also has an advantage over analyzing and handling of the fuzzy problems. In addition, it can sufficiently exploit the experts' advices, and improve the rationality of the comprehensive evaluation.

On account of the above analysis, this paper presents a fuzzy comprehensive evaluation model based on vague set. Moreover, it offers a possibility for handling the data deficiency and fuzzy decision-making of the traditional models.

II. THE RELIABILITY MODELING OF CNC MACHINE TOOLS

There are three steps in reliability prediction of CNC machine tools: firstly it is necessary to establish a relationship model among the various subsystems; secondly the predicted reliability degree of each subsystem should be calculated; and finally the predicted reliability degree of the whole machine can be obtained. Based on the principle of the GO methodology [8], GO chart is used to represent the relationship among subsystems and whole machine for the reliability analysis, as shown in Figure 1.

© 2015. The authors - Published by Atlantis Press

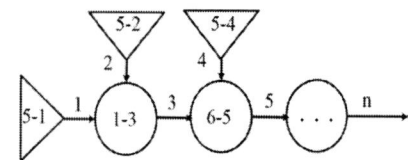

FIGURE I. THE GO CHART MODEL FOR THE SYSTEM.

The system reliability model of CNC machine tools is shown as:

$$R_S = P_S = P_S(0) \prod_{i=1}^{n} P_S(i) \prod_{i=1}^{n} P_C(i) \tag{1}$$

Where $P_S(0)$ is the reliability degree of the system input, $P_C(i)$ is the reliability degree of each subsystem, $P_S(i)$ stands for the control signal and n is the number of the subsystems.

On the basis of the reliability GO chart of the system, the reliability prediction process of CNC machine tools using VSA is shown in Figure 2.

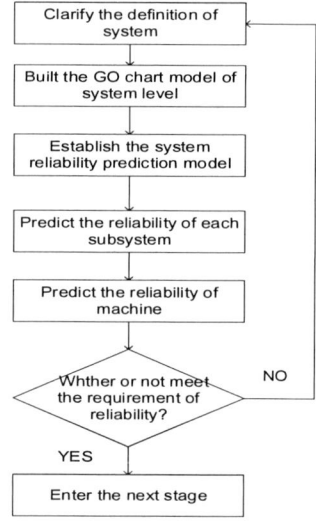

FIGURE II. FLOW CHART FOR RELIABILITY PREDICTION OF CNC MACHINE TOOLS.

III. THE RELIABILITY PREDICTION MODEL BASED ON VAGUE SET

Vague set is defined as $\hat{A}(x) = [t_A(x), 1 - f_A(x)]$, $\hat{A}(x) \in [0,1]$, where $t_A(x)$ is the lower bound of membership derived from the evidences in support of x, $f_A(x)$ denotes the lower bound of non-membership exported from the evidences against x, and $t_A(x) + f_A(x) \leq 1$ [9]. Then the fuzzy synthesis reliability prediction process based on vague set in this paper is as follows.

A. Influence Factor Set Building

In the reliability prediction of CNC machine tools, the four influence factors: design, test, manufacture and maintenance, are respectively expressed by u_1, u_2, u_3, u_4. The influence factors can be increased or decreased depending on specified object of study. The influence factor set in reliability prediction of CNC machine tools is built as follows.

$$U = \{u_1, u_2, u_3, u_4\} \tag{2}$$

B. Factor Weight set Establishment

Essentially the weights of factors just reflect the relative importance in the process of evaluation, so the weight set is considered as follows.

$$\tilde{W} = \{w_1, w_2, w_3, w_4\} \tag{3}$$

(1) Set the rating scale and its benchmark score,

$$X = \{x_1, x_2, ..., x_k\} \tag{4}$$

Where x_j is the benchmark score of the *jth* rating scale.

The weights of factors are determined by *m* experts in accordance with the integrated voting model, and expressed by vague values. The possible range of the factor u_i's weight belonging to the rating scale is as follows

$$\hat{P}_{ij} = \{[t_{ij}, 1 - f_{ij}]; \alpha_{ij}\} \tag{5}$$

Where $i = 1, 2, ..., n; j = 1, 2, ..., k$, α_{ij} stands for the possibility of nonvoters tending to cast an affirmative vote, which is named as propensity score.

(2) Calculate the satisfaction degree of $\hat{P}_{ij}$, namely to estimate the fuzzy membership, and then get the estimated value $D_{ij}(x)$.

$$D_{ij}(x) = t_{ij}(x) + \alpha_{ij} \times [1 - t_{ij}(x) - f_{ij}(x)] \tag{6}$$

(3) Calculate the basic weights of the factors. In order to synthesize the decision information of experts, the weighted average method is used here. Considering the estimated value $D_{ij}(x)$ as the weights, each rating scale x_i is weighted, by which the obtained value is used as the basic weight of factor. Thus the basic weight of factor u_i can be got.

$$z_i = \sum_{j=1}^{k} D_{ij} x_j \bigg/ \sum_{j=1}^{k} D_{ij} \tag{7}$$

(4) Normalization processing is carried out on the basic weight z_i, getting the weight set $\tilde{W} = (w_1, w_2, ..., w_n)$ and

$$w_i = z_i \bigg/ \sum_{i=1}^{n} z_i \qquad (8)$$

Since fuzzy set is the special case of vague set, the weight can also be referred as $\hat{W} = (w_1, w_2, ..., w_n)$.

C. Alternative Set Establishment

Alternative domain contains the reliability degree range which may be possessed by the subsystems, so the alternative set is the set of subsystem reliability degree. Roughly estimate the reliability degree of the object, then take this as the alternative domain. Discretize the continuous domain into m values, which are regarded as alternative elements, and the the alternative set can be obtained.

$$V = \{v_1, v_2, ..., v_m\} \qquad (9)$$

D. Comprehensive Evaluation

Comprehensive evaluation requires that experts should vote for whether the predicted object is taken as the alternative element v_i based on each sub-factor of all factors. After getting the alternative element's membership, the relationship between factor subset and alternative set will be changed to fuzzy relation from vague one.

The membership of the predicted object belonging to the alternative element v_j in the evaluation of factor u_i is set up as

$$r_{ij} \ (i = 1,2,3,4; \ j = 1,2,3,...,m).$$

So the evaluation matrix is

$$\tilde{R} = \begin{bmatrix} r_{11} & r_{12} & \cdots & r_{1m} \\ r_{21} & r_{22} & \cdots & r_{2m} \\ \vdots & \vdots & \vdots & \vdots \\ r_{41} & r_{42} & \cdots & r_{4m} \end{bmatrix} \qquad (10)$$

Then the factor u_i is comprehensively evaluated, using the weighted average algorithm.

$$\tilde{B} = \tilde{W} \circ \tilde{R} = (b_1, b_2, ..., b_m) \qquad (11)$$

Weighted average is brought forward to the comprehensive index in order to get the value of reliability prediction afterwards.

$$P_C = \sum_{j=1}^{m} b_j v_j \bigg/ \sum_{j=1}^{m} b_j \qquad (12)$$

The reliability prediction values of all subsystems can be got in accordance with the above steps. Finally the predicted reliability of CNC machine tools can be obtained with all the values calculating in Eq. (1).

IV. CASE ANALYSIS

Based on the analysis of system structure, CNC machine tool can be divided into 14 modules, including basic components, NC rotary table, pallet changer, ATC, headstock, gearing, hydraulic system, pneumatic system, lubrication system, cooling system, chip-removal system, protective system, joint interface and electrical system. The system reliability GO chart is shown in Fig. 1, with the number of subsystem $n = 14$. Here NC rotary table is taken as an example, and considering the four factors of design, test, manufacture and maintenance, the fuzzy comprehensive evaluation based on vague set is adopted to predict the reliability of the rotary table.

Set up the evaluation criterion firstly. The rating is assumed as 4 grades here: common, less important, important, more important, which are given corresponding benchmark scores.

$$X = \{x_1, x_2, x_3, x_4\} = \{2, 4, 6, 8\}$$

20 experts are invited to evaluate the weights for factors. The result of design factor u_1 can be expressed as:

$$\hat{P}_1 = \left\{ \frac{\{[0.1, 0.1]; 0\}}{x_1} + \frac{\{[0.25, 0.4]; 0.55\}}{x_2} + \frac{\{[0.45, 0.75]; 0.74\}}{x_3} + \frac{\{[0.2, 0.6]; 0.62\}}{x_4} \right\}$$

From Eq. (6), one can easily get: $D_1 = \{0.1, 0.3325, 0.672, 0.448\}$, then put the result into Eq.(7), yielding the basic weight of the design factor $z_1 = 5.89114$. Similarly the other three factors' basic weights can be calculated in this way: $z_2 = 4.85632, \ z_3 = 5.45571, \ z_4 = 4.88126$.

Normalize the all above z_i, thus obtaining the weight set of 4 factors.

$$\tilde{W} = (w_1, w_2, w_3, w_4) = (0.2794, 0.2303, 0.2588, 0.2315)$$

According to the reliability data analysis results of similar machine tools [10], the mission reliability degree of CNC rotary table should be in the range of 0.95 to 0.97. Appropriately expanding the range, the alternative set is built as follows.

$$V = \{0.94, 0.95, 0.96, 0.97, 0.98\}$$

The 4 factors of CNC rotary table are comprehensively evaluated based on the voting model, with results as follows.

$$\hat{P}_{r1} = \left\{ \frac{\{[0, 0.2]; 0.1\}}{v_1} + \frac{\{[0.25, 0.4]; 0.4\}}{v_2} + \frac{\{[0.55, 0.75]; 0.54\}}{v_3} + \frac{\{[0.2, 0.5]; 0.6\}}{v_4} + \frac{\{[0, 0.3]; 0.3\}}{v_5} \right\}$$

$$\hat{P}_{r2} = \left\{ \frac{\{[0, 0]; 0\}}{v_1} + \frac{\{[0.35, 0.45]; 0.5\}}{v_2} + \frac{\{[0.45, 0.75]; 0.45\}}{v_3} + \frac{\{[0.25, 0.65]; 0.55\}}{v_4} + \frac{\{[0, 0.4]; 0.25\}}{v_5} \right\}$$

$$\hat{P}_{r3} = \left\{ \frac{\{[0, 0.1]; 0.1\}}{v_1} + \frac{\{[0.5, 0.7]; 0.5\}}{v_2} + \frac{\{[0.35, 0.65]; 0.4\}}{v_3} + \frac{\{[0.2, 0.5]; 0.25\}}{v_4} + \frac{\{[0, 0.3]; 0.2\}}{v_5} \right\}$$

$$\hat{P}_{r4} = \left\{ \frac{\{[0, 0]; 0\}}{v_1} + \frac{\{[0.4, 0.55]; 0.5\}}{v_2} + \frac{\{[0.4, 0.65]; 0.54\}}{v_3} + \frac{\{[0.2, 0.6]; 0.55\}}{v_4} + \frac{\{[0, 0.35]; 0.4\}}{v_5} \right\}$$

The evaluation matrix is calculated by Eq. (6).

$$\tilde{R} = \begin{bmatrix} 0.02 & 0.310 & 0.658 & 0.380 & 0.090 \\ 0 & 0.400 & 0.585 & 0.470 & 0.100 \\ 0.01 & 0.600 & 0.470 & 0.275 & 0.060 \\ 0 & 0.475 & 0.535 & 0.420 & 0.140 \end{bmatrix}$$

According to Eq. (11), the comprehensive evaluation index values are calculated by Matlab.

$$\tilde{B} = (0.0082, 0.4440, 0.5641, 0.3828, 0.0961)$$

Finally, the predicted reliability of CNC rotary table can be got by Eq.(12)

$$P_C = \sum_{j=1}^{m} b_j v_j \Big/ \sum_{j=1}^{m} b_j = 0.9607$$

From the result above, the predicted reliability degree of NC rotary table during the period of periodic maintenance can be obtained as 0.9607. Similarly, the predicted reliability degrees of other subsystems can be received based on the above steps. Combined with the system GO chart and the flow chart for reliability prediction in Fig.3, the predicted reliability degree of the whole machine can be finally obtained by the calculation using Eq.(1).

V. CONCLUSION

1) A fuzzy reliability prediction method based on vague set is presented in this paper, which is suitable for the situation of data deficiency and fuzzy decision information, especially in the early design stage of CNC machine tools. It is also applicable for predicting the reliability degrees of subsystems.

2) The experts' advices can be exploited extensively by this method, and the rationality of comprehensive evaluation can be improved.

3) Establishing the reliability model among different subsystems based on GO method does not need to understand the life distribution.

4) The model in the paper is highly practical. Meanwhile, it has made up the shortcomings of existing fuzzy comprehensive evaluation.

REFERENCES

[1] S. Chatterjee, S. Bandopadhyay, Reliability estimation using a genetic algorithm-based artificial neural network: An application to a load-haul-dump machine [J]. Expert Systems with Applications. 2012(39):10943-10951.

[2] D. Zhao, W. WEN, C. Duan, A Model of Aero engine Reliability Prediction Based on Fuzzy Number [J]. Journal of Aerospace Power. 2004, 19(3):320-325.

[3] Q. Hao, Z. Yang, C. Chen, F. Chen, G. Li. Reliability prediction for NC machine tool based on interval AHP [J].Journal of Jilin University(Engineering and Technology Edition). 2012, 42(4): 845-850.

[4] Y. Chen, W. Xie, S. Zeng, Functional Analysis and Physics of Failure Associated Reliability prediction [J]. Acta Aeronautica Et AstronauticaSinica.2008, 29(5):1133-1138.

[5] G. Ren, D. Wang, X. Miao, Research on the CNC reliability prediction based on grey system theory [J].Machinery Design & Manufacture. 2010(5):191-192

[6] D. Zhao, W. Wen, C. Duan, Application of mechanical reliability prediction based on Fuzzy Theory [J].Transactions of Nanjing University of aeronautics & Astronautics. 2004, 21(1):76-80.

[7] K.C. Hung, K.Y. Gino, Peter Chu, T.J. Warren, An enhanced method and its application for fuzzy multi-criteria decision making based on vague set [J].Computer-Aided Design. 2008 (40):447-454.

[8] Z. Shen, X. Huang, Principle and application of GO methodology [M].Beijing: Tsinghua University Press, 2004

[9] D. Zhao, Fuzzy Prediction and Allocation Techniques for Mechanical System Reliability in Early Design Stage [M].Beijing: National Defense Industry Press.2010.5.

[10] G. Zhang, Z. Xu, W. He, L. Tu, Research on Reliability Enhancement Testing Method of NC Rotary Table [J].China Mechanical Engineering. 2011, 22(8):948-951.

Failure Analysis and Diagnosis of Spark Erosion Machine Tools

H. Meng

AVIC Shenyang Liming Aero-Engine Group Corporation
China

N. Xin

AVIC Shenyang Liming Aero-Engine Group Corporation
China

H. Guo

AVIC Shenyang Liming Aero-Engine Group Corporation
China

L. Zhang

AVIC Shenyang Liming Aero-Engine Group Corporation
China

Abstract—**In recent years, along with the rapid development of industrial technology, the development of spark erosion technique becomes very important and plays a key role in the molding processing technology, especially under the situation when this technology is urgently needed. Based on the first-hand data from the production site, it is introduced here regarding the structure and the working principle of hcd500k spark erosion machine tool; at the same time, the common faults and problems of the pulse power and servo feed system are analyzed in details and proposed the solution on them. The detailed analysis is also given for the problems like the frequent poor discharge. The reasons and diagnosis methods are deeply studied as well. It is very helpful and supportive on providing evidence for minimizing the machine design defects and making continuous improvement in design. This paper is meaningful in helping machine tooling design to step into a higher level and having spark erosion machine tooling been improving continuously.**

Keywords-spark erosion processing; diagnosis; design defects makeup; solution

I. INTRODUCTION

At the beginning of this century, scientific technology continued the momentum of the last century and maintained its development with high speed. Many products were updated into next generation frequently. New products emerged, from aircraft, cruise, to mobile phones, watches. Various kinds of high and new technology products, went into people's life directly or indirectly, bringing great convenience to people. However, along with these new products, market competition became more and more severe, which required higher and higher service from them. The traditional machining technology could not meet the high leveled needs for new products manufacture any more. In order to increase the competition ability, various advanced and new technology were applied to the manufacture of new products. As one of key methods in non-traditional processing technology, electrical spark machining technology, with its own characteristics and advantages, plays a more and more important role in more fields.

II. BRIEF INTRODUCTION OF HCD500K SPARK EROSION FORMING

HCD500K CNC spark erosion forming machines, equipped with MD20D power supply cabinet, is mainly used for processing various types of molding like metal parts, molders, etc. The host machine adopts vertical lathe bed, horizontal working table, in single column "C" type structure. Three axis driven power adopts ac servo motor. The driven system is composed of the precise linear rolling guide and precise ball screw and bearing. With an excellent response in ac servo motor speed, torsional resistance is very strong. The Z axis ac servo motor is equipped with braking device. When power is on, the brake will be loose, so the servo control can be realized on the spindle head. When the power is off, the brake will be tightened to lock the spindle. The machine liquid tank is installed with double doors, which is a greatly convenient structure for clamping and finding electrode. With pumping connector and movable magnetic blunt oil block, oil flushing process can be realized at any angel and any place.

A. Failure Analysis and Diagnosis of HCD500K Spark Erosion Forming Machine

Due to high utilization and flexible processing way, the HCD500K spark erosion forming machine is very important for the completion of site production. However, during the process of usage, frequently occurred problems influence the equipment application seriously. Therefore, in this paper, we discussed the data out of the equipment failures in the production process, and proposed reasonable solutions, which provided the reliable data for later machine design and improvement.

B. Abnormal Electro Discharge Machining

Poor discharge machining performances are artifact burning, over-speedy machining, seriously exceeding normal processing speed, and even failure of normal discharge. There are many reasons for this kind of failure. The main problems occur on the facts including pulse power circuit board, power supply, circuit board, dc power supply, high power diode, and even on wire, jigs, etc. In addition, the problems also come from machining conditions and processing parameters. The following items are the factors affecting the discharge machining.

(1) the influence of pulse peak electric current
(2) the influence of the pulse width
(3) The influence of the inter pulse processing
(4) the influence of the media

C. *Power Tube Connection Failure. Abnormal Current Machining*

Power tube interface signal, which is driven by U38, U39 (SN74LS245) on the interface board, isolated by the 6N137 optocoupler and again driven by 7406,4049, controls the open and the close of the corresponding power tube after deleting the grid of the power tube on the board.

In case of failure occurred, check first the working status of the IP0 - IP16 luminous tube on the pulse power board. If the tube status cannot changed regularly, this means that the control signal of electrical current works normally from the main control board to the interface board. The problem comes from the power board. Continue to check the power board. By the IP0-IP32 signal, the light coupling piece U40-U56 signal is measured, and then there is no output signal from U46, which results in no output from IP5 causing no output from the field-effect tube DIP5 (IRFP44). Replacement with a new 6N137 film solves the problem.

D. *Processing Size Exceeding Tolerance*

In processing the translation function can be used to realize the expanding function, but the machined parts do not have round holes, so this means that the motor does not move normally in the translational axis X, Y.

The backstepping detection is the way how to solve the problem. The method is to test starting from the driver control signal, till to the master plate. First, switch the control signal on the X, Y axis, and verify the point of failure. The result shows the same problem, meaning the driver works well. Second, check the driven film of the interface board, switch U21 and U13, and the result changed, meaning the problem comes from U21 and U31. Replacement with a new U13 (US26LS31), the problem is solved. This means U13 film has been damaged.

III. DATA AN ANALYSIS ON FAILURE FREQUENCY

Graph has its own characteristics in expression. Especially for something like time, space and else like abstract thinking, it can transfer info with the effects that written or verbal words cannot replace. The data listed in the columns and rows in the worksheet can be mapped into the line chart, which can show serial data changing with time. Therefore, graph is very suitable for displaying the data trend in different time intervals. In the line chart, category data and value data is distributed in uniform separately along the horizontal axis and vertical axis.

According to the analysis results summarized from onsite data, the failure rate will be high in case that the machine tool failed due to abnormal discharge processing and power failure; the failure rate will be low in case that the machine tool failed due to the deviation tolerance is exceeded in processing current and sizes. In order to make sure that the problems mentioned above can be solved by the upcoming designing, failure causes

should be taken into consideration for later improvements during designing machine tools.

Failure frequency data, for example:

TABLE I. FAILURE FREQUENCY DATA.

Failure phenomenon	Failure period(month)	Failure frequency(times)
Electro discharge machining	3	2
Power tube	2	4
Processing current	6	2
Processing size	4	5

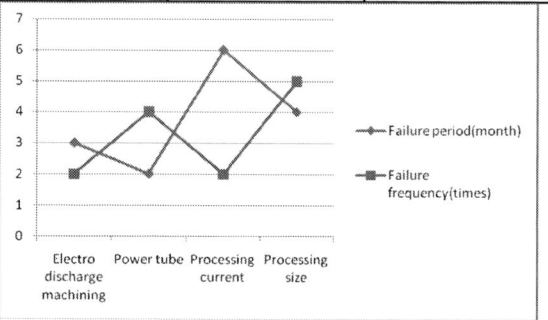

FIGURE I. FAILURE PERIOD, FREQUENCY ANALYSIS.

IV. SUMMARY

HCD500K spark erosion machine has advanced structural design with good stability, high precision, simple operation, strong control and high production efficiency, etc. According to the data of the problem occurred in the production process, it can be seen that there is high failure rate with the power tube and the over-tolerance in processing, low failure rate with current processing by discharge machining. In view of the failure frequency mentioned above and in order to ensure a more completeness of the subsequent machine design, this article proposed a series of solutions after studying and analyzing the data out of the failures and problems by spark erosion machine tool working on production site. At the same time, many examples are shown for the further improvement of equipment design as great support and assistance for improving machine design.

REFERENCES

[1] Pailong Zhu; Changzheng Wei and Zhiyuan Wen; Approach and electrode study of deep hole large depth to diameter ratio[J], journal of mechanical engineering(2006) .

[2] Xueke Luo, Yuezhong Li, CNC EDM machine tool (2004).

[3] Wansheng Zhao, EDM Technology Haierbin Industrial University press(2007).

[4] Ketokivi M.Sehroeder R Manufacturing practices.strategic fit and performance:A routine based view(2004).

[5] Mills, j. latts K.G regory M Aframework for the design of manufacturing strategy the Processes. A contingency approach(1995).

[6] Zhenhui Liu, Jiajie Yang. Special processing [M]. Chongqing University Press(2000).

[7] Jicheng Bai, Yongfeng Guo, Jinchun Liu. Special Processing Technology [M], Harbin Industrial University Press(2006).

[8] Zhen Li, Advanced Manufacturing Technology, Beijing University of Science and Technology Press(2009).

International Conference on Power Electronics and Energy Engineering (PEEE 2015)

Design and Development of Material Management System Based on Digital Pulp Pipeline

H.W. Liu

Yunnan Da Hongshan Pipeline Co., Ltd.
Kunming, China

J.D. Wu

Engineering Research Center for Mineral Pipeline
Transportation YN
Kunming, China

Z.L. Yu

Faculty of Information Engineering and Automation
Kunming University of Science and Technology
Kunming, China

Abstract-Material Management is refers to the enterprises to manage the behavior of procurement, use and storage of the necessary materials. It is an important part of the normal production and operation of enterprises, and affects the economic benefits of enterprises directly. At present, most of the material management systems in the enterprise have some problems: different kinds of materials and unreasonable classifications; procuring not timely and inventory backlog; data transformation not timely and data inaccurate, and so on. Aiming at this problem, a material management system based on B/S mode was designed. It combines procurement, inventory, audition and other processes, realizing workflow clarity, delivery of information timely, systematization and standardization of material management. The practical application shows that: the system has good practicability.

Keywords-material management system; workflow; procurement; audit

I. INTRODUCTION

Da Hong Shan Pipeline company has already built a website platform, integrated management and control systems, data acquisition and monitoring of SCADA systems, But for the lack of effective connection and integration between the various platforms, resulting in the management unable to view information timely [1]. A material management system based on B/S mode has been designed to solve these problems.

In recent years, with the development of database technology and the database management system, many enterprises have realized the importance of database management system [2]. At present, many enterprises are gradually improving their material management system, although some achievements have been made, there are still many aspects need to be improved [3].

Advanced computer technology, network technology, database technology and management concept can solve problems that the enterprises are facing [4]. The system is designed by the modular design method, and there are connections and differences between various modules, the corresponding module can be added as the development and needs of the enterprises, so it has the good adaptability. Meanwhile, the rapid development of information technology makes business processes clear at a glance, easy for users to understand and operate [5]. In order to ensure the safety of corporate information, different permissions were configured according to different departments, leaders and staff at all levels, it is better to regulate their responsibilities as well as the business related to privacy. The operation of the system shows that: the system can strengthen the links between the various departments, ensure the normal and orderly conduct of materials management activities and improve the efficiency of business operations.

II. SYSTEM DESIGN SCHEME

Through the analysis of the life cycle of materials, this paper analyzes each big enterprise's production and operation management standards, and combined with the characteristics of the enterprise, after repeated research, we divided material management process into six stages: material purchase, material inbound, material change, material allocation, material scrap, and material outbound.

A. Material Purchase

Each department needs to purchase the goods and submit the procurement plan to the relevant departments of the company, the relevant departments will come to purchase after summarizing the materials from all departments. This process is the beginning of materials coming into company, and its smooth completion is the base of the following process. Material purchase is the starting point of material life cycle.

© 2015. The authors - Published by Atlantis Press

B. Material Inbound

After the completion of the procurement of materials, supplies will be transported to the management department, and management department will make registration materials and notes after receiving the materials.

C. Material Change

Material change includes several aspects: changes of material price and quantity in material purchase and material allocation, the normal equipment changes to scrapped equipment or new equipment after maintenance. This process is relatively complex, materials will be forbidden into the next corresponding processes if the materials attribute have changed, and need to return to the former process to re-submit an application after modifying something.

D. Material Allocation

Each department can get the materials according to their procurement plan after materials come into the storage, but perhaps department A may not need some certain materials which it has purchased before temporarily, and department B just needs the materials which it hasn't purchased before, then the materials can be allocate to department B at this time. This kind of phenomenon is very common in the actual production conditions, in order to improve the work efficiency, material allocation process is very necessary.

E. Material Scrap

Material scrap generally can be divided into three aspects: First, the purchased materials failed to use or couldn't used; second, materials failed to maintain or maintenance costs are too high; scraped materials are also part of company's property, it can also bring great economic benefits for the company by rational use.

F. Material Outbound

Material will have outbound proof and note when department takes the purchased materials. We should understand the department of used materials is the department of purchased materials eventually. Material outbound is the end of the life cycle.

The following is function structure chart of material management process:

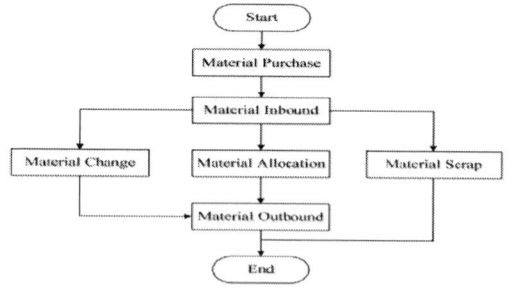

FIGURE I. THE FUNCTION STRUCTURE CHART OF MATERIAL MANAGEMENT PROCESS.

III. ANALYSIS OF THE MAIN WORKFLOW OF MATERIALS MANAGEMENT

A. Material Purchase

Material from purchase to inbound can divide into several steps actually: the unit leadership for approval, the branch leadership for approval, material procurement department for purchasing, acceptance of storage. Unit leaders for approval including leaders in charge audit and competent leadership review.

The department submits the material purchase plan after making the plan, leaders in charge will audit firstly, if the plan is reasonable, it will be submitted to the competent leadership, otherwise, it will be rejected with feedback suggestions. Competent leadership will submit the plan to the branch leaders if the plan is reasonable, otherwise, the plan will be rejected with feedback suggestions. The branch leaders will submit the plan to material procurement department if the plan is reasonable, otherwise, the plan will be rejected with feedback suggestions. Purchasing department will accept the material and material inbound process is complete.

The following is flowchart of material purchase and material inbound:

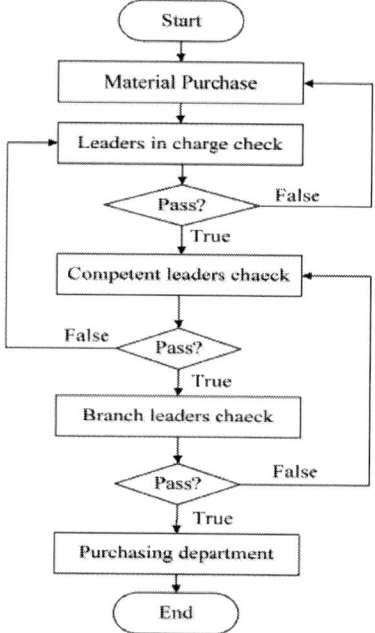

FIGURE II. THE FLOWCHART OF MATERIAL PURCHASE AND MATERIAL INBOUND.

B. Material Change

Material change include price change and quantity change, material management department submits the material change to leaders in charge, if the leaders in charge think it is reasonable, then material management department will modify material inventory record the changes, otherwise, then material management department will reject the plan and explain the reason.

C. Material Allocation

Material allocation is a material transfer process between the used department and the owned department of the material, the owned department of the material will outbound the materials after it checked the materials applied by the used department of the material, the materials will be inbounded after the used department of the material received them.

D. Material Scrap

Material scrap means material can't continue to be used, if the material has been used over its safety use year, then it can apply for scraping, after each leader checked and given some advice, the material scrap will be reported to the energy conservation center. Finally, material management department will record material scrap and Charge-offs for materials inventory.

E. Material Outbound

Material application department submits materials requisition to material management department, and then material management department will check the inventory, if the inventory is sufficient, the competent leadership will arrange materials outbound after checking the outbound plan, and material management department will hander the Charge-offs for materials outbound, then applied department can get the recipients.

IV. CONCLUSIONS

This paper discusses the design and development of material management system based on digital pulp pipeline, in addition, it has good reference to solve the problems existing in the enterprise material management. The practice shows: This system can achieve the enterprise materials management, not only ensures the timely supply of material needs and improves employee productivity, but also provides more convenient to know the need of production and the information of material inventory. With the continuous development of information technology, there will be a better material management system, and it would be helpful to provide some reference to people of engaged in developing materials management system.

ACKNOWLEDGEMENT

This work is supported by National Natural Science Foundation of China (No. 51169007), Science & Research Program of Yunnan province (No.2011CI017 & 2012CA022&2013DH034).

REFERENCES

[1] CHEN Hua, ZHANG Pengwei, CHEN Jingxia. Design and Implementation of material management system for small and medium enterprises in MIS[J]. Journal of Shaanxi University of Science and Technology,2003,21(6):78-81.

[2] WANG Jianwen, ZHANG Junming, HAN Lipeng. Design and Implementation of Materials Management System based on ExtJS[J].College of Electrical and Information Engineering,2010,23(233):5012-5014,5055.

[3] HE Nixia,YANG Shenghua. Design and Implementation of Enterprise Substances Management System based J2EE Technique System[J]. Development and application of computer, 2005, 06: 26-28.

[4] YANG Bo. Modeling and Development of Material Management System for Power Company Based on UML[J]. Central China Power,2003,16(3):67-70.

[5] ZHANG Li. Application of Information technology in Material Management System [J]. Modern economic information, 2013, (24): 114–114, 117.

International Conference on Power Electronics and Energy Engineering (PEEE 2015)

Transient Elastohydrodynamic Lubrication Analysis of Spur Gear Running-in Considering Effects of Solid Particles

X.B. Huang
School of Mechanical Engineering
Qingdao Technological University
Qingdao, China

Q. Liu
School of Mechanical Engineering
Qingdao Technological University
Qingdao, China

Y.Q. Wang
School of Mechanical Engineering
Qingdao Technological University
Qingdao, China

N. Dong
School of Mechanical Engineering
Qingdao Technological University
Qingdao, China

Abstract—**The elastohydrodynamic lubrication model considering solid particles was set up. Taking effect of solid particles into account, the Reynolds equation was deduced. The elastohydrodynamic lubrication analysis of spur gear running-in was completed considering time-variant effect. Results show that oil film pressure in region 2 where particle is located increases dramatically but oil film thickness in region 2 decreases thinking about effects of solid particles. When particle size becomes greater oil film pressure increases obviously, oil film thickness diminishes. The minimum film thickness and maximum film pressure both diminish considering effects of solid particles.**

Keywords-solid particles; spur gears; running-in; time-variant effect; elastohydrodynamic lubrication

I. INTRODUCTION

Gearing is the most common mechanical transmission. Before normal use of generated gears, they must be tested in running-in process. Qualified gears are approved to be released from factories. The efficiency and quality of running-in process have significant influence on operation efficiency and service life of gears. Therefore, doing researches about lubrication mechanism of gears in running-in process is quite meaningful. A large number of experiments and theoretical researches on gears running-in have been done by relevant scholars at home and abroad. Castro.J,et al[1] established mixed film lubrication model and analyzed the change of normal stress and shear stress at meshing points on gears meshing line. In addition, why debris may appear in lubricating oil was interpreted. Bentley.J.A,et al[2] discussed the importance of lubricating oil on industry gears in running-in process, the effects of compounded lubricating oil, extreme pressure lubricating oil and synthetic oil on gears lubrication in running-in process, respectively. The influence of solid particles on lubrication mechanism of gears in running-in process was not considered in all studies above. XIE,et al[3] set

up elastohydrodynamic lubrication model of grease lubrication under line contacts considering solid particles, lubrication equation was revised and the effects of different sizes, locations and velocities of solid particles on oil film pressure and film thickness were analyzed. However, effects of solid particles on oil lubrication of gears were not taken into account. The author will revise Reynolds equation considering solid particles. The changes of film pressure and film thickness of spur gears in running-in process will be analyzed deeply taking solid particles, time-variant effect, load and rotate velocity into consideration.

II. MATHEMATICAL MODEL

Revised Reynolds equation considering solid particle can be described as.

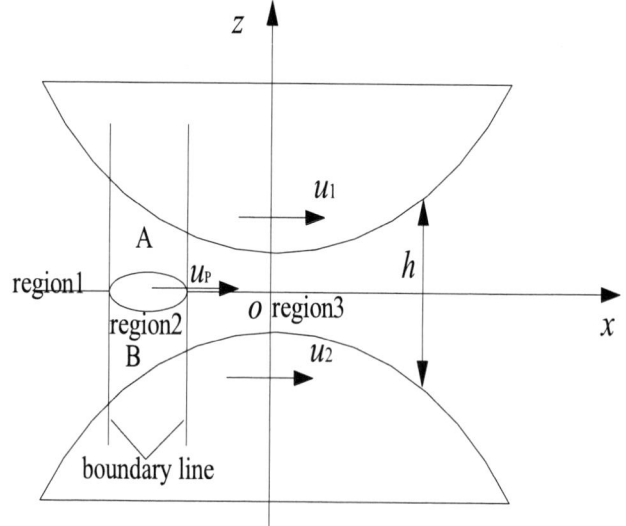

FIGURE I. CONTACT AREA CONTAINING SOLID PARTICLE ELASTOHYDRODYNAMIC LUBRICATION MODEL.

© 2015. The authors - Published by Atlantis Press

Lubrication equation in region 1 and region 3

For derivation process details see Numerical Analysis of Fluid Lubrication[4] , isothermal Reynolds equation of infinite long term contact:

$$\frac{\partial}{\partial x}\left(\frac{\rho h^3}{\eta}\frac{\partial p}{\partial x}\right) = 12\frac{\partial}{\partial x}(\rho u h) + 12\frac{\partial(\rho h)}{\partial t} \tag{1}$$

Lubrication equation in region 2

As Figure 1 shows that region 2 consists of A and B, $z0$ is half size of solid particle on z axis direction, for A and B, according to infinitesimal body of lubricating oil film force balance condition:

$$\frac{\partial p}{\partial x} = \frac{\partial \tau_x}{\partial z} \tag{2}$$

Newton law of viscosity can lead to $\tau_x = \eta\frac{\partial u}{\partial z}$, substituted into Eq.(2) Eq.(3) can be achieved.

$$\frac{\partial p}{\partial x} = \frac{\partial}{\partial z}\left(\eta\frac{\partial u}{\partial z}\right) \tag{3}$$

Since p and η is not the function of z, calculating indefinite integral about z on both sides of Eq. (3) yields

$$\eta\frac{\partial u}{\partial z} = \frac{\partial p}{\partial x}z + C_1 \tag{4}$$

Calculating indefinite integral about z on both sides of Eq. (4) yields

$$u = \frac{1}{2\eta}\frac{\partial p}{\partial x}z^2 + \frac{C_1}{\eta}z + \frac{C_2}{\eta} \tag{5}$$

Substituting boundary conditions of A described as $z = z_0, u = u_p; z = \frac{h}{2}, u = u_1$ into Eq. (5) yields

$$C_1 = \frac{\eta(u_1 - u_p)}{\frac{h}{2} - z_0} - \frac{1}{2}\frac{\partial p}{\partial x}\left(\frac{h}{2} + z_0\right), C_2 = \frac{\eta(z_0 u_1 - \frac{hu_p}{2})}{z_0 - \frac{h}{2}} + \frac{hz_0}{4}\frac{\partial p}{\partial x}$$

Oil lubricant flow in A is denoted by q_{xA} ,

$q_{xA} = \int_{z_0}^{\frac{h}{2}} u dz$,substituting C_1, C_2 into Eq.(5) yields

$$q_{xA} = -\frac{1}{12\eta}\frac{\partial p}{\partial x}\left(\frac{h^3}{8} - z_0^3\right) + \frac{hz_0}{8\eta}\frac{\partial p}{\partial x}\left(\frac{h}{2} - z_0\right) + \frac{u_1 + u_p}{2}\left(\frac{h}{2} - z_0\right)$$

Substituting boundary conditions of B described as

$z = -\frac{h}{2}, u = u_2; z = -z_0, u = u_p$ into Eq. (5) yields

$$C_1 = \frac{\eta(u_2 - u_p)}{z_0 - \frac{h}{2}} + \frac{1}{2}\frac{\partial p}{\partial x}\left(\frac{h}{2} + z_0\right), C_2 = \frac{\eta(z_0 u_2 - \frac{hu_p}{2})}{z_0 - \frac{h}{2}} + \frac{hz_0}{4}\frac{\partial p}{\partial x}$$

Oil lubricant flow in B is denoted by q_{xB} ,

$q_{xB} = \int_{\frac{h}{2}}^{-z_0} u du$,substituting C_1, C_2 into Eq.(5) yields

$$q_{xB} = -\frac{1}{12\eta}\frac{\partial p}{\partial x}\left(\frac{h^3}{8} - z_0^3\right) + \frac{hz_0}{8\eta}\frac{\partial p}{\partial x}\left(\frac{h}{2} - z_0\right) + \frac{u_2 + u_p}{2}\left(\frac{h}{2} - z_0\right)$$

Total flow in region 2 is denoted by q_x , $q_x = q_{xA} + q_{xB}$,it can be calculated as

$$q_x = -\frac{1}{48\eta}\frac{\partial p}{\partial x}(h - 2z_0)^3 + (h - 2z_0)\frac{u_1 + u_2 + 2u_p}{4}$$

Fluid mass in region 2 is denoted by m_x, $m_x = \rho q_x$,substituting m_x into equation of continuity after integral[5] yields

$$\frac{\partial m_x}{\partial x} + \frac{\partial(\rho h)}{\partial t} = 0 \tag{6}$$

The isothermal Reynolds equation considering solid particle can be expressed as follows

$$-\frac{\partial}{\partial x}\left[\frac{\rho(h - 2z_0)^3}{48\eta}\frac{\partial p}{\partial x}\right] + \frac{\partial}{\partial x}\left[\rho(h - 2z_0)\frac{u_1 + u_2 + 2u_p}{4}\right] + \frac{\partial[\rho(h - 2z_0)]}{\partial t} = 0 \tag{7}$$

III. BASIC EQUATIONS

Reynolds equation in region 1 and 3 is Eq. (1). Reynolds equation in region 2 where solid particles settles is Eq. (7). The load equation, film thickness equation, viscosity-pressure relationship and density-pressure relationship for details see Numerical Analysis of Fluid Lubrication[4].Each equation and its boundary conditions need to be nondimensionalized. The dimensionless parameters are defined as follows:

$$X = x/b \ , \quad W_0 = w_0/(ER_0) \quad U_0 = \eta_0 u_0/(ER_0) \ ,$$

$$\bar{h} = hR_0/b^2 \ , \quad P = p/p_H \ , \quad \bar{\eta} = \eta/\eta_0 \ , \quad \bar{\rho} = \rho/\rho_0 \ ,$$

$$C_{Rt} = R/R_0 \ , \quad C_{ut} = u/u_0 \ , \quad C_{wt} = w/w_0 \ , \quad \bar{t} = tu_0/b \ ;$$

75

Where w_0 is reference load and w is practical load, b is half of Hertz contact width of both rigid teeth when loaded by w_0, $b = \sqrt{8w_0 R_0/(\pi E)}$, p_H is the maximum Hertz pressure when loaded by w_0, $p_H = Eb/(4R_0)$, ρ_0 and η_0 denote lubricant environmental density and environmental viscosity respectively, C_{wt}, C_{Rt}, C_{ut} denote load time-variant coefficient, comprehensive radius of curvature time-variant coefficient and speed time-variant coefficient respectively, R_0 is reference radius of curvature, $R_0 = R_{ba}R_{bb}\tan\varphi/(R_{ba}+R_{bb})$, R_{ba} and R_{bb} denote base radius of gear 1 and gear 2 respectively, φ is reference circle angle, R is comprehensive radius of curvature of meshing point, $R = R_1 R_2/(R_1+R_2)$, $R_1 = R_{ba}\tan\varphi - s$, $R_2 = R_{bb}\tan\varphi + s$, s is the distance between meshing points and pitch points, u_0 is the entrainment velocity of pitch point, u is entrainment velocity of meshing points, $u = (u_1+u_2)/2$, $u_1 = \omega_1 R_1$, $u_2 = \omega_2 R_2$, ω_1 and ω_2 denote rotational velocity of two gears respectively, h is film thickness.

IV. NUMERICAL APPROACHES

Multi-grid method can be applied to completing pressure solution. Multiple grid integration method can be used to calculate elastic deformation. W circle is adopted in computation. There are six layers gridding, the most dense layer locates 961 nodes. The transient before meshing into point is handled as steady state. There are 120 transients from meshing into points to recess action points. The convergence criterion is that transient pressure and load's relative errors must be lesser than 10^{-3}.

V. RESULTS AND DISCUSSIONS

Table 1 is the list of lubrication relevant parameters. Figure 2 is variation diagram of load along with line of action, in which A、B、C、D、E are five transients of time-variant load. Velocity of solid particles is 0 and dimensionless centre coordinate of solid particles in x direction XC=-0.13.

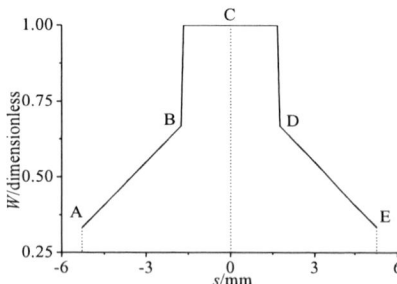

FIGUREII. LOAD ALONG WITH LINE OF ACTION VARIATION DIAGRAM.

TABLEI. LUBRICATION RELEVANT PARAMETERS LIST.

oil viscosity η_0 (Pa·S)	0.075
viscosity-pressure coefficient α (Pa^{-1})	2.19×10^{-8}
oil environmental density ρ_0 (kg·m^{-3})	870
gears density $\rho_{1,2}$ (kg·m^{-3})	7850
Young modulus $E_{1,2}$ (Pa)	2.06×10^{11}
Poisson ratio $v_{1,2}$	0.3
number of teeth z_1、z_2	35,140
module m (mm)	2
rotate velocity of pinion n_1 (r / min)	1500
tooth width B (mm)	20
gear pressure angle $\varphi(^\circ)$	20
transmitted power P (kW)	10
addendum coefficient h^*, teeth gap coefficient c^*	1.0,0.25
dimensionless position coordinate of particles X_C	-0.13

A. The Effects of Solid Particles on Oil Film Pressure and Film Thickness Comparing With None

Figure 3 is oil film pressure and film thickness distribution diagram. A significant increase of oil film pressure in region 2 where solid particles settles can be seen from Figure 4 .Film thickness decreases considering solid particles and film necking position is closed to export zone.

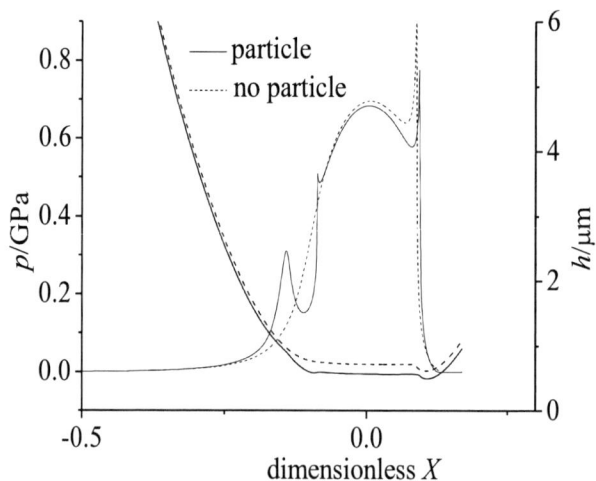

FIGUREIII. OIL FILM PRESSURE AND FILM THICKNESS DISTRIBUTION DIAGRAM.

B. Effects of Solid Particles Sizes on Oil Film Pressure and Film Thickness

Figure 4 consists of (a) and (b) that are oil film pressure and film thickness distribution diagrams considering different sizes of solid particles (studied particles are spherical particles, SR0.01 indicates that dimensionless radius is 0.01). From (a) can be seen that the minimum size of particles has negligible influence on oil film pressure, on contrary, the maximum size of particles causes the greatest rise in film pressure. Hence, a conclusion can be drawn that larger size of particles may cause more distinct rise in film pressure. From (b) can be seen that if radius of spherical particles becomes greater film thickness diminishes.

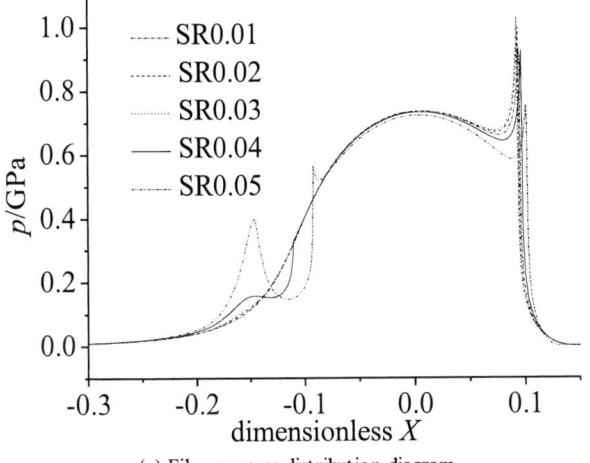

(a) Film pressure distribution diagram

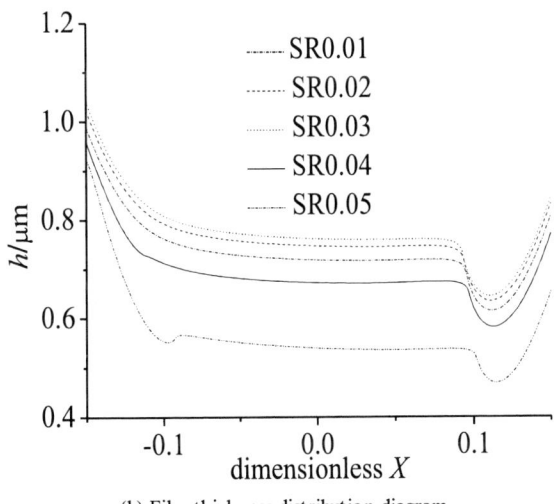

(b) Film thickness distribution diagram

FIGUREIV. OIL FILM PRESSURE AND FILM THICKNESS VARIATION DIAGRAM.

C. Effects of Solid Particles on Oil Film Pressure and Film Thickness under Time-Variant Effect

Figure 5 consists of (a) and (b) that are transient film pressure and transient film thickness distribution diagrams considering influence of solid particles. From (a) can be seen that oil film pressure in the entrance region considering solid particles becomes greater. From (b) can be seen that film necking position moves to export zone first and then come back to entrance region.

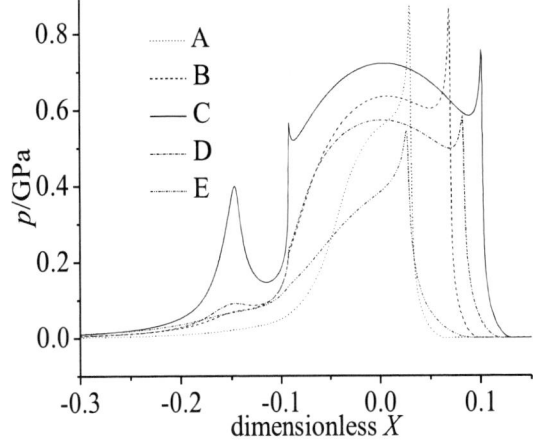

(a) Film pressure distribution diagram

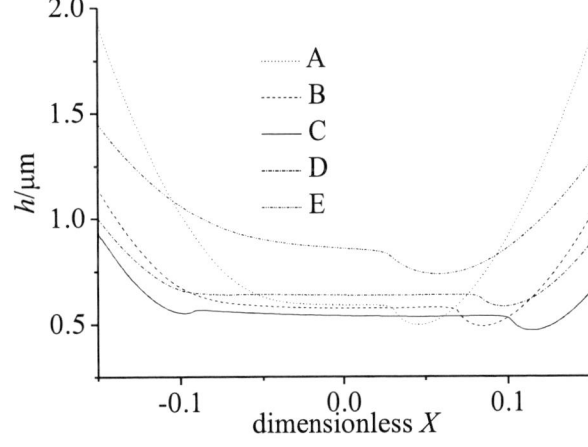

(b) Film thickness distribution diagram

FIGUREV. OIL FILM PRESSURE AND FILM THICKNESS VARIATION DIAGRAM.

D. Effects of Solid Particles on Minimum Film Thickness and Maximum Film Pressure

Figure 6 consists of (a) and (b) that are the minimum film thickness and the maximum film pressure distribution diagram considering solid particles. From (a) can be seen that the minimum film thickness decreases when solid particles are taken into consideration. From (b) can be seen the maximum film pressure also decreases considering solid particles.

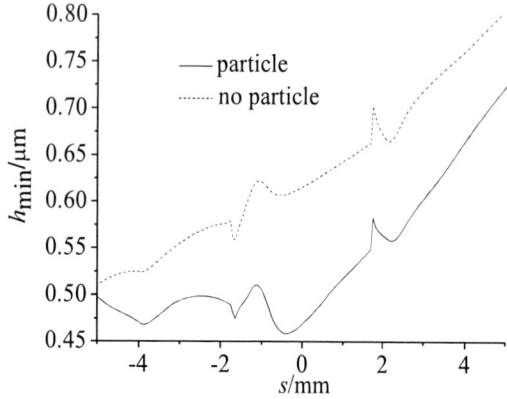

(a) Minimum film thickness variation diagram

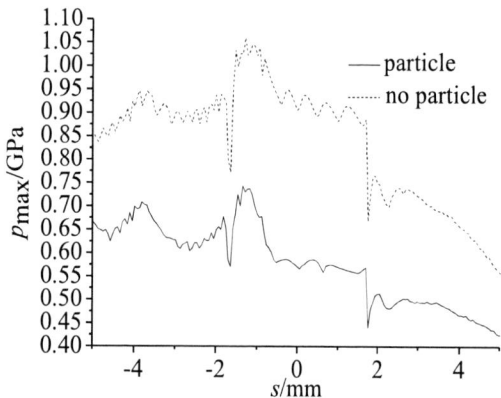

(b) Maximum film pressure variation diagram

FIGURE VI. MINIMUM FILM THICKNESS AND MAXIMUM FILM PRESSURE DISTRIBUTION DIAGRAM.

VI. CONCLUSIONS

(1) The film pressure in region 2 where solid particles are located increases dramatically and the film thickness in region 2 decreases considering sold particles. The film pressure changes remarkably with increases in size of solid particles. If radius of spherical particles becomes larger film thickness in contact area will be thinner. Moreover, increase in radius of spherical particles can result in distinct decrease in film thickness in region 2.

(2) The minimum film thickness and maximum film pressure decrease taking solid particles and time-variant effect into account.

ACKNOWLEDGEMENT

The project was supported by the National Natural Science Foundation of China (51175275) and Qingdao Technology Project(12-1-4-4-(2)-jch).

REFERENCES

[1] Castro J, Seabra J. Global and local analysis of gear scuffing tests using a mixed film lubrication model[J]. Tribology International, 2008, 41(4): 244-255.

[2] Bentley J A, Korane K J. Lubes Ensure Slick-Running Gears[J]. Machine Design, 1995, 67(9): 95-97.

[3] XIE X P, PENG Z L, CHEN S L. Numerical Analysis of Influence of Solid Particles on Elastohydrodynamic Line Contacts under Grease Lubrication[J]. Journal of South China University of Technology (Natural Science Edition),2012,7:011.

[4] YANG P R. Numerical Analysis of Fluid Lubrication[M]. Beijing: National Defence Industry Press, 1998.

International Conference on Power Electronics and Energy Engineering (PEEE 2015)

Electric Power Steering System for Agricultural Truck Matching Design

Z.G. Chen

Mechanical and energy engineering department
Shaoyang University
Shaoyang, Hunan, China

Q. Zhang

Mechanical and energy engineering department
Shaoyang University
Shaoyang, Hunan, China

W. Zhang

Mechanical and energy engineering department
Shaoyang University
Shaoyang, Hunan, China

T.M. Zhou

Zhuzhou Elite Electro Mechanical Co, Ltd
Zhuzhou, Hunan, China

Abstract—this thesis aimed at the characteristics and the structural of a type of truck. It confirmed that its chosen power-assisted form of electric power steering system is Column EPS——that is C-EPS. On the platform of a sort of car we combined with the technical requirements of a certain model car steering system performance upgrades and matched design of the motor, the decelerating device, and the transducer. Finally device of the EPS in a certain automobile model truck for the characteristics test of input and output, it have shown that the EPS has a good matching design of the booster effect.

Keywords-EPS; matching; design; test

I. INTRODUCTION

With the development of products of the automotive industry, energy efficiency and environmental protection will become a trend .automotive electric power steering system(Electric Power Steering，referred to as EPS) is the first choice for every large automobile company both at home and abroad due to its advantage of well assist characteristic, fuel economy, good low-temperature running, compact structure, and easy assembly. This thesis aimed at the characteristics and the structural of a type of car. It confirmed that its chosen power-assisted form of electric power steering system is Column EPS——that is C-EPS. In order to make sure that electric power steering possesses better handle, lower noise, and higher reliability, we combined with the technical requirements of a certain model car steering system performance upgrades and matched design of the motor，the decelerating device, and the transducer. Finally we have run testes device of the EPS in a certain automobile model car for the characteristics of input and output.

II. SYSTEM AND OPERATING PRINCIPLES OF EPS

EPS system mainly includes assisted motor, torque sensor, decelerating device, and controller (ECU).Its structures are shown following Figure 1.

Operating principles: EPS was in a working state after auto engine started. Torque sensor will detect the torque and angle size what for the rotation in the steering wheel as soon as driver manipulate and rotate the steering wheel. At the same time, torque sensor will receive signals of engine speed and vehicle speed. Then it will transform into digital signal and input control unit. After that it calculate we will get and make driving cycle so good at adapting assisting torque. Next, controlling assisted motor, outputting the corresponding size and direction, increasing torque by the decelerating device, it finally applied to the gear and rack so that it perform power-assisted steering[1].

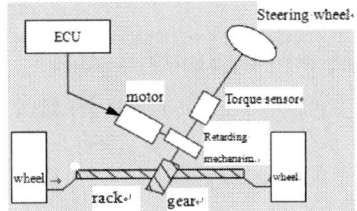

FIGURE I. THE SYSTEM STRUCTURE DIAGRAM OF EPS.

III. THE MATCHING DESIGNS OF THE MOTOR OF EPS AND THE RETARDING MECHANISM

The parameters of the entire car one type of car, see Table 1.

TABLEI. THE PARAMETERS OF THE ENTIRE CAR.

No	Project	Code	Parameter
1	Full mass(kg)	m	3 500
2	Steering gear angle ratio	I	17.2
3	Number of turns about the steering wheel(collar)	n	2.9-3.3
4	π	π	3.1416
5	Sliding friction	f	0.70
6	The max rotato speed of steering wheel(r/min)	W_h	75
7	Ldeal situ steering wheel torque (N·m)	M_{h0}	5.50
8	Transmission efficiency of wrom and gear mechanism	η_0	0.80
9	Pressure of the wheel(MPa)	P_0	0.22
10	Front axle load(Kg)	G_1	820

© 2015. The authors - Published by Atlantis Press

According to the load state of a type of farm vehicle and the calculation on relative parameters and the space size of vehicle itself, cab interior space becomes small. Space layout of the clutch pedal, the brake pedal and the accelerator pedal become small, too. To meet the needs of man-machine It is considered final what engineering, steering column is turned into form of assist. That is C-EPS structure. Motor is power source of EPS structure as instructions of ECU output suitable assisting torque. Electric machinery is key component of EPS structure. It has significant effect on performance of EPS structure. Electric machinery of C-EPS has many characteristics such as good working environment (cab's environment), little ripple of the torque, faint machinery vibration, low sealing requirement, low noise, low power, small size, low revs, connector no waterproof requirement, low requirement of surface protection, good heat dissipation surrounding, low machine-electro time constant, and so on [2].The role of decelerating device is amplification. of assisting motor output torque. Transmission of decelerating device fairly large. It requests that electrical machinery outputs lower torque and higher working speed. Meanwhile the size of decelerating device becomes very large. This lead to difficulty in spacial arrangement. Transmission ratio of decelerating device is so small that it requests the electrical machinery output higher torque and lower working speed. It makes the size of the electrical machinery increase and improves the requirement of electrical machine design. In order to facilitate machine and assemble, it mainly require higher transmission efficiency and smaller rotary inertia when the decelerating device select option. It also requires the size as small as possible. To act as the power component of the EPS system, electromechanical decelerating device is very important for matching the whole system.

IV. THE CHOICE OF TYPE ON THE DYNAMO AND THE DECELERATING DEVICE

At present, the assisted machine of EPS system mainly consists of brush dc motor and brushless dc motor at home and abroad. The brush dc motor using the mechanical commutator for reversing, it needs to maintain carbon brush regularly. Its power density is low. The brushless dc motor adopts electronic commutation, no maintenance, high power density, but the cost is higher. Generally, choosing a brush dc motor that is a good idea for lower power-assisted cars. Choosing a brushless dc motor high car is suitable for higher power-assisted car. With mature techniques, simple controller and low cost, the brush dc motor are widely used in domestic electric power steering system.

EPS commonly used is worm and worm wheel type and planetary gear type about retarding mechanism. With high efficiency, when retarding mechanism of Worm and worm wheel type are arranged, it requires less space, stable and reliable. Retarding mechanism of planetary gear type is relatively complex. It can change transmission characteristics of the steering system's angle. Usually, choosing and matching the worm and worm wheel type retarding mechanism and steering-shaft-type power system, the requirements for smaller space collocation are met.

Considering the actual needs, in this essay, assisted motor chosen was brush dc motor. In the meantime, retarding mechanism chosen was worm and worm wheel type.

V. PARAMETER DESIGN OF THE MOTOR OF EPS AND THE RETARDING MECHANISM

Because the motor act on assisted torque T of sheering shaft which is influenced by many factors, i (the speed reducing ratio of retarding mechanism), Km (electromagnetic torque constant of the motor), Ke (back electromotive force constant of the motor), w (angular velocity of steering wheel) and so on. As the ratio of retarding mechanism increases, the steering wheel torque also increases[3].At the same time, response time of yaw velocity get longer,and phase lag rise. The relationship between them is as follows:

$$T_{max}(w) = iK_m(U_m - K_e iw)/R_m \tag{1}$$

Among them: Tmax(w):the maximum assisted torque is acted on steering shaft by the electric machinery .Um：the voltage is added to two ends of motor circuit by power.

Increasing the reduction ratio will lead to reducing transmission efficiency of retarding mechanism It makes the effect of power assisting and returnability worsen. If the reduction ratio is too big,it will make its reverse efficiency is too low, even appear self-locking phenomena. At this rate, steering system wil lose the function of returnability. When the motor of EPS and retarding mechanism match, the motor of high torque constant and retarding mechanism of low reduction ratio will be chosen. For schneckenantrieb, its transmission ratio is generally from 15 to 20.This essay choose 16.5 as reduction ratio.

According to the calculating motor the parameter of the front axle load:

Situ steering torque:

$$Mr = f*((9.8*G1)3/P0)1/2/3000 = 360.1(N \cdot m) \tag{2}$$

No power the steering wheel torque:

$$Mh = Mr/(\eta*iw) = 27.9(N \cdot m) \tag{3}$$

To the maximum torque:

$$Ma = Mh - Mh0 = 22.9(N \cdot m) \tag{4}$$

Motor rated power:

$$P = 2\pi*Ma*Wh/(60*\eta 0) = 225(W) \tag{5}$$

Motor nom speed:

$$n = Wh*i = 1023.75(r/min) \tag{6}$$

Motor rated torque:

$$T = 60*P/(2\pi*n) = 1.94(N \cdot m) \tag{7}$$

In accordance with the calculation consequence, Mitsubishi brush dc motor was chosen with motor nom speed was chosen 1050 r/min，Motor rated power was chosen270W，and Motor rated torque2.4N.m.

VI. NONCONTACTING PICKUP MATCHING DESIGN

In electric power steering system, when driver control the steering wheel, torque sensor was used to measures the driner's the force of the hand, HW Hand-Wheel Angle, rotational speed and so on. Nevertheless, these informations are all ECU control signals. The accuracy of the information is very important for electric power steering system. Torque sensor is one of the key hardware electric power steering system.

Common torque sensor are mainly contact type and non-contact type torque sensor. Contact type torque sensor generally include swing arm torque sensor, dual planetary gear type torque sensor and torsion bar type torque sensor. Contact torque sensor cost is low, but easy to be resulted in temperature drift, short service life. Non-contact torque sensor includes optical torque sensor and magnetoelectric torque sensor. Compared with contact torque sensor, the non-contact torque sensor is high precision, good anti-jamming capability, but the cost is higher. The choice of torque sensor is usually should consider measurement accuracy, the cost, working environment, and service life[4].

EPS of a type of farm vehicle was used a new electromagnetic induction non-contact sensors. The sensor is wearproof, small hysteresis，small influence of follow-up attention outside factors. And it was solved the problem of the sensor potential with the temperature drift [5]. Non-contact sensor structure is shown in Figure 2.

Sensor characteristics shall meet the following requirements: the main line, output the corsspoint of auxiliary road---that is the initial work location 0, (that is, middle position of the effective electrical Angle).When Working voltage is+5±0.05V, +12±0.12V, the voltage of the main line and the auxiliary road should be 2.5±0.05V. Any other location shall meet Vmain line+Vauxiliary road=5V±0.15V .And it require 0.8V≤Vmain limit≤4.2V，0.8V≤Vauxiliary limit≤4.2V. Figure 3 meet the requirements for non-contact sensor in the characteristics of a type in the actual test.

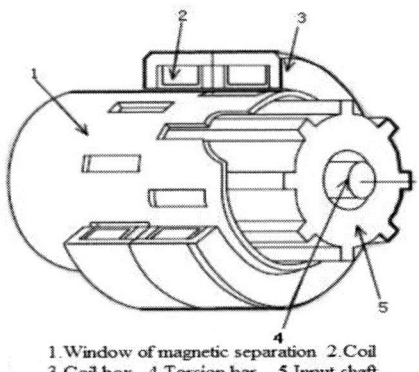

1.Window of magnetic separation 2.Coil
3.Coil box 4.Torsion bar 5.Input shaft

FIGURE II. MECHANICAL STRUCTURE OF THE SENSOR.

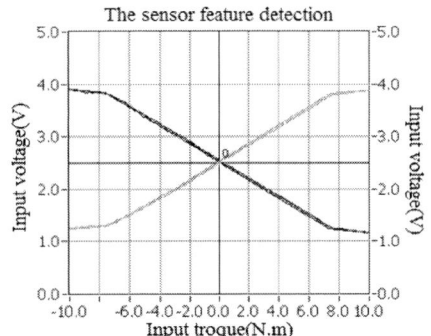

FIGURE III. THE SENSOR CHARACTERISTIC

VII. THE TEST ON EPS OF INPUT-OUTPUT CHARACTERISTICS

The matching design of EPS after input/output characteristic test, the test tools include EPS controller, torque sensor, Angle sensor, data acquisition system, computer, inverter and so on. To test under different speeds, the input and output torque/force curve that is whether the input and output characteristics comply with the design requirements and whether the curve of different vehicle speeds has good symmetry. To set the device installed on a test rack and different speed, with 20r/min~30r/min speed uniform rotation of input shaft, we record the speed of input and output torque moment/force curve. The test results as shown in Figure 4, 0~80 km/h input/output characteristic curve. Figure illustrates the input of 2 Nm, situ output is 4 Nm，20 Km/s .Output is 1 Nm, 40 ~ 80 km/s input is 0.8. Input is 3 Nm, in situ output is 6.2 N/m, 20 km/s output for 4 Nm, 40 km/s output to 2 Nm, 60 km/s output is 1.5 Nm, and 80 km/s output is 1 Nm. At low speeds, power is larger. Assisting power should weaken as the speed increase. It is easy to see after the design of EPS matches，it meet the power demand. It can make the vehicle has good steering portability and sensitiveness, too. Besides, it has a good handle and driving stability at high speeds[6].

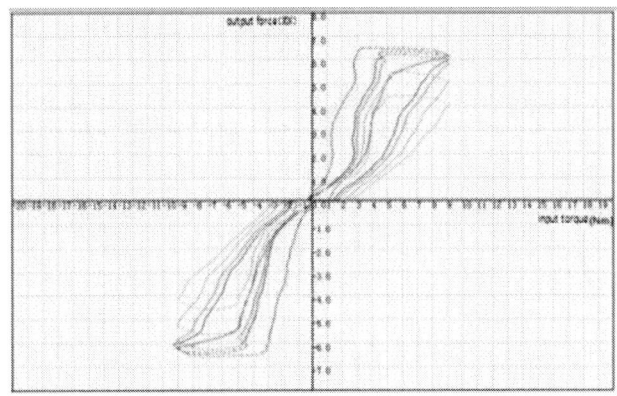

FIGURE IV. INPUT AND OUTPUT CHARACTER CURVE.

VIII. CONCLUDING REMARKS

Based on the analysis and calculation of vechicle parameter of a type of agricultural truck，a type of agricultural truck was matched and designed. After matched, EPS was done input/output characteristic test. The results show that EPS system can achieve good effect of power. Meanwhile，it has the advantage of good steering portability, sensitiveness, handle and driving stability.

ACKNOWLEDGEMENTS

This research was financially supported by Hunan Province University Innovation Platform Development Foundation (Grant No.13K109) and Research Innovation Projects of Shaoyang University graduate (CX2013SY026)

REFERENCE

[1] ZHOU Tingming, LIU Zhihui, LI Mengqi, CHEN Zhigang, TANG Ning. Electronic Power Steering System and Its Key Technologies [J]. Hydromechatronics Engineering，2012，40(7):176-179，209.

[2] XIANG jinquan. Motor Matching Technologies of Automotive Electric Power Steering System[J]. Automobile Parts，2013(02):58-61, 64.

[3] Kong Fansheng. Study on Electric Power Steering System Matching with Xiali Passenger Vehicle[D]. Journal of Jilin University Engineering and Technology Edition.2010.

[4] Liu Cai. Electronic Power Steering Matching Design on a Mini-van [D]. Journal of Hunan University.2013.

[5] Yang Guangchuan.Discussion on EPS sensor of the median voltage Key points of design and development [J]. Mechanical and electrical Technology.2013 (01):48-50.

[6] XIANG Dan, LI Wu-bo, YANG Yong. Study on Return-to-center Control and Simulation for Electric Power Steering System [J]. Machinery Design & Manufacture .2012(08).

International Conference on Power Electronics and Energy Engineering (PEEE 2015)

Research on Precision Forging Blade TENON Processing Technology

L.J. Huang

Institute of CAPP & Manufacturing Engineering Softerware
Northwestern Polytechnical University
Xi'an, China

X.T. Tian

Institute of CAPP & Manufacturing Engineering Softerware
Northwestern Polytechnical University
Xi'an, China

M. Kou

Institute of CAPP & Manufacturing Engineering Softerware
Northwestern Polytechnical University
Xi'an, China

Z.Y. Wang

Xi'an Aircraft Industry (Group) Company Limited under AVIC
Xi'an, China

Abstract—**TENON machining is an important and major processing for aero-engine precision forging blade. Traditional processing methods for TENON machining cannot meet the modern aero-engine performance, efficiency and environmental protection needs. In the paper, a "hard" clamping process method to machine blade TENON is proposed, and the key technologies of the method, fixture layout optimization and deformation controlling, are discussed in detail. Finally, we introduce specific scheme that designed a set of hydraulic special fixture with adaptive auxiliary support to realize "hard" clamping processing.**

Keywords-aero-engine forging blade; blade TENON manufacturing; fixture layout; deformation control

I. INTRODUCTION

The aero-engine blade is space curved surface, and must bear complex stresses and micro vibration in working. For this reason, there is high quality requirement for blade material and machining process. Precision forging can obtain curved surface of blade with no allowance, which solve the difficulty of mechanical processing for difficult-to-machine materials and thin blade. However, blade TENON still needs mechanical processing. Blade TENON is a standard of blade design and machining, and should be machined based on blade body. Traditionally, a low-melting alloy casting process is used to machine the TENON, shown in Table 1. The alloy with low melting point is poured a square box to transform the standard of blade body into square box, and then the TENON is processed based on the surface of the square box. However, there are many disadvantages using the production process, e.g. complex preparation, low efficiency, labor intensive, bad working conditions. In addition, the trace residue alloy can be attached on the surface of the blade in the next high temperature melting process, which would potentially impacted on the blade body surface quality.

Because of these significant technical shortcomings, a "hard" clamping process method has used to machine the TENON of precision forging blade. Compared with low melting

point alloy casting process, the hard clamping process is based on the blade body direct positioning and clamping to machine, using a special fixture tool. The blade TENON is machined using milling or grinding, which positioned with three points on the body surface of the precision forging blade, two points on the intake side and one point on the bottom of the TENON, and clamped directly on the blade body. As the processing of TENON of precision forging blades is through the free-form surface of blade body to position, the main issues to be addressed include positioning, stability, clamping deformation, and repeatability etc. In the paper, we focus on the analysis of fixture layout optimization and deformation controlling issues about the "hard" clamping process, and introduce our fixture scheme.

II. FIXTURE LAYOUT OPTIMIZATION

Fixture layout includes location layout and clamping layout. Location layout mainly limits six freedom degrees of the part to make sure the repeatability and accuracy of positioning. Clamping layout mainly chooses the appropriate clamping position and ensures the stability of the clamp.

Based on the blade body positioning, the six points positioning method is generally used, which include the three points on the body, two points on the intake side, one point on the bottom of the blade TENON. Through the six points positioning, the blade can be completely restricted to keep the blade at the determined theoretical position. Since the blade body is free curve, easily deformed, and there are shape errors of the work piece, making it difficult to select an anchor point [1].Liu WW etc. [2], Northwestern Polytechnic University, invented a hydraulic fixture for the blade root machining of precision forging blade, located by vane basin, intake side, the inside edge of the plate, using a spring washer on the dorsal to clamp. Though the fixture is simple and easy to use, the clamping force cannot be guaranteed accurately. Al-Habaibeh A [3], Afzeri A G E [4], who designed a pin-type universal fixture system, with a clamping force evenly distributed and shape reconfigurable features. But for the distribution, diameter, length, material of

© 2015. The authors - Published by Atlantis Press

thimble still need further study. Ma J [5] used topology optimization methods for complex fixture layout design. Y. Wang [6, 7] proposed a fixture layout optimization method, and considered the repeatability, stability and fixity of fixture.

TABLEI. TRADITIONAL AND NEW PROCESSING METHODS FOR TENON MACHINING.

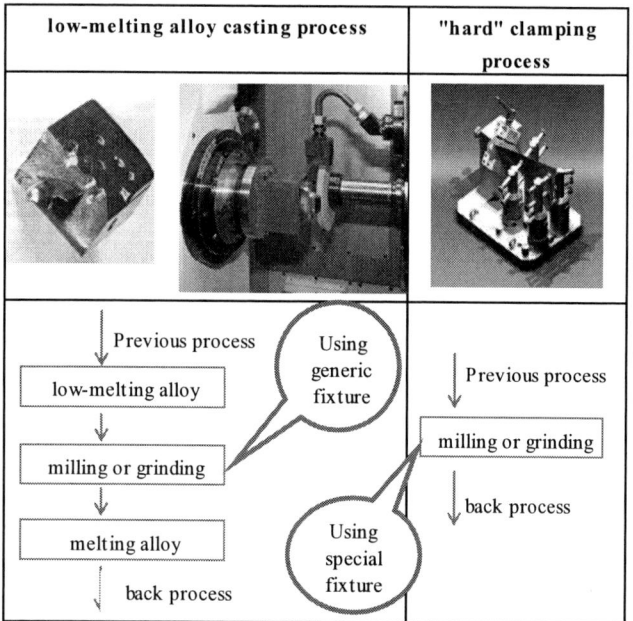

Layout optimization is based on the workpiece positioning repeatability and positioning accuracy. Clamping layout optimization provides a method for minimizing the clamping force. Clamping optimization program is as follows: 1) finding a suitable location to ensure that does not move the clamping requirements; 2) the optimal clamping scheme between suitable clamping position and stability constraints.

III. DEFORMATION CONTROLLING

In actual processing, deformation of the workpiece can be divided into two types of clamping deformation and processing deformation. The former is static deformation caused by the clamping force, another is dynamic deformation caused by cutting force.

Workpiece clamping always accompanies by varying degrees of deformation. Machining accuracy is affected by the accuracy of clamping. Especially for weak rigid thin-walled parts, clamping deformation is more prominent, seriously affecting the machining accuracy and surface quality of the workpiece [8]. 20% to 60% is caused by a machining error of the workpiece clamping [9]. Generally deformation controlling is optimizing the clamping design. In order to control the deformation of the workpiece clamping, need to consider the clamping layout, clamping force and the order of clamping.

The main factors affect the deformation of weak stiffness workpiece are the initial stress, clamping, cutting parameters and cutting paths. Machining distortion associates with the blade

structural shape, positioning and clamping layout, cutting force, cutting heat and tool path etc. In the actual process, the weak stiffness structure will exceed the difference due to the coupling effect of these factors, interaction, and error accumulating, so that difficult to control the deformation process [10]. When increased the stiffness of workpiece and fixture system, the machining deformation could be well controlled.

IV. SOLUTIONS

According to the characteristics of the blade TENON machining, we designed a set of hydraulic special fixture with adaptive auxiliary supports. The fixture comprises positioning device, the clamping device, the auxiliary supports and control system.

A. Scheme of Locating and Clamping

The fixture scheme is shown in Figure 1. It is used six points positioning scheme for the positioning device. By analyzing the processed blade, we get six anchor points. Three positioning pins contact the blade body surface, which limit three degrees of freedom; two pins contact the intake side of the blade, limit two degrees of freedom; and the last freedom degree is limited by a pin contacting the bottom of the TENON.

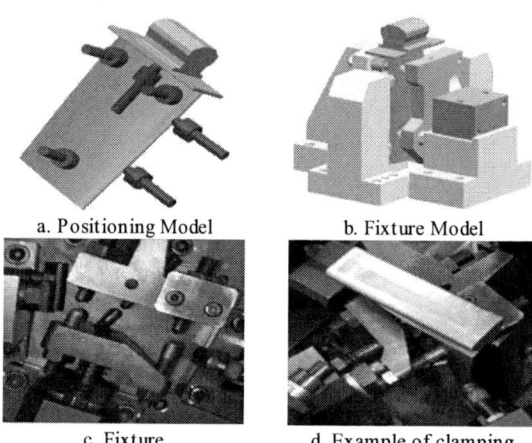

a. Positioning Model   b. Fixture Model

c. Fixture   d. Example of clamping

FIGURE I. THE SCHEME OF LOCATING AND CLAMPING.

Clamping device consists of main clamping component and lateral clamping component. Main clamping component is used for clamping the body of the blade, and the latter is used for clamping the exhaust side of the blade. Both the clamping component has a hydraulic cylinder to drive the clamping block. Connecting the clamping device using a clearance fit, the device can be automatically adjusted based on the fit clearance with the clamping force, so that all three points have access to the blade's surface.

B. Deformation Analysis

We have analyzed a blade deformation under the combined effect of clamping force and cutting force using ANSYS software, shown in Figure 2. There is large deformation on the enlarged head and tip portion of the blade, but small deformation on the body portion of the blade.

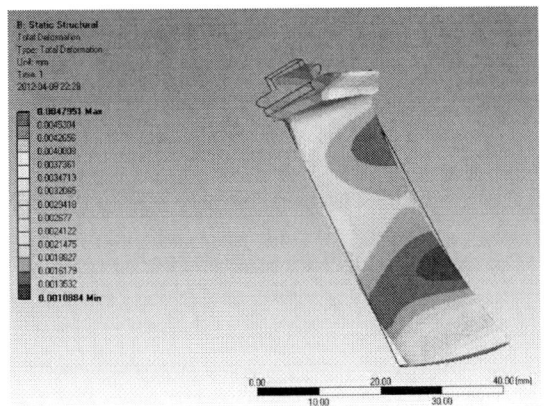

FIGURE II. DEFORMATION ANALYSIS WITH COMPREHENSIVE FORCE.

FIGURE III. COMPREHENSIVE AUXILIARY SUPPORT.

Deformation analyses are as follows:

1) Blade body deformation. The deformation of the blade comes from two aspects. One is the clamping deformation affected by the clamping force sizes, the layout of clamping points and the clamping order. The other is the process deformation, to which leaded as the reaction force from the locating and clamping pieces results local stress concentration.

2) TENON deformation caused by cutting force. When the blade TENON machined, the workpiece and fixture system is formed into a cantilever structure, and resulting in deformation under the effect of cutting force. At the same time, the change of the cutting force causes vibration, which affects the TENON processing quality seriously.

C. Deformation Controlling

To control the deformation of the blade TENON, we have designed a hydraulic auxiliary support device. The auxiliary supports provide reaction force against the cutting force, increase the workpiece stiffness and reduce the clamping force to ease clamping deformation. The auxiliary support device, shown in Figure 3, includes four auxiliary support mechanisms disposed opposite. Wherein the supports driven by cylinder adaptively contact the surface of TENON and locked. When one side of the TENON machined, the supporting head is retracted to the side, but the corresponding head supports. With this manner, the position of the blade can be maintained without excessive positioning. The stiffness of the thin-walled blade can be greatly improved, and the clamping force apply to the blade can be reduced to control clamping deformation.

In order to avoid the cutter interference with the supports, we have designed a set of controller. The controller communicates with the programmable logic controller of the machine tool and hydraulic control valves, in order to obtain the tool position data from the numerical control system in real time and to control to retract or support. Each auxiliary support head is adjusted by built-in algorithms according to the cutting tool path, to avoid the interference and make the machining process smoothly.

V. CONCLUSION

In the paper, we study machining process of the no margin forging blade TENON, propose a new type of TENON processing scheme with positioning and clamping the blade directly, multi-directional auxiliary support for the TENON with digital control of hydraulic system. The advantages of the new technology are as follows:

1) Positioning and clamping against the thin curved blade directly, and removing low melting point alloy casting and melting processes, the blade machining process chain is dramatically shorten.

2) The design of the adaptive auxiliary support device is used to remain the support in contact with the blade TENON, and the blade position can be maintained without excessive positioning. Therefor the machining deformation can be effectively reduced, and the processing quality improved.

REFERENCES

[1] Vishnupriyan S, Majumder M C, Ramachandran K P. Optimal fixture parameters considering locator errors. International Journal of Production Research, 2011, 49(21): 6343-6361.

[2] Liu W W, Wan X S, etc. Forging the blade root machining hydraulic TENON jig. China Patent, CN CN102615534B. 2013.11.06

[3] Al-Habaibeh A, Gindy N, Parkin R M. Experimental design and investigation of a pin-type reconfigurable clamping system for manufacturing aerospace components. Proceedings of the Institution of Mechanical Engineers, Part B: Journal of Engineering Manufacture, 2003, 217(12): 1771-1777.

[4] Afzeri A G E, Sutjipto A K M, Nurul Amin R M. Determination of pin configuration for clamping fixture by means of solid model contact analysis.//Proceedings of the International Conference on Mechanical Engineering (ICME), Dhaka, Bangladesh. 2005.

[5] Ma J, Wang M Y, Zhu X. Compliant fixture layout design using topology optimization method.//Robotics and Automation (ICRA), 2011 IEEE International Conference on. IEEE, 2011: 3757-3763.

[6] Wang Y, Chen X, Liu Q, et al. Optimisation of machining fixture layout under multi-constraints. International Journal of Machine Tools and Manufacture, 2006, 46(12): 1291-1300.

[7] Wang Y, Chen X, Gindy N, et al. Elastic deformation of a fixture and turbine blades system based on finite element analysis. The International Journal of Advanced Manufacturing Technology, 2008, 36(3-4): 296-304.

[8] Wu N H, Chan K C, Leong S S. Static interactions of surface contacts in a fixture-workpiece system. Internal Journal of Computer Applications in Technology, 1997, 10(3/4):133-151.

[9] Qin, G H, Wu, Z X, Zhang, W H. Analysis and control technique of fixturing deformation mechanism of thin-walled workpiece. Jixie Gongcheng Xuebao. 2007, 43(4): 211-216.

[10] Lu D. Deformation Prediction and Fixture Layout Optimization of Aerospace monolithic components. Jinan, China: Shandong University, 2007.

High Precision Work Station Research

X.M. Cheng
School of Material Engineering
Ningbo University of Technology
Ningbo, China

Y.Q. Chen
School of Material Engineering
Ningbo University of Technology
Ningbo, China

P.N. Zhang
School of Material Engineering
Ningbo University of Technology
Ningbo, China

H.C. Fan
School of Mechanical Engineering
Ningbo University of Technology
Ningbo, China

Z. Zhao
School of Mechanical Engineering
Ningbo University of Technology
Ningbo, China

C.Z. Li
School of Mechanical Engineering
Ningbo University of Technology
Ningbo, China

Abstract—a wide range of high precision work station can today be using various fields. The position precision will directly determine the precision and ultra precision processing equipment machining accuracy. It is necessary to research highly precise station system. In this paper, the development of a two-dimensional worktable with high accuracy is presented. This work station uses the linear motor as the driving source. First of all, the components of the station technology and its control system are described. The latter comprises a grating scale sensor system for position measuring. The accuracy and repeatability of work station achieved high requirements.

Keywords-high precision; work station; linear motor; close loop control

I. INTRODUCTION

With the development of advanced manufacturing technology, the machining accuracy is increased more higher in industrial fields of military, microelectronics, micro machinery. High precision work station have become a core issue, pose estimation as a research hot in the field of manufacture research has been carried on for many years. The issue of study here are the manufacturing technology, driving technology, controlling technology and measuring technology. There are a large number of studies including those technologies and some methods about the high precision station. Tokyo Institute of Technology developed a nano-positioning stage. The system uses a DC motor and air screw as the hydrostatic drive system, with hydrostatic bearings support the rotating screw. The entire table is completed using aerostatic support and guide rails. By laser interferometer as a feedback device, using closed-loop control, laser interferometer resolution up 0.3nm, positioning accuracy can be achieved with disabilities across the table 4nm [1, 2]. Compiegne University of Technology in France developed a

high-precision table. The table uses a brushless DC motor friction drive as drive and with hydrostatic support rails as the entire table. The maximum speed is up to 10mm / s and the positioning accuracy is up to 16nm [3].Shanghai University in China introduces the key techniques used for getting high precision and resolution. A control system based on macro-micro dual loop method is represented in which a macro stage is driven by ball screw and AC servo motor with metal reflect grating inspector to comprise servo control system. The linear compensation to the real error curve can reduce the position error from 76μm to 3μm. A novel micro stage holds on the macro stage's upper surface. The detected error signal from macro loop will be changed to micro loop and can fulfill large stroke with nano precision positioning [4]. Cristiana Delprete focuses on a particular case study belonging to the bio-mechanical field. The research then has started with an exhaustive analysis of the state of the art, followed by the design and the practical realization of a manipulator matching the specific implementation but whose characteristics could be employed also for other applications [5].

Research on high-precision positioning stage from the drive mode aspects, can be used in many different forms-driven approach, such as ball screws, hydrostatic lead screw, piezoelectric ceramic and linear motors. Methods and levels can be directly positioned from a combination of macro and micro aspects of the structure of the table. In addition, high-precision positioning control method of the stage also diverse.However, those systems are very complex with too many parts. This article mainly focuses on the high precision station based on linear motor and grating scale measure sensor.

II. SETTING UP OF MECHANICAL SYSTEM

High-precision work station system mainly consists of drive system, guidance system, measure system and control system

© 2015. The authors - Published by Atlantis Press

and other components. Drive system is to determine the key for the whole table positioning accuracy. Feed accuracy and resolution of the drive system directly determines the positioning accuracy of the station. High-precision station feed can include two structural forms which is single-stage feed and two-stage feed. Two-stage feed refers to the macro and micro-positioning feed combined. It gets even nanometer or sub-nanometer positioning accuracy of general application in small displacement situations. In this paper, two-dimensional work station requires precision XY axis of the effective stroke 500mm and 400mm, respectively. Positioning accuracy required to achieve micron. So the system uses a single-stage form of the feed structure.

A. Diver Motor Selection

Traditional precision drive mechanism mainly refers to use the ball screw. Ball in motion the process to rolling friction instead of sliding friction. The structure has a simple structure, low friction, high transfer efficiency, motion stability and long life, etc which widely used in precision processing areas. But the ball screw drive system drive chain length, component complexity, and there is accumulated pitch error and backlash errors, it is difficult to meet the requirements of high-precision positioning applications. Although the high level of precision ball screws can achieve micron level positioning accuracy and repeatability of positioning accuracy. But the cost is high. It's difficult to control.

Linear motor drive technology has a simple structure, get a larger movement speed and acceleration. And the stroke is not limited, to achieve the required precision positioning feed movement. Linear motor drive technology is increasingly used in precision and ultra-precision machining fields. This thesis studied precision dimensional table with linear motor drive technology. Achieve large stroke and high precision positioning requirements.

Permanent magnet linear synchronous motor mover consists of a three-phase armature windings hollow coil windings by epoxy encapsulation magnet on the inside of the support block and considerate. The stator and the overall structure of a "U" shaped groove structure. The mover reciprocates in the groove. The biggest difficulty of "U" shaped hollow permanent magnet linear synchronous motor is mounting accuracy and uniformity that the mover of the permanent magnet air gap during moving. It has a large normal suction due to the contrary in the "U" groove opposite polarity of two permanent magnets. It needs to ensure that the support block tightly close together when the installation of the permanent magnet of the permanent magnet. The length of "U" type magnet design can not be too long affected control accuracy. The length is too long to cause non-uniformity of the air gap and to affect the accuracy of the hollow type servo control of permanent magnet linear synchronous motor. In applications where a large stroke in order to ensure the uniformity of the air gap, generally a plurality of magnets fixed length splicing. The linear motor is shown in Figure1.

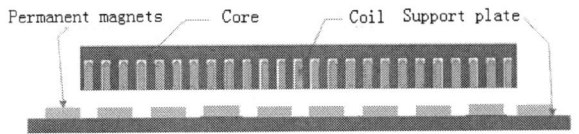

FIGURE I. "U" TYPE LINEAR MOTOR.

B. Sensor and Measure Device

Displacement measure device is an important part of the closed-loop control and semi-closed loop control, directly determines the accuracy of high hydrostatic mobile platform. There are many kinds of means for measure the displacement of the mobile platform. Here we select the grating displacement sensor to consist the measure system. Grating is mainly used for high-precision measurement of the displacement. The grating system includes a grating scale raster and a raster scanning head. The grating scale is fixed to the fixed table. The raster scanning head is fixed on the table moving parts.

C. System introduce

The work station used the marble slabs as the substrate. Marble slabs unaffected by temperature, no distortion. Marble slabs while long-term placement, it will not be deformed due to internal stresses. The work station has x, y two-dimensional directions. Each direction used permanent magnet linear synchronous motor. X-direction stroke is designed for 400mm and Y-direction stroke is 300mm. The linear guide stroke is selected as the support element. This linear guide stroke has many advantages as lower friction, good straightness and easy installation. The grating ruler is mounted in the middle of the work stage. There has a very higher parallel degree requirement between the liner guide stroke and the line motor and the grating ruler. Respectively the table cushion is mounted on both ends to protect against shock bumps. There is a standard installation flat above the table easy to install other components. The work station structure is shown in Figure2.

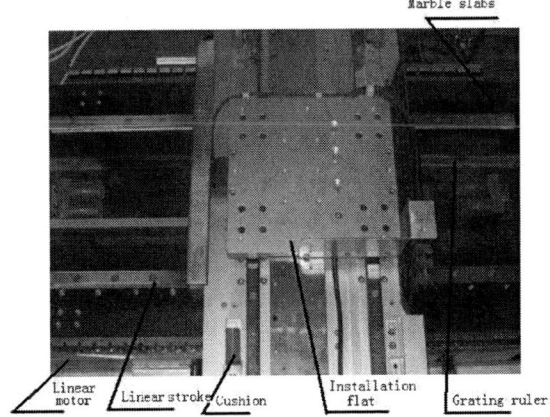

FIGURE II. WORK STATION.

D. Control System

Closed-loop control method is to achieve the actual position of the controlled object and feedback to controller which generates a position error between the detected position and the control position. By controlling principle of the deviation to reduce bias. Theoretically closed-loop servo control system can eliminate all intermediate transmission errors caused by the transmission mechanism. Positioning accuracy of the system depends on the accuracy of the measure element. It would cause great difficulties to commissioning transmission vice friction characteristics, and other nonlinear factors clearance system. These features make the system performance decline and even cause oscillation. Closed-loop control of the system has high demands to control strategy. The system achieves a direct drive to eliminate all intermediate gear. Output displacement linear motor is the actual output table displacement. Everything nonlinear factors outside interference will directly affect the positioning accuracy of the station. In order to produce the effects of various disturbances on the effective correction, direct-drive system can only use closed loop control. To prevent electromagnetic interference of linear motor itself on the position of the sensor, the nearly actual position of the measured signal is fed back to the motion control card. After the control card digital operation, the control card sends a new control signal to control the motors. The whole experiences consist of a closed loop control systems. The system can achieve high precision positioning requirements. The close-loop control system shown in Figure3.

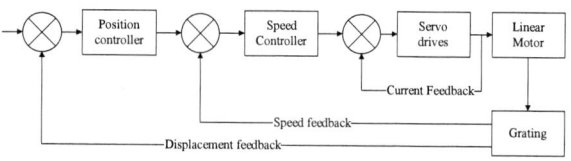

FIGURE III. SERVO PRINCIPLE STRUCTURE.

III. SYSTEM TEST

After commissioning the system is optimized to run at the optimal state. The station system is tested to meet the requirements of high-precision work station. The length of x-direction is 500mm and y direction is 400mm. The work station is running at maximum speed 1m/s. The positioning accuracy is 5μm and the repeat positioning accuracy is 1 μm. The maximum acceleration is $10m/s^2$.

IV. CONCLUSION

In this paper, our current efforts on the development of a high precision work station were presented. The linear motors are employed to perform move function. The components of the station and its control system were described. We discussed the work station components: an assembly using the marble; a sensor measure using the linear grating; a closed loop mode for the control system. After Adjusted and optimized, the station meets the requirements of high-precision. The project here presented has been a very good way for developing the high precision system in the particular field application giving the beginning of the work station.

ACKNOWLEDGEMENT

The work was supported by National Natural Science Foundation of China (NSFC, Grant Nos. 51275251 and 51075217), Zhejiang Provincial Natural Science Foundation of China (ZJSFC, Grant No. Y1100528).

REFERENCES

[1] J. H. Mao, H. Tachikawa, A. Shimokohbe. Double-inte gratoreontrol for Preeision Positioning in the Preseneeo ffrietion [J]. Precision Engineering, 2003 (27): 419-428.

[2] J. H. Mao, H. Taehikawa, A. Shimokohbe. Preeision Positioning of a DC-motor-driven aero statieslide system [J]. Preeision Engineering, 2003(27):32-41.

[3] Mekid, Samir. High Precision linear slide. Partl: Design and eonstruetion [J]. International Journal of Machine Toolsand Manufaeture, 2000, 40(7): 1039-1050.

[4] L. Z. Sun, M. M. LI, W. M. Cheng. Study on precision positioning technique. Optics and Precision Engineering. 2005, Vol.13, Nov: 69-75.

[5] Cristiana Delprete, Carlo Rosso, Cristina Scarzella. New concept for micro-manipulation systems: a practial experience. Int J Adv Manuf Technol (2014) 74:1077–1085.

International Conference on Power Electronics and Energy Engineering (PEEE 2015)

Matching Research Between Engine and Transmission of Vehicle Based on AVL-Cruise

F. Du

School of Automobile and Transportation
Tianjin University of Technology and Education
Tianjin, China

C.H. Liu

School of Automobile and Transportation
Tianjin University of Technology and Education
Tianjin, China

Z.W. Guan

School of Automobile and Transportation
Tianjin University of Technology and Education
Tianjin, China

Y.S. Wan

Collage of mechanical engineering,
Tianjin University of Technology and Education
Tianjin, China

Abstract—**Using the analysis of the design goals for a new car, the basic structure type of this new car was determined. Under the presupposition of giving power unit, the various of matching scheme of driveline and engine was selected preliminary. The vehicle model was construced by using AVL-Cruise software, and the performance simulation and theory analysis were carried out for the different combination of the power plant and driveline, the optimum matching of powertrain to meet the performance required was chosed in the eventually. Practice has proved that the development cycle of new car can be shortened, the research and development costs can be reduced by using the simulation platform of AVL-Cruise software.**

Keywords-optimization matching; power; driveline; cruise simulation

I   INTRODUCTION

The development of automobile industry so far, the requirement and the expectation to all aspects of vehicle performance is more stringent and higher, how to improve the vehicle performance has become one important subject[1]. The whole performance of vehicle depends not only on the quality of each assembly and component, but also on the parameters matching among the different assemblies one another. Usually, the assessment to the dynamics, economy and emission of vehicle depends on whether it is reasonable that matching between engine and dynamic system to a great degree, matching scheme between the engine and drive train may have a variety of programs, if all program must have real test, the development costs of new product will be increased, the design cycle will be extended, therefore, it is necessary to do the work of the power matching and performance analysis in the design phase of new product, namely there is no experimental prototype case[2]. In view of this situation, some car companies tried to predict the car performance, and study the optimum matching of power plant with driveline by using computer simulation techniques in the early stages of the design, and had achieved good effect[3]. This

approach helps the R & D department to identify design scheme quickly, not only improving the performance of new products, but also shortening the development cycle and saving the development cost, so it has been paid more and more attention to production enterprises.

At present, on the premise of the test for typical driving cycle of car, the major car companies or research institutions have developed their own car simulation programs, these simulation programs are used to simulate the power and economy performance of car, and to find the matching parameters of meeting the design requirements. Such as Austria AVL Company, it provides specially advice and technical support for the automotive power train design or manufacturing company, the CRUISE software developed by AVL Company can carry out some simulation research for power, fuel economy, emissions and braking performance of automotive, and so on[4]. The simulation platform has been successful build up a bridge between the manufacturers and parts suppliers, through its convenient general model element, intuitive data management system and modeling processes and software interfaces based on the engineering application. With the increasingly strict emission regulation and the fuel shortage, the regulations to emissions and economy of car in many countries are becoming more and more stringent; the cruise simulation software will also promote the use of more.

In this paper, a newly developed medium-sized sedan is studied with simulation technology by using Cruise software, its purpose is to find the optimum matching of driveline parameter from the different types of transmissions and final drives, to give full play the power performance of vehicle while taking into account the economy.

II   SELECTION OF ENGINE POWER

During the development process of vehicle, the design scheme of the power driveline is often a kind of combination matching among the mature engine on the market, transmission

© 2015. The authors - Published by Atlantis Press

and main reducer under the premise of design tasks identified. Based on the requirements of dynamic performance and fuel economy, the power plant parameters of vehicle were selected in accordance with the following conditions or design experience[5].

A. Power Selection Using the Expectant top Speed

The higher the top speed, the lager the engine power required, so the greater the back-up power of vehicle, the better necessarily the acceleration and grade ability. So, although the maximum speed is only one of the power index, but it also reflects the acceleration and grade ability of vehicles in essence. Therefore, it often preliminarily selects a proper engine power to ensure the expectant top speed in a design. Namely, the engine power should be approximately equal to or slightly greater than the sum of the driving resistance power when traveling at maximum speed.

$$P_e = \frac{1}{\eta_T}\left(\frac{Gf}{3600}u_{a\max} + \frac{C_D A}{76140}u_{a\max}^3\right) \tag{1}$$

In the above formula, Pe is the engine power, ηT is the transmission efficiency, G is the total vehicle weight, f is the rolling resistance coefficient, uamax is the maximum speed, CD is the air resistance coefficient, A is the car windward area.

B. Power Determination Using the Specific Power of Vehicle

Specific power of vehicle refers to the engine power of the unit total mass. The specific power of different countries and different types of automobiles (such as car and truck) are located in a statistical region, this area represents the overall level of vehicle design and manufacture. Therefore, in practical work, the engine power can be selected preliminarily according to the statistical data of specific power of some vehicles that have the same total mass or the same type.

$$P = \frac{1000 P_e}{m} = \frac{fg}{3.6\eta_T}u_{a\max} + \frac{C_D A}{76.14 m\eta_T}u_{a\max}^3 \tag{2}$$

In the above formula, P is the vehicle specific power, m is the vehicle total mass, g is the acceleration of gravity.

III   PARAMETER SELECTION OF TRANSMISSION DEVICE

A. Selection of the Gear Number and Speed Ratio of Each Gear

Theoretically, the more the number of transmission gear, the better the power and economy performance of the vehicle, but the gear number are subject to maximum, minimum speed ratio and the ratios between the every gear. If the ratios are too large, it will cause a difficult shift, so this ratio must not exceed 1.7-1.8 generally.

Usually, each gear ratio of transmission is distributed according to the geometric progression, that is, the ratio between the two adjacent gears is a constant value. But in practice, considering the economy requirement and the application frequency of each gear (car is often traveling with a higher range mainly), owing to utilization rate of each gear exists great difference, so the ratio interval between the adjacent higher gears should be smaller, each gear ratio of transmission is often allocated according to the following relationship.

$$\frac{i_{g1}}{i_{g2}} \ge \frac{i_{g2}}{i_{g3}} \ge \cdots \ge \frac{i_{gn-1}}{i_{gn}} \tag{3}$$

In the above formula, $i_{g1}, i_{g2}, ..., i_{gn}$ is gear ratio, respectively.

B. Selection of Gear ratio for Transmission

When determining the maximum gear ratio of transmission (first gear), the maximum grade ability should be considered firstly, if necessary, the checking computations should be done to attachment rate, the lowest stable speed is also the one of the factors that should be considered for off-road vehicle. Usually, the ratio for first gear of transmission should meet the following conditions.

$$i_{g1} \ge \frac{G\left(f\cos\alpha_{\max} + \sin\alpha_{\max}\right)r}{T_{tq\max}i_0\eta_T} \tag{4}$$

In the above formula, αmax is the maximum slope angle, r is the wheel rolling radiu, Ttqmax is the Maximum engine torque, i0 is the final drive ratio.

C. Selection of Gear Ratio for Final Drive

Under the condition of preliminary defining maximum engine power and its corresponding speed, the gear ratio of final drive chosen should be to ensure that the car has top speed as high as possible, which should meet the following conditions:

$$i_0 = 0.377\frac{r \cdot n_p}{i_p \cdot u_{a\max}} \tag{5}$$

In the formula, np is the engine speed corresponding to peak-power; ip is the top gear ratio of transmission.

IV   MODELING AND SIMULATION ANALYSIS OF VEHICLE PERFORMANCE USING CRUISE

This research object is a pre-precursor midsize sedan, the type of engine and vehicle-related parameters have been determined, See Table 1.

In order to obtain a reasonable power and economy, the two kinds of 5-speed manual transmission (Each speed ratio is different, see Table 2) and the three kinds of main reducer (Ratio i0 is 2.85, 3 and 3.75, respectively) was selected to match for this engine, the six kinds of driveline combinations was formed eventually, and each gear ratio and the final ratio was calculated and determined in accordance with the aforementioned theory. The following main work of this study is to build such a vehicle model by using the Cruise software, and select the optimum match from these six kinds of combinations through simulation and analysis subsequently.

Figure 1 is the vehicle model established by using Cruise software finally, the modeling process can be referred in the related literature[6]. Through setting computing tasks and matrix simulation procedure, each simulation results of the performance can be obtained and shown in Table 3. Figure 2 and Figure 3 are two kinds of instance of the performance curves by simulating.

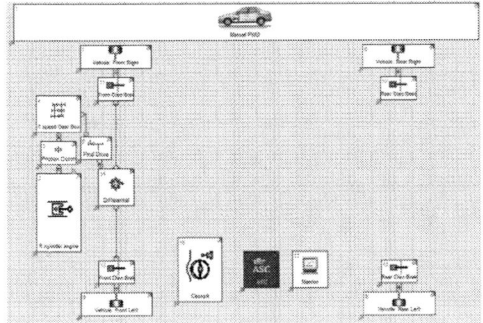

FIGURE I. MANUAL FRONT-WHEEL DRİVE MİDSİZE SEDAN MODEL BY USİNG CRUİSE.

As can be seen from table 3, if we only pursue the power performance of vehicle, then the driveline matching of A-3.75 and B-3.75 are relatively satisfactory. If we hope the maximum speed, the ability of overtaking acceleration (top gear, from 60km/h to 140km/h) and climbing grade of car are all stronger, then selecting the A-3.75 of combination; if we hope the ability of starting and acceleration (from 0 to top speed), overtaking and climbing are good, then selecting the B-3.75 of combination. If only from the perspective of economic terms, there is almost no difference in fuel consumption of constant speed driving at three speed, if referring fuel consumption of driving cycle further, then the B-2.85 of combination is the most ideal. If both the power and

economy must be taken into account, that is in the premise of ensuring power, while having the better economy as far as possible, then from the A-3.75 and B-3.75 two kinds of matching, only choosing the combination of B-3.75. Through the above analysis we can see, In pursuit of the goal of different design, the simulation calculation using CRUISE can provide technical and data support for designers according to the different design goal. If recombining the professional knowledge to analyze the forming factors of all data, it will deepen the understanding to theory and enhance the design experience in engineering.

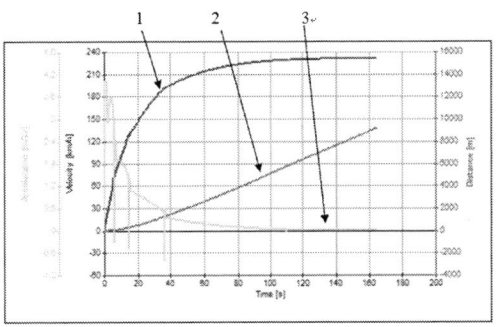

(1: Speed-time; 2: Displacement – time; 3: Acceleration - time)

FIGURE II. ACCELERATİON CURVE OF CONTİNUOUS SHİFT FROM STANDİNG START.

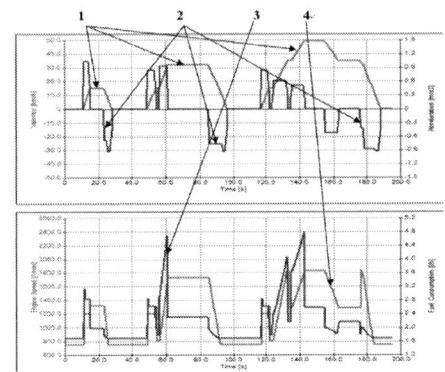

(1: Speed-time; 2: Acceleration - time 3: Fuel consumption - time; 4: Engine rotational speed - time)

FIGURE III. FUEL CONSUMPTİON CURVE OF DRİVİNG CYCLE.

TABLEI. PARAMETERS OF VEHICLE AND ENGINE.

Item	Vehicle parameter	Item	Engine parameters
Total mass / Curb Weight, (kg)	1930/1450	Engine Type	Water-cooled 4-stroke naturally aspirate
Length×width×height, (mm)	4650×1750×1505	Engine displacement, (L)	2.478
Wheelbase, (mm)	2650	Number of cylinders	6
Tread (front/rear), (mm)	1627/1618	Maximum power, (kW/r/min)	127/6000
Drive form	Front-wheel drive	Maximum torque, (N.m)	175

TABLEII. WO KINDS OF GEAR RATIO SCHEMES OF MANUAL TRANSMISSION.

Shift	1st gear	2nd gear	3rd gear	4th gear	5th gear
Scheme A	3.62	2.22	1.51	1.08	0.85
Scheme B	3.46	1.94	1.29	0.97	0.8

TABLEIII. SIMULATION RESULTS OF VEHICLE PERFORMANCE BY USING CRUISE.

Index Scheme	Top speed, (km/h)	Grade-ability, (%)	Acceleration time, (s)		Fuel consumption with constant speed, (L/100km)			Cycling fuel consumption, (L/100km)
			$0 \rightarrow v_{max}$	$60 \rightarrow 140$km/h	60km/h	90 km/h	120 km/h	
A-2.85	229.28	46	90	50	4.2	4.2	5.0	12.97
A-3	226.84	49	80	45	4.2	4.2	5.2	13.12
A-3.75	231.50	55	110	38	4.3	4.2	5.5	14.22
B-2.85	231.53	43	105	57	4.1	4.2	5.0	12.04
B-3	232.50	46	103	51	4.1	4.1	5.0	12.55
B-3.75	228.31	52	70	43	4.2	4.2	5.1	13.57

V CONCLUSION

In this paper, the vehicle modeling and performance simulation were applied to match driveline system of a mid-size sedan by using AVL-Cruise software, the optimum matching scheme of this car power-train was obtained by analyzing the calculation results, this scheme can further improve the economy of the vehicle on the premise of ensuring good dynamics. Research shows that the optimum matching of driveline and engine is able to quickly and easily implement by using simulation function of Cruise software, this technology can predict vehicle performance, shorten the development cycle, and reduce R & D costs in early development.

ACKNOWLEDGMENT

The authors gratefully acknowledge the National Natural Science Foundation of China (No.51105131) and the Key Project of Tianjin Research Program of Application Foundation and Advanced Technology (No.12JCZDJC34500) for financial support of this research work.

REFERENCE

[1] LIU Wei-xin, GE Ping, LI Wei. Study of optimal matching between automobile engine and transmission parameters. Automotive Engineering, 1991, 13(2): 65–72.

[2] YANG Lian-sheng. The optimal matching between internal combustion engine performance and transmission. Beijing, Academic Periodical Press, 1988: 12–26.

[3] XIAO Ming-wei, YANG Jing. Study on the Matching between 495QME engine and CDK6710 carriage transmission system. Hunan University. Institute of Machinery and Automotive Engineering, 2006: 50–55.

[4] LIU Zhen-jun, ZHAO Hai-feng, QIN Da-tong. Simulation and analysis of vehicle powertrain based on CRUISE. Journal of Chongqing University (Natural Science), 2005, 28(11): 8–11.

[5] LIU Qing-quan. Design & matching for city vehicle and engine. Jilin University. Institute of Automotive Engineering, 2004: 35–37.

[6] Shen Ailing, Fu Jun, Zhang Yanfa. Matching simulation and driveline optimization for CA7204 passenger car. Journal of Central South University (Natural Science), 2011, 42 (3): 677-681.

A Kind of Research on Regenerative Braking Algorithm of Hybrid Electric Vehicle

Z.L. Liao
Department of Control Engineering
Academy of Armored Force Engineering
Beijing, China

P. Fan
Bureau of Military Agent of Transport Military Agent's
Section of Nanjing Area
Nanjing, China

Q. Gao
Department of Control Engineering
Academy of Armored Force Engineering
Beijing, China

R.F. Hu
PLA Agency in No.201
Beijing, China

Abstract- **Discussed an energy recovered method of the regenerative braking system of a Parallel Hybrid Electric Vehicles (PHEV), and promoted the performance, efficiency, and reliability in a minimal cost. Discussed the generative braking algorithm based on the simulation of PHEV in MATLAB.**

Keywords-generative braking; hybrid electric; controlling algorithm

I. INTRODUCTION

The tire is driven by a vehicle electric system, which is consists of a lot of nonlinear system, such as power plant, drive system, differential system and shaft system. In addition, some other attachments such as steering system and braking system are attached to the power plant. The vehicle electric system, which is a nonlinear dynamic composite system, is formed by electrical, mechanical, chemical, and thermal equipment. It is mainly used to supply energy.

Hybrid Electric Vehicle (HEV) combined Electric Vehicle's (EV) electric system and traditional electric system. The PHEV, which included electric driven system and traditional electric system, supplied energy to the tires. Through the auxiliary power unit, it can be combined with traditional electric vehicle driven system. The HEV can prolong performance envelope and the fuel economy. Meanwhile, it can also reduce the exhaust gas emission.

Most HEVs used the traditional braking system and RBS at the same time. Traditional braking system included the friction brake drum or disc brake, and driven by the hydraulic system. RBS used the motor to supply negative torque to stop the vehicle and change the kinetic energy into electricity energy to charge the battery. In the braking time of EV or HEV, the kinetic energy loss can be recovered effectively by controlling the motor in a power generation working condition. The recovery energy can be saved in storage devices.

II. GENERATIVE BRAKING CONTROLLING ALGORITHM

The generative braking controlling algorithm adopted MATLAB to build the models and coding in auto. The code will be downloaded into the controller and test tools. Not only the hydraulic braking torque, but also the parallel braking system produced a generative braking torque to the tires. The braking order delivered by motor controller which based on coordinated controller doesn't included hydraulic brake. Hence, it can recycle the vehicle kinetic in maximum. The hydraulic braking torque was delivered by the driver when he steps on the brake to give a electric signal. The generative braking order was delivered by braking controller and coordinated controller. Meanwhile, the electric braking torque can strength the hydraulic braking torque and the parallel brake was decided by total braking demand function.

In the parallel braking system, the central controller can calculated the torque which was plus by electric braking torque and hydraulic braking torque based on a given function. The fellow equation describes the relationship between hydraulic braking torque and electric braking torque.

$$T_e = \left[\frac{(g's \cdot R_w \cdot W_v) - (2 \cdot BF_f \cdot P_f) - (2 \cdot BF_r \cdot P_r)}{g_{4\times4} \cdot g_{axle}} \right] \quad (1)$$

The braking torque of tire was decided by braking controller as fellow.

$$P_f = P_{mc}, P_r = P_{mc} \qquad for \quad P_{mc} \leq X \quad (2)$$

$$P_r = X + \delta(P_{mc} - X) \quad for \quad P_{mc} > X \quad (3)$$

The electric braking torque which is plus into hydraulic braking torque was decided by motor torque features, surface state of road and tires, brake pedal stroke and static braking force.

The static braking force was decided by the picture of static braking force. The picture including the brake shutting features in different roads, braking ratio in front and back, and the vehicle's deceleration state. The braking shutting curves is the maximum braking torque in different roads of the braking torque as the Figure1 shows.

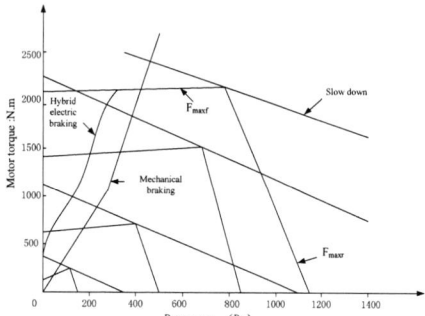

FIGURE I. STATIC BRAKING FORCE CURVE.

In Figure1, the vertical coordinate means the front braking torque, in the other hand, the abscissa means the back braking torque. And their intercepts were given as fellow.

$$F_{\max f} = \frac{\mu_p(W_v B / L)}{1 - \mu_p H / L} \tag{4}$$

$$F_{\max r} = \frac{\mu_p(W_v A / L)}{1 + \mu_p H / L} \tag{5}$$

$$slope_{f\max} = \frac{\mu_p H / L}{1 - \mu_p H / L} \tag{6}$$

$$slope_{r\max} = \frac{-\mu_p H / L}{1 + \mu_p H / L} \tag{7}$$

The braking torque was influenced by the braking force, and the value was given as fellow.

$$F_{rear} = \frac{2 \cdot BF_r \cdot P_r}{R_w} \tag{8}$$

$$F_{front} = \frac{2 \cdot BF_f \cdot P_f}{R_w} \tag{9}$$

The total braking force equals the front braking force plus the back. In addition, its value equals the vehicle's quality multiply the vehicle's deceleration. The equation was given.

$$F_r = F_{front} + F_{rear} = \left(\frac{W_v}{g}\right) a_x \tag{10}$$

If the driver given a deceleration about 0.6g on a road with a road condition about 0.8 μ, the front and back braking force would meet the braking demand in maximum based on a steady driving, only if the braking force hold on the triangle deceleration area and correspond the maximum braking torque on the road about 0.8 μ.

The optimal generative braking control strategy can choose a better ratio to fit in all the design aims, such as the brake stopping distance, the maximum braking force and so on. It can make the total brake force gets the max value on the demanding frontier and get to use the generative brake as well.

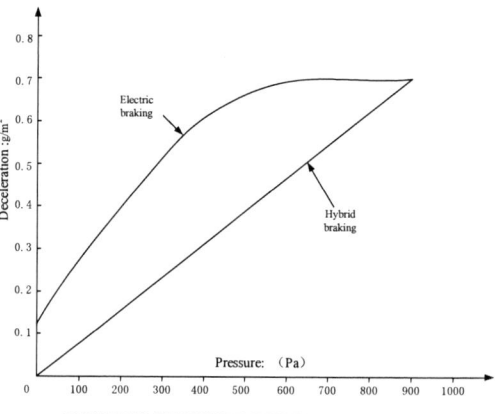

FIGURE II. PRESSURE-DECELERATION.

The picture about pressure and deceleration decided the relationship between the electric braking torque which is plus to the mechanical and the road. The mechanical curve can be the function of pressure-brake curve. It was shown as the fellow equation.

$$decel = \frac{2 \cdot BF_f \cdot P_f + 2 \cdot BF_r \cdot P_r}{W_v \cdot R_w} \tag{11}$$

In order to recover the energy, the coefficient of road adhesion must get close to 0.7 μ. Once the coefficient of road adhesion get higher than 0.8 μ, the braking force will get to the peak value too fast to recover the energy. Usually, the drivers will give a braking force about 0.2g when they meet a red light. At this moment, the braking controller has distributed the electric braking torque in maximum limitation to produce an energy feedback.

The braking force inside the tire consists of motor braking torque and hydraulic braking torque, and the relationship between electric braking torque and tire braking torque. It was shown as Figure3.

94

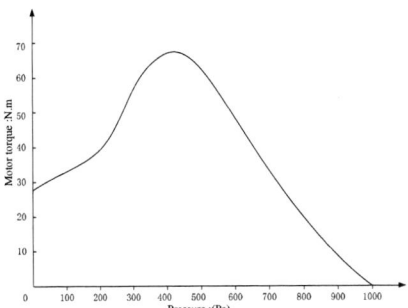

FIGURE III. GENERATIVE BRAKING TORQUE.

The value of the motor's positive and negative torque was given by the motor controller. The given value of the motor torque was shown as fellow.

$$T_{available} = \frac{P_{rated}}{\omega_m} \cdot 5252, \quad \omega_m > \omega_b \quad (12)$$

$$T_{available} = T_{rated}, \quad \omega_m \leq \omega_b \quad (13)$$

The given value of the useful feedback braking torque was given by the fellow equation.

$$T_{regenavail} = \frac{P_{rated} \cdot 5252}{\omega_m} - T_{compression}, \vdots \quad \omega_m > \omega_b \quad (14)$$

$$T_{regenavail} = T_{rated} - T_{compression}, \vdots \quad \omega_m \leq \omega_b \quad (15)$$

The motor cannot supply enough negative torque to recover the energy when the vehicle working on a low speed. It will be stopped only by electric brake when the electric braking torque can fit in and the energy was recovered in maximum limitation. However, when the energy storing device was full, the braking torque was supplied by mechanical brake.

$$T_{compression} = \frac{g's \cdot R_w \cdot W_v}{g_{4 \times 4} \cdot g_{axle}} \quad (16)$$

Motor's driven torque is a nonlinear function of the motor's rotate speed. And the loss torque produced by inverter was also a function of the motor's rotate speed. The function was shown as the equation.

$$T_m = \frac{P_{rated} \cdot 5252}{\omega_m} \quad \omega_m > \omega_b \quad (17)$$

$$T_m = T_{rated} \quad \omega_m \leq \omega_b \quad (18)$$

In these equations, there is another equation.

$$T_e - T_m \cdot 1.3558 = J_m \omega_r \quad (19)$$

The load current of inverter is a function of both the motor's rotate speed and the motor torque and the voltage of bus line. It can be shown as fellow.

$$I_{load} = \frac{T_m \cdot \omega_r}{e_{tb}} \cdot \frac{1.3558}{\eta} \quad (20)$$

The equation was shown as fellow equationwhen the motor working on the generative braking condition.

$$I_{load} = \frac{T_m \cdot \omega_r}{e_{tb}} \cdot \eta \cdot 1.3558 \quad (21)$$

The controlling algorithm needs the fellow inputs. Such as braking switch signal, braking pedal location, motor's rotate speed, the estimating value of motor's torque, the logic value of choosing mode and the value of Iq. The output signals including the demanding ratio value of hydraulic braking torque and the demanding value of Iq in electric brake.

The drivers have two choices. One is the total electric drive and another is the hybrid drive. The controlling strategy decided the working mode, which including driven by engine, driven by motor or driven by both the two, when the vehicle working on hybrid mode. The motor will work alone when the vehicle brake in the generative mode. The engine will work alone when the traditional hydraulic braking. When the vehicle brakes in hybrid driven mode, the engine's velocity slows down to the idling. The driver should shift the gear to keep the power transmission. The linear brake should be used to avoid discontinuous brake when the generative brake changed to hydraulic brake.

The accelerated position of the pedal will be auto detected by sensor, which instead of the accelerated signal of the driver. If the position of pedal performed abnormal, or if the braking signal opened, the vehicle would be stopped by the empty signal of acceleration. This order was delivered through a NOT gate signal. This signal and the braking signal became a braking logic through an OR gate. This signal is a high level when the applied brake working.

III. TEST DATA AND SIMULATION RESULT

The initial value of the test is set as 30km/h. The result is gotten through modeling and simulating by Matlab. The parallel hybrid electric vehicle brakes when the vehicle is driven by electromotor, driven by engine and electric or driven by engine.

The Figure4 indicates that when electric brake works alone, the braking torque is supplied by electromotor. When the vehicle is getting close to stop, the braking torque is supplied by mechanical braking. The Figure5 indicates that when electric and engine work together, the braking torque is supplied by electric braking and mechanical braking. The Figure6 indicates that when the engine works alone, the braking torque is supplied by mechanical braking.

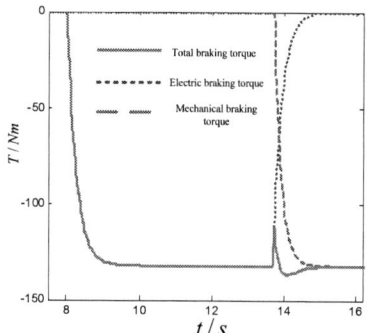

FIGURE IV. DRIVEN BY ELECTRIC.

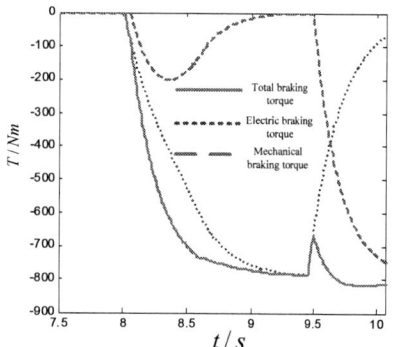

FIGURE V. DRIVEN BY ENGINE AND ELECTRIC.

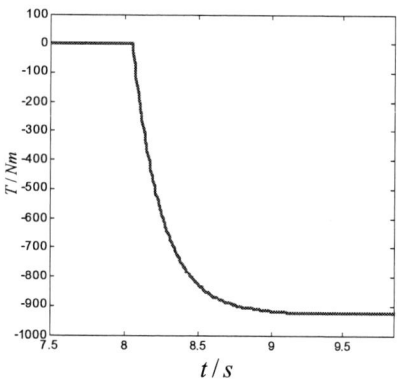

FIGURE VI. DRIVEN BY ENGINE.

IV. SUMMARY

A new improved scheme on parallel hybrid electric vehicle is discussed in this article. It can improve the fuel efficiency of the engine, and enhance the energy recovery at the greatest extent when braking. The regenerative braking effectively improves the braking through reasonable brake control strategy.

REFERENCES

[1] Xue-guo AN, Nian-yu Li.The Forward Appliance and Challenge of Military Electric/Hybrid Electric Vehicle:1st[J].Foreign Tanks. 2008(7):45-49

[2] Ke-mao Zang. Study on the All-electric Technology of Land Warfare Platform[J]. Journal ofAcademy of Armored Force Engineering. 2011, 25(1):1-7

[3] Loic.B.D, Alain.B, Olivier.P, Marie.P. Simulation model of military HEV with a highly redundant architecture [J].IEEE Trans on vehicular technology. 2010, 59(6):2654-2663.

[4] Ming-hui Hu. Study on Energy Management Strategy for Mild Hybrid Electrical b Vehicle with CVT.Chongqing University,2007

[5] Donghyun Kim, Sungho Hwang, Hyunsoo Kim. Vehicle Stability Enhancement of Four-Wheel-Drive Hybrid Electric Vehicle Using Rear Motor Control[J]. IEEE Transactions on Vehicular Technology. 2008, 57(2): 727-735.

International Conference on Power Electronics and Energy Engineering (PEEE 2015)

Study on Modeling Method of the Precision Machined Surface Geometry form Error Based on Bi-Cubic B-Spline

Z.Q. Zhang, X. Jin
School of Mechanical Engineering
Beijing Institute of Technology
Beijing, China

X. Jin
School of Mechanical Engineering
Beijing Institute of Technology
Beijing, China

Z.J. Zhang
School of Mechanical Engineering
Beijing Institute of Technology
Beijing, China

Abstract—**For precision mechanical system, the different 3d spatial distribution of form error causes different assembly contact state. Since 3d spatial distribution of form error is not taken account in modeling and evaluation method of geometric form error and it is hard to quantitatively analyze the relationship between form error and assembly accuracy of precision mechanical system, a modeling method of machined surface form error based on bi-cubic B-spline surface reconstruction is proposed. Firstly, a reconstruction algorithm of CMM date based on bi-cubic B-spline is proposed, so the mathematical model of the surface form error is established. Then, the data transaction method between Matlab and Pro/E is proposed, and the 3D solid model of parts is established, which has the surface form error. Finally, the validity and reliability of the proposed method is validated by the establishment of the typical mating surfaces form error model.**

Keywords-form errors; assembly; bi-cubic b-spline; surface reconstruction

I. INTRODUCTION

As everyone knows, the geometrical error of parts surface has been a key factor of assembly performance analysis for precision mechanical system. Because mechanical part surfaces cannot be smooth planes actually, the geometrical error has quite an effect on assembly performance. For precision mechanical system, machining errors include dimension error, geometric form error and position error, usually affect the assembly accuracy fundamentally. High precision assemblies cannot be analyzed with the assumption that form errors are negligible. The dimension error and position error are scalars, which can be relatively simply considered in assembly. However, the influencing mechanism of the form error on the assembly accuracy is more complex due to the complexity of form error distribution. The form error and assembly force result in non-uniform contact state between mating surfaces, and thus non-uniform stress in components, which will degrade the assembly accuracy. In order to make quantitative analysis on the

surface geometric error and analyze the effective rules of the geometrical error on assembly accuracy, it is necessary to establish an accurate 3d model of the geometrical error.

In the last 30 years, various methods have been developed to characterize the machined surfaces. A high order polynomial function to model form tolerance for nominal planner part feature by Turner [1]. The Fourier series is the first used in the decomposition method in order to define periods on circular by Goto and Iizuka [2].Henke et al. [3] worked on the geometrical variations of cylindrical surfaces and matched the Chebyshev polynomials with Fourier series to evaluate systematic manufacturing defects. Two dimensional Fourier is used by Capello and Semeraro [4] to define the form parameters of rectangular shapes. Serge Samper [5] proposed a new way to define form error parameters based on the eigen-shapes of natural vibrations of surfaces. Huang and Ceglarek [6] used the discrete cosine transform (DCT) to describe rectangular surfaces. Fractal theory and wavelet analysis theory is used by Srinivasan [7] to analyze the form tolerances, and a two-dimensional model of the shape error is established

The above research is limited to decomposition and reconstitution of the machined form error and the errors are too big. Furthermore, it is difficult to establish the accurate 3D model of surface form error. In order to solve these problems, this paper puts forward a new modeling method of the machined surface geometry form error for precision assembly, the three-dimensional distribution of form error can be described more accurately, assembly technology for precision mechanical system can be effectively designed, the assembly precision and precision stability can be improved.

II. METHODOLOGY

A. Modeling Method of Form Error Based on Bi-Cubic B-Spline Surface Reconstruction

The surface geometric errors of parts are mainly caused by processing technology because of force deformation and

© 2015. The authors - Published by Atlantis Press

thermal strain. The surface is consistent with the physical characteristics of shape modification of B-spline Surface. Spatial curved surface is described with B-spline Surface, which has been widely applied in practical engineering. Because of the local support property and C^2 continuous of B-spline Surface, it has been widely applied in CAD/CAE. Surface Reconstruction Based on bi-cubic B-spline surface of Points Cloud Data from CMM is proposed in this paper and solid model of the part is established using the three-dimension CAD software Pro/E. Firstly an algorithm for inverse calculation of cubic B-spline is introduced, and based on this evidence, this paper proposes an algorithm based on bi-cubic B-spline surface interpolation.

B. Reverse Calculating the Control Points of Cubic B-Spline

In order to make cubic B-spline curve through a set of data points ($P_i, i = 0, 1, \cdots, n$), the endpoint of B spline curve is made respectively with the first and last data points consistent in the process of inverse calculation. Interpolation equation of cubic B-spline is represented as Eq. (1).

$$ P_i(u) = \sum_{j=i-3}^{i} d_j N_{j,3}(u), u \in [u_i, u_{i+1}] \subset [u_3, u_{n+3}] \tag{1} $$

where $d_j \left(j = 0, 1, \cdots, n+2 \right)$ and u are control point and knots vector of B-spline, respectively. In the paper, knots vector U are calculated using the algorithm of Riesenfeld (1975).

$$ U = \left[0, 0, 0, 0, \frac{l_1 + l_2}{L}, \frac{l_1 + l_2 + l_3}{L}, \cdots, \frac{\sum_{i=1}^{n-2} l_j}{L}, 1, 1, 1, 1 \right] \tag{2} $$

U is calculated as Eq. (3).

$$ \begin{cases} l_i = \left| P_i - P_{i-1} \right|, L = \sum_{i=1}^{n} l_i, i = 1, 2, \cdots, n \\ u_0 = u_1 = u_2 = u_3 = 0 \\ u_i = \dfrac{\sum_{j=1}^{n-2} l_j}{L}, i = 4, 5, \cdots, n-1 \\ u_n = u_n = u_n = u_n = 1 \end{cases} \tag{3} $$

where U is rewritten as $U = \left[u_0, u_1, \cdots u_i, u_{n+3} \right]$.

Characteristic-points and the corresponding knots are put it into a mathematical formula (1), which can be expressed in matrix form:

$$ \begin{bmatrix} 1 & & & & \\ a_2 & b_2 & c_2 & & \\ & \ddots & \ddots & \ddots & \\ & & a_n & b_n & c_n \\ & & & & 1 \end{bmatrix} \begin{bmatrix} d_1 \\ d_2 \\ \vdots \\ d_n \\ d_{n+1} \end{bmatrix} = \begin{bmatrix} e_1 \\ e_2 \\ \vdots \\ e_n \\ e_{n+1} \end{bmatrix} \tag{4} $$

In matrix, a_i, b_i, c_i and e_i are calculated as follows

$$ a_i = \frac{\left(\Delta_{i+2} \right)^2}{\Delta_i + \Delta_{i+1} + \Delta_{i+2}} $$

$$ b_i = \frac{\Delta_{i+2} \left(\Delta_i + \Delta_{i+1} \right)}{\Delta_i + \Delta_{i+1} + \Delta_{i+2}} + \frac{\Delta_{i+1} \left(\Delta_{i+2} + \Delta_{i+3} \right)}{\Delta_{i+1} + \Delta_{i+2} + \Delta_{i+3}} $$

$$ c_i = \frac{\left(\Delta_{i+1} \right)^2}{\Delta_i + \Delta_{i+1} + \Delta_{i+2}}, i = 1, 2, \cdots, n $$

$$ e_i = \left(\Delta_{i+1} + \Delta_{i+2} \right) p_{i-1}, i = 2, 3, \cdots, n $$

$$ e_1 = p_0 + \frac{\Delta_3}{3} \dot{p}_0, e_{n+1} = p_n - \frac{\Delta_{n+2}}{3} \dot{p}_{n+1} $$

Boundary condition of B-spline is represented as Eq. (5).

$$ \begin{cases} d_0 = p_0 \\ d_{n+2} = p_n \\ \dot{p}_0 = \dot{p}(u_3) = \dfrac{3}{\Delta_3}(d_1 - d_0) \\ \dot{p}_n = \dot{p}(u_{n+3}) = \dfrac{3}{\Delta_{n+2}}(d_{n+2} - d_{n+1}) \end{cases} \tag{5} $$

where Δ_i is forward difference of u and $\Delta_i = u_{i+1} - u_i$. There are n+3 points obtained by resolving equation group (4), and cubic B-spline can be obtained through points $p_i \left(i = 0, 1, \cdots, n \right)$.

C. Surface Reconstruction Based on Bi-Cubic B-Spline

B-spline surface is of great importance in computer-aided design. Through the analysis of characteristics of the parts surface, the point clouds as characteristic points of B-spline surface are obtained using CMM, which must be in rectangular topology array. The reconstruction of the surface is that control points calculated based on characteristic points of the surface, and the B-Spline surface equation is calculated. Characteristic points of the surface is represented

98

as $P_{i,j}\left(i=0,1,\cdots,m;j=0,1,\cdots,n\right)$ and surface equation based the bi-cubic B-spline can be represented as follows

$$P_{i,j}(u,v)=\sum_{i=0}^{m+2}\sum_{j=0}^{n+2}d_{i,j}N_{i,3}(u)N_j(v)$$

$$\Rightarrow P_{i,j}(u,v)=\sum_{i=0}^{m+2}\left(\sum_{j=0}^{n+2}d_{i,j}N_j(v)\right)N_{i,3}(u) \tag{6}$$

Furthermore, $P_{i,j}(u,v)$ is rewritten as follows

$$P_{i,j}(u,v)=\sum_{i=0}^{m+2}C_i(v)N_{i,k}(u) \tag{7}$$

$$C_i(v)=\sum_{j=0}^{n+2}d_{i,j}N_{j,3}(v) \tag{8}$$

Reverse calculation method of surface control points based on bi-cubic B-spline has two procedures.

1) Fixed V value of these curves, and sectional curves are calculated using B-spline curve interpolation algorithm in U, and control points $C_i(v),i=0,1,\cdots,m+2$ are calculated;

2) Fixed U value of these curves, and sectional curves are calculated using B-spline curve interpolation algorithm in V, and control points is represented as

$$d_{i,j},i=0,1,\cdots,m+2;j=0,1,\cdots,n+2$$ are calculated.

Bi-cubic B-spline surface can be confirmed when control points are calculated; the mathematic model of form error is described by the bi-cubic B-spline surface.

D. 3d Cad Modeling Method of Form Error Based on Matlab and Pro/E

This paper proposes a modeling method of the geometry form error based on bi-cubic B-spline, and rigorous mathematical model of form error can be set up. Because of representation of three-dimensional surface based on point matrix in Matlab, the 3D solid model of form error using for assembly simulation calculation cannot be established. Data transmission method between Matlab and CAD system is proposed and it solves the problems existing in the data transmission between mathematical model point matrix and CAD system.

Firstly, coordinate measuring machine (CMM) was used to measure geometrical surface of parts to produce point clouds. Mathematical model of form error is established using the algorithm based on bi-cubic B-spline, and the reconstruction of the surface is completed. Coordinates of points in u directional

curves and v directional curves are stored in IBL document consistent with the Pro/E format. IBL document is modified, and then imported into Pro/E. Finally, the 3D solid model of form error in Pro/E is generated. In order to satisfy the requirements of Pro/E data format, "Open Arclength" and "Begin Section" are inserted beginning of the file. "Begin Curve" is inserted beginning of the each curve, and the three-dimensional coordinate of the each point is separated by space. IBL file format rewritten as follows:

Open（or Closed）Arclength（or Pointlength）

（Open Curve: Open，or Closed；different amounts in curve: Arclength，or Pointlength）

Begin Section！1   Section 1

Begin Curve！1   the first curves contained in Section 1

1 x1 y1 z1   the first point coordinate in the first curves contained in Section 1

2 x2 y2 z2

……

Begin Curve！2   the second curves contained in Section 1

……

Begin Section！2   Section 2

Begin Curve！1   the first curves contained in Section 2

……

III. CASE

To verifies the efficiency of the proposed method, four plate parts are machined on X5050 numerical control machine tool, whose size is 115cm×60m×15m, material is 45 steel, using processing techniques first rough- milling and then finish-milling. After machined, machined surfaces are measured using PMM12106G CMM, whose uncertainty measurement is (0.6 ± 1/600) um, and it is much lower than the machined error. The surfaces are measured at 5mm sampling spaces, and there are 23×11 measurement points obtained from CMM on each plate parts. Form error models based on bi-cubic B-spline surface reconstruction are showed in Figures 1.

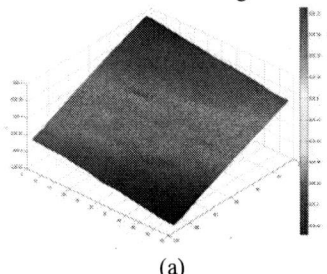

(a)

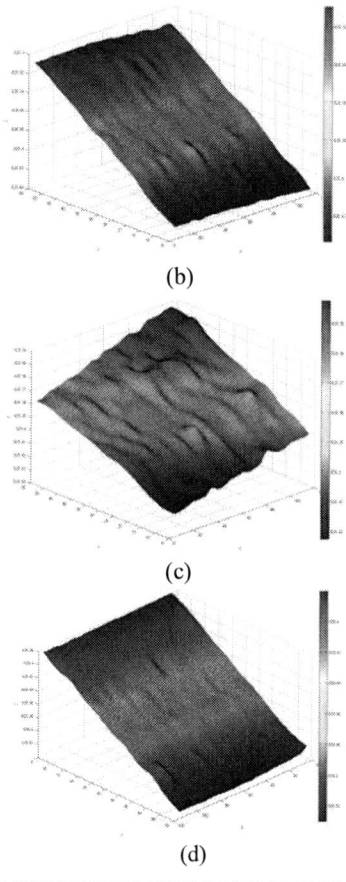

(b)

(c)

(d)

FIGURE I. FORM ERROR MODEL BASED 0N BI-CUBIC B-SPLINE
SURFACE RECONSTRUCTION.

The 3D form errors of a common flat and cylindrical surface are established. For the purpose of showing the surfaces more clearly, the surface of form error is enlarged by 4 times in Figure 2(a), and 100 times in Figures 2 (b) and (c).

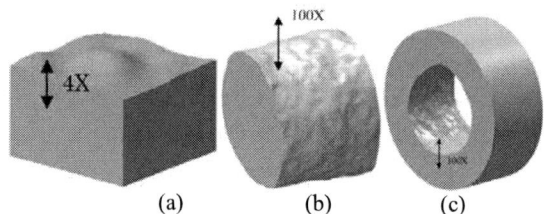

FIGURE II. CAD MODEL WITH FORM ERROR

IV. SUMMARY

The influencing mechanism of the machined error on the assembly performance is mainly reflected in contact state and assembly accuracy between mating surfaces, and different contact state would lead to different assembly error and stress distributions. The form error and assembly force result in non-uniform contact state between mating surfaces, and thus non-uniform stress in components, which will degrade the

assembly accuracy. At present, form errors of mating surfaces have significant impact on assembly accuracy, which gradually attracts the academic attentions, but assembly simulation and analysis of precision mechanical system takes into Account Plane Form Error but neglect the importance of form error and 3D distribution of form error. In order to make quantitative analysis on the surface geometric error and analyze form error and the Three dimensional distribution effective rules of the geometrical error on assembly accuracy, this paper proposes a modeling method of machined surface form error based on bi-cubic B-spline surface reconstruction, study on Data transmission method between Matlab and Pro/E, CAD solid model of parts with form error is established, it can provide theoretical foundation for the further research of effective rules of the geometrical error on assembly accuracy.

ACKNOWLEDGEMENT

This project is supported by the National Natural Science Foundation of China (Grant No. 51375054) and the Electrical-Optical Micro-Nano Manufacturing Program of Beijing Emphasis Course.

REFERENCES

[1] Turner JU and Woany MJ. Tolerances in Computer-Aided Geometric Design. The Visual Computer 1987;3:214-216.

[2] Goto M and Iizuka K. A method for evaluating form errors of cylindrical parts. Japan Soc Prec Eng 1975;41(5):.

[3] Henke RP and Summerhays KD et al. Methods for evaluation of systematic geometric deviations in machined parts and their relationship to process variables. Precision Eng 1999;23:273-292.

[4] Capello E, Semeraro A. Harmonic fitting approach for plane geometry measurements. Int J Adv Manufacturing Technol 2000;16:250–258.

[5] Samper S. and Formosa F. Form Defects Tolerancing byNatural Modes Analysis. Transactions of the ASME 2007; 7:44-51.

[6] Huang WZ and Ceglarek D. Mode-based Decomposition of Part Form Error by Discrete-Cosine-Transform with Implementation to Assembly and Stamping System with Compliant Parts. Annals of the CIRP 2002;51 (1):21-26.

[7] R.S.Srinivasan,K.L.Wood. Geometric tolerancing in mechanical design using fractal-based parameters. Journal of Mechanical Design, 1995, 117: 203-206.

[8] Grandjean J, Ledoux Y, Samper S, et al. Form Errors Impact in a Rotating Plane Surface Assembly[J]. Procedia CIRP, 2013, 10: 178-185.)

The Suppression of Narrow band Jammers Algorithm and the Implementation of Beidou B3 Signal

C.M. Li
School of information science and engineering
Hebei University of science and technology
Shijiazhuang, Hebei, China

L. Huang
School of information science and engineering
Hebei University of science and technology
Shijiazhuang, Hebei, China

X.J. Wang
School of information science and engineering
Hebei University of science and technology
Shijiazhuang, Hebei, China

Abstract—**This paper investigates the use of the transform domain adaptive filtering algorithm to suppress norrowband jammers in Beidou B3 signal, which reduces the amount of calculation and improves Convergence properties compared to the time domain least mean square algorithm. The theorem and derivation process of the timeand frequency domain adaptive filters and the convergence property is discussed. Finally, analysis the frequency spectrum properties of transform domain adaptive filtering before and after suppressing of the narrowband jammers and the convergence of weights and mean square error via matlab simulation, And compare to the time domain least mean square which show that the frequency domain adaptive filtering can effectively suppress the single-tone jammer.**

Keywords-the least mean square; tap weight convergence; adaptive filtering; single-tone jammer; matlab simulation

I. INTRODUCTION

Due to the satellite signal is weak and susceptible to interference, it is necessary to employ some type of jammer suppression technique. We proposes the transform domain adaptive filtering to suppress narrow-band jammers because of the large amount of calculation of the time domain least mean square(LMS) algorithm, which mainly studying the single-tone jammers of the narrow-band jammers. According to the characteristics of the satellite signals and jammer signals, and the energy of the jammer signals concentrated in one or some frequency bits through the Fast Fourier transform from the time domain to the frequency domain. But the useful signal has a low power spectral density in all frequency bins. The adaptive filtering should seek amplitude peaks which are indicative of a high-power jammer and suppressing jammer signals and retain the useful signal. Finally, the convergence properties of the frequency domain adaptive filter are illustrated via matlab simulation, which based on the theoretical derivation of the algorithm.

II. THE SIGNAL MODEL AND THE TIME DOMAIN LMS ADAPTIVE ALGORITHM

A. The Signal Model

The received signal ,sampled at the chip rate, can be represented as[1]:

$$x(k) = s(k) + j(k) + n(k) \tag{1}$$

Where, $s(k)$ is the B3 signal of a satellite, $j(k)$ denotes the single-tone jammers, $n(k)$ is assumed to be a sample function of a white Gaussian noise.

B. Derivation of the Time Domain Lms Adaptive Filtering Algorithm

Transform domain adaptive filtering was derived based on the time domain LMS algorithm, so it is necessary to understand the time domain adaptive filtering first. The process of the LMS algorithm is to continuously adjust these weights[2-3]:

Input signals: $x(k) = [x_1(k) \quad x_2(k) \quad ... \quad x_n(k)]^T$ and Weight signals: $W(k) = [w_1(k) \quad w_2(k) \quad ... \quad w_n(k)]^T$

Output signal:
$$y(k) = \sum_{i=1}^{n} w_i(k)x_i(k) = x^T(k)W(k) = W^T(k)x(k)$$

The error signal: $\varepsilon(k) = d(k) - y(k)$, Where, $d(k)$ represents the desired response

Performance function of mean square error:
$$\xi(k) = E[\varepsilon^2(k)]$$

© 2015. The authors - Published by Atlantis Press

LMS adaptive filteringalso do not require explicit measurements of correlation functions or matrix inversion, which is based on the method of steepest descent. Using the ∇ represents the surface of the mean square error performance:

$$\nabla = \frac{\partial \xi}{\partial W} = [\frac{\partial \xi}{\partial w_1(k)} \frac{\partial \xi}{\partial w_2(k)} \cdots \frac{\partial \xi}{\partial w_n(k)}]^T = 2E[\varepsilon(k)(\frac{\partial \varepsilon}{\partial w_1(k)} \frac{\partial \varepsilon}{\partial w_2(k)} \cdots \frac{\partial \varepsilon}{\partial w_n(k)})]^T \quad (2)$$

In order to obtain the weight iteration formula, it is necessary to further derivation, $-x(k)$ are the partial derivatives of the mean square error with respect to the weight values:

$$[\frac{\partial \varepsilon}{\partial w_1(k)} \frac{\partial \varepsilon}{\partial w_2(k)} \cdots \frac{\partial \varepsilon}{\partial w_n(k)}]^T = -x(k) \quad (3)$$

Substituting the above equation into (2):

$$\nabla = -2E[\varepsilon(k)x(k)] \quad (4)$$

We assume that the input signals and expected response can be regarded as stationary stochastic variables, the input signals and desired response are mutually uncorrelated, the mean-square error can be calculated as follows:

$$\xi(k) = E[\varepsilon^2(k)] = E[d^2(k)] + W^T(k)E[x(k)x^T(k)]W(k) - 2E[d(k)x(k)]W^T(k) \quad (5)$$

Where, the mean-square error is a quadratic function of the weight values.

We assume the $R = E[x(k)x^T(k)]$ is the autocorrelations of the input signals, and the column matrix $r = E[d(k)x(k)]$ is the set of cross correlations between the input signals and the desired response signal. The mean square error can also be defined as

$$\xi(k) = E[\varepsilon^2(k)] = E[d^2(k)] + W^T(k)RW(k) - 2rW^T(k) \quad (6)$$

By using the LMS algorithm for finding approximate solutions .The weight vector on the $(k+1)$ th iteration is:

$$W(k+1) = W(k) + \mu(-\nabla) \quad (7)$$

Where, μ is the convergence factor. According to the above equations,(7) can be expressed as:

$$W(k+1) = W(k) + 2\mu x(k)[d(k) - W^T(k)x(k)] \quad (8)$$

and then taking the expected value of both sides of (9),becomes:

$$E[W(k+1)] = (1 - 2\mu R)E[W(k)] + 2\mu r \quad (9)$$

The convergence of $E[W(k)]$ can be defined as :

$$0 < \mu < \frac{1}{\lambda_{max}} = \frac{1}{E\{|x_i(k)|^2\}_{max}} \quad (10)$$

Where λ_{max} is the maximum eigenvalue of the autocorrelations of the input signals.

III. The Derivation of the Transform Domain Adaptive FilteringAlgorithm

A. Frequency Domain Adaptive Filtering theory

Frequency domain adaptive filtering process: the input signal is processed by N points Fourier transform for conversion into the frequency domain. A signal-tap complex adaptive filter acts on each frequency bin of the transform domain signal, then obtaining the error signal by difference between the expected signal and the weights of the output signal, and then will be processed by an inverse Fourier transform to return to the time domain, which can suppress the jammers. We assume that the error signal as the output signal. Figure 1 is the block diagram of a frequency-domain adaptive filter.

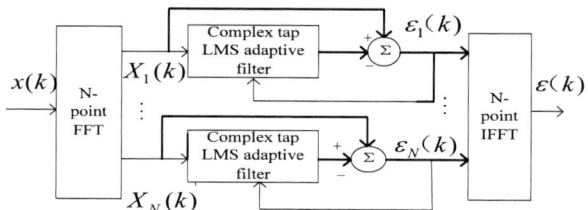

FIGURE I. THE FREQUENCY ADAPTIVE FILTERS

In the Suppression of norrowband jammers algorithm that based on FFT, in order to reduce the spectrum leakage, add the window function in every data segmentation before FFT transform in the frequency domain direct sequence spread spectrum anti-jammer algorithm.

The computation comparison: the transform-domain filter provides significant implementation advantages compared to time-domain filter[4] . Accordingly, the ratio of the frequency domain LMS for computation of time domain LMS as follows:

$$\frac{number\ of\ frequency\ domain\ multiply}{number\ of\ time\ domain\ multiply} = \frac{3\log_2(N/2)+4}{2N} \quad (11)$$

The Fourier transforms produce a set of nearly orthogonal component, which can form autocorrelation matrix that approximate to diagonal matrix, which improved the convergence properties compared to the time domain LMS algorithm.

B. *The derivation of frequency domain adaptive filtering algorithm*

As with time domain adaptive filtering, the frequency adaptive weights can be obtained by the Widrow-Hoff algorithm. As for one tap weight adaptive filter, the update weight can be described as follows:

$$W(k+1) = W(k) + 2\mu\varepsilon(k)X^*(k) \qquad (12)$$

Where, $X(k)$ are the frequency signal, and the "*"represents the complex conjugate.

The following discussion of frequency adaptive algorithm for the suppression of narrowband jammer in spread spectrum domain[5-6].The block diagram of frequency domain adaptive filter can be modified as following:

The input signal becomes one from two, and using $X(k)$ instead of the desired signal;

Introducing weight leakage factor α,($0 < \alpha < 1$)and the weight update equation becomes:

$$W(k+1) = \alpha W(k) + 2\mu\varepsilon(k)X^*(k) \qquad (13)$$

Making the error signal $\varepsilon(k)$ as the output signal

$$\varepsilon(k) = X(k) - W^T(k)X(k) = [1 - W^T(k)]X(k) \qquad (14)$$

This derivation is based on the basic Widrow-Hoff LMS algorithm. Substituting the error signal into (12), becomes:

$$W(k+1) = \alpha W(k) + 2\mu[1 - W(k)]X(k)X^*(k) \qquad (15)$$

Assuming that $X(k)$ is independent of $X^*(k)$ and $W(k)$ is independent of $X(k)$ and $X^*(k)$.Additionally, if the input samples are assumed to be zero mean, $E\{X(k)X^*(k)\}$ can be replaced by the input signal power σ^2.The expected value of (15)is given by:

$$E\{W(k+1)\} = (\alpha - 2\mu\sigma^2)E\{W(k)\} + 2\mu\sigma^2 \qquad (16)$$

Assume that the weight equal $W(0)$ at time zero,The expected value of the $(i+1)$ th weight can be described as following (17) based on the equation above:

$$E\{W(i+1)\} = (\alpha - 2\mu\sigma^2)^{i+1}W(0) + 2\mu\sum_{n=0}^{i}[\alpha - 2\mu\sigma^2]^n\sigma^2 \qquad (17)$$

Consider the first term on the right, as $i \rightarrow \infty$, $E\{W(i)\}$ tends to a certain value, and $E\{W(i)\}$ is independent of $E\{W(0)\}$,thus, the convergence factor value should be satisfied the bound :

$$|\alpha - 2\mu\sigma^2| < 1 \qquad (18)$$

Due to the convergence factor $\mu > 0$,therefore, a bound is placed on μ to ensure that weight value is convergence:

$$0 < \mu < (\alpha + 1)/2\sigma^2 \qquad (19)$$

Consider the second term of (17) on the right equal $2\mu\sigma^2/[1 - \alpha + 2\mu\sigma^2]$ as $i \rightarrow \infty$, because it is the Geometric series summation of common ratio is $\alpha - 2\mu\sigma^2$ and The first item is one.In conclusion, the first term approaches zero as $i \rightarrow \infty$,then obtaining the optimized weight:

$$W(\infty) = \sigma^2/\{[(1-\alpha)/2\mu] + \sigma^2\} \qquad (20)$$

IV. ALGORITHM SIMULATION AND PERFORMANCE ANALYSIS

Frequency domain LMS algorithm is compared with the time-domain LMS computational complexity and convergence properties simulation diagram can be obtained by the formula (11) as shown in Figure 2:

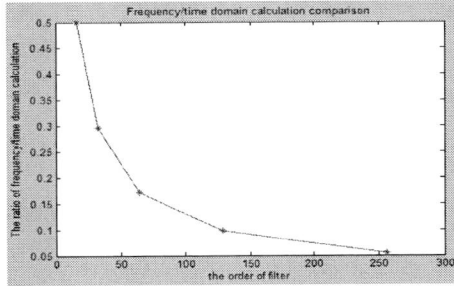

FIGURE II. FREQUENCY DOMAIN LMS COMPARED WITH TIME DOMAIN LMS COMPUTATIONAL COMPLEXITY(LEFT)

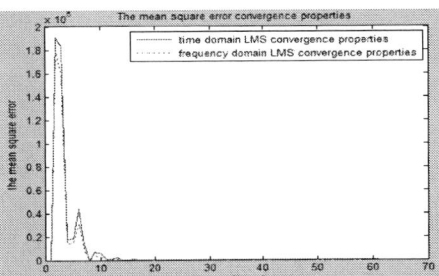

FIGURE III. FREQUENCY DOMAIN LMS COMPARED WITH TIME DOMAIN LMS CONVERGENCE PROPERTIES(RIGHT)

Figure 2 shows that the frequency domain LMS algorithm computational complexity is greatly reduced compared with the time domain LMS algorithm.

In Figure 3, the simulation results show that the time domain LMS and frequency domain LMS have similar convergence speed under the same convergence factor when the algorithm is convergent.

The analysis of the frequency domain adaptive anti-jamming propertiesas follows: assuming signal sampling frequency: $f_s = 62MHz$, carrier frequency: $f_0 = 1268.52MHz$, jammer-to-signal power: $jsr = 70dB$, carrier-to-noise ratio: $CNR = 37dB$, and the tap weight leakage: $\alpha = 1 - 2^{-14}$, By the weights of convergence condition formula(19) can calculated the convergence value: $0 < \mu < 1.26 \times 10^{-12}$. Here, we use 256 FFT points instead of 128 FFT points, As the general data points more, select the data section after FFT transform and the simulation analysis, adopting 61 iterations.

Add a single tone jammer at time zero, and Figure 4 shows that signal spectrum before and after suppressing the single-tone jammers under the same convergence factor. Figure 4(a), (b) respectively before and after the single tone jammer suppressed signal spectrum.

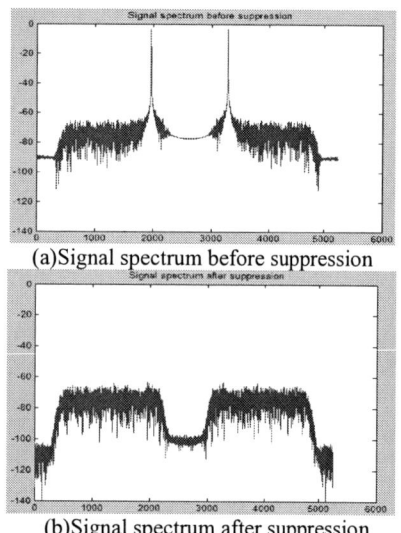

(a)Signal spectrum before suppression

(b)Signal spectrum after suppression

FIGURE IV. SIGNAL SPECTRUM BEFORE AND AFTER SUPPRESSING SINGLE-TONE JAMMERS.

Figure5 shows the curve of the amplitude of weights and mean square error convergence properties. Figure 5(a)(b) respectively show the amplitude of jammer frequency signal and non-jammer frequency signal for the times of iteration, Figure 5(c)shows the mean square error as a function of the times of iteration of different convergence factor. For the convergence factor $ata1 = 2^{-43}$, the weights of amplitude and mean square error tends to be stable after 40 iterations, and the

convergence time is about $t_2 \approx 83us$, while the convergence factor $ata = 2^{-41}$, the weights of amplitude and mean square error tends to be stable after 12 iterations, and the convergence time is about $t_2 \approx 25us$. Therefore, the bigger convergence factor results in less number of iterations within a certain range of convergence factor as show in Figure 5(a)(b)(c),and the convergence rate is consistent between the jammer frequency signal and non-jammer frequency signal.

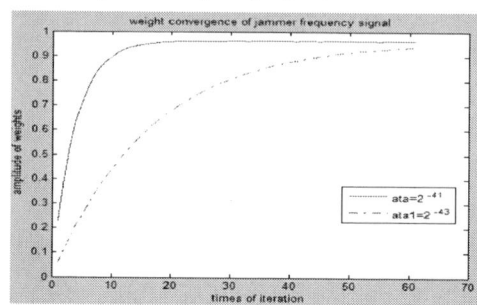

(a) Weight convergence of jammer frequency signal(left)

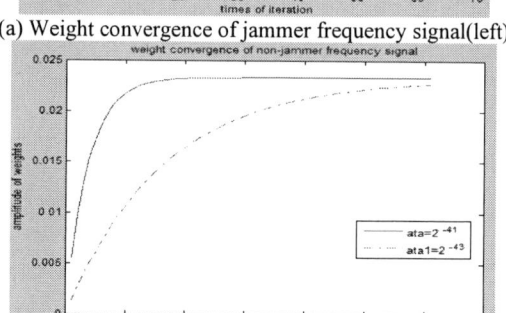

(b) Weight convergence of non-jammer frequency signal(right)

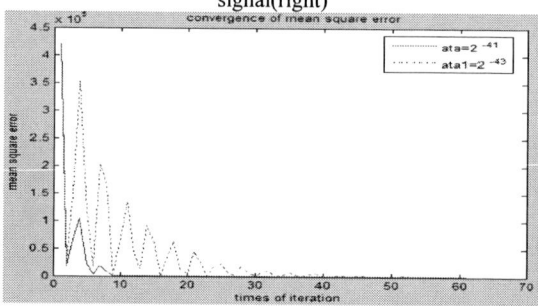

(c) Convergence of mean square error

FIGURE V. THE CONVERGENCE OF WEIGHT AND MEAN SQUARE ERROR.

V. CONCLUSIONS

This paper mainly discusses the use of the transform domain adaptive filter for suppression of signal-tone jammer which in the beidou B3 signal. Compared with time domain LMS algorithm,the simulation results show that similar convergence rate under the same convergence factor,but the amount of calculationis reduced greatly. Simulation analysis presented in this paper show the transform domain adaptive filtering can effectively suppress the signal-tone jammer. The convergence time is consistent whether the jammer frequency signal or non-jammer frequency signal, and the different convergence factor results in different time.

ACKNOWLEDGEMENT

The paper is supported by 2014 Academy of Scientific Research Foundation.

REFERENCES

[1]   J. W. Hsu and A. A. Giordano. Adaptive Algorithms for Estimating and Suppressing Narrow-Band Interference in PN Spread-Spectrum Systems[J]. IEEE Trans. Commun, vol. COM-30, no. 5, pp. 913-924, May 1982.

[2]   Yao-huan Gong. Adaptive filter(Second edition)—Time Domain Adaptive Filtering and Smart Antennas[M]. Beijing: Publishing House of Electronics Industry, 2003.

[3]   F. A. Reed and P. L. Feintuch. Acomparison of LMS Adaptive Cancelers Implemented in the Frequency Domain and the Time Domain[J]. IEEE Trans. Signal Proces, vol. ASSP-29, no. 3, pp. 770-775. June 1981.

[4]   Yan-xin Gou. Wireless Anti-Intercept and Anti-Jamming Communication[M]. Xian: Xidian University Press, 2011.

[5]   Saulnier, G. Suppression of Norrowband Jammers in a Spread-spectrum Receiver Using Transform-Domain Adaptive Filtering[J]. IEEE Journal on Selected Areas in Communication. Vol. 10. No. 4 .may 1992

[6]   Chao Tian. Study of frequency-domain block LMS Adaptive filtering Algorithm[J]. Xian: Xian university of science and technology press, 2010: 15-23.

Numerical Studies of Abrasion Wear on the Guide Vanes in a Submersible Axial Flow Pump

H.M. Zhang
Department of Engineering Mechanics
Kunming University of Science and Technology
Kunming , P. R. China

L.X. Zhang
Department of Engineering Mechanics
Kunming University of Science and Technology
Kunming , P. R. China

Abstract—**The paper presents the numerical studies of abrasion wear on the guide vanes in a submersible axial flow using OpenFOAM code, which is an Open Source CFD Package. Hashish erosion model was implemented in this code. The 3-D turbulent particulate-liquid two-phase flow equations are employed in this study. The computing domain is discretized with a full three-dimensional mesh system of unstructured tetrahedral shapes. The finite volume method is used to solve the governing equations and the pressure-velocity coupling is handled via a Pressure Implicit with Splitting of Operators (PISO) procedure. Simulation results have shown that the sand erosion rate on pressure side is more than on the suction side of the guide vanes. The abrasion wear occurs mainly at the lower part of the guide vanes.**

Keywords- CFD; Abrasion wear; Submersible axial flow pump; open FOAM

I. INTRODUCTION

A submersible axial flow pump often used in sewage pumping station. Due to sewage contains large amounts of sediment, solids and other impurities, the components of axial flow bump will be worn and results in reduced efficiency. The abrasion wear is due to the dynamic action of sediment or solid particles flowing along with water impacting against a solid surface of hydraulic components. In general, a number of factors influence the development of abrasion wear process of hydraulic machinery. These factors include mean velocity of particles, mass of the particle, concentration of the abrasive particles in a liquid flow, grain size and shape of the particles and angle of attack at which the particles collide with the surface etc. With the aid of CFD, numerical solutions for the equations governing fluid flows can be successfully obtained in order to accurately predict the different outcomes of fluid-surface interactions. Meng and Ludema[1] separated 28 erosion models out of the almost 2000 empirical models they encountered and concluding that each equation is the result of a very specific and individual approach. López at al. [2] compared the different erosion models in an Eulerian-Lagrangian frame using OpenFoam. They believe that Hashish's model for erosion would be a reasonable approach when experimentation is not feasible. They analyzed the erosion rates of a single channel of a centrifugal pump subjected to solid particle impingement using four different methods with OpenFoam [3]. Results yielded some disparities between models due to the different factors taken into consideration.

This paper presents using OpenFOAM code only to make a qualitative analysis of the abrasion wear on the guide vanes in a submersible axial flow pump. OpenFOAM is an object-oriented C++ library of classes and routines of use for writing CFD codes. It has a set of basic features similar to any commercial CFD solver, such as turbulence models and discretization schemes. With OpenFOAM it is easy to add any modification to any part of the implementation. Hashish erosion model was implemented here.

II. GOVERNING EQUATIONS

The flow model used for the numerical simulations is based on the generalized homogeneous multiphase flow model with the additional sources of momentum for the effects of the Coriolis and centrifugal accelerations in a steady rotating frame of reference. The governing equations are described below[4]

Continuity equation of liquid:

$$\frac{\partial}{\partial t}(\rho) + \frac{\partial}{\partial x_j}(\rho \boldsymbol{u}_j) = S = -\sum n_p \dot{m}_p \tag{1}$$

Momentum equations of liquid:

$$\frac{\partial(\rho u_i)}{\partial t} + \frac{\partial}{\partial x_j}(\rho u_j u_i) = -\frac{\partial p}{\partial x_i} + \frac{\partial}{\partial x_j}[\mu_e(\frac{\partial u_i}{\partial x_j} + \frac{\partial u_j}{\partial x_i})]$$
$$+ \sum \rho_p C_D \frac{3}{4} \frac{R_{ep}\mu}{\overline{\rho}_p d_p^2}(u_{pi} - u_i)$$
$$+ \sum C_D \frac{3}{4} \frac{R_{ep}\mu}{\overline{\rho}_p d_p^2}(\frac{v_p}{\sigma_p}\frac{\partial \rho_p}{\partial x_j}) + F_{cj} \tag{2}$$

Where F_{cj} represents the Coriolis force. $P = P^* + 0.5\rho(\omega r)^2$ is total pressure, P^* is static pressure, ρ_p is the volume density of the particle phase, n_p is the number of particles per unit volume, m_p is the particle mass, C_D is the drag coefficient of unsteady flow.

Continuity equation of particle:

© 2015. The authors - Published by Atlantis Press

$$\frac{\partial}{\partial t}(\rho) + \frac{\partial}{\partial x_j}(\rho \boldsymbol{u}_j) = \frac{\partial}{\partial x_j}(\frac{v_p}{\sigma_p} \cdot \frac{\partial \rho_p}{\partial x_j}) \tag{3}$$

Momentum equations of particle:

$$\frac{\partial(\rho_p \boldsymbol{u}_{pi})}{\partial t} + \frac{\partial}{\partial x_j}(\rho_p \boldsymbol{u}_{pj} \boldsymbol{u}_{pi}) = \frac{\partial}{\partial x_j}[\rho_p \boldsymbol{u}_p(\frac{\partial u_{pi}}{\partial x_j} + \frac{\partial u_{pj}}{\partial x_i})]$$

$$+ \sum \frac{m_p C_D \frac{3}{4} \frac{R_{ep} \mu}{\overline{\rho}_p d_p^2} + m_p}{m_p} \frac{v_p}{\sigma_p} \frac{\partial \rho_p}{\partial x_i}$$

$$+ \frac{\partial}{\partial t}(\frac{v_p}{\sigma_p} \frac{\partial \rho_p}{\partial x_i}) + F_{pci} + F_{pli} + F_{ppi} \tag{4}$$

Where F_{pci} is the Coriolis force, F_{pli} is the particles centrifugal force, F_{ppi} is the particle phase pressure.

Erosion model:

Hashish erosion model is used here [5]. It modified Finnie's model for erosion by solid particle impingement in ductile materials taking into account particle shape and including no empirical constants. The equation, taken from [2], and used for the simulation is equation (5).

$$W = \frac{7}{\pi} \frac{M}{\rho_p} (\frac{V}{C_k})^{2.5} \sin(2\alpha)\sqrt{\sin\alpha} \tag{5}$$

Where M is the mass of the abrasive particles, ρ_p is the density of the particles, V is the velocity magnitude of the particles, α is the angle of impingement, C_k is a coefficient which can be obtained from equation (6).

$$C_K = \sqrt{\frac{3\sigma_f R_f^{\frac{2}{5}}}{\rho_p}} \tag{6}$$

Where R_f is the roundness factor of the particulate phase.

Numerical Algorithm

In the OpenFOAM , the spatial discretization is performed using a cell centered co-located finite volume method for unstructured meshes with arbitrary cell-shapes, and a multi-step scheme are used for the time derivatives. For the simulations presented in this paper, a second order implicit time scheme is used combined with second order linear interpolation in space, except for the convective terms. The time step is equal to 1 degrees of runner revolution. The iterative solvers are considered converged when the residuals have been reduced by

a factor of 10-5. The pressure-velocity coupling is handled via a Pressure Implicit with Splitting of Operators (PISO) procedure. The $k-\omega$ SST turbulence model is adopted for simulation of the unsteady flow through a submersible axial pump. It has been implemented in the OpenFOAM libraries.

Grid, Boundary and Initial Conditions

Figure 1 shows the calculation domain of the submersible axial flow pump. The parameters are listed in Table 1.

The computational grid of the complete axial flow pump is generated using ICEM CFD, including inlet, guide vane, impeller and outlet with 23,600,000 elements was used. After importing the meshed geometry in OpenFOAM, it is divided into four domains namely inlet, vanes, impeller and outlet, in which impeller is rotating domain and others are stationary domains. The element number of impeller domain is 13,600,000.

In the algorithm, the length of the time step was equal to one of the impeller revolution. Rotation speed of the runner was 580 r/min. Thus, the time step was 0.0036236s. Inlet bulk mass flow rate and fixed outlet pressure boundary conditions are applied. The calculation is done on a SUGON high-performance computers using 160 CPU core.

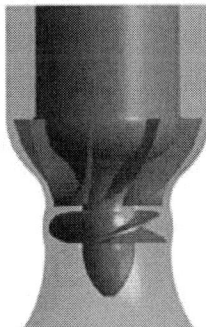

FIGURE I. SUBMERSIBLE AXIAL FLOW PUMP

TABLE I. PARAMETER OF SUBMERSIBLE AXIAL PUMP

Diameter of impeller (m)	0.165
Number of impeller blade	3
Number of stay vane	8
head (m)	5.85
Rotational speed (rpm)	580
Rated flow rate (m³/s)	0.981

III. ANALYSIS OF NUMERICAL CALCULATION RESULTS

The sand volume fractions and the sand erosion rate are the important feature for abrasion wear in hydraulic turbine. The region of higher sand volume fraction and sand erosion rate with the region of abrasion wear is consistent qualitatively. Figures 2 show the pressure distribution for the guide vanes on the pressure side and section side. The pressure is relatively high at the lower part of the guide vanes. Figures 3 show the volume fraction of sand distribution for the guide vanes on the pressure side and the section side. It is clearly seen that the sand volume fraction at the lower part of the pressure side is higher than others. Compared with the sand erosion rate distribution of Figure 4, the maximum of sand volume fraction is at same location with the maximum of sand erosion rate. Figures 5 show the wall stress distribution for the guide vanes on the pressure side and section side. Its maximum position is also consistent with the maximum position of the sand erosion rate and sand volume fraction. Figures 6(a) show the sand erosion rate distribution on guide vanes and Figures 6(b) show the sand volume fraction distribution on guide vanes. The sand erosion rate and sand volume fraction in the lower part of the guide vane is relatively high.

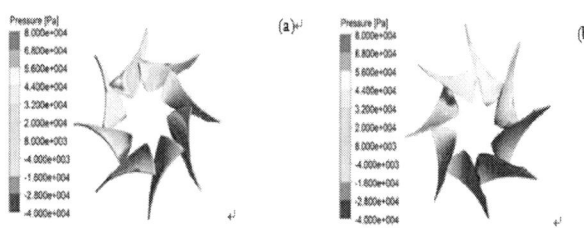

FIGURE II. THE PRESSURE DISTRIBUTION FOR THE GUIDE VANES ON THE PRESSURE SIDE AND SECTION SIDE

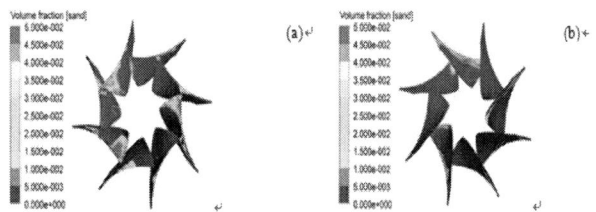

FIGURE III. THE SAND VOLUME FRACTION DISTRIBUTION FOR THE GUIDE VANES ON THE PRESSURE SIDE AND SECTION SIDE

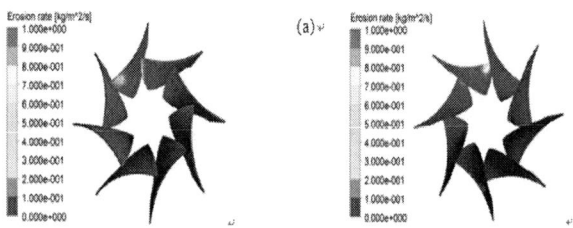

FIGURE IV. THE SAND EROSION RATE DISTRIBUTION FOR THE GUIDE VANES ON THE PRESSURE SIDE AND SECTION SIDE

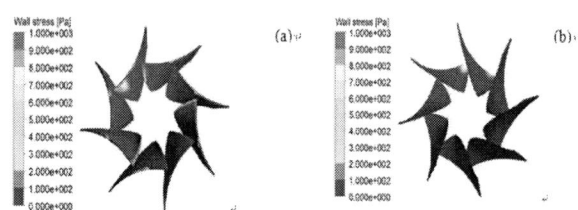

FIGURE V. THE WALL STRESS DISTRIBUTION FOR THE GUIDE VANES ON THE PRESSURE SIDE AND SECTION SIDE

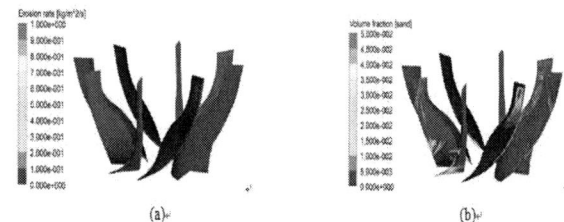

FIGURE VI. (a) THE SAND EROSION RATE DISTRIBUTION ON GUIDE VANES (b) THE SAND VOLUME FRACTION DISTRIBUTION ON GUIDE VANES

IV. CONCLUSION

In this paper, the numerical studies of abrasion wear on the guide vanes in a submersible axial flow pump has been carried. Simulation results have shown that the sand erosion rate on pressure side is more than on the suction side of the guide vanes. The abrasion wear occurs mainly at the lower part of the guide vanes. Hashish erosion model implemented in OpenFoam is able to simulate the wear problem well.

ACKNOWLEDGEMENTS

This work was financially supported by the National Natural Science Foundation of china (No. 51279071) and the Natural Science Foundation of Kunming University of Science and Technology (No.20149x16).

REFERENCES

[1]. Meng H C, Ludema K C. Wear models and predictive equations: their form and content[J]. Wear, 1995, 181: 443-457.

[2]. López A, Stickland M T, Dempster W M. Comparative study of different erosion models in an Eulerian-Lagrangian frame using Open Source software[J].

[3]. Lopez A, Stickland M, Dempster W. Modeling erosion in a centrifugal pump in an Eulerian-Lagrangian frame using OpenFOAM®[J].

[4]. Tang Xue-Lin, TANG Hong-Fen,WU Yu-Lin. Simulating and erosion prediction of 3D two phase flow through a turbine runner,JOURNAL OF ENGINEERING THERMOPHYSICS, Vol.22, N0.1, jan.,2001.

[5]. Hashish M. Modified model for erosion. Proceedings of the 7th International Conference on Erosion by liquid and solid impact, 1987, Cambridge, UK, pp. 461–480

International Conference on Power Electronics and Energy Engineering (PEEE 2015)

The Design of Rice Direct Seeding Machine

S.F. Ding, Q.X. Li, Q. Huang

School of Mechanical and Materials Engineering
Jiujiang University
551 Qianjing East Road Jiujiang, Jiangxi, China, 332005

S.F. Ding

Key Research Lab for Numerical Control of Jiangxi Province
Jiujiang University
551 Qianjing East Road Jiujiang, Jiangxi, China, 332005

Abstract—**In order to change the existing status of big turning radius of the large direct seeding machines with diesel and gasoline as fuel and improve the low efficiency and time-consuming condition of the machine driven by human or animal, an electric rice planter was designed. The mechanical parts and circuit design had been finished. The rear wheel shaft, seeding wheel, seeding frame and other key parts had been designed. The charging circuit of the maintenance-free storage battery and the power supply circuit of the DC motor had been given. The main technical parameters of the rice planter as follows, L * B *H: 1700 * 1200 *1200 (mm), row distance* planting distance*width = 250 * 200 *1000 (mm). The rice planter driven by DC motor and maintenance-free battery has a compact structure, a small turning radius, and it can be used in hills and mountainous regions.**

Keywords-rice direct seeding machine; turning radius; maintenance-free storage battery; DC motor

I. INTRODUCTION

Rice direct seeding technology, which makes rice planters sow the rice seeds directly into the field, is a relatively extensive planting pattern; it eliminates the seedlings, seedling up and transplanting, simplifies the operation process, decreases the time, the cost and the seedling bed, and increases the production[1]. The applicable machinery for hills, mountains, which cover more than 45% of China's territory, is a severe lack where the mechanization level is only about a third of the national average level and most still in the undeveloped stage. This seriously affects the normal labor transfer conditions of agricultural production and restricts the development of comprehensive mechanization of agricultural production [2]. In the south, the situation is similar.

The rice planter was normally operated by a single person, or driven by a large motor traction machine with diesel and gasoline engine as power source. The planter with a diesel often makes the turning radius very big, which makes it not suitable for small, scattered growing areas in hills and mountain; the rice planter operated by a single person could work in the hills and mountains, but it's time-consuming and low-efficient.

To solve the single person-manipulated rice planter's engine problem, electric traction was used to replace human or animal traction with appropriate change to the whole machine structure. The maintenance-free battery was chosen as the power supply and the DC motor as the traction; it changed the time-consuming and low efficiency status. It accords with the requirements of the ecological agriculture development. This

development has broken the traditional dynamic design, it is low-carbon, energy-saving, and easy to use.

II. MECHANICAL PARTS DESIGN

A. The Main Technical Parameters

The planter's configurations are listed below.

The motor power: 0.35 KW

The body length: 1700 mm or less

The body width: 1200 mm or less

The body height: 1200 mm or less

Row space: 250 mm

Seeding planting distance: 200 mm

Working width: 1000 mm.

B. Working Principle

When the DC motor is turned on, it drives the small chain wheel to rotate. The latter turns the big chain wheel as the big chain wheel is fastened to the rear shaft. The shaft rotates along with the walking wheel and the seeding wheel. When the walking wheel rotates, the rice planter goes straight and when the seeding wheel rotates, the filling, cleaning, protecting and casting processes are finished from the uniformly distributed sowing boxes in the rack [3]. When the DC motor is turned off, the rice planter stops working. The farmer turns the DC motor on or off by steering the front switch in order to implement the seeding operation and to stop the operation. The rice planter turns left or right through the steering wheel.

C. Structure Diagram

The host general structure diagram was shown in Figure 1, including 1 front wheel, 2 front shaft, 3 steering wheel, 4 body, 5 maintenance-free battery,6 DC motor, 7 small chain wheel, 8 chain, 9 big chain wheel, 10 rear axle, 11 seeding wheel, 12 sowing rack .

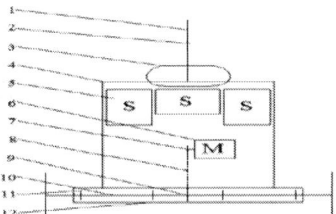

FIGURE I. STRUCTURE DIAGRAM.

© 2015. The authors - Published by Atlantis Press

The shape structure of the rice planter was shown in Figure 2.

FIGURE II. SHAPE DIAGRAM.

III. KEY COMPONENTS DESIGN

A. Rear Axle

The rear axle structure was shown in Figure 3. The power was transferred to the big sprocket in the rear axle. The evenly tightened wheel finished the seeding. The seeding frame was connected with the rear axle. The rear walking wheels were connected on both ends of the shaft which was a stepped one and its material was 45# steel. There were some operations in the heat treatment of the shaft: the tempering after hardening could improve its hardness and strength, and the plasticity and impact toughness as well [4].

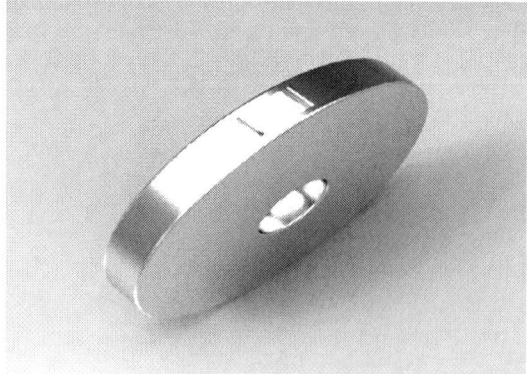

FIGURE III. REAR WHEEL SHAFT.

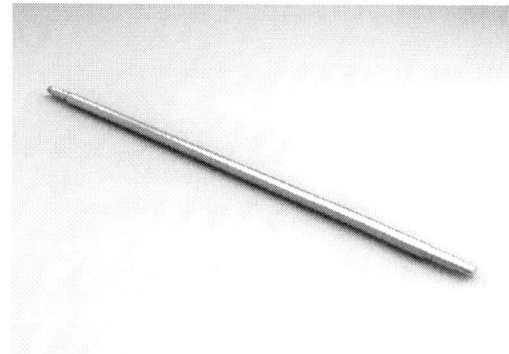

FIGURE IV. SEEDING WHEEL.

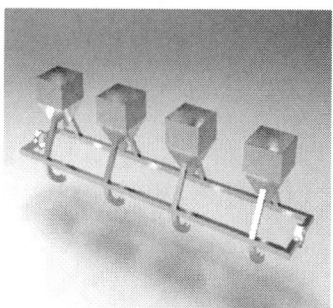

FIGURE V. PLANTING FRAME.

B. Seeding Wheel

The seeding wheel's main function is filling seeds' and cleaning; the wheel structure was shown in Figure 4. The device has a simple structure and a low cost. The wheel was fastened in the rear axle and its inner diameter is the same to the rear axle; there are two eye sockets uniformly distributed on the cylindrical; the seeding row distance of main technical parameters is 200 mm, so the outer diameter is 127 mm; the seed spacing of main technical parameters is 250 mm and the seeding wheels are uniformly located in the rear axles.

C. Seeding Frame

There were evenly distributed seeding boxes and seeding grooves. The main function of the seeding boxes was storing seeds and releasing seeds, while the main function of the seeding grooves was protecting seeds and casting seeds. Its structure was shown in Figure 5. This device is connected to the rear axle. When the seeding wheel rotates, the seeds in the boxes fall to the eye sockets and slip to the soil with the seeding wheel rotates under gravity.

IV. THE CIRCUIT DESIGN

The motor type was DC motor [5] and it was the power source of the rice planter. The DC motor drove the small chain wheel rotate and made the big chain wheel rotate by chain transmission.

DC motor power came from maintenance-free battery; maintenance-free battery charging circuit was shown in Figure 6.

FIGURE VI. MAINTENANCE-FREE BATTERY CHARGING CIRCUIT.

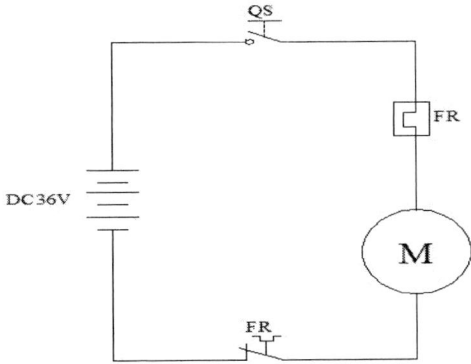

FIGURE VII. THE POWER SUPPLY CIRCUIT OF THE DC MOTOR.

DC motor power supply circuit was shown in Figure 6; the power supply was a 36 v maintenance-free battery and the load was a DC motor. The switch was set between the source and load. A thermal relay was set for the DC motor overload protection [6], and the thermal relay type was JR20 10-14-16A.

V. SUMMARY

The rice planter was combined with the DC motor and maintenance-free battery. The DC motor acted as the execution device and the maintenance-free battery as the power unit. The high energy consumption, heavy pollution and high cost condition caused by using diesel and gasoline as the energy source have been improved. The low efficiency condition caused by using man power or animal power as the traction source has also been improved. It accords with the requirement of the ecological agriculture development.

ACKNOWLEDGEMENTS

This work was financially supported by the Jiangxi Provincial Key Technology R&D Program of the department of technology (20122BBF60129).

REFERENCES

[1] Li-qiang Wang, Chong-you Wu, Lian-xing Gao, An-fu Tu, Cheng-qian Jin. The Paddy Planting Status and the Research of Developing Paddy Mechanical Direct Seeding in China. Journal of Agricultural Mechanization Research,2006(3):28-30.in Chinese

[2] Huang-zhen Lv, Xiong-wei Wang, Yun-tao Fan, Bing-nan Yang. Review on Agricultural Machinery Research Progress in 11th Five Year Plan and Direction of Future Development. Agricultural Engineering, 2011,1(2):1-7.in Chinese

[3] Mei Jin, Chun-hua Xia, Chong-you Wu, Wen-yi Zhang. The Present Situation and Trends in Development of Seeding Apparatus of Paddy Planter. Chinese Agricultural Mechanization, 2010, (5):39-42.in Chinese

[4] Zong-ze Wu, Sheng-guo Luo. Course Design Manual for Mechanical Design. Beijing: Higher Education Press, 2006.5. in Chinese

[5] Ze Wang.1GD-40 Type DC Electric Multi-functional Micro Tillage Machine. Journal of Agricultural Equipment Technology, 2007 (2) : 56. in Chinese

[6] Li-jiang Yang. Electric Control Technology in Machine Tool. Beijing: Beijing Institute of Technology Press, 2008.6. in Chinese

Designing A Refrigeration and Heat-Removal System for Rapid Detector Based on Freezing Point

X.T. Yu
International College Beijing
China Agricultural University
Beijing, China, 100083

J. Leng
Foreign Language Department
China University of Geosciences (Beijing)
Beijing, China, 100083

X.B. Guo
School of Engineering and Technology
China University of Geosciences (Beijing)
Beijing, China, 100083

Abstract—**Based on a detection method of freezing point, the paper proposed a simple, reliable, portable, and low-cost refrigeration and heat-removal system. Semiconductor refrigeration was chosen through comparing operating principles and features of three refrigeration solutions. Forced air cooling was then selected through comparing three heat dissipating methods for hot ends of semiconductor chilling. After determining the refrigeration and heat-removal solution, semiconductor refrigeration and heat dissipating by air cooling for the hot ends were selected as erection schemes, with stirring apparatus and condensate water processing unit. Mechanical compression was chosen for fixation through comparing features of three installation methods. It is safe to say that this refrigeration and heat-removal system can be used to rapidly detect the categories and quantities of adulterated milk with accuracy.**

Keywords-freezing point detection; refrigeration; heat-removal; systematic design

I. INTRODUCTION

Milk is of rich nutrition and has become a popular dairy product, and its demand increases with people's raising health consciousness. To make profit, some lawbreakers adulterate water, salt, sucrose and preservative with fresh milk. Such adulterated milk is closely involved with milk drinker health, especially for the young and the aged with weak immunity. Through inspecting freezing point of milk sample to be detected, freezing-point detector can be used to qualitatively and quantitatively detect milk adulteration. A bottle-neck issue for the national products is complex preprocessing, long testing period, and difficult to carry [1]. And the counterpart for the foreign like product is high-cost and inconvenient maintenance [2]. The crux of the matters above lies in its refrigeration and heat-removal system. This paper thus attempts to design a refrigeration and heat-removal system with three characters of simple and reliable structure, portable operating and low-cost, so as to make a rapid and efficient detection of the milk quality possible for freezing- point rapid detector.

II. TESTING PRINCIPAL AND METHOD

A. Testing Principals

The freezing point of pure water is $0\,^{\circ}\text{C}$, which is higher than milk. When water or other impurities are adulterated with milk, its freezing point will change. The more substance gets adulterated with milk, the more will the freezing point be fluctuated. By testing the freezing point of samples and comparing them with database, we can find out whether milk is adulterated. We can even figure out the category and content of impurities [3]. Our national recommendatory standard is that the freezing point of milk can reasonably fluctuates from $-0.546\,^{\circ}\text{C}$ to $-0.508\,^{\circ}\text{C}$. Samples whose freezing point is above $-0.508\,^{\circ}\text{C}$ or below $-0.546\,^{\circ}\text{C}$ can be judged as disqualification.

B. Testing Methods

Refrigerate the liquid samples until their temperature drops below freezing. Then the samples will crystallize, curdle, and give off heat. After temperature rise, the samples will remain constant within a short period, and the temperature is read as the freezing point. Processor can output picture signals of the temperature curve on the display. The samples will be discharged at last through liquid waste processing unit.

III. DESIGN OF REFRIGERATION SYSTEM

Three main refrigerating methods available to date for miniature instruments are: vapor compression refrigeration, absorption refrigeration and thermoelectric refrigeration; and whose working diagrams are shown in Figure 1. The cyclic process of vapor compression refrigeration system for small sized device whose refrigerating capacity is less than 25 kw is achieved by the flow of refrigerating fluid, as shown in Figure 1a. Refrigerating fluid 1 flows out from condenser 2, and then flows in the closed pipeline 3. When the refrigerating fluid goes through throttle valve 4, thermal insulation isenthalpic expansion will occur. The cooling is conducted as the evaporation of the refrigerating fluid absorbs heat when it goes through evaporator 5. After that, the refrigerating fluid will be compressed by a piston compressor 6 and enter into condenser 2 to cool itself. Vapor compression refrigeration has large refrigerating capacity, good refrigeration effect, and high cost performance. However, the refrigerating fluid, such as Freon, may destroy the ozone layer and pollute the environment.

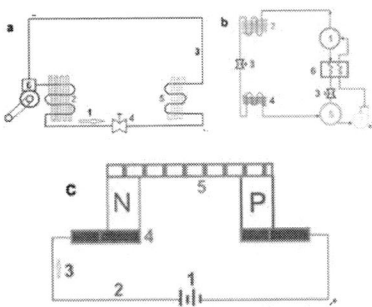

FIGURE I. SCHEMATIC DIAGRAM OF WORKING PRINCIPLE OF COOLING SYSTEMS.
(A) VAPOR COMPRESSION REFRIGERATION (1. REFRIGERATING FLUID; 2. CONDENSER; 3. CLOSED PIPELINE; 4. THROTTLE VALVE; 5. EVAPORATOR; 6. COMPRESSOR.); (B) ABSORPTION REFRIGERATION (1. GENERATOR; 2. CONDENSER; 3. THROTTLE VALVE; 4. EVAPORATOR; 5. ABSORBER; 6. HEAT EXCHANGER; 7. SOLUTION PUMP.); (C) THERMOELECTRIC REFRIGERATION (1. POWER SUPPLY; 2. ELECTRIC WIRE; 3 ELECTRON CURRENT; 4. COLD END; 5. HOT END.)

As shown in Figure 1b for absorption refrigeration system, a refrigeration cycle can be formed in this way. Refrigerating fluid is heated in generator 1, and the isolated refrigerating vapor enters condenser 2 through pipelines, where the vapor condenses into liquid. After the liquid refrigerating fluid is depressurized in throttle valve 3, the cooling is occurred as the evaporation of the refrigerating fluid absorbs heat when it goes through evaporator 4, where the refrigerating fluid changes from a liquid to a vapor, and then enters the absorber 5. The refrigerating fluid flowing out from generator 1 passes through heat exchanger 6 and is depressurized by throttle valve 3. After that, it enters the absorber 5 to absorb the vapor of refrigerating fluid. The solution pump 7 allows the refrigerating fluid to reenter generator 1 after it absorbs heat to warm up in heat exchanger 6. The refrigerating fluid is natural water or ammonia, which is harmless to the environment and atmospheric ozonosphere. Its driven energy can be residual heat, waste heat, and solar energy. However, its shortcomings are small refrigerating capacity, high cost, large energy consumption and bulky set.

Thermoelectric refrigeration system shown in Figure 1c works under action of Peltier Effect: current carriers form an electric current in the conductor. Current carriers are in different energy levels for different materials, so they release energy when moving from high energy levels to low ones. Vice versa, they absorb energy when moving from low energy levels to high ones. In this case, if power supply 1 turns on, the electron current 3 will form in electric wire 2. The thermocouple is composed of P-semiconductor and N-semiconductor materials with conspicuous pyroelectric effects. When the electric current

goes through it, the flow direction is N→P from the thermocouple to layup in upper end. The heat is absorbed and temperature drops, which forms the cold end 4. If the flow direction is N→P in the layup of lower end, the heat is released and temperate rises, which forms the hot end 5. Since there is no mechanical motion in thermoelectric refrigeration, it is efficient to work and convenient to use.

The above comparative investigation allows us to choose semiconductor refrigeration. By altering the supply voltage of refrigerator, we can achieve continuous adjustment of refrigerating capacity. By means of wide adjustable temperature range, light weight, and small volume, such refrigeration is also non-hazardous and environmentally friendly, and thus becomes a good candidate applicable for miniature instruments like freezing-point rapid detector.

IV. DESIGN OF HEAT-REMOVAL SYSTEM

During thermoelectric refrigeration, heat-removal effectiveness of the hot end directly affects the semi-conductive refrigeration performance. Only if the heat in hot end is driven away without delay and ensures that the temperature is not too high, the cold end can be continuously refrigerated. Therefore, removing the heat of hot end is a critical point for semiconductor refrigeration. Three main heat-removal methods possible for semiconductor refrigeration are air natural convection, air cooling and water-cooled radiator, and whose operating principles are shown in Figure 2. Figure 2a shows the air natural convection, which installs heat sink 3 at the hot end of semiconductor refrigeration piece 2. In order to refrigerate, it employs air natural convection to remove the heat of objects under refrigeration into surroundings. With simple structure, the heat-removal efficiency of the method is not so good. Air cooling, shown in Figure 2b, is based on natural convection to remove heat. It installs an axial flow fan at the end of a cooling fin. With simple structure, small volume and low cost, the system is more effective in removing heat. As shown in Figure 2c, water-cooled radiator connects a cold plate 2 to the hot end of semiconductor refrigeration piece 1. Under action of pump 5, water in tank 3 enters the water-cooled tube through water pipe 4. It uses the circular flow of water to remove heat from the hot end of object under refrigeration 6. Its heat-removal effectiveness is good enough, but suffers from bulky structure and noise from running pump.

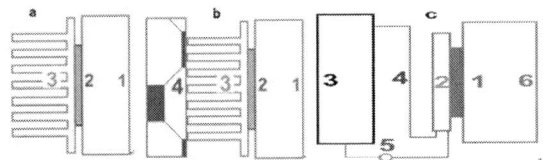

FIGURE II. SCHEMATIC DIAGRAM OF WORKING PRINCIPLE OF HEAT-REMOVAL SYSTEMS.
(A) AIR NATURAL CONVECTION (1. OBJECT UNDER REFRIGERATION 2. SEMICONDUCTOR REFRIGERATION PIECE 3. HEAT SINK); (B) AIR COOLING (1. OBJECT UNDER REFRIGERATION 2. SEMICONDUCTOR REFRIGERATION PIECE 3. HEAT SINK 4. A COOLING FAN); (C) WATER-COOLED RADIATOR (1. SEMICONDUCTOR REFRIGERATION PIECE 2. COLD PLATE 3. TANK 4. WATER PIPE 5. PUMP 6. OBJECT UNDER REFRIGERATION)

As a result, we can see air cooling is superior to the other heat-removal methods applicable for small sized devices like freezing-point rapid detector.

V. ASSEMBLY OF COOLING AND HEAT-REMOVAL MODULE

Figure 3 exhibits an assembly drawing of cooling and heat-removal system. The core component is semiconductor refrigeration piece 1, and whose cold end can cool samples in cooling pool 2. The hot end of semiconductor refrigeration piece employs air cooling, where heat sink 3 cooperates with cooling fan 5. There are clapboards 4 around cooling pool, and heat sink is separated with cooling pool by insulation boards 6. Below the cooling pool, there is a collecting tank 7 used to collect condensed water, produced during the cooling process. There is an air-out fan on the back wall of the cabinet 8, used for removing heat from the hot end.

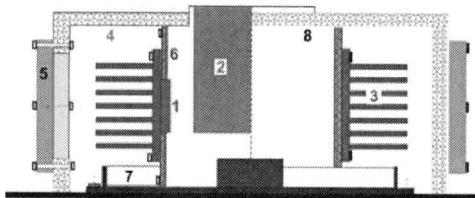

FIGURE III. ASSEMBLY DRAWING OF COOLING AND HEAT-REMOVAL SYSTEM.

(1. SEMICONDUCTOR REFRIGERATION PIECE; 2. COOLING POOL; 3. HEAT SINK; 4. CLAPBOARD; 5. COOLING FAN; 6. INSULATION BOARD; 7. COLLECTING TANK FOR CONDENSED WATER; 8. CABINET.)

Three common installation methods are welding, bonding and mechanical compression fixation. The welding method suffers from high cost, inconvenient maintenance, and easy damage of the refrigeration piece during operation. Bonding is simple and convenient to install, but is also difficult to disassemble and maintain, and whose tack coat is easy to age and shed. The reliable mechanical compression fixation is simple to install and convenient to disassemble, and thus becomes our option.

VI. CONCLUSION

Aiming at current issues for freezing-point rapid detector to milk, this work offers a design for refrigeration and heat-removal system applicable for rapid detector, after introducing testing principle and method of the rapid detector. A comparative investigation of three refrigeration solutions allows us to pick up semiconductor refrigeration in view of rapid refrigeration and simple structure. After comparing working principles and features of three heat-removal methods for hot ends of semiconductor refrigeration piece, air cooling is chosen due to its heat-removal character and working condition. In case that the refrigeration and heat-removal solution is determined, the final design belongs to semiconductor refrigeration and air cooling for removal heat around the hot ends, with stirring apparatus and condensate water processing unit.

ACKNOWLEDGMENT

This work was supported by college students' scientific research and entrepreneurial action plan project of Beijing.

REFERENCES

[1] Y.B. Li, Z.J. Zhang, J.S. Li, H.G. Li, Y. Chen, Z.H. Liu, Simple, stable and sensitive electrogenerated chemiluminescence detector for high-performance liquid chromatography and its application in direct determination of multiple fluoroquinolone residues in milk, Talanta. 84(2011) 690-695.

[2] S. Qu, X. Yu, X.J. Liu, P.F. Yu, A design of turbidimetry based rapid microbiology detector, Advanced Materials Research. 662(2013) 758-761.

[3] P.A. Lieberzeit, F.L. Dickert, Rapid bioanalysis with chemical sensors: Novel strategies for devices and artificial recognition membranes, Analytical and Bioanalytical Chemistry, (391)2008 1629-1639.

International Conference on Power Electronics and Energy Engineering (PEEE 2015)

Investigation of Negative Influences on Ride Comfort Performance of In-Wheel Motor Vehicles with High Unsprung Mass

T.Z. Shi, D.F. Wang, S.M. Chen

State Key Laboratory of Automotive Simulation and Control
Jilin University
Changchun, China, 130022

Abstract—**The in-wheel motor electric vehicle is directly driven by motors integrated in wheels. This unique design provides actuation flexibility, energy efficiency and performance potentials. Yet the introducing of in-wheel motor caused massive increment of unsprung mass, which shows negative effect on vehicle ride comfort performance. In this paper, a rigid-elastic coupling multi-body dynamic model of a typical in-wheel motor vehicle is built. Based on the model, vehicle ride comfort performance with different unsprung mass including root mean square of weighted body acceleration, wheel dynamic load, suspension deflection is investigated. The amplitude-frequency characteristics of vehicle is analyzed based on quarter vehicle system.**

Keywords-in-wheel motor vehicle; ride comfort; unspung mass; rigid-elastic coupling; suspension parameters

I. INTRODUCTION

With potentials in emissions and fuel consumption reduction, the electric vehicles including hybrid electric vehicles, plug-in hybrid vehicles and pure electric vehicles are considered as promising vehicle architectures[1-3]. The in-wheel motor electric vehicles (IMEV) which are driven by four independently actuated in-wheel motors are one of those unique designs. With motors directly actuating wheels, the in-wheel motor vehicle is designed without transmission systems or mechanical links[4]. The driving torque of each motor is controlled independently. With the control flexibility and quick response of motors, the vehicle performance can be improved promisingly[5]. The EV and IMEV have achieved impressive driving performance along with the improvements of motors and batteries.

With in-wheel motors integrated in wheels, the unsprung mass of IMEV is increased dramasticly. Studies have shown that with larger unsprung mass, the ride comfort performance of vehicle tends to come down[6,7]. Adjusting the suspension parameters is a effective way to alleviate the negative effect of ride comfort[8]. With different types of in-wheel motor integrated, the unsprung mass are different. In this paper, a rigid-elastic multi-body dynamic model of typical IMEV is built. Based on the model, the influence on vehicle ride comfort performance of different in-wheel motor mass is investigated.

II. FULL VEHICLE MODELING

A rigid-elastic coupling muti-body dynamic model of an IMEV is built. The model is composed of double pivot front suspension system; rear suspension system with a flexible body twist beam; steering system; anti-roll bar system; tire system; in-wheel motors; vehicle body and battery cells. Major parameters of the multi-body dynamic model is shown in Table 1.

TABLE I. MAJOR PARAMETERS OF IMEV MODEL.

Parameter	Value
Number of moving parts	58
Degree of freedom	74
Curb weight (kg)	1540
In-wheel motor weight (kg)	54
Battery cell weight (kg)	280
Wheel base (mm)	2389

The mass and geometry parameters were obtained by physical measurements. The bushing, spring and damping characteristics are obtained by disassembling the suspension parts and measuring on hydraulic actuator test bench. The in-wheel motor with motor controller is installed in wheels respectively.

III. RIDE COMFORT ANALYSIS

A. Quarter Vehicle Modeling

The vehicle ride performance can be considered as vibration response of vehicle-road system. In order to analysis the vibration response of vehicle and road interact system, a quarter vehicle system is built. In this model, m1 and m2 are the unsprung and sprung mass of vehicle respectively; Z1 and Z2 are vertical displacement of wheel and body; K is the suspension spring stiffness; Kt is the tire vertical stiffness; C is damping coefficient of damper system; q represents for the irregularity of road. The dynamic equation of the quarter vehicle model is:

$$\begin{cases} m_2\ddot{z}_2 + C(\dot{z}_2 - \dot{z}_1) + K(z_2 - z_1) = 0 \\ m_1\ddot{z}_1 + C(\dot{z}_1 - \dot{z}_2) + K(z_1 - z_2) + K_t(z_1 - q) = 0 \end{cases} \quad (1)$$

Solve Eq. (1) to obtain the transfer function of body and wheel displacement in response of q:

$$H(j\omega)_{z_1-q} = \frac{K_t(-\omega^2 m_2 + j\omega C + K)}{U} \quad (2)$$

© 2015. The authors - Published by Atlantis Press

$$H\left(j\omega\right)_{z_2-q} = \frac{K_t\left(j\omega C + K\right)}{U} \tag{3}$$

$$U = \left(-\omega^2 m_2 + j\omega C + K\right)\left(-\omega^2 m_1 + j\omega C + K + K_t\right) - \left(j\omega C + K\right)^2 \tag{4}$$

B. Body Acceleration Response

The body acceleration response is often considered as an evaluation index for ride comfort performance. The amplitude-frequency characteristics $\left.|H\left(j\omega\right)|\right|_{\ddot{z}-\dot{q}}$ of body acceleration in response of road irregularity is:

$$\left.\left|H\left(j\omega\right)\right|\right|_{\ddot{z}-\dot{q}} = 2\pi f \frac{K_t}{K} \left[\frac{1 + 4\xi^2\left(f/f_0\right)^2}{V}\right]^{\frac{1}{2}} \tag{5}$$

$$V = \left[1 - \left(f/f_0\right)^2 \left(1 + \frac{K_t}{K} - \frac{m_1}{m_2}\left(f/f_0\right)^2\right) - 1\right]^2 + 4\xi^2\left(f/f_0\right)^2\left[\frac{K_t}{K} - \left(\frac{m_1}{m_2} + 1\right)\left(f/f_0\right)^2\right]^2 \tag{6}$$

where $f_0 = \sqrt{k/m_2}/2\pi$ defines the natural frequency of vehicle body, $\xi = C/2\sqrt{K \times m_2}$ is the damping ratio of quarter vehicle system. The root mean square of body acceleration is given as:

$$\sigma_{\ddot{z}} = \sqrt{4\pi^2 G_q\left(n_0\right)n_0^2 v \int_0^\infty \left.\left|H\left(j\omega\right)\right|\right|_{\ddot{z}-\dot{q}}^2 df} \tag{7}$$

Eq. (2), (3) and (7) show the vibration characteristics of quarter vehicle system. It can be seen that the factor m1/m2 is a key factor influences acceleration response. Figure 1 shows the relationship between body acceleration and mass ratios in different speeds. The result is obtained by adjust the motor mass of multi-body full vehicle model on B class pavement. With larger m2 value, the body acceleration tends to increasing slightly.

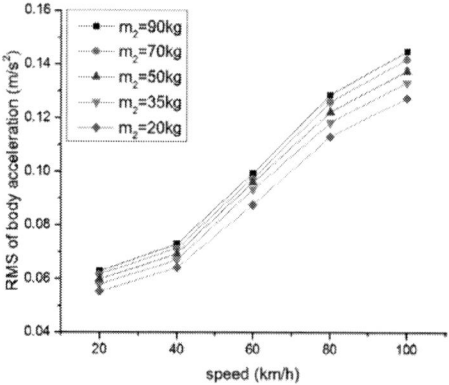

FIGURE I. BODY ACCELERATION RESPONSE.

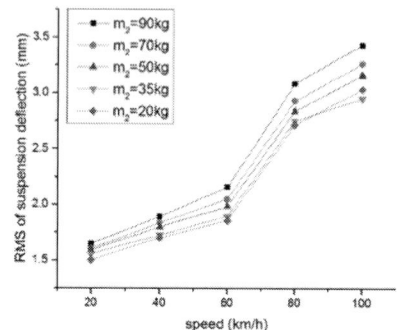

FIGURE II. SUSPENSION DEFLECTION RESPONSE.

C. Suspension Deflection Response

The root mean square of suspension deflection σfd is described as:

$$\begin{cases} \sigma_{fd} = \sqrt{4\pi^2 G_q\left(n_0\right)n_0^2 v \int_0^\infty \left.\left|H\left(j\omega\right)\right|\right|_{z-\dot{q}}^2 df} \\ \left.\left|H\left(j\omega\right)\right|\right|_{z-\dot{q}} = \frac{f K_t/K}{2\pi f_0^2 \sqrt{B}} \end{cases} \tag{8}$$

Adjust the motor mass of multi-body full vehicle model and run the models on B class road, relationship between suspension deflection and mass ratios is obtained as in Figure 2. With higher speed, the influence of higher unsprung mass is greater. The increment is 11.65% at maximum.

D. Wheel Dynamic Load Response

The amplitude-frequency characteristic of wheel dynamic load σFd in response of speed and the root mean square of wheel dynamic load is given as:

$$\begin{cases} \left.\left|H\left(j\omega\right)\right|\right|_{F_d-\dot{q}} = 2\pi f \frac{K_t}{K}\sqrt{\frac{\left[\left(f/f_0\right)^2 - \left(1 - m_1/m_2\right)\right]^2 + 4\xi^2\left(f/f_0\right)^2}{\left(1 - m_1/m_2\right)V}} \\ \sigma_{F_d} = \sqrt{4\pi^2 G_q\left(n_0\right)n_0^2 v \int_0^\infty \left.\left|H\left(j\omega\right)\right|\right|_{F_d-\dot{q}}^2 df} \end{cases} \tag{9}$$

Figure 3 shows the relationship between wheel dynamic load and mass ratios on B class pavement of different speed. With higher speed and higher unsprung mass, the wheel dynamic load is higher. The greatest increment is 40.4%, which means this index is heavily influenced.

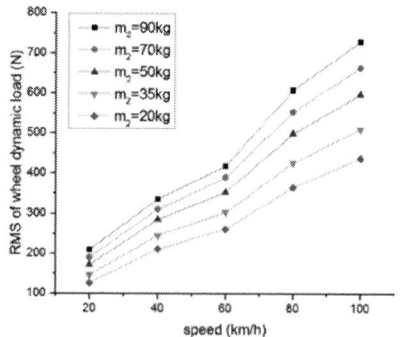

FIGURE III. WHEEL DYNAMIC LOAD RESPONSE.

IV. CONCLUSIONS

A rigid-elastic coupling multi-body dynamic model of a in-wheel motor vehicle was built. The ride comfort performance of different mass ratios is investigated. The vibration characteristics of vehicle system is analyzed based on the quarter vehicle model.

Negative influence on vehicle ride comfort is investigated. The body acceleration is influenced by the change of unsprung mass. The largest increment on suspension deflection is 11.65%. The wheel dynamic load is influenced heavily in a increment of 40.4%. For ride comfort performance of vehicle with high unsprung mass, a higher damper coefficient is suggested.

ACKNOWLEDGMENT

This research work was supported by National Natural Science Foundation project of China (No. 51205152). The authors would like to express their appreciations for the above fund support.

REFERENCE

[1] R. Wang, Y. Chen and D. Feng et al., Development and performance characterization of an electric ground vehicle with independently actuated in-wheel motors , J Power Sources 196 (2011) 3962-3971.

[2] K. M. Rahman, K. M. Rahman and N. R. Patel et al., Application of Direct-Drive Wheel Motor for Fuel Cell Electric and Hybrid Electric Vehicle Propulsion System, Ieee T Ind Appl 42 (2006) 1185-1192.

[3] H. Alipour, M. Sabahi and M. B. B. Sharifian, Lateral stabilization of a four wheel independent drive electric vehicle on slippery roads , Mechatronics (2014).

[4] L. Rambaldi, E. Bocci and F. Orecchini, Preliminary experimental evaluation of a four wheel motors, batteries plus ultracapacitors and series hybrid powertrain , Appl Energ 88 (2011) 442-448.

[5] W. Kim, K. Yi and J. Lee, Drive control algorithm for an independent 8 in-wheel motor drive vehicle, Journal of Mechanical Science and Technology 25 (2011) 1573-1581.

[6] G. Nagaya, Y. Wakao and A. Abe, Development of an in-wheel drive with advanced dynamic-damper mechanism , Jsae Rev 24 (2003) 477-481.

[7] D. Hrovat, Influence of unsprung weight on vehicle ride quality , J Sound Vib 124 (1988) 497-516.

[8] S. Chen and D. Wang, Optimization of Vehicle Ride Comfort and Controllability Using Grey Relational, Journal of Grey System 23 (2011) 369-380.

Electrical Power Management for Distributed Automotive System

X.F. Zhang, M.H. Luo, Y. Shen
Clean energy automotive engineering center
Tongji University
Shanghai, China

J.D. Cao
China Academy of Transportation Sciences
Beijing, China

Abstract—**The electrical power management for distributed automotive system is proposed. The electrical power requirement is analyzed and the steady power source and electrical safe power source are independently proposed. Power supply channel with three-level current protection is designed to construct an electrical safe power supply which includes selective overload protection, fast short current protection and fuse backup protection. Prototype of electrical power management system is developed based on intelligent relay. The test result showed that the electrical management works well and the protection method is effective.**

Keywords-electrical power management; steady power source; electrical safe power source

I. INTRODUCTION

With the rapid development of automotiveelectronictechnologies, electronic/electrical devices (EEDs) such asvehicle information systemshave become much more complicated than everbefore (Gu et al. 2010), at the same time more and more optimization work were carried out to simplify harness wires in automotive system(Zhang & Shen 2014). Available surveys (Toerngren et al. 2008, Giovanni et al. 2005)revealed that the embedded control with network connections, such as CAN (Control Area Network), LIN (Local Interconnection Network), has become a trend.At the same time, more and more intelligent EEDs come into reality in Automotive, which have over current protection and self-diagnostic function (Russell 2005). In such asystem, the wires which transfer digital signals can be separated from the wires that transfer the electrical power, inducing the independence of communication network and electrical power network, which facilitates the electrical power distribution and management design.

In this paper, an electrical power management method,which is suitable for distributed electrical system in automotive is proposed.Three-level protection method, includingsoftware overload protection, fast short current protection and fuse backup protection,are adopted to meet over current protection requirement. Electrical power management system (EPMS) and its prototypeis developed and tested,which is proved to be effective and came into usage.

II. DISTRIBUTION OF AUTOMOTIVE ELECTRICAL POWER SUPPLY

A. Electrical Power Requirement

The requirements for electrical power can be divided into two types. First, the power forelectricaldevices,such as electrical motors, electrical heaters, electrical vales and lights, etc. The voltage is always 12/14 V in passenger cars and 24/28 V in trucks or buses. Second, the power for circuits in ECU (Electric Control Unit) and communication network. The circuits inside an ECU usually works under 5 or 3.3 V, sometimes 2.5 V or even lower. Astabilized power of 8-18V is also asked by LIN bus system as a reference voltage.

System voltage perturbation appears frequently when the electronic circuits are instable, eg.at the starting time by the starter, and sometimes by the burst into usage of high power devices or cutoff of inductive loads while vehicle's working. System voltage perturbation could cause digital communication problems, such as CAN error frames and LIN communication errors.

In general, in a distributed automotive electrical system, an electrical-safe power source (ESPS) as well as stable-neat power source (SNPS) is independently necessaryfordifferent requirements. The former is for lower voltage stability and higher energy consumption, and the latter provides an uncontaminated stable power source and lower energy consumption.

B. System Layout

A typical electrical system is as Figure 1 shows (Gu et al. 2010).d and δ, namely the intelligent EEDs with CAN and LIN digital interface, are composed of the original EEDs and an on-site ECU. The intelligent EEDs are networked and organized into one system, which features a hierarchical structure. The δs are firstly grouped as several local networks (L in Figure 1) according to a certain criteria, e.g. function similar criteria, location adjacent criteria, etc. secondly Ls and ds are connected through gateways. Finally, Ls and ds together compose the system.

© 2015. The authors - Published by Atlantis Press

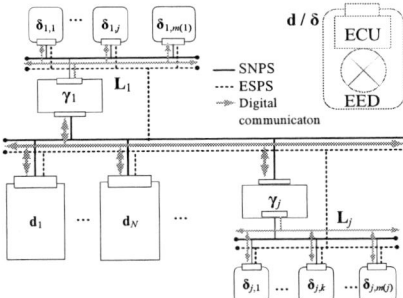

FIGURE I. DISTRIBUTED ELECTRICAL SYSTEM LAYOUT.

As referred in last section, SNPS and ESPS compose an independently electrical power network in an electrical power supply. SNPS is usually composed of a constantly-on-source part and aswitchable part. Some EEDs need slight electrical power in a parking vehicle, e.g. remote sensor, clock, memory inside odometer and instrument to keep the number of miles and individual configuration, anti-thief system, keyless entrance system, etc. The EEDs mentioned above need sustained but much limited power supply, e.g. 10mA@12V or even lower. In contract, most of electrical circuits inside ECUs are shut down while parking, when there is no consumption and energy requirement.

So normal-open SNPS andswitchable SNPS are figured out. The former is to power the constantly working EEDs while the latter is for all other ECUs'and to minimize the transient current during parking, which has a great effect on battery's state of charge (SOC).

C. Electrically Safe Power Source

The ESPS network is composed of several power supply channels (PSCs). PSC is a switchable electrical transfer wire with over current protection. Grounding points are necessary for less wires and connectors. A high power volume EED has its dedicated PSC, e.g. starter, while several EEDs in the adjacent area or same local network share one PSC to reduce the harness complexity. It is the problem to keep electrical safety after the removal of traditional fuse box. The EPMS is introduced for PSC current protection as well as battery SOC monitoring.

III. THREE-LEVEL OVER CURRENT PROTECTION

A. Current Protection of Power Supply Channel

Three-level over current protection is proposed to ensure PSC's electrical safety. The current protection points A~C along the PSC are as Figure 2 shows.

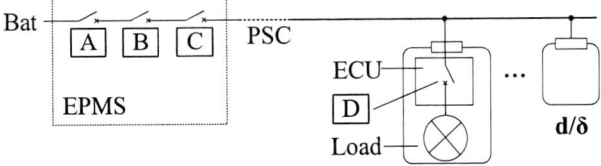

FIGURE II. PROTECTION POINTS OF POWER SUPPLY CHANNEL.

As shown in Figure 2, "A" is software over current protection based on MCU (Micro-Control Unit).The time delayed before current break is calculated according to the magnification

of the actual current to the allowed maximum. Itobtains the tolerance of current pulse and make it flexible(Gu& Li 2005). In most case, protection will be activated at point D as the PSC should not be broken down frequently because it is responsible to several EEDs. As a result, time delay function and current pulse tolerance are necessary. "B" is fast break-out protection in case of short circuit, e.g. PSC wire short to ground. "C" is arecoverable fuse as backup protection. It will cut off the PSC in case of "A" and "B" failure. "D" is intelligent power electronics inside the on-site ECU of d/δ rather than a traditional fuse. And A~C are set for the possible failure at point "D".

B. Software Current Protection

Nowadays, intelligent circuit breaker, three-segment protection method is adopted (Kosack 1999), namely the long time delay, short time delay over current protection and short circuit protection, is used to obtain excellent tolerance.

The real time current or normal current IR equals to the sum of each running EEDs' normal power divided by nominal system voltage. Then the nominal delay time features proportion to the magnification factor K. The relation between action time T and K can be as in Equation (1):

$$\begin{cases} KT = K_1 t_1 & K_1 < K < K_2 \\ KT = K_2 t_2 & K_2 < K \end{cases}, K = I_R/I \quad (1)$$

where, I= the real time current from the sensor;K1= the lower limit of long time delay protection over current factor;K2= the lower limit of short time delay protection over current factor;t1, t2= the action time under K1 and K2.

The software current protection is implemented by MCU. As is shown in Figure 1, all the EEDs' status can be monitored and the reference current on a specific PSC can be obtained with the additional rated power of the working EEDs. An individual current sensor is laid on PSC and the real time over current factor ki can be achieved at each sample time.

Left time of protection τ is defined as Equation (2):

$$\tau_i = \begin{cases} \tau_{i-1} + t_R & k_i < K_1 \\ \tau_{i-1} - \dfrac{T_i}{T_{i-1}} T_s & k_i > K_1 \end{cases} \quad (2)$$

where, τi and τi-1= the left time at the ith and (i-1)th sample time; Ti, Ti-1 = the action delay time according to Equation (2); tR = the recoverable factor;Ts = MCU's sample time. τ has its up limit of Tu, which is also the initial value of τ. When τ≤ 0, the PSC will be cut off and the software current protection happens.

C. Fast Break-out Protection

In case of short circuit current on PSC, which exceed the PSC's maximum allowed current, a fast break out protection should be taken place. Software protection cannot react within one millisecond with a slow 8 bit MCU. For that a circuit for fast break out is designed. As Figure 4 shows, U1 is from the current sensor and U0 is the reference voltage, namely the maximum allowed current on PSC. In case of the short circuit current on PSC, U1 is higher than U0, the comparator's output will be zero and the R&D trigger will give out a low voltage level on Q terminal, too. So second coil powered by the MOSFET will be shut down because of the gate electrode voltage drop, thus the relay is broken out.

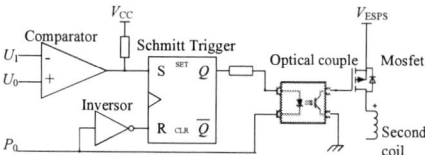

FIGURE III. FAST BREAK-OUT CIRCUIT.

After the clearance of short current status sensor voltage U1will drop and the comparator output will be recovered to the TTL high voltage. Although, thanks to status lockage by the R&D trigger, the output of Q terminal remains lower, and the MOSFET remains closed, thus keeps the relay closed. R&D trigger can be reset through port P0 by MCU.

The delay time of the fast break out protection consists of:

– Response time of Hall sensor, equal to $O(10^{-6})$s;

– Delay by first order transfer function of RC filter. The delay can be calculated as Equation (3):

$$t_{RC} = -T_{RC} \cdot \ln(1 - \frac{1}{K}) \qquad (3)$$

– where, TRC= the RC constant;K=theratio of current magnitude to maximum allowed current of PSC, and K is always more than 1.

– Response time of R&D trigger, equal to $O(10-9)$s;

– Response time of optical couple, about $50 \times 10-6$s;

– Response time of MOSFET, equal to $O(10-9)$s;

– Response time of Relay, about $2 \times 10-3$s.

– According to analysis above, the total time delay can be within 3 milliseconds. Thus the delay can be equal to $O(10-6) \sim O(10-3)$s.

D. Back up Fuse Selection

Backupfusemay work in case of failure of fast break out protection and software protection. It is very important to make sure that backup fuse wouldn't act before the software protection or fast break out protection. Protection curve of backup fuse should be selected compared with that of time delay protection. The dashed line is obtained from Equation (3) and reasonable margin between the two lines. So the backup fuse's time delay protection curve should always be above that of the software protection.

IV. POWER MANAGEMENT SYSTEM PROTOTYPE

A. System Layout

In the EPMS, each PSC is controlled by IRs (Intelligent Relays), while the number of IRs depends on system requirement. The gateway (GW) is used to generate the SNPS as well as to transfer digital signals between the local network and its

superior. The BMU (Battery ManagementUnit) monitors the total current consumed. System layout is as shown in Figure 4.

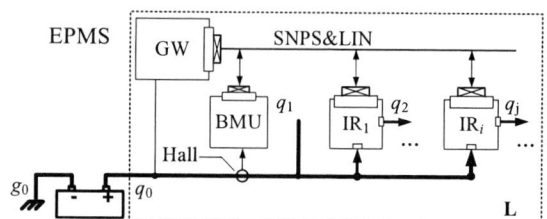

FIGURE IV. EPMS SCHEMATIC ANDLAYOUT.

Contact-type relays were chosen and the schematic of IR is designed. LIN interfacecircuit and relay contact diagnostic circuit are also designed except the fast break-out circuit mentioned above. The controllers of IR are designed and implemented. The EPMS system was also developed as one assembly for a travel bus electrical system, as shown in Figure 5.

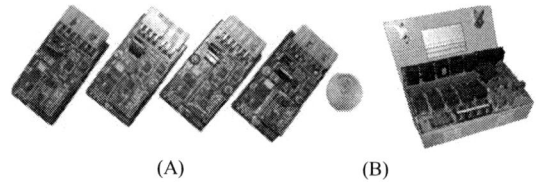

(A)   (B)

FIGURE V. IR'S CONTROLLER AND EPMS PROTOTYPE.

B. Over Current Protection Test

The CAN bus test tool Canalyzer was used to simulate the digital communication network. Gateway was also used to set the reference current and the actual load was performed by electrical load. The software protection test process is as Figure 6 shown.

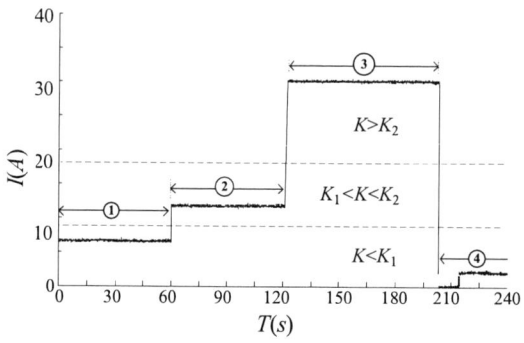

FIGURE VI. SOFTWARE PROTECTION TEST.

K1 were set to 1.15and K2 were 2. Andt1, t2 were set to140 and 90s. Four segments were performed:

– the reference current was set to 8A and the load was about 10A. EPMS worked well and no shut down warning signal was sent;

– the actual load was set to about 12A, so slight over load happened. After around 73 seconds, "Channel will shut down in 60 seconds" warning was sent out and recorded by Canalyzer;

120

– the actual load was set to about 30A, so serious over load happened. After around 80 seconds, PSC was closed;

– the state of IR was reset and a current of 3A is loaded. The IR and PSC recovered.

The proposed electrical power distribution method as well as the EPMS has been applied to a prototype vehicle, a mid-size concept car and a twelve-meter long city shuttle bus.

In total of 10 PSCs are created to power more than 100 EEDs. More than twenty thousand kilometers road test was done and it was proved to be safe and reliable.

V. CONCLUSIONS

(1) Steady-neat power supply and electrical safety power supply can fulfill the electrical power requirement of distributed electrical system in automotive.

(2) Power supply channel based electrical power distribution method is feasible. Software over current protection obtain good flexibility, fast break out protection is also work well.

(3) The three-level current protection method can ensure electrical safety in automotive system.

ACKNOWLEDGMENT

This work was supported by the and National High Technology Research and Development Program of China (2013AA12A026)and National Instrument Special Funding on Automotive fuel cell system testing (2012YQ150256).

REFERENCE

[1] Christopoulos, C. &Wright, A. 2005.Electrical power system and its protection. Beijing: electrical power press of China.

[2] Giovanni, T., Joseph, N. & Gary, B. 2005. Evolutionand trends in automotive electrical distributionsystems. In Proceedings of the IEEE Vehicle Powerand Propulsion Conference: 812–818.

[3] Gu, S.Q. & Li, F.R. 2005.Electrical power system and its protection. Beijing: electrical power press of China.

[4] Gu, Z.M.,Yang, D.G. &Zhang, X.F. 2010. Distributed vehicle body electric/electronic system architecture with central coordination control. International journal of automotive engineering 224(2): 189-199.

[5] Kosack, K. 1999. Low voltage electrical device and switch datasheet: a select criteria and design guide.Beijing:China Machine Express.

[6] Russell, M.E. 2005. Integrated automotive sensors.IEEE Transaction on Microwave Theory andTechniques 50(3): 674-677.

[7] Toerngren, M., Johansson, K.H., Andersson, G.,Bodin, P. &Purdue, D.2008. Assessment of high-integrity embedded automotive control systems using hardware in the loop simulation. Journal of Systems and Software 81(7): 1163-1183.

[8] Zhang, X.F. & Yong, S. 2014. Networking optimization for Holo-distributed automotive body electrical system. Journal of computers 9(1): 112-117.

International Conference on Power Electronics and Energy Engineering (PEEE 2015)

Novel 0.1 Hz Exponential Wave Generator Based on Semiconductor Switch

Z. Hou, C.B. Yang, H.J. Li, S.C. Ji

Xi'an Jiaotong University
Xi'an, Shaanxi Province, P. R. China, 710049

J. Li, X. Chen

Guangxi Electric Power Research Institute
Nanning, Guangxi Zhuang Autonomous Region, P. R. China, 530023

Abstract—**Very low frequency (VLF) high voltage technique has been used to some extent in the past for the field test of XLPE power cables. Equipment available has used 0.1 Hz voltage having a variety of waveforms such as square waves, triangular waves, etc. This paper describes a novel laboratory test system conducted with the objective to develop an exponential wave generator as the complement of existing VLF test waveforms. The generator consists of a high voltage (HV) 50 Hz AC source, a HV semiconductor switch unit based on the series-connected IGBTs, and shape-control resistors. The proposed HV switch can block 20 kV rated voltage and has the current capacity of 40A. Finally, the new designed approach permits the cable load of up to 1μF to be tested at a withstand voltage level of 20 kV. Preliminary experiments are performed in the laboratory using the proposed generator and the output waveforms are presented. The experiment results show that the exponential wave appears to be a satisfactory alternate to the waveforms used in the traditional VLF test and the generator is a utility installation for the diagnostic test of power cables.**

Keywords-**VLF; test systems**

I. INTRODUCTION

Recent publications in the technique community have shown that Very Low Frequency (VLF) test of XLPE power cables at 0.1 Hz has increasingly gained interests during the past 10 to 15 years [1]. The basic idea in using VLF test for the diagnostic test is to take advantage of the low charging current required to charge the specimen to a high voltage over a relatively long time interval [2]. Moreover, the dissipation factor (tan δ), which is an important parameter for identifying the water tree content in a cable or the presence of other defects in the insulation or terminations, obtained during VLF test is much larger than at power frequency test, giving advantage of reducing the sensitivity requirement of the testing system.

The two most commonly used VLF technologies differ in the wave shapes of 0.1 Hz AC voltage. The sinusoidal waveform technology features a continuously sinusoidal waveform with a period time of 10 seconds, while the cosine-rectangular waveform technology generator a 0.1 Hz rectangular waveform with a cosine-shaped rising and falling edge which lasts 2 to 6 milliseconds. The topology of the former waveform generator is presented in [3], which uses a low-pass filter to convert the high frequency sinusoidal-wave-enveloped voltage waveform into 0.1 Hz waveform and has the voltage level of only 2.4 kVp-p. The latter waveform generator is described in [4],

which includes a complex software for controlling the digital signal processor (DSP).

This paper proposes a new approach to the VLF test. We developed a 20 kV rated 0.1 Hz exponential waveform generator, which has the advantage of lower cost, equipment size and simpler in structure over other technologies. Preliminary experiments on the capacitive specimen are performed using the generator in the laboratory, showing that the deliberate engineering development effort is applicable on the tan δtest.

II. OPERATION PRINCIPLE OF PROPOSED EXPONENTIAL WAVER GENERATOR

A detailed design and diagram of the generator is described in this section.

A. Circuit Description

Figure 1 shows the circuit diagram of the exponential wave generator. The input power is directly obtained from the normal 220V power frequency source. The output voltage amplitude is controlled by the variable transformer T1, as depicted in fig. 1.Controlled using optical fibres, P1, the polarity HV switch, consists of two major units, the forward conducting switch Q1 and the reverse conducting switch Q2. R1 is the current limiting resistor, preventing the step up transformer from transient over-current. C1 is the output voltage filter of P1, and R2 acts as the shape regulator of the output voltage on the capacitive specimen. S1 and S2 are AC switch pairs, turning on and off periodically to coordinate with P1 to protect the transformer core of T2 from saturation.

Operating principle of our schematic is depicted in Figure 2, where vg is the gate control voltage of the gate-controlled equipment and high level of vg means turning on the corresponding device. As can be seen, Q1 and Q2 perform alternately conducting process which lasts only half of the whole period of time (t0~t4). Turning on the forward switch Q1 charges the specimen to the preset positive voltage (t0~t1) through the shape-controlling resistor R2, then Q1 is turned off and Q2 is turned on, discharging the specimen (t1~t2) and recharging in a negative way (t2~t3). Finally, Q1 is turned on again and Q2 is turned off, and the generator operates in discharging process at the negative voltage (t3~t4). S1 is turning on during the charging process, whereas S2 is conducted during the discharging process to offer a reverse current path in case of the saturation of the core.

© 2015. The authors - Published by Atlantis Press

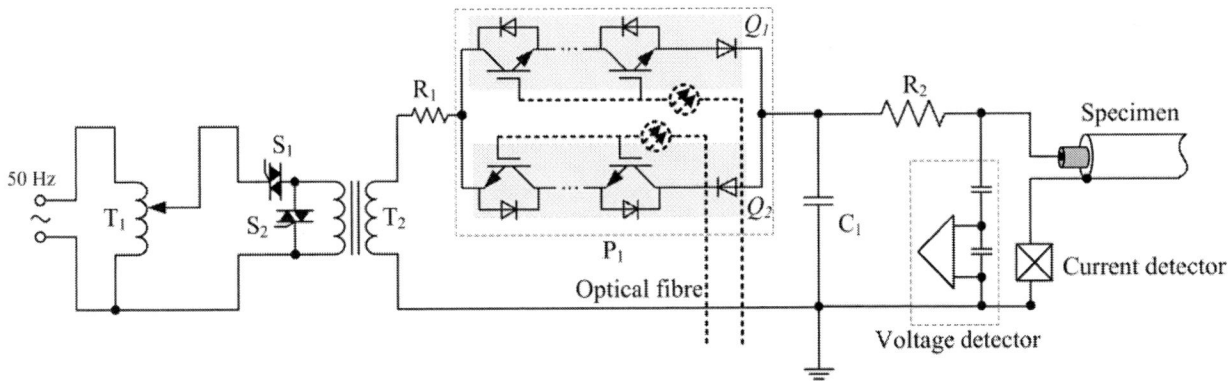

FIGURE I. THE STRUCTURE OF PROPOSED EXPONENTIAL WAVE GENERATOR.

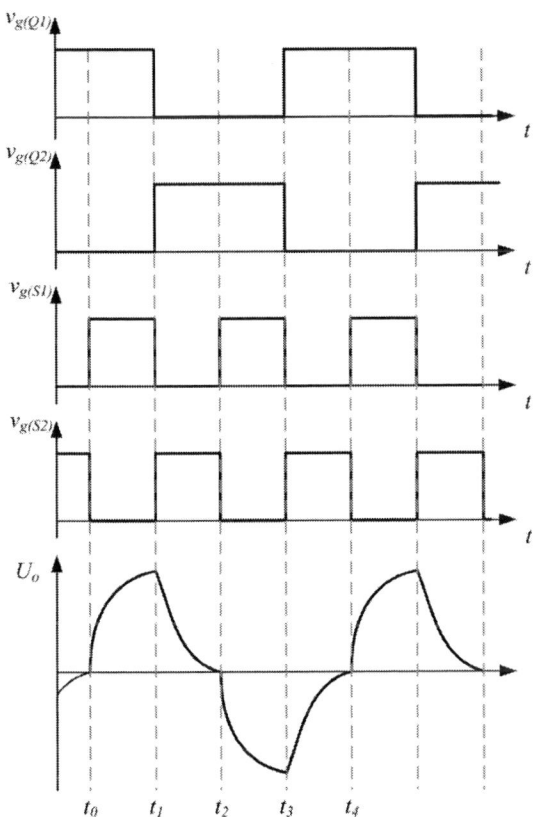

FIGURE II. TYPICAL WAVEFORMS OF THE PROPOSED GENERATOR.

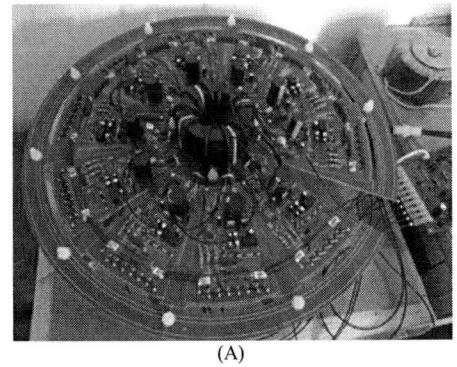

(A)

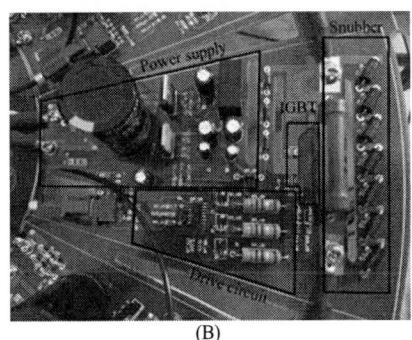

(B)

FIGURE III. THE APPEARANCE OF PROPOSED HV SWITCH: (A) OVERALL DESIGN; (B) DETAILS.

B. Design of the High Voltage Switch

The Structure of the proposed HV switch is shown in Figure 3. It consists of several main assemblies: a switch stack, a power supply system and the trigger unit. Each stack is made up of several series connected IGBT switch units (SUs), namely the IGBT chips, their accessory drive circuits, power-supply outputs and the snubbers. Thanks to the voltage balancing methods, which are paralleled statistic voltage sharing resistors and a series of dynamic voltage sharing transient voltage suppressors, each IGBT is rated only for an identical fraction of the full blocking voltage. To reach the blocking voltage requirement of 20 kV, nine IGBTs should be used as a basis if each of them carries the anticipate voltage of 2.5 kV, margin reserved for reliability operation. Considering the size and shape of the stack, redundancy concept is used. So the developed stack is configured as ten IGBT SUs connected in total, with each rated for 2 kV.

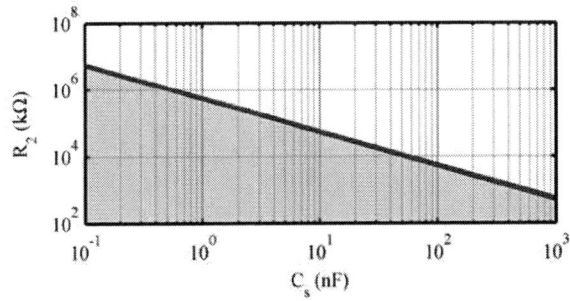

FIGURE IV. PERMISSIBLE REGION FOR R2 AS A FUNCTION OF CS

123

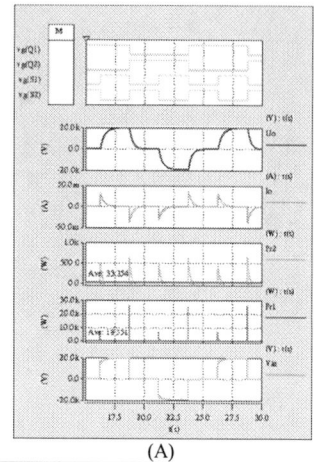

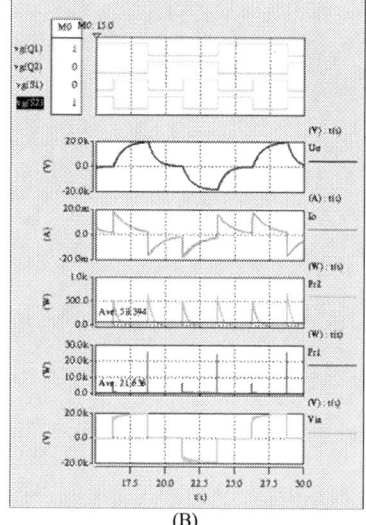

FIGURE V. FIGURE 5. THE SIMULATION WAVEFORMS OF THE
EXPONENTIAL WAVE GENERATOR AT (A) CS =500 NF AND (B) CS =
1 MF.
GATE CONTROL VOLTAGE OF Q1, VG(Q1); GATE CONTROL
VOLTAGE OF Q2, VG(Q2); GATE CONTROL VOLTAGE OF S1, VG(S1);
GATE CONTROL VOLTAGE OF S2, VG(S2); OUTPUT VOLTAGE ON
THE SPECIMEN, UO; CURRENT FLOWING THROUGH THE
SPECIMEN, IO; POWER DISSIPATION ON R2, PR2; POWER
DISSIPATION ON R1, PR1; VOLTAGE ON C1, VIN.

C. Parameter Specification and Simulation Analysis

As can be seen from Figure 2, voltage on the specimen Uo
in one period of time could be deduced as

$$
U_0 = \begin{cases}
V_{in}(1-e^{-\alpha(t-t_0)}), & t_0 \leq t \leq t_1; \\
V_{in}e^{-\alpha(t-t_1)}, & t_1 \leq t \leq t_2; \\
V_{in}(1-e^{-\alpha(t-t_2)}), & t_2 \leq t \leq t_3; \\
V_{in}e^{-\alpha(t-t_3)}. & t_3 \leq t \leq t_4;
\end{cases}
\tag{1}
$$

Where Vin is the voltage on C1, and α is the parameter defined as

$$
\alpha = \frac{R_2 + R_s}{R_2 C_s R_s}
\tag{2}
$$

where Cs and Rs respectively are the equivalent capacitive
and resistive value of the specimen. In most cases, the value of
Rs is very large (thousands of MΩ) and can be neglected.

It should be mentioned that the shape of the generator output voltage is varied with the capacitive value of the specimen,
which is linearly related to the length of the cable to be tested.
Assuming that the specimen is charged to the voltage level no
less than kVin as defined above during time interval of t0~t1,
we could get

$$
V_{in}(1-e^{\frac{-\alpha T}{4}}) \geq kV_{in}, \qquad 0.9 \leq k < 1
\tag{3}
$$

Where T is the period time of the exponential wave (T= 10
s), derived from (3),

$$
R_2 C_s \leq -\frac{T}{4\ln(1-k)},
\tag{4}
$$

Figure 4 shows the operation range for selecting R2, where
the parameter k is specified as 0.99. From Figure 3 we can see
to guarantee that the circuit is properly operated under the
maximum load condition of 1μF as mentioned above, the minimum value of R2 is 540 kΩ. In the following paragraph we
chose R2 = 500 kΩ to meet the worst-case requirement.

The performance of the generator designed was evaluated
using SABER Sketch simulation package. A number of simulations were run for varying output voltage and load conditions.
Various waveforms obtained demonstrated that the output of
the generator is consistent with the theoretical waveforms.
Some sample waveforms obtained with the variation of Cs are
shown in Figure 5. We can see that because of the reduced
value of load, the charge and discharge speed in Figure 5(a) is
much faster than in Figure 5(b), results in the variation of the
wave shape.

III. EXPERIMENTAL RESULTS AND APPLICATION

To verify the aforementioned analysis, a laboratory exponential wave generating system is built. To refine the voltage
shape under the specified load condition, the specifications of
the system are as follows: R1 = 15 kΩ, R2 = 6 MΩ, C1 = 200
nF. T1: 220V input, 500W varitran; T2: 220V to 20kV, 500W
step up transformer; S1, S2: 1 kV, 50A AC switch; Q1, Q2:
proposed HV switch series connected with 40kV, 1A silicon
stack.

The experimental setup is depicted in Figure 6(a) and the
results are given in Figure 6(b) the specimen parameters are Cs
= 50 nF and Rs = 1660 MΩ. In Figure 6(a) we can see the amplitude of output voltage is 20 kV, and the wave shape is similar with the simulating results, which is the proposed exponential wave.

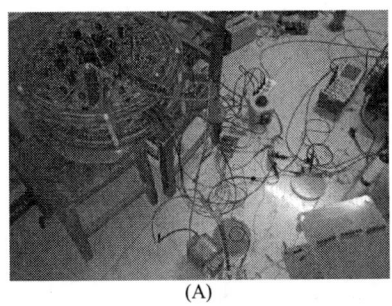

(A)

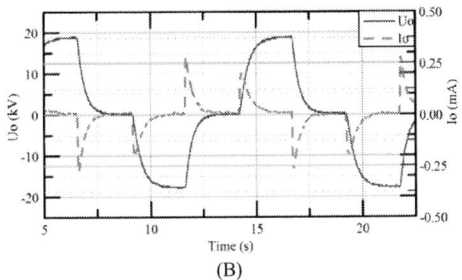

(B)

FIGURE VI. (A) EXPERIMENT SETUP AND (B) OUTPUT
WAVEFORM.

IV. CONCLUSION

On the basis of the development of a 20 kV double module switch built up with commercially available IGBTs, a novel 0.1 Hz exponential wave generator is designed and tested in the laboratory. Simulating results are proposed, showing that the system could generate expected waveform under various load condition. To refine the wave shape under the worst case, system parameters are calculated and specified. We setup a prototype and the feasibility of this scheme is proved by the produced exponential wave of anticipated rate of 20 kV. It should be mentioned that the further application of the generator is to calculate the dissipation factor (tan δ) and dielectric spectrumby using the excitation voltage Uo and the response current Io. Compared with traditional sinusoidal 0.1 Hz VLF generator, the exponential wave generating system allowsanalyses of harmonics as well as the fundamental frequency. Further research on this kind of application will be proposed later.

REFERENCE

[1] J. C. Hernandez-Mejia, R. Harley, N. Hampton, and R. Hartlein, "Characterization of Ageing for MV Power Cables Using Low Frequency Tan delta Diagnostic Measurements," Ieee Transactions on Dielectrics and Electrical Insulation, vol. 16, pp. 862-870, Jun 2009.

[2] R. Reid, "High voltage VLF test equipment with sinusoidal waveform," in Transmission and Distribution Conference, 1999 IEEE, 1999, pp. 8-12 vol.1.

[3] S. Seesanga, W. Kongnun, A. Sangswang, and S. Chotigo, "A new type of the VLF high voltage generator," in Electrical Engineering/Electronics, Computer, Telecommunications and Information Technology, 2008. ECTI-CON 2008. 5th International Conference on, 2008, pp. 929-932.

[4] D. Pepper and W. Kalkner, "PD-pattern of defects in XLPE cable insulation at different test voltage shapes," in High Voltage Engineering, 1999. Eleventh International Symposium on (Conf. Publ. No. 467), 1999, pp. 313-316 vol.5.

International Conference on Power Electronics and Energy Engineering (PEEE 2015)

Dynamic Analysis of Aerial Work Platform Working Device Based on Virtual Prototype

J. Q. Guo, D.W. Liu
Automobile Engineering Department
Qingdao University
Qingdao, Shandong, China

J.G. Jia
Qingdao Jite Auto Technology Co. Ltd.
Qingdao, Shandong, China

Abstract—**In order to analyse dynamic response of aerial work platform working device, a virtual prototype was established by using dynamics software of ADAMS. Got the force of joint between components with the change of luffing angle which was carried by the dynamic analysis of work divece. A practical method of dynamic analysis of aerial work platform was created, and provided a reference on finite element analysis about aerial work platform working device parts in the future.**

Keywords-aerial work platform; working device; virtual prototype

I. INTRODUCTION

Working device is one of the most important parts of aerial work platform, the dynamic response of working device is directly related to the safety of staff, so it is significant to analyse the dynamic response of working device. In recent years, a great number of scholars and enterprises make lots of studies on aerial work platform [1, 2, 3, 4, 5, 6]. In this paper, Folding boom type aerial work platform was dynamic analysed by using virtual prototype, and got the force of joint between components. Provided the references of finite element analysis on working device in the future.

II. WORK PRINCIPLE OF AERIAL WORK PLATFORM WORKING DEVICE

Figure 1 demonstrates the aerial work platform working device. There were two oil cylinder to make the lower arm and the upper arm to work. The lower arm was moved by the lower oil cylinder. The lower arm range of angle was form 0° to 75°. The upper oil cylinder do not work until the lower oil cylinder stop to work. When the lower oil cylinder stop to work, the upper arm oil cylinder began to work. The upper arm range of angle was form 0° to 68°. In the process of work, the work platform kept horizontal all the time to ensure the safety of staff.

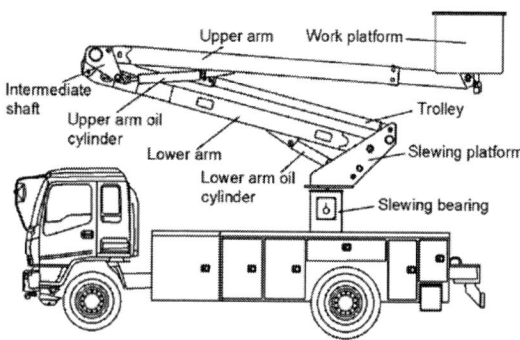

FIGURE I. AERIAL WORK PLATFORM.

III. ESTABLISH THE VIRTUAL PROTOTYPE OF WORKING DEVICE

This article is based on the hypothesis as follow as modeling:

(1) Assume the various components are rigid-body, and each kinematic pair is ideal constraints without considering the friction force between kinematic pair.

(2) In the working process of the aerial work platforms, consider only the effects on the capability of the motion components and work load, not the effect of the wind.

(3) Assume the slewing bearing as rigid-body which is fixed to the frame of aerial work platform, without considering the effects on the capability of the frame, and the suspension, the tires and ground deformation.

Because 3D geometric modeling tool in the software of ADAMS is difficult and complex, and there is no guarantee of the dimensional accuracy of the model and the setting position. For this reason, 3D modeling of working device's mechanical system were completed using the physical design software in this paper, then import 3D modeling is built to ADAMS environment, add constraint and load to the geometric model of prototype under ADAMS environment. In the model, the interrelation between the slewing platform and the lower arm, slewing platform and trolley, trolley platform and the push rod of lower arm cylinder, the tube of lower arm cylinder and the lower arm, the intermediate shaft and the lower arm, the intermediate shaft and trolley, the intermediate shaft and the upper arm, the intermediate shaft and the push rod of upper arm cylinder, the tube of upper arm cylinder and the upper arm

© 2015. The authors - Published by Atlantis Press

are defined the constraint of the plane rotation pairs restriction, the interrelation between the piston rod and the cylinder is defined the constraint of the plane moving pairs. In this way, the simulation model is a bit close to the actual operation of the working device.

In the working device, the luffing angle of the lower arm is 0°, and the luffing angle of upper arm is 0°, as shown in Figure 2; The upper arm always keeps horizontal while the lower arm luffing, as shown in Figure.3; The luffing angle of lower arm is 75°, and the luffing angel of upper arm is 0° as shown in Figure 4; The luffing angle of lower arm is 75°, and the luffing angel of upper arm is 68° as shown in Figure 5.

FIGURE II. LOWER ARM 0°AND UPPER ARM 0°.

FIGURE III. LOWER ARM AND UPPER ARM LUFFING.

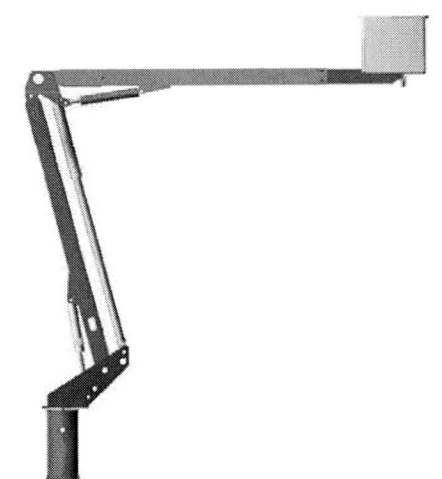

FIGURE IV. LOWER ARM 78°AND UPPER ARM 0°.

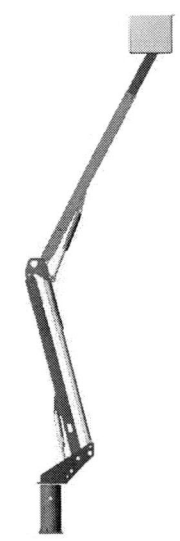

FIGURE V. LOWER ARM 78°AND UPPER ARM 68°.

IV. THE FORCE OF WORKING DEVICE

In the simulation calculation, the work platform was loaded at 200kg, the luffing process was simulated by translation joint between hydraulic cylinder and cylinder rod. The force of lower arm and upper arm changing by the luffing angle was showed as follows:

Figure 6 demonstrated the force of lower arm oil cylinder changing with the luffing angle. As shown in Figure 6, the force of lower arm was reducing with the luffing angel increasing. The maximum force appeared at the initial position. The curve had a sudden change when the lower arm luffing angle at 75°, the reason of these suddenly change was the lower arm cylinder stop to work.

Figure 7 demonstrated the joint force between lower arm and slewing platform changing with the luffing angle. As shown in Figure 7, the force of the joint force between lower arm and slewing platform was reducing with the luffing angel increasing. The maximum force appeared at the initial position, and this was the maximum force in this working device, reaching up to 270000N. The curve had sudden change when the lower arm luffing angle at 75°, the reason of these suddenly change was the lower arm cylinder stop to work.

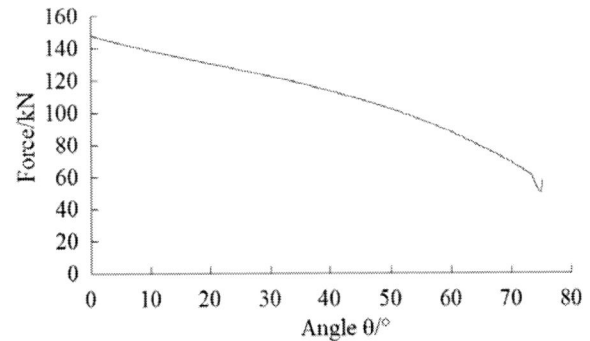

FIGURE VI. LOWER ARM OIL CYLINDER FORCE.

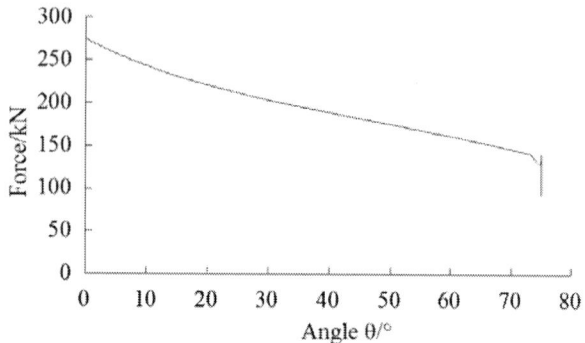

FIGURE VII. LOWER ARM FORCE.

Figure 8 demonstrated the joint force between upper arm and intermediate shaft changing with the luffing angle. As shown in Figure 8, the force of the joint force between upper arm and intermediate shaft was reducing with the luffing angel increasing. The maximum force appeared at the initial position.

Figure 9 demonstrated the joint force between upper arm oil cylinder and intermediate shaft changing with the luffing angle. As shown in Figure 9, the force of the joint force between upper arm oil cylinder and intermediate shaft was reducing with the luffing angel increasing. The maximum force appeared at the initial position.

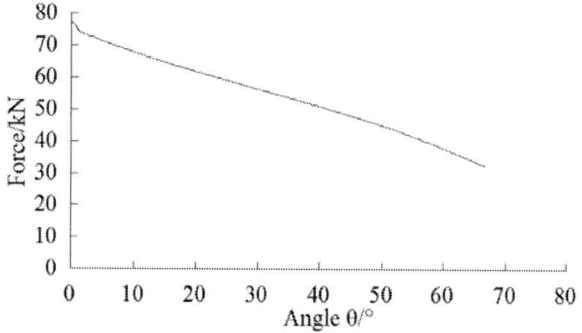

FIGURE VIII. UPPER ARM FORCE.

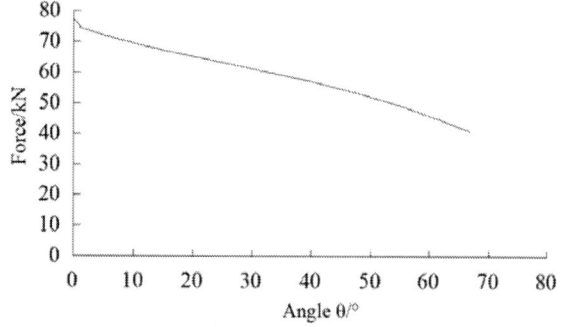

FIGURE IX. UPPER ARM OIL CYLINDER FORCE.

V. SUMMARY

(1) According the structural parameters of aerial work platform, this paper established the virtual prototype of aerial work platform by the Multi-body dynamics analysis software of ADAMS.

(2) The force of oil cylinder changed with the luffing angle during work process of working device. The maximum force occured at the initial position. The maximum force of lower oil cylinder was 1480000N and the maximum force of upper oil cylinder was 75000N.

(3) The maximum force of aerial work platform occured at the joint between lower arm and slewing platform in the initial position, size of 270000N. Finally the force stabilized at 88000N.

REFERENCES

[1] Wu Shengdi, How to improve the technology of our country in aerial working platform,J. Construction Mechanization.1996(5)36-38.

[2] Tian Liminng, Guo Weibin, The production and development of aerial work platform in China, J.Construction Machinery and Equipment.2003(2)20-22.

[3] Wang Zongjun, The Development of Macro Vision of Aerial Work Mechanical,J.Construction Machinery.1997(11)28-30.

[4] Wang Fei, Research of Electro-Hydraulic Control System of Aerial Work Platform,D. Xinan：Chang'an University, 2009.

[5] Wang Hui, The Finite Element analysis and Improve Research on New Crank Arm Type Aerial Work Platform,D. Nanjing:Southeast University, 2008.

[6] Chen Huabo, The Virtual Prototype and Structural Analysis of Supeer Structures of Aerial Work Platforms,D. Changsha:Hunan University, 2011.

International Conference on Power Electronics and Energy Engineering (PEEE 2015)

TFT Substrate Glass Geometrical Parameter Measurement and Data Processing

H. Guo

Xi'an University of Technology

NO.5 South Jinhua Road, Xi'an, Shaanxi, China

Abstract—**This paper has analyzed various sensors for the measurement of the TFT substrate glass geometric parameters and ascertain one of optical sensor for this task. The influence of measuring accuracy due to measuring platform has been studied, and then a solution for guarantee high precession measurement has been proposed.**

Keywords-TFT glass; spectral confocal sensor; measuring platform

I. INTRODUCTION

There are many factors which influence the quality of TFT substrate glass and the LCD panel such as glass optical refractive index difference, manufactory stress and fault. The geometric parameters are also important factor, special the thickness and warpage. A new instrument consists of optical sensor, high-precision granite platform, CNC automatic controller was developed. It can measure the thickness, warpage, length and width, parallelism and perpendicularity of four edges of TFT substrate glass. The suitable sensor and the support platform on which the TFT substrate glass is lay when measuring is important faction for measuring accuracy. The study of material and the layout of the platform are presented in this paper.

II. MEASURING TRANSDUCER

The traditional measurement of glass thickness on the production line, including the detection of average thickness and the variation in thickness in both the flow and non-flow directions, is mostly done by the digital micrometer to manually measure the multi-point thickness on the glass edges, manually calculate the average thickness value and the variation in thickness in different directions. The thickness of TFT substrate glass is only 0.3-0.7mm, the required accuracy is high, in a large area of 2000*18000mm of sixth generation version, the uncertainty for the thickness measurement should be controlled within 1.5μm. The measuring speed of the traditional measuring methods is slow, and it is difficult to get the thickness value of whole part of glass. With the development of ultra-thin glass substrate towards the trend of new generation, the glass area is becoming larger and larger that it is more and more difficult for the thickness parameter of the edge part to reflect the thickness condition of the whole plate. The traditional method can no longer be used to handle the warpage measurement of substrate glass of large area.

New measuring instruments and equipment with different principles instead of the traditional mechanical measuring instruments in the glass thickness measurement field are studied, such as the ultrasonic thickness gauge, capacitance thickness gauge and various optical thickness gauges.

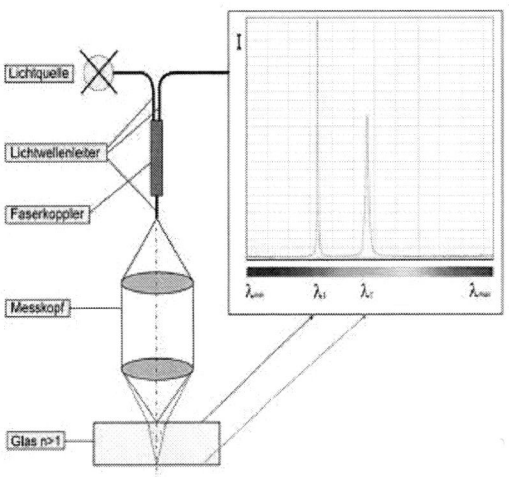

FIGURE I. PRINCIPLE OF SPECTRAL CONFOCAL MEASUREMENT.

The prime advantage of the optical measurement method is of its non-contact nature, which can avoid the precision loss caused by measuring force on the thickness, especially on the warpage. It can also realize the measurement of both the thickness and warpage simultaneously and improve the measurement efficiency. It can realize the CNC automatic measurement to conduct the objective assessment on the geometrical parameters of the whole plate by means high-speed and high-density sampling and data processing, so that optical sensor is very suitable for the measurement of TFT substrate glass. this paper has adopted the spectral confocal optical measurement sensor.

The spectral-confocal displacement sensor is developed based on in the confocal microscopy according the principle of spectral-confocal measurement. Its light source is not a single frequency laser, but the polychromatic light, after emitted, the polychromatic light shall go through the fiber coupler, after strict chromatic aberration correction through the lens group in the optical system, radiate the beam onto the surface to be measured to produce a spectral dispersion due to the action of color difference, forming a series of focus discrete in space. The monochromatic light focused on the object measured is reflected back to reach the sensor controller through optical fiber coupler, you can determine the frequency of the monochromatic light through the spectral analysis by

© 2015. The authors - Published by Atlantis Press

spectrometer, as each frequency corresponds to a distance value, the displacement can be reckoned by the frequency so detected, so that you can accurately measure out the distance from the optical probe to the surface of object to be measured, of course, the refraction coefficient of the object to be measured can be substituted into the distance value in the calibration process.

If both the top and bottom surfaces of glass under test are placed within the measuring range of sensor probe, there shall have two different spectral peaks, λ_1 and λ_2, appeared on the spectrogram of the sensor detector, which correspond to the distances of two surfaces, S_1 and S_2, respectively.

The light ray shall enter into the top surface at an angle , if there is no refraction (n=1), the beam shall intersect the optical axis at the point S0 away from the sensor. In case of refraction (the refractive index of glass usually n>1), the beam shall converge with the optical axis at a refraction angle β at the point S2 away from the sensor.

According to the law of refraction, the relative refractive index n is a function of wavelength λ :

$$\sin\alpha / \sin\beta = n(\lambda) \tag{1}$$

By the triangular relationship

$$\frac{s_2 - s_1}{s_0 - s_1} = \frac{d_2}{d_0} = \frac{\tan\alpha}{\tan\beta} = \frac{\sin\alpha}{\sin\beta} \bullet \frac{\cos\beta}{\cos\alpha} = n \cdot \frac{\cos\beta}{\cos\alpha} \tag{2}$$

Through comprehensive considerations of all the different beams, the real thickness dw can be determined by the measured thickness.

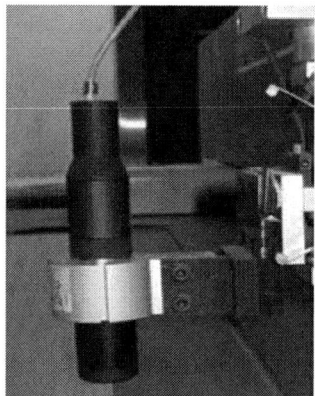

FIGURE II. EXPERIMENTAL INSTALLATION.

The spectral confocal sensor can measure the transparent or translucent workpiece with thin coating, as the light of different frequencies is reflected from the contact surface of different materials, therefore we can analyze on the distribution of spectrum detected by spectrometer to determine the thickness of each material layer through detecting the wavelength corresponding to each wave crest.

Based on this special optical principle and structural design, the spectral confocal sensor has a high accuracy in the measurement of displacement and thickness.

III. IMPACT OF GLASS SUPPORT CONDITION ON MEASUREMENT

As mentioned above, the measuring principle of sensor is based on the reflection of color light on the top and bottom surfaces of glass, when the glass is placed on different carriers, because the refractive index of supporting material itself, uniformity of material, state of contact interface and contact stress would affect the optical characteristics of the contact interface. The impact on the reflection on the top and bottom surfaces is different, which would greatly interfere the high precision measurement, produce the measurement noise and bring adverse influence to the measurement and evaluation. In order to study the impact of different supporting materials on the sensor, we select the optimum material to be used as the measuring platform.

In the experiment we placed the glass on different materials, such as rubber, plastic, granite and suspension (air) to measure same substrate glass sampled. The measurement results are shown in Figure 3. The abscissa represents the number of measuring points, while the ordinate represents the corresponding value of thickness. In order to reflect most of the measuring points, the range of thickness value on the ordinate is adaptively adjusted.

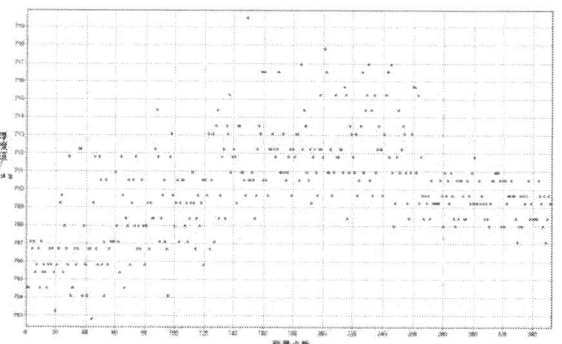

FIGURE III. (A) RUBBER.

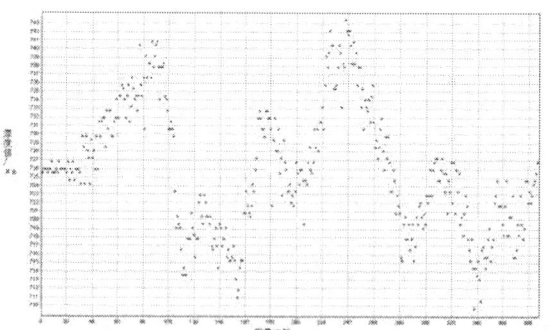

FIGURE III. (B) PLASTIC.

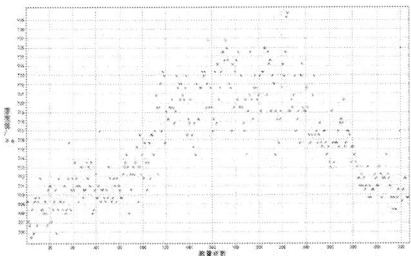

FIGURE III. (C) GRANITE.

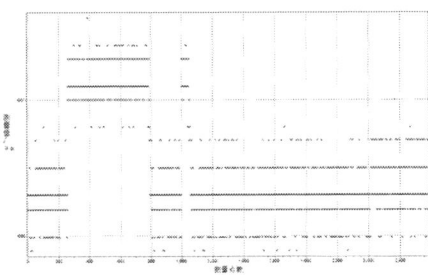

FIGURE III. (D) AIR.

The rubber surface is rougher that the diffuse reflection is stronger; the plastic surface is relatively rough, the reflectivity is moderate; because the surface of granite platform is finished, the surface is relatively smooth with a certain mirror reflection, and also with a certain degree of diffuse reflection.

It can be seen from the measuring point-thickness curve in the experiment, if the glass is placed on the surface of different carriers, the reflectivity of the carrier itself would affect the measurement results to some extent. The diffuse reflection of rubber is strong, there is no regular pattern to follow as to the impact on the measurement results. The impact of plastic carrier with moderate reflectivity on the measurement results is weaker in the regular pattern that the positions of the maximum and the minimum measured values have a certain repeatability, but the difference between the maximum and minimum values measured exceeds $30 \, \mu m$, and the stability is poor. The impact of granite carrier on the measurement results is similar to plastic, the difference between the maximum and minimum

measured values is smaller, about $20 \, \mu m$, but because the distribution regularity of measuring points is not regular, it is impossible to be used to achieve the glass measurement.

The measuring point - thickness curve in the suspended state is shown as a line segment, the thickness value is mostly changing with the changes in the measuring points, the changes in the stable measurement data are within the $1.5 \, \mu m$ linear deviation. Therefore, in order to ensure the stability of the measurement accuracy, it is necessary to measure the glass in the suspended state.

IV. TOPOLOGY OPTIMIZATION DESIGN OF MEASURING PLATFORM

In order to realize the warpage measurement of flat glass, there must have a measuring datum. Due to its excellent stability and vibration reducing performance, the granite is the preferred material for the mechanical measuring platform of large-sized precision measurement equipment. The automatic measuring equipment for TFT the substrate glass has adopted the granite platform, which is also used as the plane datum for the measurement of high precision guide and warpage pair in the measurement equipment. But when the granite material is used as the background of optical measurement, it would have an impact on the optical measurement.

In order to further study the impact of granite platform background on the results of the dynamic thickness measurement, it is necessary to make the further experimental researches. Fix the probe in the vertical direction of the glass plane and move along the glass plane direction, to write down the measurement results under the same high granite carrier background and suspended background respectively. Measure the thickness for 20 times under each of the two working conditions whenever the probe is moving a displacement L by the preset step length, and write down the mean value and standard deviation. The distribution of thickness values within the measuring range is shown Figure 4.

The distribution of standard deviation under the two carrier backgrounds is shown Figure 5.

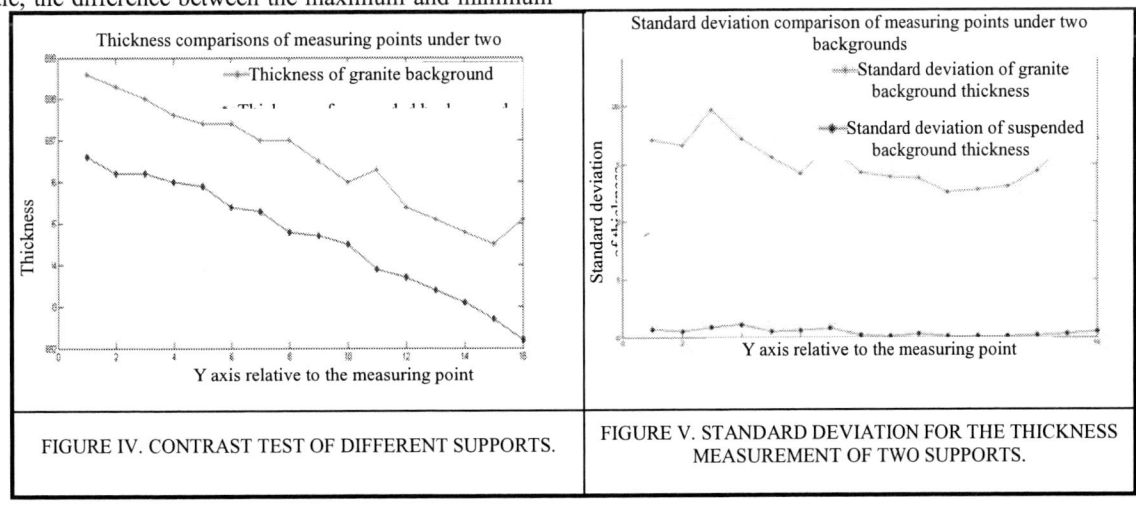

| FIGURE IV. CONTRAST TEST OF DIFFERENT SUPPORTS. | FIGURE V. STANDARD DEVIATION FOR THE THICKNESS MEASUREMENT OF TWO SUPPORTS. |

It is proved by the experimental analysis that, the difference in the measurement results under the granite background and suspended state is quite great, of the measurement under the granite carrier background, the distribution of measurement data is scattered, with great obviously random error; in contrast, with the dynamic measurement in the suspended state, the repeatability of measurement data is very good, the standard deviation is less than 1μm, which can meet the thickness measurement requirements.

Therefore, in order to use the stable granite platform as the warpage measurement datum, while realizing the suspended measurement of ultra-thin substrate glass, this paper proposed to provide the equidistant measuring grooves in the platform along the non flow direction of TFT substrate glass, to avoid the impact of medium fluctuation between granite surface and glass interface on the probe data, which can also improve the measurement precision of TFT substrate glass thickness and warpage. Seeing from the angle of the glass thickness measurement, we hope the suspended glass area should be as large as possible. But for the glass with thickness to be 0.3-0.7mm, the plate support of equidistant grooves would produce additional deformation of thin glass. Therefore, the physical dimension of the groove section should be designed in a reasonable manner, the slot width should be as wide as possible under the condition that it would not produce the additional warpage deformation. In this paper we have

conducted analysis and experiment by the section design of supporting platform for the 2000*1800mm glass with thickness to be 0.5 mm.

In order to ensure the reliable operation of warpage and thickness optical measurement sensor, we should ensure the stability and reliability of original measured data. Based on the theoretical analysis and experimental results, the bottom surface of glass should be suspended for more than 3mm. By which the slots are opened in the granite measuring plate.

However, the width of slot and the shape of cross section shall have a certain impact on the glass warpage. In addition, the slot structure is closely related to the layout of measuring positions, wider the slot to open, the more favorable for the layout and arrangement of measuring positions shall be, but the wider slot would have more obvious impact on the glass warpage.

For this purpose, we conducted the in-depth analysis on the impact of slot structure on the warpage. For the TFT substrate glass with thickness to be 0.5mm (with elastic modulus to be 4.62*109, Poisson's ratio to be 0.245 and density to be 2500), there are multiple supporting ribs across the glass, the supporting spacing is ranged from 90mm to 50mm that the in-depth analysis is made under various changed conditions. The results are as follows.

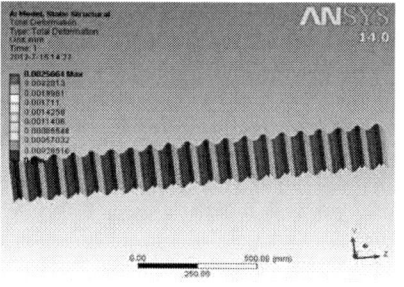 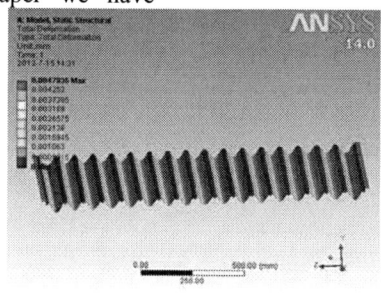 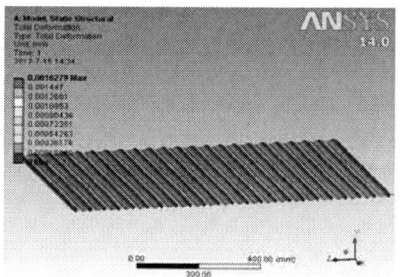

FIGURE VI. (A) CENTER DISTANCE OF SUPPORTS IS 90 MM, THE MAXIMUM DEFORMATION IS 4.9M.

FIGURE VI.(B) CENTER DISTANCE OF SUPPORTS IS 80 MM, THE MAXIMUM DEFORMATION IS 2.6M.

FIGURE VI.(C) CENTER DISTANCE OF SUPPORTS IS 70 MM, THE MAXIMUM DEFORMATION IS 1.6M.

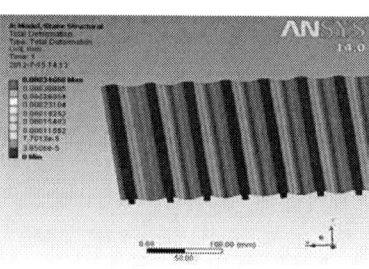 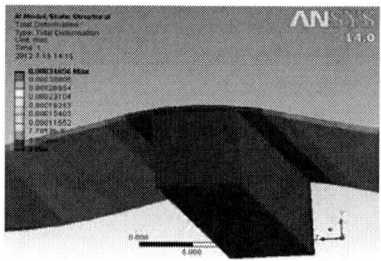

FIGURE VI.(D) CENTER DISTANCE OF SUPPORTS IS 50 MM, THE MAXIMUM DEFORMATION IS 0.3M.

FIGURE VI.(E) DEFORMATION ALONG GLASS EDGE.

FIGURE VI.(F) DEFORMATION AT THE SUPPORT.

The analysis results show that, for the 0.5 mm thick TFT substrate glass, smaller the support spacing is, the less impact of slotting on the warpage would be. When the slot spacing is 50mm, the impact of slotting on the warpage is only 0.3μ.

Taking into account the limits by the granite slotting process, it shall also consider the comprehensive influence of sampling interval and measurement layout of thickness measurement. The slot spacing is 50mm. The width of supporting edge plane is 5mm, the slot width is 45 mm, the depth is 3 mm.

V. SUMMARY

The spectral confocal sensor is an ideal sensor for the thickness and warpage measurement of TFT substrate glass, in order to accurately measure the thickness and warpage of substrate glass, you should correctly set the conditions for use. Through theoretical analysis and experimental verification, the measurement of spectral confocal sensor should be carried out on the slotted high-precision granite plate. These conclusions have been proved in the TFT substrate glass measurement and successfully applied in engineering.

REFERENCES

[1] Yang Cheng ,WangYutian Glass Thickness measuement system of double Beam path Compensation Based on CPLD Thechnology. Yanshan University

[2] WangXingying,ChangXuguang Instrument of Glass thiekness Measurement, Jounal of transducer Thecknology, 1995, 02 : 29-33.

[3] Song Chen1,Liu Cen,Guo Qi, a princple for Glass Thhickness Measuemrnt Based of Laser Doppler , ACTA PHOTONICA SINICA, 2008, 37（08） : 1635-1638.

[4] WANG Wei,WANG Zhao-ba, Capability Research of Glass Thickness Detection

[5] Based on CCD Displacement Sensor, Instrument Technique and Sensor, 2006, 9 : :44-45.

[6] KATHRYNJ W, Richard R C, GUN Young-Lee. Quantitative in situ measurement of transfer film thickness by a Newton's rings method[J]. Article in Press, 2007, 10(4):6-56.

[7] Maru Yama H, Inoue S,Mitsu Yama T, et al. Low-coherence interferometer system for the simultaneous measurement of refractive index and thickness[J]. Appl Opt, 2001, 41(3):1315-1322.

[8] Sirat, Gabriel Y., Paz, Freddy, Agronk, Gregory, Wilner, Kalman.Conoscopic holography, SPIE of the International Society for Optical Engineering[J], USA, 2005, 5972: 021-026.

[9] Totzeck, M., Tiziani H.J. Phase-shifting polarization interferometer for microstructure line width measurement[J]. Opt Lett.1999, 24(5): 294-29.

International Conference on Power Electronics and Energy Engineering (PEEE 2015)

Electrical Retail Tariff Model Based on the Load Characteristics

J.J. Wu, H.T. Huang, C. Gao, F. Yu
College of Electric Power Engineering
Shanghai University of Electric Power
Shanghai, China

J. Pan
Jinhua Electric Power Company, State Grid Zhejiang Electric
Power Company
Jinhua, Zhejiang Province, China

Abstract—According to the recent policy requirements in China, which present to built the users classification of electrical retail tariff base on the users load characteristics, this paper put forward the consumers classification according to the voltage grade and load factor, and design the pricing system considering of the load characteristics. Combining with the power system and electricity price system environment in China, this paper deeply analyses the system cost including the marginal capacity cost and marginal electricity cost; then apply the marginal cost pricing and cost-based pricing to build the load factor electrical retail tariff model. This model is more fair compared with the traditional model in capacity cost allocation; finally, its effectiveness is proved in the example.

Keywords-electrical retail tariff model; consumers classification; voltage grade and load factor

I. INTRODUCTION

The reasonable cost allocation guarantees the electrical retail tariff well operating, but current consumers classification system in China is primarily based on the trade, which is defective to reflect the different affect of power cost between different users. This price system have rebelled against the fair burden pricing principle. Therefore, National Development and Reform Commission of China put forward the retail tariff reform twice in 2003 and 2013. The reform clearly requires to simplify the price classification into three categories including resident living, agricultural production and trade and other users. The latter tranche is further classified according to the voltage grade and load factor, then form the classification system based on the users load characteristics.

In the early 50's of last century, pricing mechanism based on the load characteristics classification has already been applied in some developed countries such as France. In China, it is still in the theoretical research stage and achievement is rare. The main achievement so far include: literature[1-2] has formed mature marginal cost pricing model. That method classified users according to the voltage grade and load factor and established a load factor pricing model. However, that model uses simultaneity factor to calculate the marginal capacity cost of each class users, which has deviation in the capacity users occupied at the system peak time. In addition, it isn't accord with the power system and electric price system environment in China; literature[3-4] discuss the necessity, theory evidence, policy design and foreign experience of the

load factor pricing, but didn't construct the load factor pricing models suitable for China.

Therefore, this paper proposes a user classification method based on the voltage grade and load factor, and designs a retail tariff system according to the load characteristics. Combining with the electric power system and electricity price system environment in China, this paper deeply analyses the system cost including the marginal capacity cost and marginal electricity cost in servicing users; then apply the marginal cost pricing and cost-based pricing to build the load factor retail tariff model. This model is more fair compared with the traditional model in capacity cost allocation; finally, its effectiveness is proved in the example.

II. BASIC FRAMEWORK OF LOAD FACTOR PRICING

In addition to some political tariffs such as the preferential price for the residents, electricity price shall reasonably reflect the real cost of power supply. However, nearly all consumers are different in the cost of power supplying, so that variant tariff should be set for each consumer in theory, but it is not practical. Therefore, consumers should be classified into several classes according to the load characteristics, and then tariff should be set for each class of users.

Combining with the advantages of marginal cost pricing and cost-based pricing, this paper classifies consumers considering the load characteristics and build a hybrid load factor pricing method. The procedure is described briefly below: 1) Carry out analysis of the users load characteristics, acquire users load factor, simultaneity factor, system simultaneity factor indexes and their statistical relationship; 2) Select the load factor and voltage grade as the users classification indexes, formulate the consumers classification system according to the load factor; 3) Account the system financial cost of the power supply; 4) Calculate users marginal capacity cost and marginal electricity cost of each class; 5) Allocate the total cost to each class users according to the marginal cost, and set tariff to each class. The basic framework is showed in Figure 1.

© 2015. The authors - Published by Atlantis Press

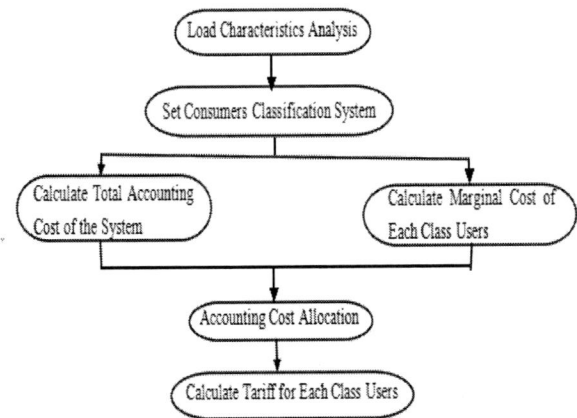

FIGURE I. BASE FRAMEWORK OF LOAD FACTOR PRICING.

III. CONSUMERS CLASSIFY MODEL

A. Consumers Classification Index

Allocation of the system cost refers to not only the cost of the power generation, transmission and distribution, but also the capacity and energy cost. In that case, multiple indexes should be utilized to reflect it synthetically. In case of the complexity of the consumers classification, load factor and voltage grade are chosen as the most typical indexes to classify users.

Consumers of the different voltage grade have difference in the cost of the power transmission and distribution, because of the distinctions of the facilities and lines users occupied. High voltage consumers occupy only the facilities of the high voltage, while low voltage consumers occupy the facilities of the low voltage in addition. Therefore low voltage consumers bear more power transmission and distribution cost. Load factor refers to the ratio of the average load within a certain period and the maximum load. It reflects the consumers utilization of the power supply facilities, consumers of high load factor utilize the power generation, transmission and distribution facilities more effectively, so improving the consumers load factor could reduce the fixed cost, and then achieve the system load curve smoothing, reducing the unit cost of power supply. In consequence, consumers of high load factor should bear less unit cost, and vice versa.

Because of the limited space, this paper only discusses the problems of consumers classification pricing according to load factor of the same voltage grade.

B. Load Factor Classify Method

Basic principle of setting load factor classification is to ensure that consumers within the same class share similar electricity price, while the weight of consumers in each class is not too small. According to the international experience, users will usually be classified into 3-6 files according to the load factor. The specific formulation method and procedure is showed in Figure 2: first collect all or sample users load factor indexes to statistic and analysis; second classify the consumers according to the load factor by 0.05 interval into 20 gears and calculate users amount and the theoretical price ratio; then

merge certain consumers on the base of the principle, on that base confirm file number and level; finally, on the base of the elementary load factor classification, confirm the final classification considering to simplify the catalogue price.

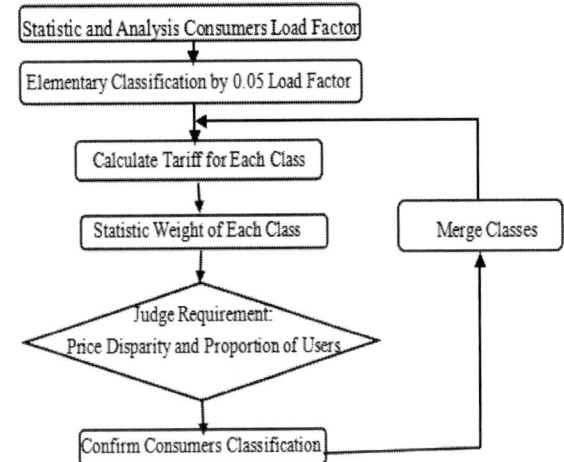

FIGURE II. CONSUMERS CLASSIFY PROCEDURE.

IV. MARGINAL COST OF POWER SUPPLY

Power marginal cost can be divided into marginal electricity cost, marginal capacity cost and consumer cost. In this paper, the load factor pricing is only aimed for large consumers, consumer cost take a little of its cost, so it can be ignored.

C. Marginal Energy Cost

Marginal electricity cost is associated with the electricity consumption, it is the operating cost for providing additional kWh. Under the traditional vertical integration system, it mainly consists of the fuel cost of the corresponding marginal plants at system operation time. Power generation companies and power companies in China are independent of each other, so the purchase price isn't Tou tariff and is single electricity price. In that case the purchase cost per kWh at any time is uniform, namely the average purchase tariff and the cost of loss. The later take a very small proportion, it is relevant with the users voltage grade.

D. Marginal Capacity Cost

Marginal capacity cost is increment capacity cost attribute to an incremental demand increase. It is relevant with the maximum demand of consumers. It shall generally include two parts, the marginal cost of generation and the marginal cost of transmission and distribution (MCT&D). However in China the power purchase price is signal tariff, so the marginal cost of generation capacity is void. Therefore, the marginal capacity cost shall be MCT&D.

MCT&D consist of two parts, one is the MCT&D of the exactly voltage grade (represented as "MCT&DB"), the other is the MCT&D of the higher voltage grade (represented as "MCT&DG"). For the higher voltage grade, incremental capacity investment is needed to meet the incremental load at

135

system peak time, on the contrary, incremental demand at off-peak time cost nothing. Therefore, according to the peak load responsibility law, at peak time, MCT&DG can be calculated by the average transmission and distribution investment at a certain period. For the exactly voltage grade, users dispersion rate is high, there is deviation between the time when users load is maximum and the system peak time, at any time load increment will cause incremental transmission and transformer capacity cost. Therefore, MCT&DB will emerge at any time, it can still be calculated according to the average transmission and distribution investment at a certain period.

V. METHOD OF LOAD FACTOR CLASSIFICATION TARIFF

Capacity and energy cost is approximatively proportion to the consumers maximum power demand and consumption. Therefore, the consumers could pay the bill by means of basic tariff and electricity tariff and take different billing methods, which is called two-part electricity price.

A. Basic Tariff

In a certain period, MCT&DG and MCT&DB were calculated as MC1sr and MC0sr. If the class i consumers load factor is fi; system simultaneity factor is dsi, which represents the ratio of the comprehensive maximum load of the users at the system peak time and the sum of all the maximum load of every user; simultaneity factor is di, which represents the ratio of the comprehensive maximum load of the users and the sum of all the maximum load of each user. When sum of the maximum demand of all of the class i users increase 1kWh, the system maximum capacity increased dsi kWh, so MCT&DG increment is (dsi * MC1sr) according to the peak load responsibility law. For the reason that incremental demand at any time will occupy transmission and distribution facility of that voltage grade, so MCT&DB should be calculated according to the users maximum comprehensive load, ie (di* MC0sr). As mentioned before, marginal capacity cost of class i is MCsr:

$$MC_{sr,i} = d_{si} \times MC_{sr}^1 + d_i \times MC_{sr}^0 \qquad (1)$$

The proportion of MC0sr in MCsr is k. Then the theoretical based price ratio of class j and class i users could be calculated as follow:

$$b_{j,i} = \frac{d_{sj} \times (1-k) + d_j \times k}{d_{si} \times (1-k) + d_i \times k} \qquad (2)$$

After financial accounting, the transmission and distribution capacity cost a certain voltage grade bear is expressed as Csr, system capacity is Pmax, the average system capacity cost is Csr/Pmax. Total capacity of cost of power system is allocated to all load factor classification users, then the basic price of class i users will be:

$$q_i = \frac{C_{sr}}{\sum_1^l (b_{i,1} \times P_{\max i})} \times b_{i,1} \qquad (3)$$

where qi=basic price of class i users; Pmax=maximum load of class i users; l=number of users classes; bi,1=let reference users be the first class of users, bi,1 is the price ratio of class i users and the reference users.

B. Electricity Price

Total annual power purchase price for power company is Csd, purchases of electricity is Qs, the average power purchase price is denoted as Csd/Qs, the average loss rate of that voltage grade is s. As mentioned, marginal electricity cost is the sum of average power purchase price and network loss cost:

$$MC_{sd,i} = \frac{C_{sd}}{Q_s} + \frac{C_{sd}}{Q_s} \times \frac{s}{1-s} \qquad (4)$$

Because the marginal cost is calculated as the average cost, allocation of financial cost is similar to that of marginal cost.

VI. CASE STUDY

A. Example Description

Example takes nearly 700 sample consumers from a certain province of 2013 in China, whose voltage grade is 35kV and capacity is beyond 315kVA. Dates collected include the load date, electricity consumption and the catalogue price of each class consumers. July18 is the maximum load day, so it's be chosen as the representative day. Example distills the date of that day to analysis users load characteristics. The average cost of 35kV is 0.6346 ￥/kWh, MCT&D proportion is 20%. MCT&DB take 20% of MCT&D. Example aims to classify the sample consumers by load characteristics, and calculate the tariff of each class.

B. Load Characteristics

Obtain the consumers load factor indexes according to users real load curve, then calculate the simultaneity factor and the system simultaneity factor and fitting the curve. Figure 3 displays the load factor-simultaneity factor curve and the load factor-system simultaneity factor curve of the sample consumers of 35kV.

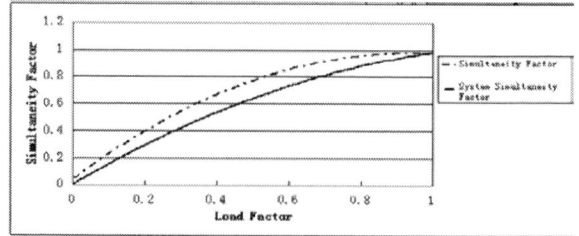

FIGURE III. LOAD FACTOR-SIMULTANEITY FACTOR CURVE & LOAD FACTOR-SYSTEM SIMULTANEITY FACTOR CURVE.

C. Result of the Load Factor Classification Tariff

For the capacity of the sample consumers is all beyond 315kVA,and it's belong to China present two-part tariff implementation range, so all sample consumers could be set load factor classification and formulate two-part tariff for.

1) Load factor classification

Distribution of the sample users load factor is showed in Figure 4. According to the principle that customers within the same class share similar electricity price, while the weight of customers in different class is not too small, and file number should be appropriate, preliminarily classify the consumers by 0.05 interval of load factor, finally classify the consumers into 3 classes after merging and adjusting. The classification is showed in Table 1.

TABLE I. CONSUMERS CLASSIFICATION BASED ON THE LOAD FACTOR.

Load Factor Classification	Load Factor	Simultaneity factor	System Simultaneity Factor
1st Class: 0-0.45	0.33	0.60	0.48
2nd Class: 0.45-0.75	0.61	0.83	0.72
3rd Class: 0.75-1	089	0.97	0.96

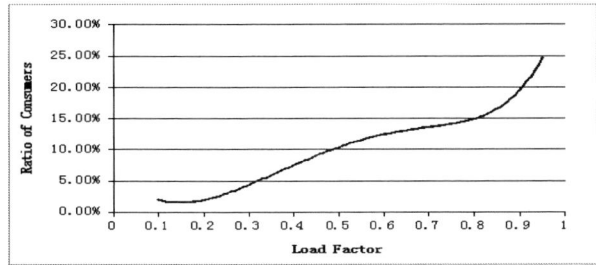

FIGURE IV. STATISTIC OF LOAD FACTOR DISTRIBUTION.

2) Load factor classification tariff

According to the classification in 5.3.1, obtain the consumers simultaneity factor and system simultaneity factor indexes, adopt the method mentioned in 5th quarter and the method from literature[1] to calculate the two-part tariff of each class. The result is showed in Table 2. The method mentioned in this paper is called "Method 1", the method taken from literature[1] is called "Method 2".

TABLE II. TWO-PART TARIFF BASED ON THE LOAD FACTOR CLASSIFICATION.

	Method 1*			Method 2*		
	Basic Price	Electricity Price	Average Price	Basic Price	Electricity Price	Average Price
1st Class	40	0.5232	0.6406	48	0.6220	0.7142
2nd Class	59	0.5232	0.6097	60	0.5971	0.6218
3rd Class	76	0.5232	0.5964	76	0.5154	0.5872

* Unit of basic price is ¥/kW/mouth, unit of electricity price and electricity price is ¥/kWh

D. Result Analysis

Figure 5 shows the load curve of the 1st and 2nd class and the 35kV system. As showed in the Figure 5, 1st class consumers load factor is lower, the load fluctuate according to the time obviously, peak-valley is larger, so the operating factor of the facility is lower. This means that the average price of the low load factor users is higher for that they should allocate more capacity cost per unit. The calculate results of the two method showed in the Table 1 are correspond to the theoretical analysis. This means two method all could better reflect the fair burden principle.

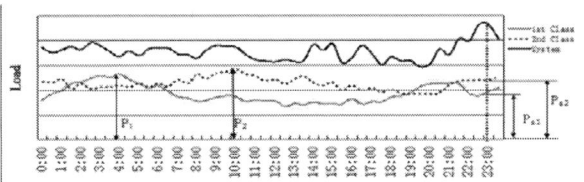

FIGURE V. DAILY LOAD CURVE OF LOAD FACTOR CLASSIFICATION CONSUMERS.

Besides, dates of the first two classes consumers are showed in Figure 5 and Table 3. Method 1 allocates the capacity cost according to the compositive load Ps1 and Ps2 at peak time of system, while method 2 use the compositive maximum load P1 and P2 of each class. Method 2 doesn't reflect peak load shifting reasonably, so the capacity cost consumers bear is more than method 1. In conclusion, compared with method 2, method 1 utilizes the system simultaneity factor to reflect peak load shifting, which will better reflect the real capacity occupation at peak time of system. In that case, method 2 could allocate cost more scientifically.

TABLE III. COMPARISON OF TWO METHOD IN CAPACITY COST OF EACH CLASS USERS.

Load factor Classification	Method 1 (¥/kW/Mouth)	Method 2 (¥/kW/Mouth)
1st Class: 0-0.45	40	51
2nd Class: 0.45-0.75	59	62
3rd Class: 0.75-1	76	78

VII. CONCLUSION

According to the recent policy requirements in China, which presents to built the users classification of electrical retail price classification structure base on the users load characteristics, this paper design the retail price system based on the load characteristics, and build the load factor pricing model suit for China. On the one hand, the model is closely connected with the electric power system and price policy in China, on the other hand, compared with the simultaneity factor, this model utilizes the system simultaneity factor to reflect peak load shifting, which will better reflect the real capacity occupation at peak time of system. System simultaneity factor could reflect the occupation of capacity at peak time of system. and it will make capacity cost allocation more scientific.

ACKNOWLEDGEMENT

This work is supported by National Natural Science Foundation of China 71203137.

REFERENCES

[1] L.S, Zhao. 1992. *Electric Pricing Design--Marginal Cost Pricing Application.* Beijing: Hydraulic and Electric Power Press.

[2] World Bank Staff Working Paper No.340. 1979. *Electric Power Pricing Policy.* Washington, DC.

[3] Z.Y, Tan. 2013. The Theory Basis, Calculation Method and Policy Options for Load Rate Tariff. Changsha: Hunan University.

[4] Z, Ye. & S, Yao. 2014. Research of Load Factor Tariff Implement in China. *Price Theory and Application5*: 44-46

Ultrasonic Liver Image Denoising Based on A Hybrid Threshold Method

H.J. Zhu

College of Information Science & Technology
Beijing University of Chemical Technology
Beijing, China, 100029

L. Rao

Jiangxi Normal University
Nanchang, China, 330027

Abstract—**The objective of this paper is to investigate a hybrid threshold denoising algorithm based on wavelet transform for the ultrasonic liver image. A novel hybrid threshold function first is discussed. The hybrid threshold denoising algorithm based on the wavelet transform is then performed for ultrasound image of the liver. Only is one parameter selected in the proposed image denoising algorithm. Several metrics such as correlation coefficient (CoC), edge preservation index (EPI), and structural similarity index (SSI) are measured to quantify the denoised results of ultrasound liver image. Experiments show that the wavelet-based hybrid threshold denoising algorithm is effective and feasible.**

Keywords-ultrasonic liver image; the hybrid threshold function; denoising; wavelet transform

I. INTRODUCTION

Ultrasound image plays an important role in the medical diagnosis and medical clinic. Unfortunately, some existential speckle noises and grey level discontinuity degrade the ultrasound images. It is difficult for the observer to identify clearly the interest details. Therefore, medical ultrasound image denoising is considered as one of the major problems in the medical ultrasonography.

Because speckle noise degrades the details and contrast resolution of medical ultrasound images, ultrasound image denoising is a key technique for medical diagnosis and medical clinic. In the reported literatures, speckle reduction approaches for medical ultrasound image include spatial filtered method [1,3,5,8,9,11] and multiscale denoising methods [2,4,6,7,10,14,16].

In the filtered methods some prominent edges of tissues or organs were preserved or enhanced after speckle noises removed. Koo [1] explored a homogeneous region growing mean filter for medical ultrasonic images speckle noising reduction. This method reduced speckle noise with edge preservation, but a proper seed region was difficult to determine at a larger speckle noise size. Damodaran [9] presented a discrete topological derivative image denoising approach to improve hyperechoic regions resulting in ultrasound medical images diagnosis. It is difficult to obtain good effect in the weak hyperechoic regions of ultrasound medical images.

Multiscale denoising methods have attracted more and more researchers' attention due to their excellent localization property.

Harmoko [14] combined a wavelet multi-scale strategy and a warping optical flow to generate a high-accuracy velocity vector from two consecutive frames of poor-quality ultrasound images. Gupta [16] provided a generalized Nakagami probability density function to denoise the medical ultrasound images. Essentially, most of these reported literatures using wavelet strategy needed to adjust multiple parameters to denoise the ultrasound images.

In medical ultrasound image, the abdominal or the liver ultrasound is one of the common ways to diagnose liver disorders and diseases. The objective of this paper is to investigate the hybrid threshold denoising algorithm based on the wavelet transform for the ultrasonic image of the liver. The proposed method constructs a hybrid threshold function with one selected parameter.

II. PROPOSED APPROACH

It is well known that the hard function is discontinuity and the soft function has bigger bias in Donoho's algorithm [12]. Some improved methods [7,13,15] were proposed in succession, but multiple parameters were tuned to denoise in these methods.

To reduce the drawbacks of thresholding function in Donoho's algorithm and the adjustment of more parameters in the thresholding function, we investigate a hybrid threshold function. The hybrid threshold function is constructed as

$$\hat{X} = \begin{cases} sgn(X)\left(|X| - \dfrac{T}{\alpha^{-|X|-T}}\right) & |X| > T \\ 0 & |X| \le T \end{cases} \quad (1)$$

Where X is the wavelet coefficients, $\hat{X}$ is the corrected wavelet coefficients, and α is one selected parameter. The hybrid threshold function only needs to adjust one free parameter.

An analytical express for the hybrid threshold function is discussed in the following. While the parameter α is very close to 0, we can get

$$\lim_{\alpha \to 0} sgn(X)\left(|X| - \frac{T}{\alpha^{-|X|-T}}\right) = sgn(X)|X| = \frac{X}{|X|} \cdot |X| = X \quad (2)$$

© 2015. The authors - Published by Atlantis Press

The function Eq. 2 is approximate to the hard-threshold function. When the parameter α is very close to 1, we have

$$\lim_{\alpha \to 1} sgn(X)\left(|X| - \frac{T}{\alpha^{-|X|-T}}\right) = sgn(X)(|X| - T) \quad (3)$$

The function Eq. 3 is close to the soft-threshold function. From Eq. 1, Eq. 2 and Eq. 3, we can see that the hybrid threshold function is a feasible threshold selection between the soft threshold and the hard threshold function. The parameter $\alpha \in (0,1]$ is selectable.

Assuming that a function $f(X)$ is equal to the first part in Eq. 1, that is

$$f(X) = sgn(X)\left(|X| - \frac{T}{\alpha^{-|X|-T}}\right) \quad (4)$$

For $X < 0$, we get $\lim_{X \to -\infty} \frac{f(X)}{X} = 1$, while $X > 0$, we have $\lim_{X \to +\infty} \frac{f(X)}{X} = 1$. And we have

$$\lim_{X \to +\infty}(f(X) - X) = \lim_{X \to +\infty}\left(sgn(X)\left(|X| - \frac{T}{\alpha^{-|X|-T}}\right) - X\right) = \lim_{X \to +\infty}\left(-\frac{T}{\alpha^{-X-T}}\right) = 0 \quad (5)$$

Eq. 5 can be used to explain the hybrid threshold function approximated gradually the line $\hat{X} = X$. That is to say, this hybrid threshold function can reduce the bigger bias between the original wavelet coefficients X and the estimated wavelet coefficients $\hat{X}$ in the soft- threshold function. Besides, the hybrid threshold function is continuous at $|X| = T$ and can decrease the bigger variance because of the discontinuity of the hard-threshold function.

The ultrasound image denoising algorithm based on the hybrid threshold function is briefly summarized as follows.

Step 1: Estimation of wavelet coefficients for original image.

Step 2: Computation of threshold by the hybrid threshold function $T = \sigma\sqrt{2log(M)}$.

Step 3: Computing the parameter α from Eq. 6.

$$\alpha = E_{I_s} / E_{I_o} = \sum_{i=0}^{W-1}\sum_{j=0}^{H-1} I_s^2(i,j) \Big/ \sum_{i=0}^{W-1}\sum_{j=0}^{H-1} I_o^2(i,j) \quad (6)$$

Where I_o be an original image and I_s represent the smoothed result using weighted median filter.

Step 4: Denoising the original image through estimating the corrected wavelet coefficients from Eq. 1 and the computed parameter α in Step 3.

To evaluate the proposed method, several metrics such as correlation coefficient (CoC) [4], edge preservation index (EPI) [1,4], and structural similarity index (SSI) [4] were calculated from the diagnosed results. Correlation coefficient and structural similarity index are measured of similarity between the original and denoised images, and EPI denotes to restore or preserve the edges after a speckle reduction method. These quality metrics defined as follows:

$$CoC = \frac{\sum(I_o - \bar{I}_o)(I_d - \bar{I}_d)}{\sqrt{\sum(I_o - \bar{I}_o)^2 \sum(I_d - \bar{I}_d)^2}},$$

$$EPI = \frac{\sum(\Delta I_o - \Delta \bar{I}_o)(\Delta I_d - \Delta \bar{I}_d)}{\sqrt{\sum(\Delta I_o - \Delta \bar{I}_o)^2 \sum \Delta(I_d - \Delta \bar{I}_d)^2}},$$

$$SSI = \frac{(2\bar{I}_o\bar{I}_d + C_1)(2\sigma_{I_oI_d} + C_2)}{(\bar{I}_o^2 + \bar{I}_d^2 + C_1)(\sigma_{I_o}^2 + \sigma_{I_d}^2 + C_2)}$$

Where I_o is original image, I_d is denoised image, $\bar{I}_o$ shows the mean of I_o, $\bar{I}_d$ shows the mean of I_d, $\Delta(I_o)$ is the high-pass filtered of I_o using the discrete Laplacian operator, and $\sigma_{I_o}^2$ the variance of I_o, $\sigma_{I_d}^2$ the variance of I_d, $\sigma_{I_oI_d}$ the covariance of I_o and I_d, $C_1 = (k_1L)^2$ and $C_2 = (k_2L)^2$ two variables to stabilize the division with weak denominator, $L = 225$ the dynamic range of the pixel values, $k_1 = 0.01$ and $k_2 = 0.03$.

III. EXPERIMENTAL RESULTS AND DISCUSSION

The first experiment tested the influence of the parameter α on the proposed method. One real ultrasound image corrupted by speckle noise with variance $\sigma = 0.01$, $\sigma = 0.05$, $\sigma = 0.3$ and $\sigma = 0.5$ was first generated. The proposed method was then applied to eliminate the speckle noise of the corrupted ultrasound image. The experimental results are shown in Figure 1.The peak signal to noise ratio (PSNR) of the denoised results and the parameters were shown in Table 1.

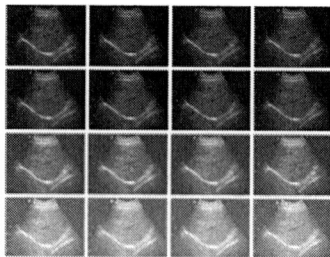

FIGURE I. FROM LEFT TO RIGHT: THE CORRUPTED ULTRASOUND IMAGE; DENOISING WITH THE SOFT-THRESHOLD FUNCTION, DENOISING WITH THE HARD-THRESHOLD FUNCTION; AND DENOISING WITH THE PROPOSED METHOD.

In the second experiment, the parameter α is first estimated for three real ultrasound liver images, and these images are denoised using the proposed method and Damodaran's method [9]. The experimental results are shown in Figure 2. For the three ultrasound liver images, the parameter α is 0.9984, 0.9742 and 0.9854, respectively. We can see that the denoised effect of the proposed method is better than that of Damodaran's method [9] in Figure 2.

Here, we use COC, EPI, and SSI quality metrics to better evaluate the proposed denoising algorithm. Table 2 shows these metrics of the second image in Figure 2 which is added speckle noise with variance $\sigma = 0.1$ and $\sigma = 0.5$. From the result, it can be seen that the key parameter α plays an important role in the proposed wavelet-based hybrid thresholding method.

TABLE I. THE ESTIMATED PARAMETERS α AND PSNRS OF THE DENOISED RESULTS.

Speckle Noise Variance	The parameter α	PSNR with the soft-threshold method	PSNR with the hard-threshold method	PSNR with the proposed method
0.010	0.3308	20.7587	20.9452	21.2978
0.050	0.5903	20.2555	20.4035	20.7602
0.300	0.8651	19.5840	19.6993	20.0564
0.500	0.9034	19.7443	19.8224	20.1806

TABLE II. THE ESTIMATED COC, EPI, AND SSI OF AN ULTRASOUND IMAGE DENOISING USING THE PROPOSED METHOD AND OTHER METHODS.

Parameter / Method	$\sigma = 0.1$			$\sigma = 0.5$		
	CoC	EPI	SSI	CoC	EPI	SSI
Speckle noise image	0.8351	0.4481	0.8630	0.6518	0.3656	0.7278
The soft-threshold function	0.9314	0. 9277	0.9293	0.8325	0.8990	0.8316
The hard-threshold function	0.8338	0.8057	0.8852	0.7958	0.8605	0.8188
Damodaran's method[9]	0.9273	0.9155	0.9242	0.8589	0.8633	0.8043
Adaptive method[16]	0.8488	0.8343	0.8816	0.7579	0.8546	0.7593
Our method with $\alpha = 0.20$	0.9120	0. 9261	0.9173	0.8964	0.8922	0.8486
Our method with $\alpha = 0.50$	0.9314	0. 9259	0.9293	0.9267	0.9024	0.8523
Our method with $\alpha = 0.80$	0.9547	0.9242	0.9380	0.8565	0.8433	0.8370

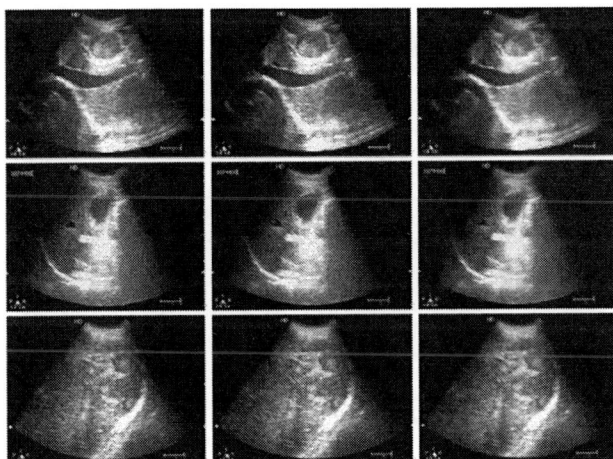

FIGURE II. DENOISING WITH DAMODARAN'S METHOD[9] AND THE PROPOSED METHOD FOR THREE LIVER IMAGES. FROM LEFT TO RIGHT: INPUT ULTRASOUND IMAGE, DENOISING WITH DAMODARAN'S METHOD [9] AND DENOISING WITH THE PROPOSED METHOD.

IV. SUMMARY

In this work, we have investigated a hybrid threshold for ultrasound liver image denoising via the wavelet transform. We applied real medical ultrasound liver image to test the proposed method. Besides, a few metrics such as correlation coefficient, edge preservation index, and structural similarity index were estimated to evaluate the proposed method. The experimental results validate the wavelet-based hybrid thresholding technique for ultrasound liver image denoising.

ACKNOWLEDGMENTS

This work was supported in part by the Beijing University of Chemical Technology Interdisciplinary Funds for "Visual Media Computing" and Beijing Training Programme Foundation for the Talents No.2012B009016000004.

REFERENCES

[1] J.I. Koo, S.B. Park, Speckle reduction with edge preservation in medical ultrasonic images using a homogeneous region growing mean filter (HRGMF), Ultrasonic Imaging, 13(1991)211-237.

[2] B. Aiazzi, L. Alparone, S. Baronti, F. Lotti, Multi resolution local statistics speckle filtering based on a ratio Laplacian pyramid, IEEE Trans. Geosci. Remote Sense, 36(1998)1466-1476.

[3] A. Donka and M. Lyudmila. Contour segmentation in 2D ultrasound medical images with particle filtering, Machine vision and applications, 22(2012)551-561.

[4] H. Rabbani, M. Vafadust, P. Abolmaesumi, S.Gazor, Speckle Noise Reduction of Medical Ultrasound Images in Complex Wavelet Domain Using Mixture Priors, IEEE Transactions on Biomedical Engineering, 55(2008)2152-2160.

[5] K. Thangavel, R. Manavalan, I.Laurence Aroquiaraj, Removal of Speckle Noise from Ultrasound Medical Image based on Special Filters: Comparative Study, ICGST-GVIP Journal, 9(2009)25-32.

[6] M.I.H. Bhuiyan, M.O. Ahmad, M.N.S. Swamy, Spatially adaptive thresholding in wavelet domain for despeckling of ultrasound images, IET Image Processing, 3(2009)147-162.

[7] A. Khare, M. Khare, Y. Jeong, H. Kim, M. Jeon, Despeckling of medical ultrasound images using Daubechies complex wavelet transform, Signal Processing, 90(2010)428-439.

[8] A.M.L. Lanzolla, G. Andria, F. Attivissimo, G.Cavone, N.Giaquinto. Improving B-mode ultrasound medical images. Proceedings of IEEE International Instrumentation and Measurement Technology Conference, 1(2011)1704-1708.

[9] N. Damodaran, S. Ramamurthy, S. Velusamy and G.K. Manickam. Speckle noise reduction in ultrasound biomedical B-scan images using discrete topological derivative, Ultrasound in medicine & biology, 38(2012)276-286.

[10] J.M. Bioucas-Dias, M.A.T Figueiredo, Multiplicative noise removal using variable splitting and constrained optimization, IEEE Transactions on Image Processing, 19(2010)1720-1730.

[11] R. Sivakumar, M.K. Gayathri, D. Nedumaran, Speckle filtering of ultrasound B-scan images - A comparative study between spatial and diffusion filters, 2010 IEEE Conference on Open Systems, 1(2010)80-85.

[12] D.L. Donoho, I.M.Jonestone. Ideal spatial adaptation via wavelet shrinkage. Biometrika, 81(1994)425-455.

[13] A.G. Bruce, H.Y. Gao, Understanding WaveShrink: Variance and bias estimation. Biometrika, 83(1996)727-745.

[14] S.A. Harmoko, M.M. Marzuki, H. Aini, M. Oteh and N.F. Moha. Enhancement of myocardial boundary tracking using wavelet-based motion estimation, Journal of Information and Computational Science, 8(2011)1779-1792.

[15] H. Yu, L. Zhao, H. Wang, Image Denoising Using Trivariate Shrinkage Filter in the Wavelet Domain and Joint Bilateral Filter in the Spatial Domain, IEEE Trans. on Image Processing, 18(2009)2364 -2369.

[16] S. Gupta, R. Chauhan, S. Saxena. Homomorphic wavelet thresholding technique for denoising medical ultrasound images, Journal of Medical Engineering Technology, 29(2005)208-214.

International Conference on Power Electronics and Energy Engineering (PEEE 2015)

Study on Aerodynamic Performance of Offshore Wind Turbine with Floating Platform Motion

X.M. Ding
Institute of Ocean Renewable Energy System
Harbin Engineering University
Heilongjiang, China

L. Zhang
Institute of Ocean Renewable Energy System
Harbin Engineering University
Heilongjiang, China

Y. Ma
Institute of Ocean Renewable Energy System
Harbin Engineering University
Heilongjiang, China

Abstract—In order to study the effect of the floating platform motion on the aerodynamic performance of offshore wind turbine, this paper analyzes the aerodynamic performance under different amplitudes of tossing movements on unsteady Blade Element Momentum Theory (BEM).The numerical model for aerodynamic performance of floating wind turbine under control system is established, which is based on the model of 5MW offshore wind turbine of National Renewable Energy Laboratory (NREL). Due to the limitation of the unsteady Blade Element Momentum Theory itself, the dynamic wake model and the yawing correction are added into the unsteady BEM theory with the Prandtl blade tip and hub loss correction, the Glauert correction and three-dimensional rotation correction. The change of thrust, torque and power in six degrees of freedom are received by calculation and the laws about aerodynamic performance of the offshore wind turbine under the six degrees of freedom platform motion is summarized.

Keywords-BEM;offshore wind turbine;NREL

I. INTRODUCTION

The bottom of ordinary stationary wind turbine is fixed with the ground rigidly. The external force from the bottom can make tower produce the structural issues such as yield and fatigue, but it will not affect the aerodynamic performance of the wind turbine. The Floating Offshore Wind Turbine (FOWT) in complex sea environment, based mainly on mooring chain. Under the action of the wind, wave and flow, the wind turbine will produce six degrees of freedom movement such as yaw, tilt, which affects the relative wind speed of wind turbine and the aerodynamic performance of the Floating Offshore Wind Turbine.

In this paper, the National Renewable Energy Laboratory (NREL)-5mw horizontal axis three blade variable speed variable pitch offshore wind turbine is used for the prototype of theory and the theory of unsteady BEM [1-2] is calculated based on revised theory. Six degrees of freedom movement producing by floating foundation influences the inflow velocity of wind turbine respectively with the main control system. It provides references for complex nonlinear coupling problem.

II. CORRECTION OF BLADE ELEMENT MOMENTUM THEORY

Because of the complex vortex system structure of the wind turbine [3-4], simplifying the wake flow structure and for the hypothesis which the number of rotor blades is boundless, Prandtl blade tip and hub loss correction is used. Define induced velocity correction factor of blade tip as F_1.

$$F_1 = \frac{2}{\pi} \arccos e^{-f_1} \tag{1}$$

$$f_1 = \frac{B}{2} \frac{R-r}{r \sin\phi} \tag{2}$$

In equation: B is the blade number of wind turbine; R is the whole radius of wind turbine; r is the local radius of wind turbine; ϕ is the inflow Angle.

Analogizing blade tip, define induced velocity correction factor of hub as F_2.

$$F_2 = \frac{2}{\pi} \arccos e^{-f_2} \tag{3}$$

$$f_2 = \frac{B}{2} \frac{r-R_{hub}}{r \sin\phi} \tag{4}$$

In conclusion, define Prandtl blade tip and hub loss correction factor as F

$$F = F_1 \cdot F_2 \tag{5}$$

When the axial induced factor a_n more than 0.4, the accuracy of data is low because of the simple theorem of momentum. The Glauert correction is used based on local aerodynamic. Define the axial induced factor as a_n

© 2015. The authors - Published by Atlantis Press

143

$$
a_n = \begin{cases} \dfrac{1}{\dfrac{4F\sin^2\phi}{\sigma C_n}+1} & a_n \leq a_c \quad (6) \\[2em] \dfrac{1}{2}\left\{2+k\left(1-2a_c\right)-\sqrt{\left[\left(k\left(1-2a_c\right)+2\right)^2+4\left(ka_c^2-1\right)\right]}\right\} & a_n > a_c \end{cases}
$$

$$
k = \frac{4F\sin^2\phi}{\sigma C_n} \quad (7)
$$

In equation: C_n is normal load coefficient; σ is solidity; a_c is axial induced velocity factor at critical time, about 0.2.

Due to the lift force and drag force coefficient of wind turbine acquired by wind the tunnel experiment of two-dimensional airfoil, the blade unit is regarded as two-dimensional airfoil. In the program separation factor model is used to calculate the Blade Element Momentum Theory. ΔC_l is increment of lift coefficient; ΔC_d is reduction of drag coefficient, so define the corresponding three-dimensional lift and drag coefficient as $C_{l,3d}$ and $C_{d,3d}$.

$$
C_{l,3D} = C_{l,2D} + \Delta C_l \quad (8)
$$

$$
C_{d,3D} = C_{d,2D} - \Delta C_d \quad (9)
$$

In equation: $C_{l,2d}$ and $C_{d,2d}$ is two-dimensional lift and drag coefficient [5-8].

III. PERFORMANCE OF OFFSHORE WIND TURBINE WITH FLOATING PLATFORM MOTION

Conceptual model parameters of the projects such as WindPACT, RECOFF and DOWEC are referenced. Integrated NREL 5 MW wind turbine main parameters are as follows:diameter of wind turbine is 126m; hub height from sea level is 90m;cut-inwindspeed, ratedwindspeed and cut-out windspeed are 3.0m/s, 11.4m/s and 25m/s;coneangle, tilt angle of main shaft, initial yaw angle and initial pitch angle are $0°$.

A. The Platform Motion

Under the action of the wind, wave and flow, the floating foundation of wind turbine will produce six degrees of freedom movement, such as yaw, pitch, roll, surge, sway, heave. This paper usesthe angle of yaw, pitch, roll$\beta_{yaw},\beta_{pitch},\beta_{roll}$ and the displacement of surge, sway, heave$S_{surge},S_{sway},S_{heave}$ to describe the motion.

In the study of six degrees of freedom movement, the wind speed is assumed as 11.2m/s less than rated wind speed11.4 m/s; the rated rotate speed of wind turbine is 12.1rpm.Six kinds of motion are sinusoidal movement:

$$
\gamma = \gamma_A \sin \omega t \quad (10)
$$

In order to obtain the aerodynamic performances when the wind turbine runs smoothly,thedata after 400s are selected in the following graphs to be studied. The thrust, torque and power of oscillation period are studied based on the sinusoidal motion condition of floating foundation. Circular frequency is elected as ω=0.1πrad/s based on Offshore Floating Wind Turbine Design Standard from DNV. It conforms to typical wave period 5-25s in China. The amplitudesof angle motion with floating foundationare 6°,8°,10°, and amplitudesof linear motion are 6m,8m,10m.

In actual cases, the wind turbine control system will affect the change of the aerodynamic performance.Assume that wind turbine speed measuring device can only be induced horizontal inflow velocity, so the control system will not affect its aerodynamic performance specially for movement can't change the horizontal inflow such as yaw and heave.

B. Aerodynamic Performance of Wind Turbine with Angle Motion of Floating Foundation

This paper only gives the aerodynamic performance changing with six degrees of freedom movement in a cycle of 20s. When the cycle is not same, the overall trend of aerodynamic performances' change is similar in each degree of freedom movement. When the inflow velocity of hub is convection, thrust, torque and power are 597.40kN,4256.67kN•m,5393.67kW.

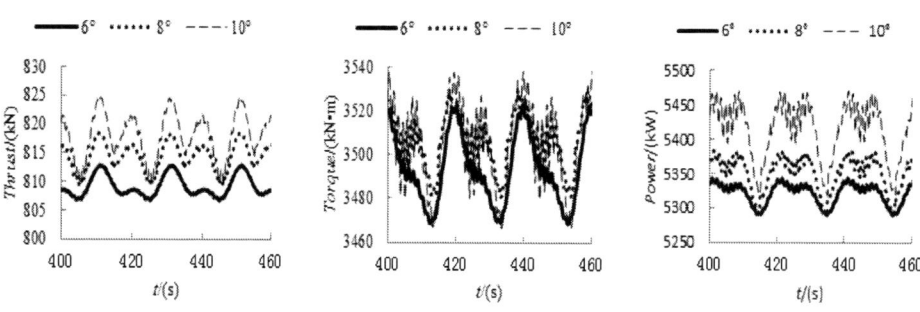

(a)Thrust (b)Torque (c)Power

FIGURE I. AERODYNAMIC PERFORMANCE IN YAW.

That can be seen from Figure 1, the average and amplitude of thrust; amplitude of torque and the average and amplitude of power increase when the angle of yaw increases, but the change is small. Due to the large rotor diameter, slight yaw will affect the inflow velocity of wind turbine to be uniform.

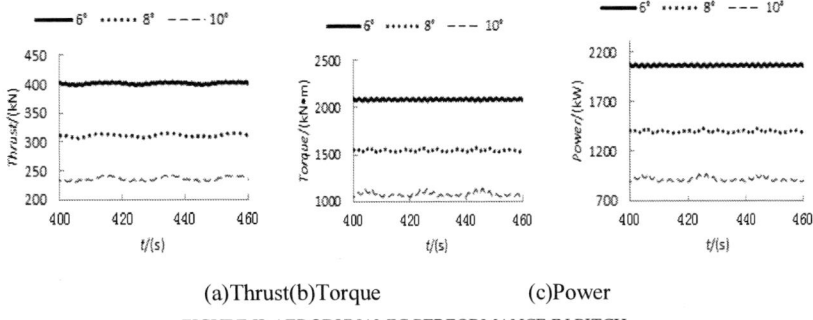

(a)Thrust(b)Torque   (c)Power

FIGURE II. AERODYNAMIC PERFORMANCE IN PITCH.

That can be seen from Figure 2, the average thrust; the average torque and the average power decrease when the angle of pitch increases. The amplitude of thrust; the amplitude of torque and the amplitude of power increase when the angle of pitch increases. The wind turbine's pitch is similar to yaw, because both of them don't change the inflow velocity of hub. But it will affect the tilt angle of wind turbine.

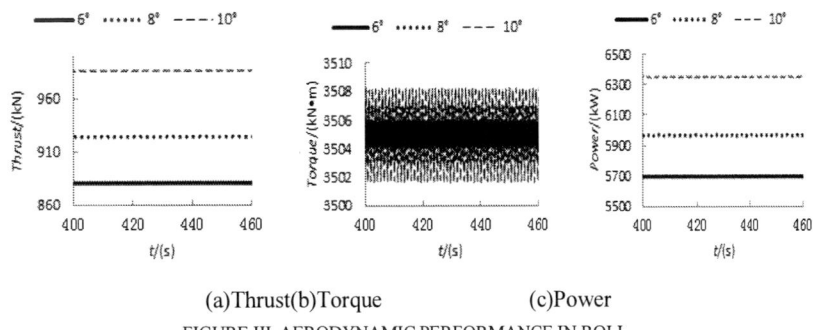

(a)Thrust(b)Torque   (c)Power

FIGURE III. AERODYNAMIC PERFORMANCE IN ROLL.

That can be seen from Figure 3, the average power decrease when the angle of roll increases. The average and amplitude of thrust; the amplitude of torque and the amplitude of power increase when the angle of roll increases. The average torque almost doesn't change. Because of the height of tower, a small roll can cause a greater change of inflow velocity. Roll occurs when wind and wave are in different directions, and can lead to inflow velocity increases. But the control system of wind turbine can be adjusted by the pitch angle to maintain the rotate speed of wind turbine staying at a rated rotate speed.

C. *Aerodynamic Performance of Wind Turbine with Linear Motion of Floating Foundation*

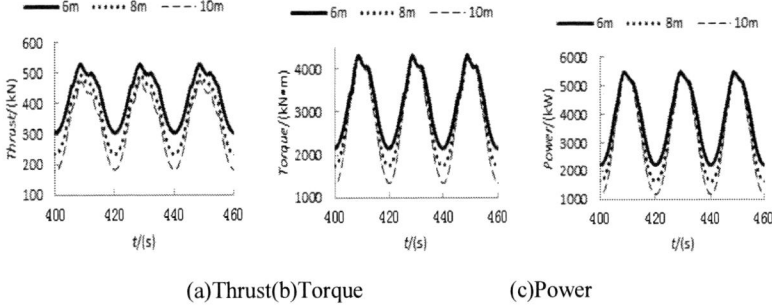

(a)Thrust(b)Torque   (c)Power

FIGURE IV. AERODYNAMIC PERFORMANCE IN SURGE.

That can be seen from Figure 4, the average thrust; the average torque and the average power decrease when the distance of surge increases. The amplitude of thrust; the amplitude of torque and the amplitude of power increase when the distance

of surge increases. The huge pneumatic thrust of upper wind turbine will generate larger surge movement when it works. Different from pitch, surge only changes the inflow velocity of hub not the tilt angle.

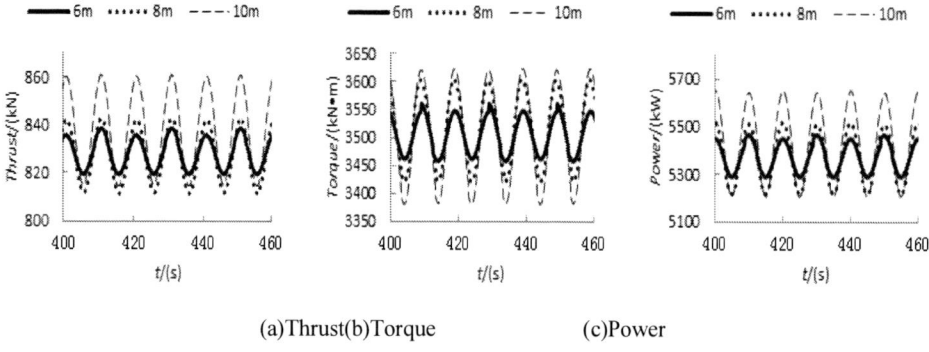

(a)Thrust   (b)Torque   (c)Power

FIGURE V. AERODYNAMIC PERFORMANCE IN SURGE.

That can be seen from Figure 5, the amplitude of thrust and the amplitude of power increase when the distance of sway increases. Sway occurs when wind and wave are in different directions, and it makes wind turbine always yaw. Sway changes horizontal wind speed and influence yaw angle of wind turbine at the same time, so the inflow velocity of hub will increase. When the speed no longer yaws, namely the inflow velocity of

hub is the right convection, and wind speed reaches minimum value at this time. After that wind turbine will yaw, and the wind speed will change too. Compared with the roll, sway belongs to the linear movement and it can't change the rotate speed.

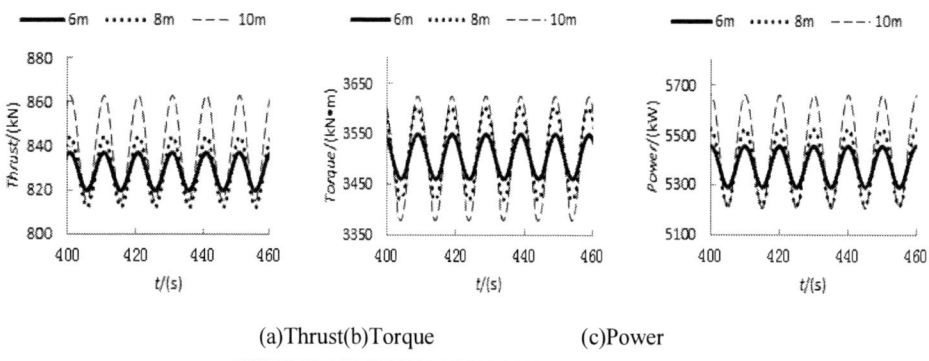

(a)Thrust   (b)Torque   (c)Power

FIGURE VI. AERODYNAMIC PERFORMANCE IN HEAVE.

That can be seen from Figure 6, the amplitude of thrust and the average and amplitude of power increase when the distance of sway increases. Heave and sway are same, they change horizontal wind speed and the yawing state of wind turbine. However, sway affects yaw angle and heave affects pitch angle. In the case of the surge, problem of the blade stall should be paid special attention to.

REFERENCES

[1] Moriarty P. J. & Colorado G., 2005, AeroDyn Theory Manual. National Renewable Energy Laboratory, pp.2-3.

[2] Leishman J. G. & Beddoes T. S., 1989, A Semi-Empirical Model for Dynamic Stall. Journal of the American Helicopter Society, Vol.34, NO.3, pp.3-17.

[3] Glauert, H., 1935, Airplane Propellers form Aerodynamic Theory, Vol. 4, Division L edited by Durand W. F. , Julius Springer, pp.169-360.

[4] Chaviaropoulos, P. K. & Hansen, M. O. L., 2000, Investigating Three-dimensional and Rotational Effects on Wind Turbine Blades by Means of a Quasi-3D Navier-Stokes Slover. Journal of Fluids Engineering, Vol. 122, pp.330-3336.

[5] Hansen M. O. L., 2008, Aerodynamics of Wind Turbines, Second Edition, Earthscan: UK and USA, pp.85-102.

[6] Glauert H., 1926, A General Theory of the Autogyro. ARCR R&M NO.1111, pp.1-6.

[7] Spera D. A., 1994, Wind Turbine Technology, First Edition, ASME Press: New York, pp.55-61.

[8] Øye. S., 1991, Dynamic stall, simulated as a time lag of separation. Proceedings of the 4th IEA Symposium on the Aerodynamics of Wind Turbines, Harwell Laboratory, Harwell, UK, pp.1-6.

International Conference on Power Electronics and Energy Engineering (PEEE 2015)

A Visualization Experimental Study of Icing on Blade for VAWT by Wind Tunnel Test

Y. Li
Engineering College
Northeast Agricultural University
Harbin, China

Q.D. Liu
Engineering College
Northeast Agricultural University
Harbin, China

J. Tang
Engineering College
Northeast Agricultural University
Harbin, China

S.L. Wang
Engineering College
Northeast Agricultural University
Harbin, China

F. Feng
College of Sciences
Northeast Agricultural University
Harbin, China

Abstract—**Blade icing is problematic to the safe and efficient operate of wind turbine. It is important to prevent and remove ice from the blade surface. To invest the characteristics of blade surface icing of wind turbine, wind tunnel tests were carried out on the wind turbine blade with NACA0018 and NACA7715 airfoil. The tests were made in winter. The outside cold air was induced into the wind tunnel, and a water spray was set up before the output of the wind tunnel to supply the icing condition. A CCD video camera was used to record the ice accretion in 20 minutes. By image processing software the icing distributions of the two kinds of airfoils were obtained and analyzed. Furthermore, the icing area increasing rate was also discussed.**

Keywords- wind turbine; blade; deicing

I. INTRODUCTION

The wind energy is one of the most used renewable energy in the world now. The power performance is not the most important point concerned by researchers benefiting from the long term study and development. The problems such as safety, noise, the effects from extreme weather to wind turbine have become the hot research fields. Generally, many wind farms were built up in cold regions, such as North Europe, North America. When the wind turbine works under the condition of wet and low temperature, icing will happen on the wind turbine, especially on blade which is the most important part to produce aerodynamics. The icing on blade changes the shape of blade airfoil and leads to a lot of serious problems. A dramatic decline of performance and safety of wind turbine will be caused. Therefore, the icing problems have received a lot of attention in the world. For example, Neil Bose researched icing on the small horizontal-axis wind turbine [1, 2]. Andrea G. Krai researched the phases of icing and ice adhesion force on wind

turbine blade by wind tunnel tests [3, 4]. The effect of temperature and droplet size variation on ice accretion of wind turbine blades was researched by Matthew C. Homola [5]. Olivier Parent proposed critical reviews about anti-icing and de-icing techniques for wind turbines in 2011 [6]. In addition, the authors also carried out some tests on wind tunnel and numerical simulations on the blade icing both for HAWT and VAWT [7,8].

Based on the past researches mentioned above, the icing problems on wind turbine have been more and more known to people. However, almost all the wind tunnel tests on blade icing just took photos at one time only and made analysis on icing characteristics of this time. The duration of icing was not recorded yet and no researches on the icing development have been carried out. Therefore, in this study, a CCD video camera was adopted and icing of 20 minutes were recorded. This makes the icing development visualization.

II. EXPERIMENTAL DETAILS

A. Test Blade Airfoil

The NACA0018 airfoil and NACA7715 airfoil were selected in this study. Based on the experimental condition, the chord of the blade was 0.22 meter, and the width of the blade is 0.1 meter.

B. Test System

Figure 1 shows the experimental system. The wind tunnel used in this study is a small scale open type one with the outlet of 0.4m×0.4m. The sensitivity of wind speed sensor is ±5%. To supply the icing condition, a water spray nozzle was set near the outlet of wind tunnel. The flow discharge was con-

© 2015. The authors - Published by Atlantis Press

147

trolled by a flow controller and measured by a flow meter. The average diameter of water droplet is 0.2mm. During the test, the cold air outside was sucked into the wind tunnel, and at the same time, the water was sprayed from the nozzle and mixed with the cold wind. The test blade is fixed at 0.3m downstream from the outlet. The wind speed used in this study was 6m/s. The flow discharge of water sprayed with wind which can be represented as the humidity of air was 0.5 L/min. The environmental temperature was -15 degree. A video CCD camera was used to record the ice accretion in 20 minutes. By using image processing software the icing photos were processed to be clearly visualization. The icing distributions on blade at every minute were obtained.

The test blade is in the water spray. Each test was carried out for 20 minutes. Then the test blade was photographed every minute. At last, data were recorded through computer. In this study, the humidity of air was 0.3L/min. The environmental temperature was -10 degree centigrade. The wind speed was 6m/s. The attack angle was 0 degree.

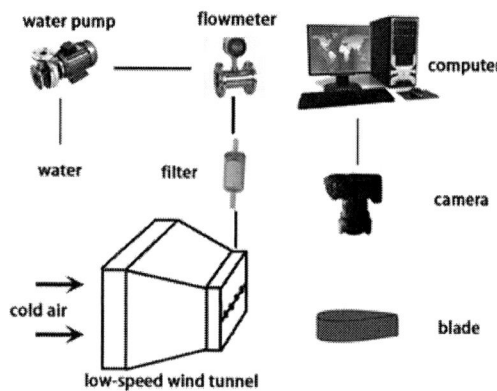

FIGURE I. EXPERIMENTAL SYSTEM.

III. RESULTS AND DISCUSSION

A. Distribution of Icing on Blade Airfoil Surface

Figure 2 shows icing development at every minute on blade surface on NACA0018 airfoil and NACA7715 in 20 minutes.

According to Figure.2, icing occurs at the leading edge part of airfoil and development along the up and down side of airfoil surface. When the ice reaches to the point near the largest thickness, it does not develop along the chord direction. Therefore, the icing of NACA0018 airfoil almost remains at the front part of airfoil at the attack angle of 0 degree. The icing on NACA7715 airfoil is not similar to NACA0018. At beginning, the icing occurs both at the leading edge and trailing edge because the NACA7715 is an airfoil. The icing development is not balance on the upside and downside. The icing on upside is many than icing on downside. Similar with the results of NACA0018 airfoil, the icing distance from leading edge of blade to the point about the first third of the length of blade chord. The icing length from trailing edge is also about first third of blade chord.

Based on the test results, it can be concluded that the icing on blade surface mainly depend on the airfoil type. For the symmetry airfoil, the de-icing and anti-icing device should be installed in the leading edge part. For the asymmetric airfoil, they should be installed both in the leading edge part and trailing edge part.

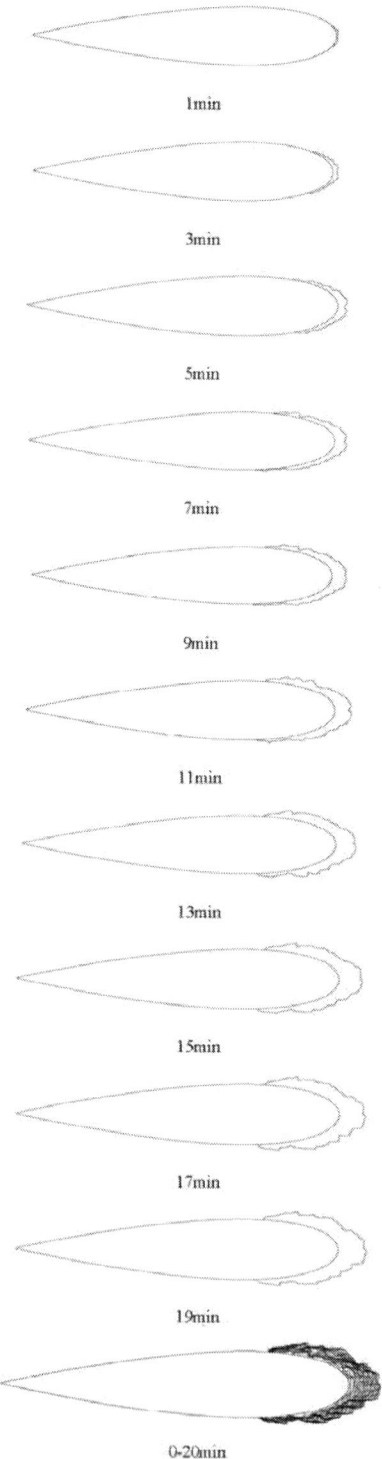

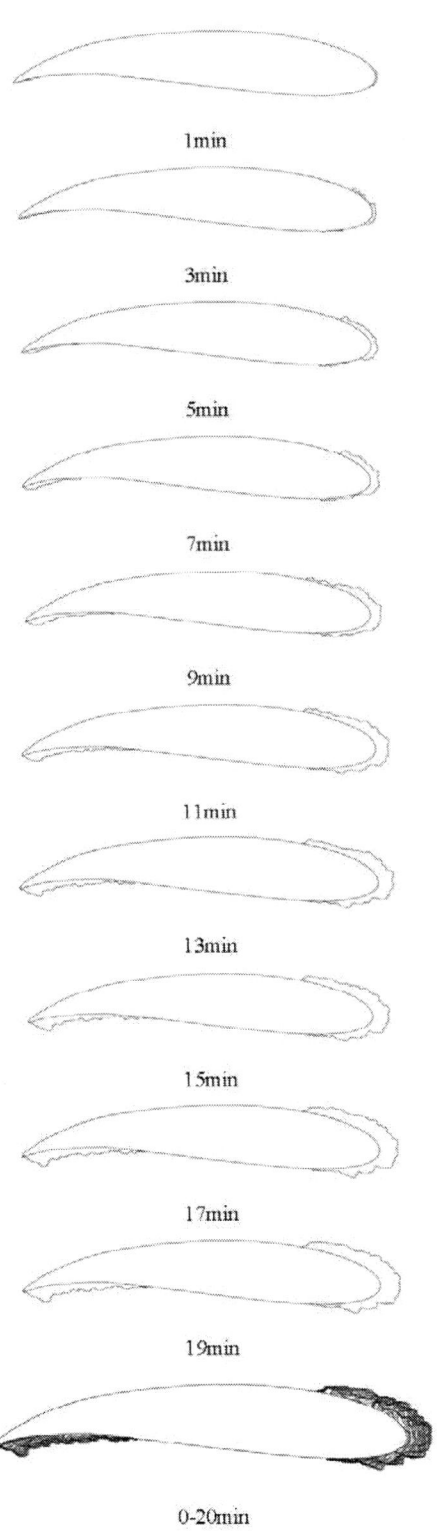

FIGURE II. ICING DISTRIBUTIONS ON NACA0018 AND NACA7715 AIRFOIL.

B. Icing Area

1) Total Icing Area Increasing Rate (TIAIR)

To clearly describe the icing rule on blade surface, icing area was analyzed. Figure 3 shows the definition of icing area at every minute. The icing area of blade surface at 1 minute is S_1, the new increasing icing area from 1 minute to 2 minutes is S_2. Finally, the new increasing icing area from 19 minutes to 20 minutes is S_{20}.

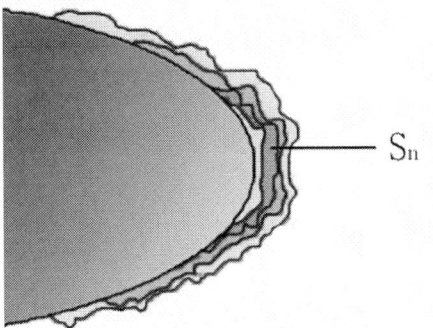

FIGURE III. DEFINITION OF ICING AREA.

Total icing area increasing rate (γ) was defined as shown in Table 1 where, S is the area of airfoil.

TABLE I. TOTAL ICING AREA INCREASING RATE.

T		Icing area rate
1min	r_1	$S_1/S*100\%$
2min	r_2	$S_1+S_2/S*100\%$
......	......	......
20min	r_{20}	$S_1+S_{2+...+}S_{20}/S*100\%$

Figure 4 shows the total icing area increasing rate for both the NACA0018 and NACA7715 airfoil. According the Figure4, the TIAIR shows a growth trend for both the two types of airfoil. During the 20 minute, the TIAIR of NACA0018 airfoil is 18.78% and NACA7715 airfoil is 20.62%. Between first minute and fifteenth minute, the TIAIR of NACA0018 airfoil is larger than that of NACA7715 airfoil. Between sixteenth minute and twentieth minutes, the TIAIR of NACA7715 airfoil is larger than that of NACA0018 airfoil.

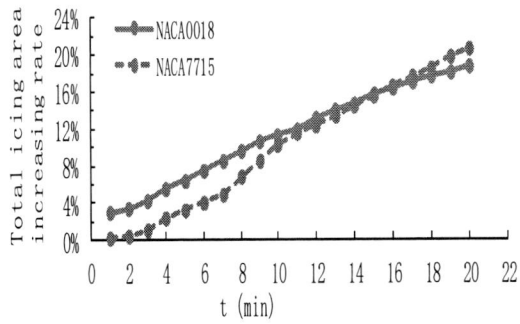

FIGURE IV. TOTAL ICING AREA INCREASING RATE.

2) Total Icing Area Increasing Rate (TIAIR)

Furthermore, the net icing area increasing rate (δ) during every one minute was defined shown in Table 2 like the definition of total icing area increasing rate.

TABLE II. NET ICING AREA INCREASING RATE.

T	Net icing area rate	
1min	δ_1	$S_1/S*100\%$
2min	δ_2	$S_2/S*100\%$
......	......	......
20min	δ_{20}	$S_{20}/S*100\%$

Figure 5 shows the net icing area increasing rate for both the NACA0018 and NACA7715 airfoil. According the Figure 5, the NIAIR shows a fluctuation trend on both two types of airfoil. NIAIR of NACA0018 airfoil is between 0.45% and 1.5%. NIAIR of NACA7715 airfoil is between 0.18% and 1.79%. In order to analyze NIAR of two types of airfoil better, trend line is introduced. There is a gradually decrease on the NIAIR of NACA0018 airfoil. Nevertheless, the NIAIR of NACA7715 airfoil increases first and then decreases.

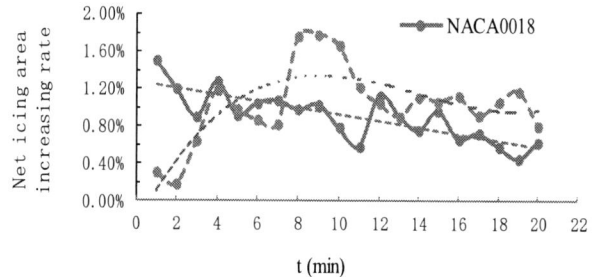

FIGURE V. NET ICING AREA RATE.

According to Figure.4 and Figure.5, it can be concluded that although the net icing area increasing rate is quite different between NACA0018 and NACA7715 airfoil during every minute, the total icing area increasing rate keeps the liner increasing speed.

IV. CONCLUSIONS

Under the conditions of this experiment, conclusions can be summed as below:

1) The icing development on both NACA0018 and NACA7715 airfoil were well recorded and the icing distributions were clearly visualization.

2) The icing on blade surface mainly depended on the airfoil type when the attack angle was fixed. For the symmetry airfoil, the icing distributions were mainly at the leading edge part of blade. For the asymmetric airfoil, the icing occurred at both the leading edge part and trailing edge part of blade. This can be a reference for installing of de-icing and anti-icing device on blade.

3) The total icing area increasing rate kept linearly increasing during the test time.

ACKNOWLEDGMENTS

This study was sponsored by Specialized Research Fund for the Doctoral Program of Higher Education of China (No.: 20132325110003) and was also done as the research for the Project 10702015 supported by National Natural Science Foundation of China (NSFC). The authors would like to express thanks to all of their supporters.

REFERENCES

[1] Neil Bose, Icing on a small horizontal-axis wind turbine - Part 1: Glaze ice profiles, *Journal of Wind Engineering and Industrial Aerodynamics,* 1992, 45(1): 75-85.

[2] Neil Bose, Icing on a small horizontal-axis wind turbine - Part 2: Three dimensional ice and wet snow formations, Journal of Wind *Engineering and Industrial Aerodynamics*, 1992, 45(1):87-96.

[3] Andrea G. Kraj, Eric L. Bibeau, Phases of icing on wind turbine blades characterized by ice accumulation, *Renewable Energy*, 2010, 35(5): 966-972.

[4] Andrea G. Kraj, Eric L. Bibeau, Measurement method and results of ice adhesion force on the curved surface of a wind turbine blade, *Renewable Energy*, 2010, 35(4): 741-746.

[5] Matthew C. Homola, Muhammad S. Virk, Tomas Wallenius, Per J. Nicklasson, Per A. Sundsbø, Effect of atmospheric temperature and droplet size variation on ice accretion of wind turbine blades, *Journal of Wind Engineering and Industrial Aerodynamics*, 2010, 98(12): 724-729.

[6] Olivier Parent, Adrian Ilinca, Anti-icing and de-icing techniques for wind turbines: Critical review, *Cold Regions Science and Technology,* 2011, 65(1): 88-96.

[7] Yan Li, Kotaro Tagawa and Wei Liu, Performance effects of attachment on blade on a straight-bladed vertical axis wind turbine. *Current Applied Physics*, 2010, 10(2): S335-338.

[8] Yan Li, Kotaro Tagawa, Fang Feng, Qiang Li, Qingbin He, A Wind Tunnel Experimental Study of Icing on Wind Turbine Blade Airfoil, *Energy Conversion and Management*, 85(2014)591-595.

International Conference on Power Electronics and Energy Engineering (PEEE 2015)

Kinematics Analysis on the Cross Step Skills of Chinese Female Javelin Thrower Lv Huihui

M.R. Zhou
Southwest University for Nationalities
Chengdu, China

Abstract—**The paper use the documents the mathematical statistics and the three-dimension picture analysis on the cross step skill of Lv Huihui, the champion of China Long Open in 2013 and get the kinematics parameters and summarize her skill features . The results show that Lv made a proper leaning back during the cross step stage. Her elbow angle is 146.4°, while the angle of world outstanding female javelin athletes is 160.2°, it has a big gap of 13.8°. In the horizontal velocity during the cross step, Lv Huihui lags behind world outstanding female athletes. The conclusion shows that Lv did not fully extend her elbow joint in the cross step stage, the elbow stress is not enough, throwing arm shoulder too loose; the speed of Lv need to be improved. Because the step size is too long, will effect postpone time of the force, thus affect the coherence of action, reduce the running speed and force effect, ultimately affect the performance in the competition; the long cross step time is the main reason for the loss of horizontal speed.**

Keywords- biomechanics; the three-dimension picture analysis; javelin thrower; the cross step skills

I. INTRODUCTION

The paper aims to use the three-dimension picture analysis on the cross step skill of Lv Huihui in China Long Throw Open in 2013 and get the kinematics parameters and summarize her skill features. Additionally, the paper also makes a comparison of her related kinematics parameters and other elite javelin throwers to find her advantages and drawbacks. All these results provide coaches and athletes the kinematics quantitative indicators in order to improve Lv Huihui' skills and enrich javelin technological theories.

II. METHODS

The main method is three-dimension picture analysis. We used two JVC9800 Synchronized cameras at the right side and back side at the game site. machine height is 1.2m，The main optical axis camera into 90 Angle（As shown in Figure 1）.The shooting frequency is 50 frames per second. The record analysis used 3-DSignalTec system and series analysis. The anthropometric dummy is Japanese Song jing xiu zhi phantom(21 articulation points, 16 segments and additional 2 testing points and 1 segment, i.e. Javelin's two endpoints and javelin link). It pass the original data filter and the cutoff frequency is 8Hz.

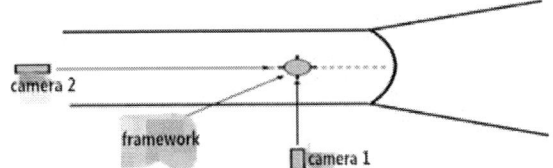

FIGURE I. . THE MAIN OPTICAL AXIS OF CAMERA.

III. RESULTS

Throwing steps phase is usually set foot on his left foot marker 2 step right leg began to finally force left foot landing, a total of four steps. （As shown in Figure 2）.The third throwing step is cross step that start from the left foot landing and the right foot actively swing ,and go to the right foot landing. When the left foot landing it should actively swing , thigh leg drive positive and powerful forward swing out, ,make the lower limbs forward acceleration to surmount apparatus to be a good post .when the right foot landing , the left leg should in the in the front of the right leg to speed up the landing time . This is a critical step to connection the run-up and the last exertion.

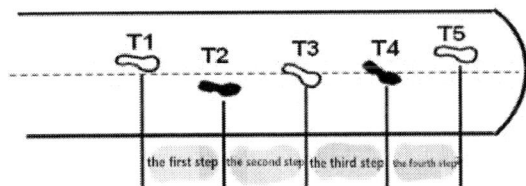

FIGURE II. THE THROWING STEPS PHASE

A. *The Analysis Of The After Trunk Angle Of Lv During The Cross Step*

Angle of trunk leaning back is the angle between the axis of javelin throwers' trunk and the vertical plane in the process of throwing. It reflects the degree of trunk leaning back. Proper leaning back can not only maintain the horizontal velocity of throwers' center of gravity, but also increase the effective distance working with javelin. Therefore, it is one of the most important elements that will increase the throwing speed. The main function of the cross step is to keep throwers' balance between velocity and the posture which form forces as a whole, therefore, the proper trunk leaning back, no doubt, makes the basis for increasing distance for the last strength. This moment, the average angle of trunk leaning back of world outstanding

© 2015. The authors - Published by Atlantis Press

female javelin athletes is 23.3°, while Lv 23.6°. It can be known form the data that there is a little gap between their angle, which means that Lv made a proper leaning back during the cross step stage.

B. *The Analysis Of Shoulder Angle And Elbow Angle Of Lv During The Cross Step*

Proper angle of right shoulder of a thrower during the cross step is the most important indication to judge his quality of directing the javelin, and an important hallmark to tell if a javelin athlete's throwing arm makes full use of arm length to increase kinetic energy produced by whiplash in the process of throwing heavily as well. Besides, proper angle of right shoulder can keep a thrower's body in moderately tense readiness, which further rushes muscles to fully prepare for whiplash.

This moment, her shoulder angle is 92.5 , compared with the world's outstanding female javelin athletes' 95.4, it has little gap, her elbow angle is 146.4, while the angle of world outstanding female javelin athletes is 160.2, it has a big gap of 13.8. According to above data, Lv shoulder and arm's muscle is loose and muscle tensity should be strengthen.

C. *Analysis Of Horizontal Velocity Of Lv Huihui In Cross Step*

The performance is in positive correlation with the horizontal velocity of body center of gravity after the cross step, that is the speed after the cross step is more important than the speed before the cross step; the horizontal velocity after the cross step reflects horizontal momentum before the final force. Therefore, the bigger the horizontal velocity, the larger the chance for athletes to get good performance; at the same time, the difference of horizontal velocity between before the cross step and after shows the amount of lost speed, which indirectly shows the effect of the cross step. Therefore, a bad cross step will affect the effect of the final force. So, the less loss of the horizontal velocity before and after the cross step, the better performance. The horizontal velocity of Lv Huihui before the cross step is 6.6m/s, and after the cross step, the horizontal velocity is 5.79m/s. The difference is 0.81m/s. This shows that in the horizontal velocity during the cross step, Lv Huihui lags behind world outstanding female javelin athletes.

D. *Analysis Of Step Size And Time Of Lv Huihui During The Cross Step*

Run-up gives athletes and instrument system a certain horizontal velocity and makes preparation for withdraw javelin

[6] iversity, 2012,14 (1).

and the final force. The rhythm of throwing steps is very important to successfully complete a series of actions from withdraw javelin to throwing in a run-up with higher speed. The rhythm of throwing steps is a guarantee for right throwing steps. Only with appropriate rhythm can throwing steps be successfully completed. According to the study of Peter Cheyne, the size and time of throwing steps affect the coordinated cooperation between the trunk and the action of upper limb. They are the conditions affecting the throwing of javelin. In theory, the average size of throwing steps of Lv Huihui should be 1.46m-1.63m, while the actual average size is 1.68m. In theory, the size of cross step is 1.61m—1.87m, while the actual average size is 2.19m. Too long size of steps postpones the moment of the final force, which affects the coherence of the whole skilled movement, reduces the speed of run-up and the effect of force. Therefore, the throwing performance is affected. The time of cross step of Lv Huihui and the world's outstanding athletes is 0.44s and 0.36s respectively. Longer time of the cross step reduce more of the horizontal velocity.

IV. CONCLUSIONS

The after trunk angle of Lv in the cross step stage is suitable; while the elbow joint angle has a big gap with world excellent athletes. This shows that Lv did not fully extend her elbow joint in the cross step stage, the elbow stress is not enough, throwing arm shoulder too loose; the speed of Lv need to be improved. Because the step size is too long, will effect postpone time of the force, thus affect the coherence of action, reduce the running speed and force effect, ultimately affect the performance in the competition; the long cross step time is the main reason for the loss of horizontal speed.

REFERENCES

[1] National sports college athletics teaching textbook committee group. Track and field teaching materials [M]. Beijing: People's sports press, 2006

[2] Zhan Yongshun. Research on [2] throwing step of Chinese Elite Women Javelin Athletes [J]. Shaanxi: Journal of Shaanxi University of Science and Technology, 2009,27 (2)

[3] Liu Shengjie, Li Jianying. Technology of throwing step of Chinese Elite Female Javelin Athletes Kinematics Study of [J]. Chinese sports science and technology, 2007,43 (4)

[4] Li Wencong, Zhuo Jiannan, Zhu Zhenjie. Technology of throwing step analysis [J]. Zhejiang sports science of our country outstanding female javelin athlete, 2008,30

[5] Fu Jianqiang cross step kinematics of Chinese Elite Male Javelin Athletes analysis [J]. Journal of Hengshui Un

Microstructure and Mechanical Properties of Micro-Alloying Modified Al-Mg Alloys

J. Zhang
College of Materials Science and Engineering
Chongqing University
Chongqing, China

J.J. Zhao
College of Materials Science and Engineering
Chongqing University
Chongqing, China

R.L. Zuo
College of Materials Science and Engineering
Chongqing University
Chongqing, China

Abstract—Er and Sr were added to 5052 Al alloy to investigate their possible effects on the microstructure and mechanical properties of the alloy. The results show that while Sr addition has a moderate grain refinement effect and has no positive effect on alloy plasticity, Er addition refines both the grain size and Al6Fe intermetallics and obviously improves the tensile elongation from 12% for 5052 alloy to 20% for Er-containing alloy. It is also revealed that Er either dissolves in the matrix or combines with Fe and Si and that Er can purify the alloy by eliminating impurity elements Fe and Si from the melt and Al-matrix.

Keywords-5052 Al alloy; micro-alloying; microstructure; mechanical properties

I. INTRODUCTION

With the increased demand of energy saving and pollution reduction, Al-Mg series (5XXX) alloys is increasingly used in automotive and transportation industry due to their high specific strength, wieldable, good workability and corrosion resistance [1]. Al-Mg alloys are non-heat treat hardening but derive their strength from solid solution strengthening and strain hardening. It has been shown that micro-alloying is an effective way to further strength the alloys. Alloying elements such as Sc, La, Ce, Y, Nd, Zr, Er, Sr have been added to high-Mg-containing Al-Mg alloys, especially 5083 aluminium alloy [2-9]. However, the role of alloying addition on low-Mg-containing Al-Mg alloys has not been studied. Comparatively, low-Mg-containing Al-Mg alloys have better formability and are more suitable for large wrought products. Demand in light rail train, coal conveyor, refrigerated truck, etc., requires larger width Al sheets, which in turn, raises higher formability requirement in order to ensure a homogenous deformation along width direction. In consideration that rare earth addition is beneficial to both microstructure and properties [10], micro-alloying elements Er and Sr have been added to 5052 Al alloy, a typical low-Mg-containing alloy, and their modification role on the as-cast microstructure and mechanical properties are investigated.

II. EXPERIMENTAL PROCEDURES

Commercial 5052 Al alloy was used as base alloy. Minor Er and Sr were added. The chemical composition was analyzed by inductively coupled plasma emission spectrometer (ICP) and is shown in Table 1. The alloys were prepared by mold casting. Samples sectioned from the center of the ingots were polished and etched for microstructure observation, by optical microscope and TESCAN VEGA II scanning electron microscope (SEM) equipped with INCA Energy 350 energy dispersive X-ray spectrometer (EDX). The grain size was measured by the mean liner intercept method. The phase transformations during the solidification of these alloys were characterized by employing a NETZSCH STA449C simultaneous thermal analyzer. In the DSC testing, samples of 30 mg were heated to 700 °C for 5 min and then cooled at a controlled speed of 15 K/min under flowing argon. Microhardness was measured using a Vickers micro-hardness tester with a load of 50g and duration of 10s. Cylindrical samples of 56 mm gauge length and 13 mm diameter were cut from the ingots for tensile mechanical testing.

TABLEI. CHEMICAL COMPOSITIONS OF THE EXPERIMENTAL MATERIALS(WT%)

Alloy	Si	Fe	Cu	Mn	Mg	Cr	Ti	Er/Sr	Al
5052	0.070	0.417	<0.1	<0.1	2.25	0.14	0.006	/	Balanced
5052+Er	0.081	0.665	<0.1	<0.1	2.27	0.10	0.020	0.30	Balanced
5052+Sr	0.070	0.460	<0.1	<0.1	2.41	0.14	0.006	0.028	Balanced

© 2015. The authors - Published by Atlantis Press

III. RESULTS AND DISCUSSION

A. Microstructure

The as-cast microstructures of 5052 base alloy and Er/Sr modified alloys were shown in Figure1. The microstructures manifest dendritic morphology with the second phase distributed along grain boundaries. The grain size of 5052 alloy is within 200~300 μm. With Er/Sr addition, the grain sizes are much reduced. As the same time, the amount of the secondary phases is obviously increased with Er addition while Sr addition does not influence the amount of the particles. The volume fraction of the particles was measured to be ~2%, 3% and 2% for the base alloy, the alloy with Er and Sr additions, respectively.

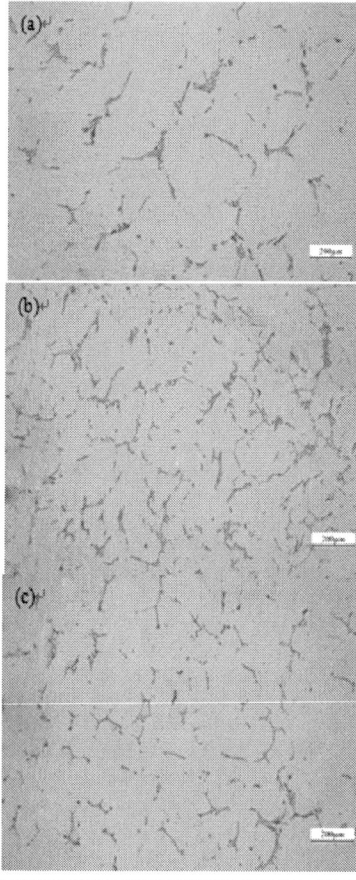

FIGUREI. AS-CAST MICROSTRUCTURES OF 5052 ALLOYS: (A) BASE ALLOY, (B) WITH ER ADDITION, (C) WITH SR ADDITION

Thermal Analysis

DSC curves of the as-cast alloys are shown in Figure 2. The addition of Er and Sr does not change the phase transformation characteristics. The exothermal peak on the cooling curves corresponds to the crystallization of the α-Al matrix. However, it is noted that the addition of Er and Sr does change the characteristic temperatures. The initial temperature TN, the peak temperature TM and the end temperature TR of the exothermal

peaks are listed in Table 2. The solidification temperature range ΔT（ΔT=TN-TR）is also given in Table 2. It can be seen that ΔT decreases by micro-alloying and it decreases the most with Er addition. The smaller the solidification temperature range, the better the fluidity and filling performance of the alloy. Therefore, the addition of alloying elements, especially Er, is beneficial to improve the casting performance of the alloy, which is important for large billet production. Furthermore, it is noted that the initial temperature increases incrementally with Sr and Er addition. This means that nucleation can occur at a relatively low undercooling temperature, a phenomenon which is normally associated with lower nucleation energy. From this point of view, the grain size of Er-containing alloy should be the finest, which accords with the microstructure observation results.

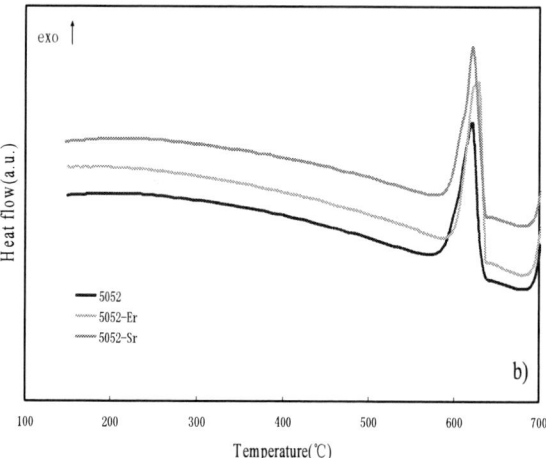

FIGUREII. DSC COOLING CURVES OF THE AS-CAST ALLOYS

TABLEII. THE CHARACTERISTIC TEMPERATURES (INITIAL TEMPERATURE TN, PEAK TEMPERATURE TM AND END TEMPERATURE TR) OF THE EXOTHERMAL PEAKS ON THE COOLING CURVES (°C)

Alloy	T_N	T_M	T_R	ΔT
5052	630.6	619.7	590.4	40.2
5052+Er	635.0	625.9	606.2	28.8
5052+Sr	631.8	622.3	595.7	36.1

B. Phase Constitution

To determine the effect of micro-alloying addition on the phase constitution, SEM and EDS analysis were performed. In 5052 alloy, Mg is almost dissolved in α-Al matrix. Impurity elements Fe and Si exist in the form of intermetallic Al6Fe and Mg2Si phases. With Sr addition, the phase constitution does not show detectable change. After adding Er, no Mg2Si particles were detected, instead, Si combined with Er, as illustrated by arrow A in Figure3 whose composition is also inserted in the figure. Al6Fe phase still exist, however, the size is much

154

reduced, being ~100 μm in 5052 alloy and ~20 μm with Er addition, respectively. Moreover, it is interesting to note that there is always minor Er detected in the Al6Fe phase. The results suggest that Er can purify the alloy by eliminating impurity elements Fe and Si from the melt and Al-matrix, thus diminishing their harmful effects. Furthermore, Er segregates in the front of the advancing solid/liquid (S/L) interface during the solidification and inhibits the growth of the secondary phase [11], resulting in a size reduction of the Al6Fe phase.

EDS results also reveal that there is certain amount of Er dissolved in the α-Al matrix. Besides, it is found [12] that Al and Er have a stronger tendency than Mg-Er and Mn-Er to combine together, therefore there might be some Al3Er particles which can act as effective nuclei for the Er-containing alloy.

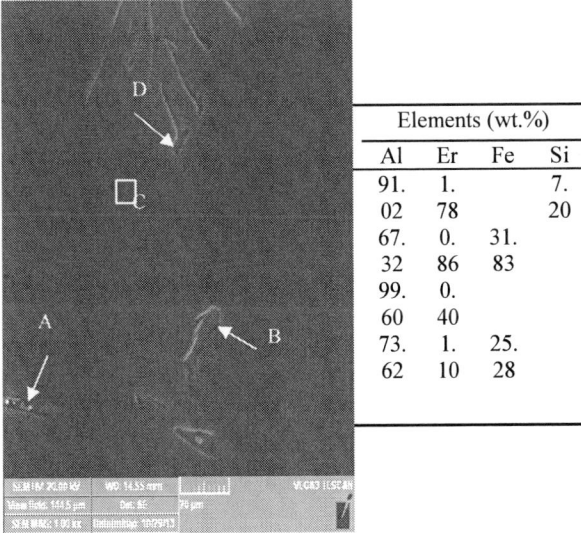

Elements (wt.%)			
Al	Er	Fe	Si
91.02	1.78		7.20
67.32	0.86	31.83	
99.60	0.40		
73.62	1.10	25.28	

FIGUREIII. SEM MORPHOLOGY AND EDS ANALYSIS RESULTS OF THE ER-CONTAINING ALLOY

C. Mechanical Testing

Hardness testing shows that HV value remains almost unchanged with alloying element addition, being in the range of 47~50. Mechanical testing shows (Figure4) that tensile strength decreases a bit with alloying element addition, while the elongation is obviously improved with Er addition, increasing from 12% for 5052 alloy to 20% for Er-containing alloy. The improvement of elongation is beneficial to subsequent plastic deformation processing.

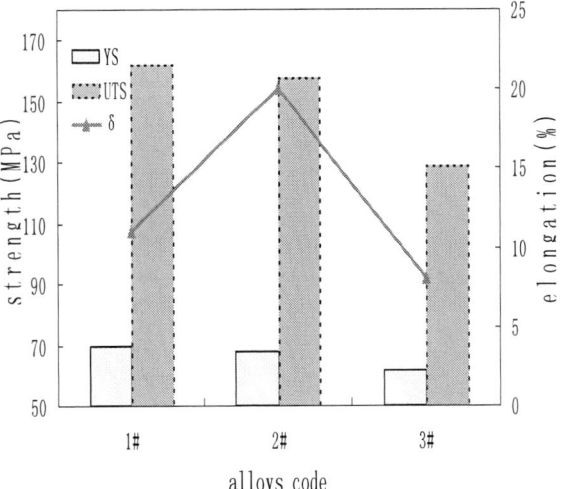

FIGUREIV. TENSILE MECHANICAL PROPERTIES OF THE ALLOYS.
1#:5052 ALLOY #2:5052+ER; 3#:5052+SR

IV. CONCLUSIONS

1) The as-cast microstructure of 5052 alloy has a typical dendritic morphology. The main phases are α-Al matrix, intermetallic compounds Al6Fe and Mg2Si.

2) Er addition refines both the grain size and the Al6Fe phase. Er either dissolves in the matrix or combines with Fe and Si; Er can purify the alloy by eliminating impurity elements Fe and Si from the melt and Al-matrix.

3) Tensile elongation is obviously improved by Er, increasing from 12% for 5052 alloy to 20% for Er-containing alloy. The improvement of elongation is beneficial to subsequent plastic deformation processing.

4) Sr has a moderate grain refinement effect. Moreover, it does not change the phase constitution and has no positive effect on alloy plasticity.

ACKNOWLEDGEMENTS

The authors are grateful for the financial support of the International Technical Cooperation Project (2011DFR50950) by the Ministry of Science and Technology of China and sharing fund of Chongqing University's large scale equipment.

REFERENCES

[1] G.C. Blaze, Alcoa Green Letter: The 5000 Series Alloys Suitable for Welded Structural Applications, Aluminum Corporation of America, New Kensington, PA, 1972.

[2] K.E. Knipling, D.C. Dunand, D.N. Seidman, Criteria for developing castable, creep-resistant aluminium-based alloys-a review, Z Metallk 97 (2006) 246–265.

[3] Yang FB, Liu EK, Xu J, Shi LK. Effects of Er on the microstructures and mechanical properties of as-cast Al–Mg–Mn–Zn–Sc–Zr–(Ti) filler metals. Acta Metall Sin 2008;44(8):911.

[4] J. Royset, N. Ryum, Scandium in Aluminum Alloys, Int. Mater. Rev. 50 (2005) 19-44.

[5] E. A. Marqis, D. N. Seidman, M. Asta, C. Woodward, Composition Evolution of Nanoscale Al3Sc Precipitates in an Al-Mg-Mn-Sc Alloy: Experiments and Computations, Acta Mater. 54 (2006) 119-130.

[6] H.Z. Li, H.J. Wang, X.P. Liang, Y. Wang, H.T. Liu, Effect of Sc and Nd on the Microstructure and Mechanical Properties of Al-Mg-Mn Alloy, J. Mater. Eng. Perform. 21 (2012) 83-88.

[7] A. K. Lohar, B. Mondal, D. Rafaja, V. Klemm, S. C. Panigrahi, Microstructural investigations on as-cast and annealed Al-Sc and Al-Sc-Zr alloys, Mater. Charact. 60 (2009) 1387-1394.

[8] Z.X. Liu, Z.J. Li, M.X. Wang, Y.G. Weng, Effect of complex alloying of Sc, Zr and Ti on the microstructure and mechanical properties of Al-5Mg alloys, Mater. Sci. Eng. A 483-484 (2008) 120-122.

[9] Q. Chen, Q.L Pan, Y. Wang, Z.Y. Zhang, J. Zhou, C. Liu, Microstructure and mechanical properties of Al-5.8Mg-Mn-Sc-Zr alloy after annealing treatment, J. Cent. South Univ. 19 (2012) 1785–1790.

[10] Z. R. Nie, T. N. Jin, J. X. Zou, J. B. Fu, J. J. Yang, T. Y. Zuo, Development on Research of Advanced Rare-earth Aluminum Alloy, Trans. Nonferrous Met. Soc. China, 13 (2003) 509-514.

[11] M.A. Easton, D.H. StJohn, A model of grain refinement incorporating alloy constitution and potency of heterogeneous nucleant particles, Acta Mater. 49 (2001) 1867–1878.

[12] J Zhang, C. Fang, F. Yuan, Grain refinement of as cast Mg–Mn alloy by simultaneous addition of trace Er and Al, Int. J. Cast Metal. Res. 25 (2012) 335-340.

International Conference on Power Electronics and Energy Engineering (PEEE 2015)

Comparative Study on Mechanical Properties of V-Shaped Insulator String under Fluctuating and Calm Wind

J.J. Huang, X.M. Chen, Y. Xiong
State Grid Hubei Electric Power Research Institute
Wuhan Hubei 430077, China

C.L. Liu
China Electric Power Research Institute
Beijing 100055, China

Y. Wang
Beijing Guowang Fuda Science & Technology Development Co., Ltd
Beijing 100070, China

Abstract—**In order to define mechanical properties of V-shaped insulator strings under dynamic wind load, a numerical commutating method is adopted. By carrying out comparative analysis of dynamic and static calculation results, this paper provides a theoretical basis for engineering design.**

Keywords-v-shaped insulator; mechanical analysis; calm wind; buckling

I. INTRODUCTION

With the rapid development of national economy of China, the electricity demand of the entire society has greatly increased. Since most energy bases in China are located at western part, the optimal configuration [1] of energy resources has to be realized on a national scale through electricity transmission from the west to the east of China, the north and the south power exchange and nationwide interconnection. A number of built power lines in China have to cross high-wind areas, so the V-shaped insulator string is adopted to improve windage yaw of insulators so as to reduce insulator windage yaw of the straight-line tower. The current design specification calculates mechanical properties of the V-shaped insulator string as per static wind load and then multiplies the security coefficient. In order to define mechanical properties of V-shaped insulator strings under dynamic wind load, it is necessary to carry out comparative study on mechanical properties of V-shaped insulator strings under fluctuating and calm wind.

II. THEORETICAL BASIS OF CALCULATION

Literature [2, 3] deduced fundamental formulas such as catenary balance equation and created an iterative algorithm. On this basis, Literature [4, 5] deduced a catenary element stiffness matrix with a finite element method, which was used to analyze planar and spatial structure of suspension cables. The basic motion equation of dynamics is as follows:

$$M\ddot{u} + C\dot{u} + Ku = F.$$ (1)

where, M is structural mass matrix, K is stiffness matrix, C is damping matrix, $\ddot{u}$ is acceleration vector, $\dot{u}$ is velocity vector, u is displacement vector, and F is load vector.

The motion balance equation is a second-order linear ordinary differential equation system with constant coefficients, which can be in theory solved with a typical method for ordinary differential equation systems. Two numerical solutions are generally adopted during actual finite element dynamic analysis, i.e. modal superposition method and immediate integration [6].

III. CALCULATION OF WIND LOAD

A wind time-history curve generally contains two components: one is long-period part whose value is always above 10min; the other is short-period part whose value is often only several seconds. The fluctuating wind is caused by wind irregularity, whose intensity changes over time as per randomness rules; since its period is short, it has a dynamic function and vibration may be caused. We adopt the down-wind horizontal fluctuating wind velocity spectrum used in Chinese specifications, which is a Davenport spectrum. Thus, the fluctuating wind velocity time history under high wind conditions is obtained through calculation.

The lateral wind load produced by wind action on conductors should take into account influences of shape factor of conductors, wind-pressure uneven factor related to wind velocity, wind-load adjustment factor related to voltage level and wind velocity, wind velocity-height change factor related to average height of conductors, angle between wind direction and conductor axial direction, etc. The wind load on unit length of conductors should be calculated by formula (2) according to standards of the power industry in China.

$$g_H = 0.625 \times \alpha \times \mu_{sc} \times (d + 2\delta) \times (K_h \times v)^2 \times 10^{-3}.$$ (2)

© 2015. The authors - Published by Atlantis Press

where, α is uneven factor of conductor wind pressure, μ_{sc} is shape factor of conductor, d is outside diameter (mm) of conductor, δ is thickness (mm) of conductor icing, K_h is wind velocity-height change factor at the average height h of conductors, v is design wind velocity (m/s) at reference height h_s of lines.

The wind load on bundled conductors is specified in IEC60826 Code for Design of Overhead Transmission Line that as follows [7]: the wind load on bundled conductors is the number of bundles multiplying wind load of a single conductor, with shielding effect of bundled conductors neglected.

IV. NUMERICAL CALCULATION MODEL

The conductor is a flexible structure, and the geometric nonlinear problems about big displacement and small deformation should be considered during analysis. At the same time, the initial static balance position of conductor system constitutes an initial condition of static and dynamic analysis, which is of great significance for results of calculation. Therefore, we should firstly carry out static balance analysis on the conductor system, and then establish a calculation model according to assembly drawing of insulator strings.

Conduct windage yaw analysis on the insulator-hardware fitting-conductor system under calm and high wind conditions. Select horizontal span of 500m and vertical span of 500m, with a difference of 0m. The length of the entire composite insulator string should be 11.0m and the model of V-shaped insulator strings is shown in Figure 1. A conductor of 6×JL/G3A-1000/45-72/7 is adopted for mechanical calculation. The starting calculation height of reference wind velocity should be 10m and the category of ground roughness is regarded as Category B. The design basic wind velocity is 33m/s (at 10m height), the average height of conductors is regarded as 30m, and the wind velocity here is converted into 39.34m/s.

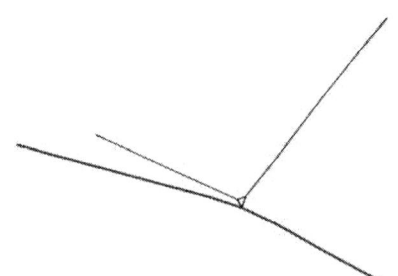

FIGURE I. NUMERICAL CALCULATION MODEL OF V-SHAPED INSULATOR STRING.

V. STATICS CALCULATION OF V-SHAPED INSULATOR STRING

When there is only one V-string with included angle of 100°, the force imposed on single-string insulators at windward side of V-string under calm wind is 148.94kN and that imposed on such insulators at its leeward side is -0.08kN.

When there is only one V-string with included angle of 110°, the force imposed on single-string insulators at windward side of V-string under calm wind is 151.27kN and that imposed on such insulators at its leeward side is 7.01kN.

When there is only one V-string with included angle of 120°, the force imposed on single-string insulators at windward side of V-string under calm wind is 159.03kN and that imposed on such insulators at its leeward side is 22.57kN.

VI. MECHANICAL PROPERTIES OF INSULATOR STRING UNDER FLUCTUATING WIND

Under high and fluctuating wind conditions, we calculate mechanical properties of V-string at fluctuating wind. After buckling deformation is caused on the insulators at leeward side under fluctuating wind, the deformation of V-shaped insulator string is shown in Figure 2.

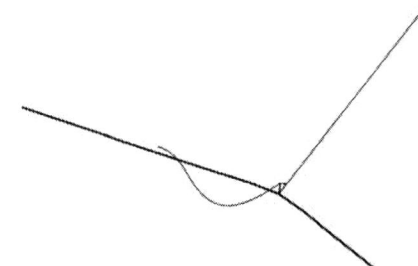

FIGURE II. BUCKLING DEFORMATION OF V-SHAPED INSULATOR STRING UNDER HIGH WIND.

A. Calculation Results When There Is Only One V-String with Included Angle of 100

The force time-history curve of single-string insulators at windward side of V-string under fluctuating wind is shown in Figure 3, and the maximum value is 256.21kN which is 1.72 times of static calculation value.

The force time-history curve of single-string insulators at leeward side of V-string under fluctuating wind is shown in Figure 4, and the maximum value is 87.35kN. The insulators have buckling deformation.

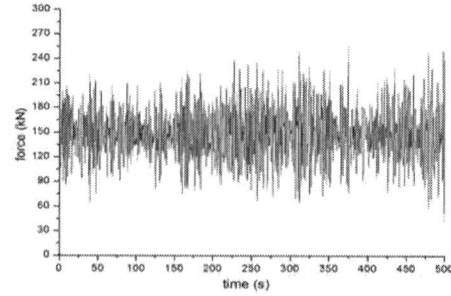

FIGURE III. FORCE TIME-HISTORY CURVE OF SINGLE-STRING INSULATORS AT WINDWARD SIDE OF V-STRING.

158

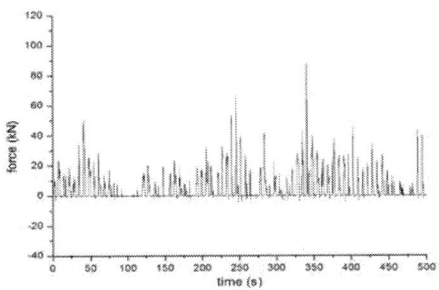

FIGURE IV. FORCE TIME-HISTORY CURVE OF SINGLE-STRING INSULATORS AT LEEWARD SIDE OF V-STRING.

B. Calculation Results When There Is Only One V-String with Included Angle of 110°

The force time-history curve of single-string insulators at windward side of V-string under fluctuating wind is shown in Figure 5, and the maximum value is 252.81kN which is 1.67 times of static calculation value.

The force time-history curve of single-string insulators at leeward side of V-string under fluctuating wind is shown in Figure 6, and the maximum value is 81.31kN. The insulators have buckling deformation.

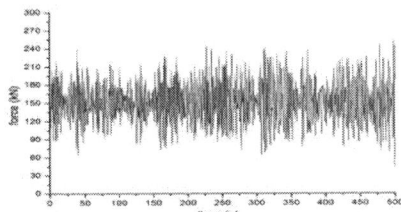

FIGURE V. FORCE TIME-HISTORY CURVE OF SINGLE-STRING INSULATORS AT WINDWARD SIDE OF V-STRING.

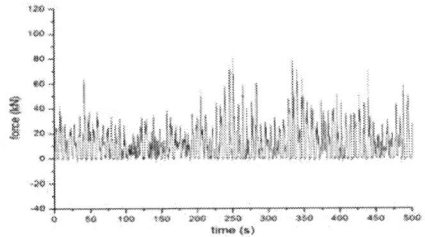

FIGURE VI. FORCE TIME-HISTORY CURVE OF SINGLE-STRING INSULATORS AT LEEWARD SIDE OF V-STRING.

C. Calculation Results When There Is Only One V-String with Included Angle of 120°

The force time-history curve of single-string insulators at windward side of V-string under fluctuating wind is shown in Figure 7, and the maximum value is 269.12kN which is 1.69 times of static calculation value.

The force time-history curve of single-string insulators at leeward side of V-string under fluctuating wind is shown in

Figure 8, and the maximum value is 98.02kN. The insulators have buckling deformation.

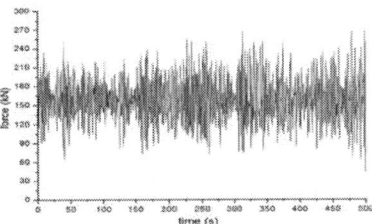

FIGURE VII. FORCE TIME-HISTORY CURVE OF SINGLE-STRING INSULATORS AT WINDWARD SIDE OF V-STRING.

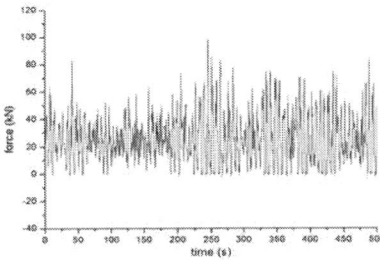

FIGURE VIII. FORCE TIME-HISTORY CURVE OF SINGLE-STRING INSULATORS AT LEEWARD SIDE OF V-STRING.

VII. SUMMARY

The following conclusions are obtained according to calculation results.

The numerical calculation can effectively simulate the force-receiving process of V-shaped insulators under calm and fluctuating wind, and corresponding mechanical properties are obtained.

The maximum value of forces on single-string insulators at leeward side of V-string under fluctuating wind is about 1.7 times of the force imposed under calm wind.

The dynamic response of insulators cannot be reflected under calm wind. However, the dynamic calculation under fluctuating wind can reflect buckling deformation of insulators.

REFERENCES

[1] Z.Y. Liu, Ultra-high voltage power grid[M]. Beijing: China Economic Publishing House, 2005.

[2] W. Terence, O. Brien, A. J. Francis, Cable movements under two dimensional loads [J]. Journal of the Structural Division, ASCE, 1964(90):89-123.

[3] W. Terence, O. Brien, General solution of suspended cable problems [J]. Journal of the structure Division Engineering, ASCE, 1967, 93(1):1-26.

[4] H. M. Irvine. Cable structures [M].The MIT Press.1981.

[5] A. Peyrot, J. Lee, H. Jensen, J. Osteraas, Application of cable elements concept to a transmission line with cross rope suspension structures [J]. IEEE transactions on power apparatus and systems. 1981, 7(100):3254-3262.

[6] X.T. Zhang, Z.P. Wang, B.C. Huang. Mechanics of Structural Vibration [M]. Shanghai: Tongji University Press, 1994.

[7] Design Criteria of Overhead Transmission Lines (IEC60826) [S].2003.

International Conference on Power Electronics and Energy Engineering (PEEE 2015)

The Internal Stress Analysis Method of Cement Paste under Core Restricted Conditions

Y. Li, X.F. Liu, W.L. Bai

The Key Laboratory of Urban Security and Disaster Engineering
MOE
Beijing Key Lab of Earthquake Engineering and Structural Retrofit
Beijing University of Technology
Beijing, 100124, China

Abstract—**Concrete cracking is a major cause affecting the durability and applicability of concrete; it draws increasing attention from researchers in the field of civil engineering. However, previous studies of early micro-cracks are insufficient. The presence of early micro-cracks is the basis of macroscopic cracks. Therefore, a calculation method of internal stress within cement paste caused by core restricted shrinkage under the core restricted condition of rigid body was studied, and a mechanical model of internal stress was established. Based on the model, the factors that affect internal stress within cement paste were further analyzed, and the law of strain variation of cement paste under the rigid elastic body confinement was discussed. The results show practical significance in the control of crack development and in the improvement of concrete durability.**

Keywords-large elastic body; micro-crack; internal stress; elastic mechanics; model

I. INTRODUCTION

Shrinkage cracking is a issue which has gotten more and more attention in the field of civil engineering. With the rapid development of the modern concrete technique, including the use of the admixture, mineral addictive, low water-gel ratio and pump concrete with better performance, it widely increases the possibility of the generation of cracks. Structures using concrete with cracks will greatly reduce its working performance, including applicability and durability. Focusing on concrete structures with cracks during their working time, previous studies include: To solve this practical problem, a technique was developed to determine shrinkage stresses by using rubber models, and a new ring test was designed for the crack resistance of concrete (Carlson R.et al., 1988). In addition, state-of-the-art research on mechanisms was reviewed which caused complex cracking and newly developed methodologies to control cracking at early stage (Mihashi H. et al., 2004). Some tests were designed to simulate restrained shrinkage cracking by using a ring-type specimen, and a theoretical model was developed (Grzybowski M. et al., 1990). J-H Moon (2006) established the strain formulas of autogenously shrinkage and restricted shrinkage that correlate with relative humidity and property parameters of concrete by using the restrained ring. Additionally, according to the comparison of finite element analysis software, restricted classification was studied, and the influence of materials type on shrinkage cracking was investigated.

In the restrained shrinkage experiment, a steel ring enveloped by concrete is created, within which hoop stress arise to restrict the shrinkage of concrete. The restrained ring test has recently become a popular method to assess a mixture's susceptibility to retrained shrinkage cracking (Krause et al., 1995; Grzybowski 1989; Lim et al., 1999). Shah (1992) has verified that the free shrinkage of the restrained shrinkage ring is the same with the axial restrained one. As of now, many experiments and methods have been conducted to reduce concrete cracking. However, there has been no simpler integrated calculation model with respect to stress field which generates micro-cracks. There has not been enough corresponding theoretical direction, either.

In this paper, aggregates and unhydrated cement particles were simplified as rigid body, because their elasticity module was considerably larger than the hydration products around them. Thus, theoretical formulas and calculation method about the internal stress in cement paste under the core restriction of large elastic modulus body (aggregates and unhydrated cement particles) were established in this paper. Then, the factors that affect internal stress in cement paste were analyzed by the formulas, and micro-cracks can be reduced by changing environmental factors and material parameters.

II. EXPERIMENTATION

When the fresh cement paste began to harden, the dry shrinkage of volume occurs. With the increase of age, cement paste is restricted by rigid bodies (such as unhydrated cement particles, aggregates, and rebar). Additional strain occurs due to the restraint effect of internal rigid bodies, which can be taken as $\varepsilon=\varepsilon(\rho)$. In order to simplify the case, the radical displacement of the cementitious material, and an axisymmetric mechanical model of cement paste under the core restricted conditions are shown in Figure 1. An element of elastomer is taken and analyzed by means of equilibrium condition and polar coordinates, as shown in Figure 2. In this work, self-weight of cement paste, inertia force and the nonlinear properties of the cement paste are not taken into account. Referring to the restrained shrinkage experiment of steel ring, the paper took the stress in single plane of concrete into consideration and the concrete ring was simply under the stress along the direction of ring and radial, plane axisymmetric assumption fulfilled. Thus, the model is a case of axisymmetric plane strain.

© 2015. The authors - Published by Atlantis Press

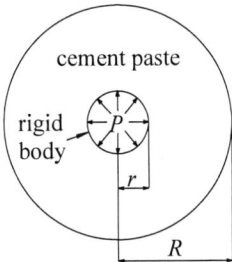

FIGURE I. THE MECHANICAL MODEL OF CEMENT PASTE.

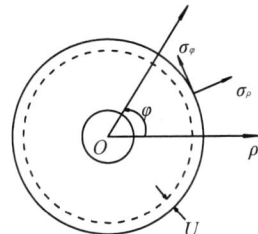

FIGURE II. THE MODEL IN POLAR COORDINATES.

Where, ρ and φ denote the coordinate direction, U ($\times 10$-3mm) is the radial displacement, and φ is the circumferential displacement.

Thus, the internal stress becomes:

$$
\begin{cases}
\sigma_\rho = A\left[\frac{2}{3}\left(1-\frac{3}{4}m\right)\frac{r}{\rho^2}+m\cdot\ln\frac{\rho}{r}\cdot\frac{r}{\rho^2}-\frac{1}{r}+\frac{\rho}{3r^2}+\frac{m}{2r}\right]-\left(\frac{1}{r^2}-\frac{1}{\rho^2}\right)\cdot A\cdot B-2\left(1-\frac{m^2r^2}{\rho^2}\right)\cdot A\cdot C \\
\sigma_\varphi = -A\left[\frac{2}{3}\left(1-\frac{3}{4}m\right)\frac{r}{\rho^2}+m\cdot\ln\frac{\rho}{r}\cdot\frac{r}{\rho^2}-\frac{1}{r}+\frac{\rho}{6r^2}+\frac{m}{2r}\right]-\left(\frac{1}{r^2}+\frac{1}{\rho^2}\right)\cdot A\cdot B-2\left(1+\frac{m^2r^2}{\rho^2}\right)\cdot A\cdot C \\
\quad +A\left(\frac{mr}{\rho^2}-\frac{1}{\rho}+\frac{\rho}{2r^3}+\frac{m}{r}\right)
\end{cases}
\quad (1)
$$

Where,

$$A=\frac{E\cdot U}{\left(1-\upsilon^2\right)\cdot\left(m^2-1\right)}$$

$$B=\frac{m\left(\ln m-\frac{3}{2}\right)+\frac{1}{6}\left(5m^3+4\right)}{m^2-1}\cdot r$$

$$C=\frac{\left(1-\upsilon\right)\cdot\left[m\left(\ln m-\frac{3}{2}\right)+\frac{1}{6}\left(5m^3+4\right)\right]}{\left(m^2-1\right)\cdot\left[\left(2\upsilon-1\right)\cdot\left(m^2-1\right)k+\left(2\upsilon-1\right)-m^2\right]\cdot r}$$

$$k=\frac{E}{E'}$$

$$R=m\cdot r$$

where, σ_ρ(GPa) represents the radial normal stress in the direction of ρ, σ_φ(GPa) represents the circumferential normal stress in the direction of φ, ε_ρ is the radial linear strain, ε_φ is the circumferential linear strain, υ is the passion ration, and E and E' are the elastic modulus of the cement paste and the rigid body, respectively.

III. EXPERIMENTAL RESULTS

In order to verify the formulas above, the microstructure of the model of cement paste was examined by imaging the

interfacial transition zone between the aggregate and the hydration products in a scanning electron microscope (SEM, FEI Quanta 200, Holland). The parameters of SEM were listed below: the accelerating voltage is 200V~30KV, the magnifying power is X25~X200 000, and the distinguishability is 3.5nm. In order to gather the first image, the specimen was put into the SEM for 60min with 80% of relative humidity.

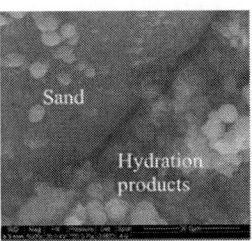

a) The relative humidity of 80%

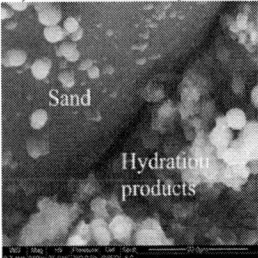

b) The relative humidity of 40%

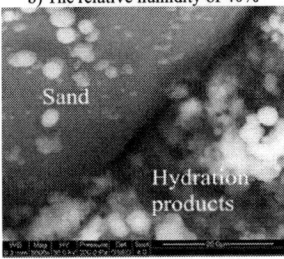

c) The relative humidity of 10%

FIGURE III. MICROGRAPHS OF CEMENT PASTE FOR VARIOUS RELATIVE HUMIDITY.

Figure 3 shows that sand was enclosed in the hydration products, which is as same as the microstructure of Figure 1. It can be seen that, with the decrease in relative humidity, the hydration products begins to shrink. Under the restriction of sand, the radial and hoop stress exerted on hydration products result in the increasing gap between sand and hydration products. Then, to find the laws which cracks obey, the relative humidity in the concrete is specified by the corresponding parameters in the Equation1.

Under the condition of the same raw materials and curing environment, the variation law of the shrinkage cracking stress under the restraining of the rigid body was explained with the Equation1 and the micrographs from Figure 3. Meanwhile, under the same raw materials (Figure 3) and curing environment in a related experiment, using the nano-indentation method, the test results were calculated from the curve of the force to the indentation depth. The elastic modulus of the sand and the hydration products are 107.06

±4.95GPa and 32.53GPa, respectively. In order to calculate σ_ρ and σ_φ in Equation1, E and E' were assumed as 33.58GPa and 110GPa, respectively. υ and υ' were assumed as 0.28 and 0.17, respectively. U was assumed as $-0.04 \times 10\text{-3mm}$. Then, k=0.305, m=10. And Table 1 lists the results of circumferential normal stress of Equation1.

TABLE I. THE RESULTS FROM EQUATION1.

NO.	r=1mm		r=2mm		r=3mm	
	ρ (mm)	σ_φ (MPa)	ρ (mm)	σ_φ (MPa)	ρ (mm)	σ_φ (MPa)
1	3	2.64	5	2.05	10	1.86
2	4	2.81	6	2.64	11	1.89
3	5	2.54	7	2.82	12	1.88
4	6	2.14	8	2.81	13	1.83
5	7	1.68	9	2.71	14	1.77
6	8	1.20	10	2.54	15	1.70

In Table 1, the circumferential normal stress (σ_φ) from Equation1 with different diameters of the rigid body are between 1MPa and 3MPa, and interrelated test was about the tensile stress test of cement paste with different mineral admixture, and the result indicated that the tensile stress was 1~3 MPa (Li Y., Yan Q.,2012) , which was in accord with the calculation results obtained by the model formula to some extent. Thus if, the parameters of the Equation1 are changed, the value of the shrinkage stress can be in control. Therefore, it's effective to control the cracks of the cement paste with the changed mineral admixtures and changed relative humidity.

IV. CONCLUSIONS

(1)A theoretical model for the internal stress analysis of cement paste under core restricted conditions was established based on the elastic mechanics and the restrained shrinkage experiment.

(2)According to this model, the main factors that influence the cracks were further analyzed. By using testing results of various components in hydrated cement paste and some assumptions, the parameters in the model are determined and internal stress were calculated.

(3)With the results from Equation1, the parameters of Equation 1 can be changed, and then the value of the shrinkage stress can be in control to some extent.

(4)With the results from Equation1, it's effective to control the cracks of the cement paste with the changed mineral admixtures and changed relative humidity.

ACKNOWLEDGMENTS

The authors would like to acknowledge the financial support provided by Program for New Century Excellent Talents in University (NCET-12-0605); The Importation and Development of High-Caliber Talents Project of Beijing Municipal Institutions（CIT&TCD20150310）; National Natural Science Foundation of China (51278014).

REFERENCE

[1] Carlson R W, Reading T J. Model study of shrinkage cracking in concrete building walls[J]. ACI Structural Journal, 1988, 85(4), 395-404.

[2] Mihashi H, Leite J P B. State-of-the-art report on control of cracking in early age concrete[J]. Journal of Advanced Concrete Technology, 2004, 2(2): 141-154.

[3] Grzybowski M, Shah S P. Shrinkage cracking of fiber reinforced concrete[J]. ACI Materials Journal, 1990, 87(2). 138–148.

[4] Moon J H. Shrinkage, residual stress, and cracking in heterogeneous materials[M]. ProQuest, 2006.

[5] Krause, P. D., Rogalla, E A., Sherman, M. R, Mcdonald, D. B., ,et al.(1995), "Transverse Cracking in Newly Constructed Bridge Decks" , NCHRP380, Project 12.37.

[6] Grzybowski, M., (1989a), "Determination of Crack Arresting Properties of Fiber Reinforced Cementitious Composites", Ph.D. Thesis, Royal Institute of Technology, Stockholm, Sweden.

[7] Lim, Y.M., H.C. Wu and V.C. Li, (1999), "development of Flexural Composite Properties and Dry Shrinkage Behavior of High Performance Fiber Reinforced Cementitious Composites at Early Age", J. of Materials, Vol.96, NO.1, American Concrete Institute, pp.20-26.

[8] Shah S.P Effect of shrinkage-reducing admixtures on restrained shrinkage cracking of concrete. ACI Material Journal, 1992,89(3).

[9] Tazawa E, Miyazawa S. Experimental study on mechanism of autogenous shrinkage of concrete[J]. Cement and Concrete Research, 1995, 25(8): 1633-1638.

[10] Li Y, Yan Q, Du X. Relationship between Autogenous Shrinkage and Tensile Strength of Cement Paste with SCM[J]. Journal of Materials in Civil Engineering, 2012, 24(10): 1268-1273.

[11] Qianqian Yan. Properties Research of Early-age Shrinkage and Crack of the Cementitious materials. Beijing: Beijing University of Technology, 2012.

International Conference on Power Electronics and Energy Engineering (PEEE 2015)

Research and Development of the Tension Pay-Off Equipment for Wire Rope Maintenance

W.H. Cui, L. Tang, C. Fu
School of mechanical engineering
University of Jinan
Jinan, China, 250022

K.Q. Gong
Shandong Electrical Power Supply & Transformation
Engineering Co., Ltd
Jinan, China, 250118

Abstract—**Traction wire rope is used in the construction of the power. The wipe rope cleaning, inspection and maintenance integration system can detect the damages on the traction wire ropes and maintain the ropes. The integration system consists of a variety of equipments and the tension pay-off equipment is one of them. According to the role of the tension pay-off equipment in the integration apparatus, the functional requirements of the equipment were determined and the structure of the equipment was designed. This paper used SolidWorks to build a three-dimensional model of the tension pay-off equipment and did finite element analysis of its key components with COSMOS Works. The structural optimizations and improvements about the tension pay-off equipment were carried out according to the results of the analysis. The tension pay-off equipment has been put into use and it works well.**

Keywords-Wire rope maintenance; Tension pay-off; Finite element analysis; Structural optimization

I. INTRODUCTION

Wire rope used for tension stringing work in relatively poor conditions, it easily produce wire rope wear, fatigue, broken wires, corrosion and deformation, thereby reduce the strength of the wire rope and shorten the service life of the wire rope, easily lead to accidents of construction[1-2]. According to DL/T1079-2007 "Twist Proof Steel Wire Ropes for Line Stringing Under Tension in Overhead Transmission Line", wire rope used for tension stringing should be regularly cleaned, tension detection, oil maintenance and other measures every year. "environment-friendly wire rope cleaning, detecting and maintaining integrated system" is a company research project, it is referred to as "wire rope maintenance line", and its solution is shown in Figure 1. The maintenance line can complete the detection and maintenance for traction wire rope of the tension stringing work, to ensure the Safety of tension stringing construction. The tension pay-off equipment（hereinafter it is referred to as tensioner） belongs to a part of the maintenance line.

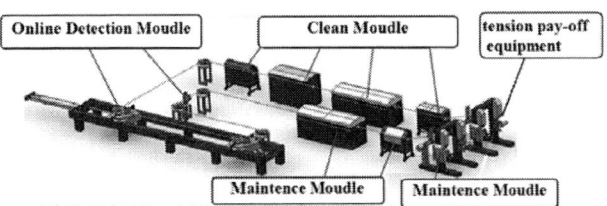

FIGURE I. ENVIRONMENT-FRIENDLY WIRE ROPE CLEANING, DETECTING AND MAINTAINING INTEGRATED SYSTEM.

According to the data access[3-4], the existing tensioner are not suitable for the wire rope maintenance line. Therefore, in this paper, according to the demand of the wire rope maintenance line, the design and improvement of tension pay-off machine are processed the objective is to develop a suitable tensioner for the wire rope maintenance line, and ensure the reasonableness of the design and use of safety.

II. STRUCTURAL DESIGN OF THE TENSIONER

A. Functional Requirements

In the maintenance of the ropes, the ropes undergoes dirt brushing, soaking, cleaning, drying, grease immersion and rope collection after being unwound from the rope plates. The wipe rope cleaning, inspection and maintenance integration apparatus has traction equipment and the tensioner doesn't need to provide power. The tensioner should have the following functions. (1) The clamping of the rope plate. (2) The lifting of the rope plate. (3) The applying of a tension to the wire rope. (4) The fixing of the rope expanding direction.

B. Structural Design of the Tensioner

In summary, the tensioner consists of frame, clamping mechanism, lifting mechanism, braking mechanism and rope arrangement mechanism. The general structure of the tensioner is shown in Figure 2. The braking device uses magnetic powder brake as the braking tool. The tension can be easily controlled with the magnetic powder brake and this provides conditions for automatic, coordinated control of the entire wipe rope cleaning, inspection and maintenance integration apparatus. To ensure the smooth passage of the joints, the rope arrangement mechanism uses a spring structure.

© 2015. The authors - Published by Atlantis Press

163

III. FINITE ELEMENT ANALYSIS AND STRUCTURAL OPTIMIZATION ABOUT KEY COMPONENTS OF THE TENSIONER

A. Analysis of Tensioner Working Conditions

Tensioner provides a tension to the wire rope to ensure the rope hold taut during maintenance. The tensioner's working principle is: Firstly, the brake device (with brake core clamper) and clamping device (with clamping core clamper) work together to clamp the rope plate. Secondly, the lifting mechanism enhance the rope plate some distance off the ground and the control system controls the magnetic brake to provide a stopping power that based on the size of traction power. Then, the brake spindle and shifting fork act on the rope plate and a tension will generate in the rope.

The brake core clamper (with shifting fork) is an important working mechanism foe it not only provides the necessary brake tension, but also bears the weight of the rope plate. As the transmission and weight bearing part, the spindle in the brake core clamper bears the twisting and bending moments. The spindle also bears cycle bending moments during the rotation of the rope plate. The performance of the spindle will directly affect the working of the tensioner. The core clamper slide carriage connects the brake core clamper and the lifting mechanism. The structure and force condition of the slide is complex. There is no analytical solution to the stress and strain that the slide bears. In summary, the spindle and the slide carriage are the key parts affecting the performance of the tensioner.

The tensioner's three-dimensional model built by SolidWorks is shown in Figure 3.

B. Finite Element Simulation Tool Selection

COSMOSWorks is fully integrated in the design and analysis system of SolidWorks [5-6]. It provides a complete means of analysis for design engineers. This paper chooses COSMOSWorks as the finite element analysis tool.

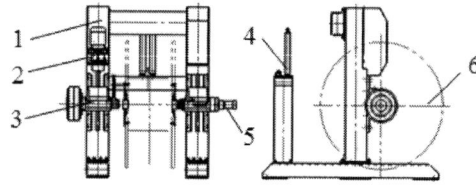

1-frame 2-lifting mechanism 3- magnetic powder braking mechanism 4- rope arrangement mechanism 5-clamping mechanism 6-rope plate

FIGURE II. THE TENSIONER.

FIGURE III. THREE-DIMENSIONAL MODEL OF THE TENSIONER.

IV. FINITE ELEMENT ANALYSIS AND STRUCTURAL

A. Optimization about the Spindle

The simplified model of the brake mechanism is shown in Figure 4. In the model, the magnetic brake, core clamper slide, bearings and so on are removed and it mainly consists of spindle, key and shifting fork. The spindle is connected with the shifting fork by the key. Brake torque generated by the spindle passes to the shifting fork through the key. The shifting fork acts on the rope plate and applies brake torque to the plate. The structure of the spindle is shown in Figure 5.

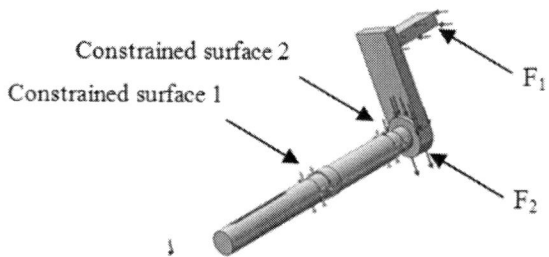

FIGURE IV. SIMPLIFIED MODEL OF THE BRAKE MECHANISM.

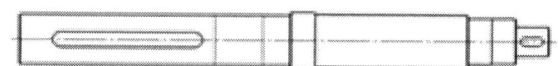

FIGURE V. STRUCTURE OF THE SPINDLE.

The spindle is supported on the core clamper slide carriage by bearings installed on constrained surface 1 and constrained surface 2. In the simulation analysis, constraints will be fixed at constrained surface 1 and constrained surface 2. The force F1 exerted on the shifting fork is determined to 2400N based on the maximum tension in the rope and the diameter of the rope plate. The weight of the rope plate is endured by two spindles and each spindle bears half of the weight. The gravity of the rope plate is 16000N and each spindle will bear 8000N. The constraints and forces applied to the brake mechanism are shown in Figure 3. The safety factor distribution map obtained by simulation is shown in Figure 6. The minimum safety factor at the key connection is 0.2599. It means there are serious safety hazards in the spindle and the design doesn't meet the requirements. The red region is the area where the safety factor is less than 1.

To ensure the safety of the brake mechanism torque transmission, the torque transfer mode must be radically improved. As shown in Figure 7, the structure of the spindle segment where bears the force is changed and the key connection is removed. The improved spindle can greatly enhance the torque transfer capacity. And keep the diameter of the spindle segment invariant and optimize the section distance. In the optimization, the optimization constraint is that the minimum safety factor should be bigger than 1.5 and the optimization objective is to minimize the weight of spindle. The optimization results are shown in Table 1.

164

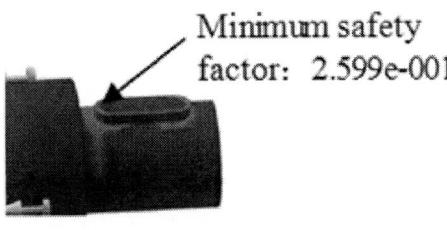

FIGURE VI. SPINDLE FINITE ELEMENT ANALYSIS CLOUD

FIGURE VII. IMPROVED SPINDLE STRUCTUR E

TABLE I. COMPARISON OF PARAMETERS THAT BEFORE AND AFTER THE OPTIMIZATION.

	The initial values	Values after comparison
Section distance （mm）	30	32
Minimum safety factor	1.39	1.53
Minimum weight （g）	1730	1752

V. FINITE ELEMENT ANALYSIS AND STRUCTURAL OPTIMIZATION ABOUT THE CORE CLAMPER SLIDE CARRIAGE

According to the tensioner's working condition and the previous design experience, the minimum safety factor of the core clamper slide carriage （hereinafter it is referred to as slide） takes 6. The three-dimensional model of the preliminarily designed core clamper slide is shown in Figure 8.

Set the material of the slide to HT200. According to the slide's force condition, impose fixed constraints to the surface at where the slide contacts with the turnbuckle, impose sliding constraints to the surfaces at where the slide contact with the guide rail and apply downward a force which is 1200N to where the bearings support the slide. Mesh the slide model and do the static analysis. The safety factor distribution map is shown in Figure 9. The minimum safety factor of the slide is 12.31 and the maximum stress appears at the surface where the slide contacts with the turnbuckle. The simulation analysis shows that the structural performance is far greater than the design requirements.

Figure 9 shows that stress at the slide's thread, through which the slide connects with the screw, is larger than other parts of the slide. The above and behind parts of the slide almost bear no force. Parts with a relatively larger potential of optimization can be found through Figure 9. This paper will optimize the dimensions and shapes of these parts.

FIGURE VIII. THE MODEL OF THE SLIDE.

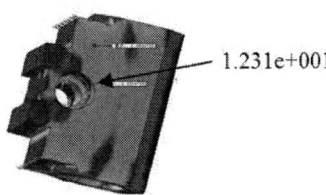

FIGURE IX. THE SAFETY FACTOR DISTRIBUTION MAP OF THE SLIDE.

The optimization purpose of this paper is to reduce the weight of the slide on the basis ensuring the slide's structural strength and stiffness. Take the slide's weight as the optimization goal. Parts that bear the most force are the spindle hole and the thread. Take the wall thickness shown in Figure 10 as the optimization parameter. The optimization results: the wall thickness is reduced from 35mm to 20mm and the minimum safety factor is reduced from12.3 to 10. Parameters before and after the optimization are compared in Table 2.

TABLE II. COMPARISON OF PARAMETERS THAT BEFORE AND AFTER THE OPTIMIZATION.

	The initial values	Values after comparison
Wall thickness （mm）	35	20
Minimum safety factor	12.31	10
Minimum weight （g）	6819	5985

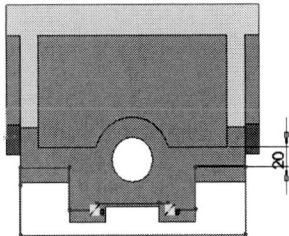

FIGURE X. THE SLIDE'S OPTIMIZATION PARAMETER.

VI. CONCLUSION

This paper designed the tension pay-off machine based on the tensioner's work condition and function requirements. In order to ensure tensioner's design rationality and safety, this paper made the finite element simulation analysis about the key parts of the tensioner. The structures of these parts were optimized based on the simulation results. The tension pay-off machinet has been put into use in a company and it works well.

REFERENCES

[1] Yun Sun, Hong-xue Chen, Dong-yao Shu, Hubei Electric Power. 34(2010) 79-81.

[2] Fang-mao Kang, Coal Mirie Machinery. 30(2009) 154-156.

[3] Hai-ping Jiang, Tension Stringing Equipment and Application, China Electric Power Press Beijing, 2005.

[4] Hai-ying Cui. Electric Wire & Cable. 6 (2003) 45-46.

[5] Yan-ping Yao, Lei Qi, Hong-fu Yue. Hoisting and Conveying Machinery. 12 (2012)47-50.

[6] Lei Qi, Zhi-de Zhang, Hao Wen, Gang Han. Hoisting and Conveying Machinery. 4 (2011) 30-33.

International Conference on Power Electronics and Energy Engineering (PEEE 2015)

Convex-Concave Property for Parabola Fitting of Dr. Bridge

J. Huang

Faculty of civil and Transportation Engineering
Guangdong University of Technology
Guangzhou, Guangdong, China, 510006

S.X. Ding

School of Civil and Transportation Engineering
South China University of Technology
Guangzhou, Guangdong, China, 510640

H.G. Gan, S.F. Zhang

Faculty of civil and Transportation Engineering
Guangdong University of Technology
Guangzhou, Guangdong, China, 510006

Abstract—**A new method is presented to determine the fitting direction while using Dr. Bridge's linear editor to model various parabolas in this paper. By considering convex-concave property at monotone interval, parabolas' different shapes are determined. Based on the method, two set of symbols are introduced to describe the monotonicity and convex-concave property of parabola, and the judgment of parabola fitting type is determined. With the presented method, a quadratic parabola can be fitted by means of two controlling points as well as three controlling points. The method is illustrated by its simplicity and convenience, but efficient to program realization.**

Keywords-Convex-concave propert; fitting direction

I. INTRODUCTION

Bridge structure usually contains hundreds of or even thousands of units (Yao 2008), it is not advisable to deal with information of units individually because of exhausting input. In Dr. Bridge system, except for some special units, a great deal of input work can actually be reduced by using quick editor. There are ten types of quick editors in Dr. Bridge system such as line, arch, cable, parallel, symmetry, offset, insert, unit, section, and coordinate, which are provided for users to select the most convenient way to model various bridge structures. Among them, linear editor has the capacity of formation of a group of units, whose top edge or midpoint height is located at the same line. When using linear editor, sections at each controlling point are first defined, and then the fitting type of each controlling section is selected. Finally the other sections can be fitted by line or by parabola based on the controlling section which has already been defined.

Dr. Bridge Specifications Instruction indicates when using linear editor to fit a parabola, a parabola must be determined by three points, the parabola fitting type at the first point is not limited, the type at the second point must be 'backward parabola', the type at the third point must be 'forward parabola'. In this way, Dr. Bridge system will regard the three points as controlling points and fits a parabola. However, the case is usually occurred when known two points to determine a parabola

(Huang 2012). At this time, only two controlling sections are defined. Moreover, the user's guide of Dr. Bridge does not offer a way how to fit a parabola based on the two controlling points.

By introducing two set of symbols to consider the monotonicity and convex-concave property of a certain parabola in this paper, a new method is presented to determine the fitting direction of parabola when using Dr. Bridge linear editor. Whether two or three controlling points are known, this presented method is suitable for identification of the parabola fitting type. The application of the method is illustrated by the modeling of three-span continuous box girder with variable cross sections, which verified the simplicity and efficiency of the approach in this paper.

II. PARABOLA FITTING METHOD CONSIDERING MONOTONICITY AND CONVEX-CONCAVE PROPERTY

Monotonicity and convex-concave property of function are characteristics of a certain curve (Kohlmann 2001, Chen & Cerone 2003, Tawarmalani et al. 2013). When there is parabola at monotonous interval, we can examine its convex-concave property as well as the parabola fitting direction.

A. Identification of Parabola Fitting Type by Two Controlling Points

When using linear editor to fit parabolas, it will commonly encounter the condition that the controlling sections are only defined at two points. The fitting type at the first point (starting point) is not limited, but it is not mentioned the fitting type at the second point (end point) whether to select 'forward parabola' or to select 'backward parabola' in user's guide of Dr. Bridge, which is always a technical difficulty in modeling process. In fact, by considering monotonicity and convex-concave property of a quadratic curve, the parabola with two controlling points fitted only appears the following four different shapes, as shown in Figure 1.

© 2015. The authors - Published by Atlantis Press

167

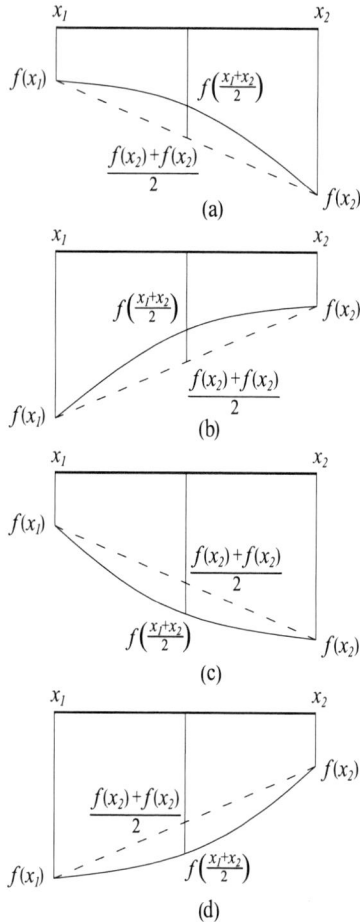

FIGURE I. PARABOLA SHAPES FITTED BY TWO CONTROLLING POINTS.

Definition 1 Suppose a parabola $f(x)$ is fitted by two controlling points, x_1 and x_2 at its monotone interval. If the relation (1) is satisfied at interval $[x_1, x_2]$, as shown in Figure 1 (a) and (b), thus we define the parabola $f(x)$ is convex. If the relation (2) is satisfied at interval $[x_1, x_2]$, as shown in Figure 1 (c) and (d), thus we define the parabola $f(x)$ is concave.

$$f\left(\frac{x_1 + x_2}{2}\right) > \frac{f(x_1) + f(x_2)}{2} \qquad (1)$$

$$f\left(\frac{x_1 + x_2}{2}\right) < \frac{f(x_1) + f(x_2)}{2} \qquad (2)$$

Theorem 1 Two set of symbols are introduced. The first set of symbols marks parabolic monotonicity, and the second set of symbols marks parabolic concave-convex property. In the first set of symbols, it is marked $+$ if the parabola is decreasing function from the first point (starting point) to the second point (end point), and it is marked $-$ if the parabola

is increasing function from the first point (starting point) to the second point (end point). Meanwhile, in the second set of symbols, it is marked $+$ if the parabola is convex, otherwise it is marked $-$.Thus, if the two set of symbols are the same signs, the parabola fitting type at the second point (end point) is 'forward parabola', and if the two set of symbols are opposite signs, the parabola fitting type at the second point (end point) is 'backward parabola'.

According to definition 1 and theorem 1, with regard to four shapes shown in Figure 1, the procedure of determination of parabola fitting direction by two controlling points is given in Table 1.

TABLE I. IDENTIFICATION OF PARABOLA FITTING TYPE BY TWO CONTROLLING POINTS.

Parabola shape	First set of symbols	Second set of symbols	Parabola fitting type at end point
(a)	$+$	$+$	'Forward parabola'
(b)	$-$	$+$	'Backward parabola'
(c)	$+$	$-$	'Backward parabola'
(d)	$-$	$-$	'Forward parabola'

B.   *Identification of Parabola Fitting Type by Three Controlling Points*

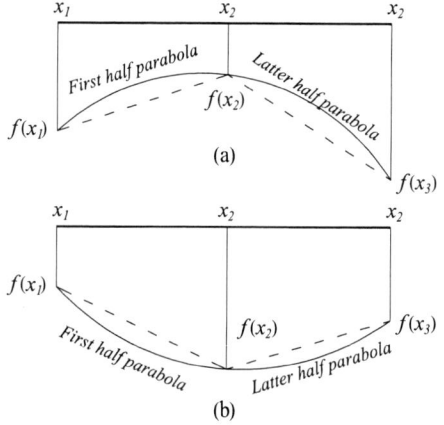

FIGURE II. PARABOLA SHAPES FITTED BY THREE CONTROLLING POINTS.

On the other hand, if a parabola is determined by three controlling points, the proposed parabola fitting method by considering monotonicity and convex-concave property is also suitable for the determination of the parabola fitting direction. At this time, the whole parabola is divided into the first half parabola (from the first point to the second point) and the latter half parabola (from the second point to the third point), as shown in Figure 2. Thus, the midpoint (the second point) is regarded as not only end point of the first half parabola but also starting point of the latter half parabola. Then let theorem 1 apply respectively at the first half parabola and the latter half parabola to determine their fitting direction. The procedure of determination of parabola fitting direction by three controlling points is given in Table 2.

TABLE II. IDENTIFICATION OF PARABOLA FITTING TYPE BY THREE CONTROLLING POINTS.

Parabola shape		First set of symbols	Second set of symbols	Parabola fitting type at end point
(a)	First half	−	+	Second point 'backward parabola'
	Latter half	+	+	Third Point 'forward parabola'
(a)	First half	+	−	Second point 'backward parabola'
	Latter half	−	−	Third Point 'forward parabola'

The user's guide of Dr. Bridge indicates that if a parabola is determined by three points, the parabola fitting type at the first point is not limited, the type at the second point must be 'backward parabola', the type at the third point must be 'forward parabola'. In this way, Dr. Bridge will regard the three points as controlling points and fits a parabola. As can be seen in Table 2, the results of the presented method are in agreement with the criterion given in the user's guide of Dr. Bridge, which verified the presented method is applicable to fit parabola by three controlling points.

III. APPLICATION EXAMPLE

A three-span continuous box girder is 100m in length, with span arrangement 30m+40m+30m. There are several segments of straight line being located at side spans, midspan join, and pier tables. The length of line segment at side span is 10m, and 4m line segments are located at piers, and 2m line segment is located at midspan join. The element arrangement of three-span continuous box girder is shown in Figure 3. The girder is divided into 100 elements, with each element 1m in length. The configurations of controlling sections are shown in Figure 4, with vertical web, single box and single room. The controlling sections at side spans and midspan are both 2.5m in height, and at piers are 5.0m high.

FIGURE III. ELEMENT ARRANGEMENT OF THREE-SPAN CONTINUOUS BOX GIRDER.

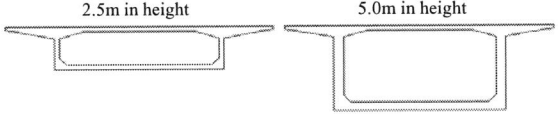

FIGURE IV. CONFIGURATIONS OF CONTROLLING SECTIONS.

We can notice that the top edge of the box girder is all located at the same line, and the bottom edge of the box girder is various from lines to parabolas, as shown in Figure 3. Thus, the linear editor of Dr. Bridge can be used to complete the model of the box girder. Let add ten controlling sections, 0m, 10m, 28m, 32m, 49m, 51m, 68m, 72m, 90m, 100m respectively to the column of distance from controlling point to starting point on the interface. Then input the size of each

controlling section, and determine the parabola fitting direction at each controlling section. Based on the presented identification of parabola fitting type by two controlling points in this paper, select at 0m 'linear interpolation', at 10m 'linear interpolation', at 28m 'backward parabola', at 32m 'linear interpolation', at 49m 'backward parabola', at 51m 'linear interpolation', at 68m 'forward parabola', at 72m 'linear interpolation', at 90m 'backward parabola', at 100m 'linear interpolation', respectively. Thereby, the model of three-span continuous box girder with variable cross sections whose span is 100m is established, as shown in Figure 5.

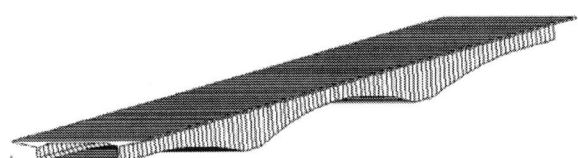

FIGURE V. MODEL OF THREE-SPAN CONTINUOUS BOX GIRDER.

IV. CONCLUSIONS

Based on monotonicity and convex-concave property of a quadratic curve, four different parabola shapes, with two controlling points fitted, are provided in this paper. By introducing two set of symbols to describe parabola characteristics, the method of determination of parabola fitting direction is presented, which explore a new approach to fit a parabola by two controlling points when linear editor of Dr. Bridge is used. The presented method is suitable for determining the parabola fitting direction whether parabola is fitted by two or by three controlling points. The method is illustrated by its simplicity and convenience, but efficient to program realization, which greatly promotes the modeling efficiency.

ACKNOWLEDGEMENTS

Financial support for this study provided by Training Programs of Innovation and Entrepreneurship for Undergraduates of Guangdong Province, China (Projects No.1184513039 and No.201411845197) is gratefully acknowledged.

REFERENCES

[1] Chen, C.P. et al. 2003. Monotonicity of sequences involving convex and concave functions. *Mathematical Inequalities & Applications*. 6:229-239.

[2] Dr. Bridge V3.2 Specifications Instruction. *Hao Shanghai with Civil Engineering Consulting Co., Ltd.*

[3] Huang, J. 2012. *Analysis of Time-dependent effects on PC cable-stayed bridge under long-term load*. Beijing: China Communications Press.

[4] Kohlmann, P. 2001. The convex-concave principle and uniqueness for hypersurfaces in space forms. *Journal of Geometric Analysis*. 11(2): 295-310.

[5] Tawarmalani, M. et al. 2013. Explicit convex and concave envelopes through polyhedral subdivisions[J]. *Mathematical Programming*. 138(1-2): 531-577.

[6] Yao, L.S. 2008. *Bridge Engineering*. Beijing: China Communications Press.

International Conference on Power Electronics and Energy Engineering (PEEE 2015)

TEM Investigation on Ceramic Strengthening NiAl-Based Composite Prepared by Thermal Explosion and Hot Extrusion

L.Y. Sheng, C. Lai, T.F. Xi

Shenzhen Key Laboratory of Human Tissue Regeneration and Repair Shenzhen Institute
Peking University
Shenzhen, China

Abstract—The NiAl-based composite strengthened by TiC and Al2O3 was synthesized by thermal explosion and hot extrusion. Its microstructure and mechanical properties were investigated by XRD, OM, TEM and HRTEM. The results reveal that the NiAl-TiC- Al2O3 composite can be densified with less porosity. The NiAl grain of the composite matrix has been well refined by the TiC and Al2O3 particles. In addition, stacking fault and microtwins formed in some TiC particles and thin amorphous layer also observed along NiAl/TiC interface. Moreover Ti2AlC particle with intergrowth TiC plate inside is formed along the NiAl grain boundary.

Keywords-NiAl-based composite; TEM; ceramic particle; microstructure

I. INTRODUCTION

NiAl and its alloys have been studied widely and thought as one of most promising candidates to substitute the superalloy used in aeroengine, because of the good physical properties and excellent high temperature corrosion resistance of NiAl [1–4]. However, the room temperature (RT) brittleness, low room temperature fracture toughness and relative poor high temperature strength have restricted its applications [5–7]. Therefore, to overcome the shortcomings of the NiAl, many techniques[8–11] have been adopted, among which fabricating NiAl based composites by adding stiffness particles is a convenient way to obtain improvement in high temperature properties. For example, boride strengthening NiAl based composite, oxide strengthening NiAl based composite and nitride strengthening NiAl based composite have been investigated extensively to incorporate the ceramics particles into the NiAl matrix [12,13]. Generally, NiAl based composite strengthened by ceramic particles can be obtained by reaction synthesis, which is an easy, quick and economic method [14–16]. However the porosity generated during the reaction synthesis is a big problem [17–19]. The recent study reveals that the thermal explosion and hot extrusion could fabricate composite with fine microstructure and less porosity [16]. Nevertheless, the poor RT ductility of the NiAl based composite still handicap its application as high temperature structure material. Fortunately, the excellent oxidation and wear resistance of NiAl make the NiAl based composite still has widely application prospect in environmental resistant

fields [19]. Therefore, in the present paper the ceramic particles strengthening NiAl-based composite is fabricated thermal explosion and hot extrusion and its microstructure characteristics are studied as well.

II. EXPERIMENTAL PROCEDURE

Elemental powder mixtures including Ni(98%, <1μm), Al(98%, <13μm), Ti(99%, <75μm), TiO2(98%,<10μm) and C(99.9%, <1μm) for composites of NiAl-15vol TiC+2vol Al2O3 were dry mixed in a ball milling for 15 h. The mixed powders were put into the thermal explosion and hot extrusion system [16]. Firstly, the mixed powders were pressed into compact and degassed; then the inductor heat reaction puncheon to 800 K to start the reaction synthesis. The temperature of the compact was detected by a thermal pair in the thermal explosion and hot extrusion synthesis system. When the reaction synthesis began, the temperature of the powder would increases dramatically. Then two second later, a force of 500 MPa was imposed on the reaction puncheon in order to densify the synthesized composite.

The samples for microstructure observation and compression test were cut from the thermal explosion and hot extrusion synthesized NiAl-TiC-Al2O3 composite by electro-discharge machining (EDM). The resultant phases in the composite were characterized by X-ray diffraction (XRD) with a Cu radiation at 40 kV and 40 mA. Microstructural characterization of sample was carried out on an OLYMPUS GX41 Optical microscope (OM). Samples for OM observations were prepared by conventional methods of mechanical polishing and chemical etching with an acidic mixture (CH3COOH/HNO3/HCl=8:4:1). The samples for transmission electron microscope (TEM) observation were cut by EDM. The slices were mechanically ground from both sides to 30 μm and then thinned by ion milling. The thin foils were examined on a JEOL-2010 high-resolution transmission electron microscope with a point resolution of 0.19 nm and operated at 200 kV.

III. RESULT AND DISCUSSION

The NiAl-TiC composite with dispersed Al2O3 oxides were successfully fabricated by thermal explosion and hot extrusion. The x-ray diffraction analysis proves that the elemental

© 2015. The authors - Published by Atlantis Press

powders have been transformed to the NiAl phase after thermal explosion and hot extrusion processing, as shown in Figure. 1(a). The peaks of NiAl matrix and TiC phase are obvious, but the peak of Al_2O_3 is so weak that it can not be found in the XRD pattern. The typical microstructure of NiAl-TiC-Al_2O_3 composite is shown in Figure.1 (b). From the OM micrograph, it can be seen that the composite is mainly composed of NiAl Matrix and white TiC. In addition, there is still some porosity inside. The average grain size of the NiAl matrix is about ten microns. Additionally, the NiAl phase is elongated along the extrusion direction, which indicates that the composite experiences deformation during hot extrusion. The TiC particles with several microns mainly distributed along the NiAl grain boundaries, which can contribute to the grain refinement.

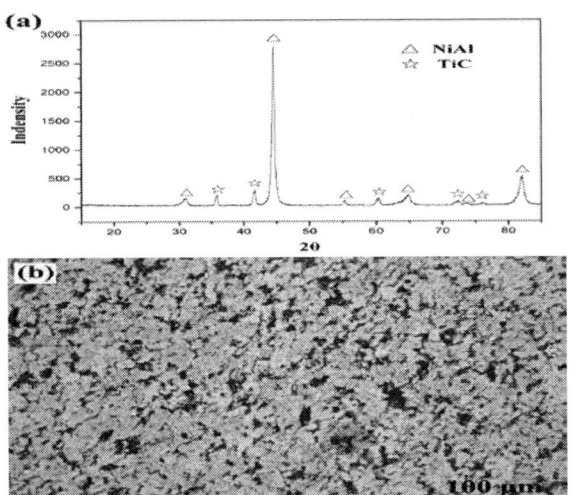

FIGURE I. (A) XRD PATTERN OF THE NIAL-TIC-AL2O3 COMPOSITE, (B) OPTICAL MICROGRAPH OF THE NIAL-TIC-AL2O3 COMPOSITE.

The TEM observation on the NiAl-TiC- Al_2O_3 composite is shown in Figure.2. The NiAl matrix has dual-grain structure, as shown in the Figure.2 (a). Except the large NiAl grain shown in Figure. 1(b), there are much fine grains with hundreds of nanometers. The TiC particles exhibit two kinds of morphologies. The TiC particles along the NiAl grain boundary agglomerate into big ones, as shown in Figure. 1(b). The TiC particle in NiAl grain exhibits polyhedron characteristics, as shown in Figure. 2(b). The inset selected area electron diffraction (SAED) pattern gives that TiC particle has an orientation relationship with NiAl matrix of $[100]_{TiC}//[100]_{NiAl}$ and $(020)_{TiC}//(0-11)_{NiAl}$, which was reported in the former research [12]. However, further observation on NiAl and TiC interface finds that an amorphous transition layer with several nanometers exists between NiAl and TiC, as shown in Figure. 2(c). According to the recent research [19], the amorphous structure contains high energy and so unstable, especially to the amorphous obtained by mechanical alloying. The heat treatment will lead to the amorphous crystallization. In the present paper, the relative ball milling and rapid cooling during synthesis process may contribute much to the formation of amorphous. In addition, the crystal parameters of NiAl and TiC are 0.356 nm and 0.435 nm, respectively. The big difference

between them will lead to great interface stress. Based on the recent research [20], the high interface stress can result in the formation of amorphous film. Therefore it can be concluded that the formation of the amorphous layer should be attributed to the synthesis procedure and crystal difference. In addition, along the NiAl grain boundary, the Al_2O_3 particles with irregular shape were found, as shown in Figure.2 (d). By the SAED pattern, it can be confirmed that most of particles are α-Al_2O_3.

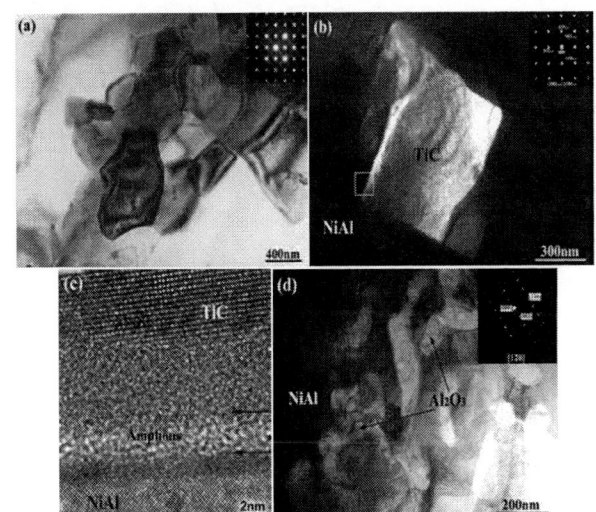

FIGURE II. (A) BRIGHT FILED TEM MICROGRAPH OF FINE NIAL MATRIX, (B) MORPHOLOGY OF TIC PARTICLE IN NIAL GRAIN, (C) HRTEM OF THE THIN AMORPHOUS LAYER ALONG NIAL/TIC PHASE INTERFACE, (D) MORPHOLOGY OF A-AL2O3 PARTICLES ALONG NIAL BOUNDARY (INSET PICTURES SHOWING THE CORRESPONDING SAED PATTERNS).

Further observations on the TiC particles find that some TiC particles have the laminate characteristic, as shown in Figure.3 (a). The inset SAED pattern confirms that stacking faults and microtwins exist inside. The corresponding HRTEM image is shown in Figure.3 (b), which reveals that the atoms on the twin boundary are different from the one in the twin crystal. The EDS test also shows that there are more Al and Ti elements in the TiC particle that contains microtwins and stacking faults defects. According to Yu's study [21], the segregation of Al and Si in TiC can obviously increase the formation chance of microtwins and stacking faults, so it may be drawn that the Al segregation leads to the formation of such defects. Moreover, some Ti_2AlC ceramic particles are observed, which has hexagonal crystal with lattice parameters of a=0.34 nm, c=1.36 nm, as shown in Figure.3 (c). Inset SAED pattern reveals that there are stacking fault inside. The corresponding HRTEM image shows that except the stacking fault and microtwins inside there are some special intergrowth plates, as shown in Figure.3 (d). According to the recent study [22,23] the intergrowth plate should be TiC. The HRTEM image also exhibits that the atoms in the plate are different from the Ti_2AlC. It can be inferred that the growth of Ti_2AlC phase leads to the lack of Al elements in neighbor area, which promote the TiC plate to form. And then it is reasonable to understand that the special microtwins in TiC phase, which may be the Ti_2AlC

plates. Such ceramic particle inside should be attributed to the ball milling and the high temperature during SHS/HE, which result in the Al elements segregate in the TiC.

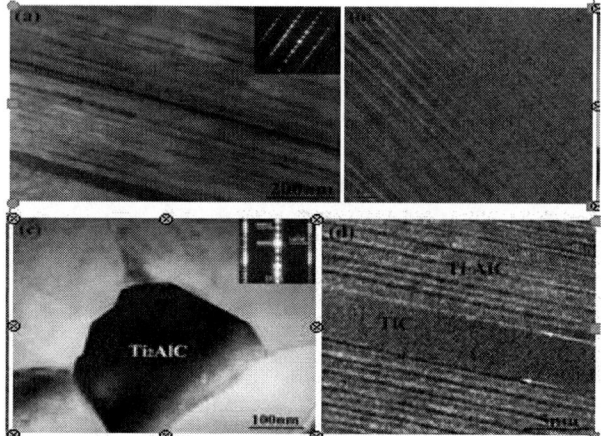

FIGURE III. (A) BRIGHT FIELD TEM MICROGRAPH OF TIC PARTICLES WITH STACKING FAULT AND MICROTWINS INSIDE, (B) HRTEM IMAGE OF THE STACKING FAULT AND MICROTWINS, (C) MORPHOLOGY OF THE TI2ALC PARTICLE, (D) HRTEM IMAGE SHOWING THE INTERGROWTH OF TIC AND TI2ALC (INSET PICTURES SHOWING THE SAED PATTERNS).

IV. CONCLUSIONS

(1) An in-situ NiAl-TiC-Al$_2$O$_3$ composite has been successfully fabricated from elemental powders using thermal explosion and hot extrusion. In the composite, TiC particles along NiAl grain boundary tend to agglomerate and grow, but the TiC particles in NiAl grain still keep polygonal and faceted. The NiAl grain of the composite matrix has been greatly refined by the in situ formed TiC particles. In addition, the Ti$_2$AlC particle with hexagonal crystal is formed along the NiAl grain boundary.

(2) The interface between NiAl and TiC is flat and sharp, and almost no interfacial precipitate, but, in some cases, thin amorphous layer is observed at the interface. The stacking fault and microtwins are also observed in many TiC particles, and moreover TiC is intergrowth within Ti$_2$AlC in some particles.

ACKNOWLEDGMENT

The authors are grateful to the Strategic New Industry Development Special Foundation of Shenzhen (JCYJ20140419114548515 and JCYJ20130402172114948), the Shenzhen International Cooperative Research Project (GJHZ20140419114548516) and the Shenzhen Technology Innovation Plan (CXZZ20140419114548507 and CXZZ20140731091722497).

REFERENCES

[1] R. Darolia, NiAl alloys for high-temperature structural applications, JOM 43(1991) 44–49.

[2] L.Y. Sheng, W. Zhang, J. T. Guo, F. Yang, Y. C. Liang, H. Q. Ye, Effect of Au addition on the microstructure and mechanical properties of NiAl intermetallic compound, Intermetallics 18(2010)740–744.

[3] L.Y. Sheng, Y. Xie, T. F. Xi, J. T. Guo, Y. F. Zheng, H. Q. Ye, Microstructure characteristics and compressive properties of NiAl-based multiphase alloy during heat treatments, Mater. Sci. Eng. A 528(2011)8324–8331.

[4] L. Y. Sheng, F. Yang, T. F. Xi, C. Lai, J. T. Guo, Microstructure and elevated temperature tensile behaviour of directionally solidified nickel based superalloy, Mater. Res. Innov. 17(2013) S101–S106.

[5] S. C. Deevi, V. K. Sikka, C. T. Liu, Processing, properties, and applications of nickel and iron aluminides, Prog. Mater. Sci. 42(1997)177–192.

[6] L. Y. Sheng, F. Yang, T. F. Xi, J. T. Guo, H. Q. Ye, Microstructure evolution and mechanical properties of Ni3Al/Al2O3 composite during self-propagation high-temperature synthesis and hot extrusion, Mater. Sci. Eng. A 555(2012)131–138.

[7] L. Y. Sheng, J. T. Guo, W. L. Ren, Z. X. Zhang, Z. M. Ren, H. Q. Ye, Preliminary investigation on strong magnetic field treated NiAl–Cr(Mo)–Hf near eutectic alloy, Intermetallics 19(2011)143–148.

[8] L.Y. Sheng, W. Zhang, C. Lai, J.T. Guo, T.F. Xi, H.Q. Ye, Microstructure and mechanical properties of laves phase strengthening NiAl base composite fabricated by rapid solidification, Acta Metall. Sin. 49(2013):1318-1324.

[9] L. Y. Sheng, W. Zhang, J. T. Guo, Z. S. Wang, H. Q. Ye, Microstructure evolution and elevated temperature compressive properties of a rapidly solidified NiAl–Cr(Nb)/Dy alloy, Mater. Design 30(2009)2752–2755.

[10] L.Y. Sheng, C. Lai, F. Yang, Q.L. Wang, T.F. Xi, Microstructure and wear behaviour of ceramic particles strengthening NiAl based composite, Mater. Res. Innov. 18(2014):S544–S549.

[11] L.Y. Sheng, L.J. Wang, T.F. Xi, Y.F. Zheng, H.Q. Ye, Microstructure, precipitates and compressive properties of various holmium doped NiAl/Cr(Mo, Hf) eutectic alloys, Mater. Design 32(2011):4810-4817.

[12] L. Y. Sheng, J. T. Guo, T. F. Xi, B. C. Zhang, H. Q. Ye, ZrO2 strengthened NiAl/Cr(Mo,Hf) composite fabricated by powder metallurgy, Prog. Nat. Sci.-Mater. Int. 22(2012)231–236.

[13] H. Choo, P. Nash, M. Dollar, Mechanical properties of NiAl–AlN–Al2O3 composites, Mater. Sci. Eng. A 239–240(1997)464–471.

[14] L. Y. Sheng, T. F. Xi, C. Lai, J. T. Guo, Y. F. Zheng, Effect of extrusion process on microstructure and mechanical properties of Ni3Al-B-Cr alloy during self-propagation high-temperature synthesis, Trans. Nonferrous Met. Soc. China, 22(2012)489–495.

[15] W. J. Ye, D. Feng, H. L. Luo, The microstructure and its properties of Ni3Al based composite, Mater. Res. Innov. 2(1999)321–324.

[16] L. Y. Sheng, W. Zhang, J. T. Guo, Z. S. Wang, V. E. Ovcharenko, L. Z. Zhou, H. Q. Ye, Microstructure and mechanical properties of Ni3Al fabricated by thermal explosion and hot extrusion, Intermetallics 17(2009)572–577.

[17] O. Ozdemir, S. Zeytin, C. Bindal,A study on NiAl produced by pressure-assisted combustion synthesis, Vacuum 84(2009)430–437.

[18] L. Y. Sheng, F. Yang, T. F. Xi, J. T. Guo, H. Q. Ye, Microstructure evolution and mechanical properties of Ni3Al/Al2O3 composite during self-propagation high-temperature synthesisand hot extrusion, Mater. Sci. Eng. A 555(2012)131–138.

[19] L. Y. Sheng, F. Yang, T. F. Xi, J. T. Guo,Investigation on microstructure and wear behavior of the NiAl–TiC–Al2O3 composite fabricated by self-propagation high-temperature synthesis with extrusion, J. Alloys Compd. 554(2013)182–188.

[20] L.Y. Sheng, F. Yang, T.F. Xi, Y.F. Zheng, J.T. Guo, Microstructure and room temperature mechanical properties of NiAl–Cr(Mo)–(Hf, Dy) hypoeutectic alloy prepared by injection casting, Trans. Nonferrous Met. Soc. China 23(2013)983–990.

[21] R. Yu, L.L. He, H.Q. Ye, Effects of Si and Al on twin boundary energy of TiC, Acta Mater. 51(2003)2477-2484

[22] L. Y. Sheng, F. Yang, J. T. Guo, T. F. Xi, H. Q. Ye, Investigation on NiAl–TiC–Al2O3 composite prepared by selfpropagation high temperature synthesis with hot extrusion, Compos. B-Eng. 45(2013)785–791.

[23] L.Y. Sheng, F. Yang, J.T. Guo, T.F. Xi, Anomalous yield and intermediate temperature brittleness behaviors of directionally solidified

nickel-based superalloy, Trans. Nonferrous Met. Soc. China 24 (2014) 673-681.

Flexural Behavior of Reinforced Concrete Beams Strengthened with BFRP Bars

G.N. Yang, B.R. Huo

College of Architectural and Civil Engineering Shenyang University
No 21 Wanghua South Street Dadong District, Shenyang, China

M.X. Zheng

Tianlang Real Estate Co Ltd Liaoning
No 227 Qingnian Street Shenhe District, Shenyang, China

Abstract—In order to strengthen buildings against various damages and prolong their service life, BFRP is considered serving as reinforcing materials instead of steel. Through static load tests on six simple supporting beams reinforced with BFRP rebar and six reinforced with steel rebar, the paper compares their flexural behavior, finds the load-deformation relationship of the BFRP reinforced concrete beam and finally concludes if the BFRP bar bonds well enough to concrete, it will be great strengthening materials, which is expected to improve the bearing capacity of beam specimens significantly. But in the experiment, the two materials fail to bond as well as assumed, which hinders the BFRP producing advantages as reinforcing materials. The BFRP rebar debonding from concrete restricts its application severely, so the first concern should be how to improve the technology for debonding-resistant BFRP rebar. Under the same loads, the deflections of all beam specimens are large, mainly due to the fact that the elastic modulus of the BFRP rebar is small, while its tensile strength is high, but not fully utilized, so the emphasis in design should be put on the use of pre-stressing tendons. Based on the experimental results, recommendations on relevant design and applications are of great importance to engineering practice and help improve the design guidance for fiber composite concrete structure.

Keywords-BFRP rebar; flexural; load-deformation relationship; debonding

I. INTRODUCTION

In recent years, the use of basalt fiber reinforced polymer (BFRP) as a means of rehabilitating or strengthening reinforced concrete (RC) beams has generated much interest in the construction industry. Compared with tendon, the BFRP textile fiber has higher tensile strength, higher corrosion resistance, lower density, higher fatigue resistance, better insulation and some other merits. Obviously, if the BFRP is used as replacement for tendon in reinforced concrete structures, reinforcing bar corrosion can be avoided thoroughly. High stiffness-to-weight and strength-to- weight ratios of these materials combined with their superior environmental durability have made them a competing alternative to the conventional strengthening and repair materials. In addition, it is a now type of green fiber composite material, pollution-free, non-carcinogenic and eco-friendly. Basalt fiber is a relative newcomer to fiber reinforced polymers (FRPs).At present, numerous research studies have been performed on carbon fiber and glass fiber reinforced concrete, while studies on concrete slabs reinforced with BFRP are limited. The mechanical properties of BFRP rebar, such as linear elastic stress-strain relationship and low elastic modulus, differ from those of steel slab greatly. It is significant to study the flexural behaviors and design theory of BFRP reinforced concrete beams, which helps design concrete beams with higher load capacity and durability, avoiding disastrous damages significantly. Experimental studies on BFRP reinforced concrete beams have made some progress, but there is still a long way ahead. The main contents of the present thesis are briefly summarized as following: 1. six simple-supported beams reinforced with BFRP rebar and six reinforced with steel rebar were tested on the three-point static load, and analyzed the results. 2. BFRP reinforced beams were tested on the ultimate load. 3. based on the test results, the load-deformation relationship of BFRP reinforced beams were analyzed.

II. EXPERIMENTAL PROGRAM OF FLEXURAL BEHAVIOR OF BFRP REINFORCED CONCRETE BEAMS

The experiment consisted of two groups of beams. Beams of group one, M1,M2 and M3, were strengthened with steel rebar, while those of group two, M4,M5, and M6, were strengthened with BFRP rebar. Two test specimens were made from each sample beam with sectional dimension 200 x 300 mm, beam length 2060mm, and clear span 1800 mm, simple supported at both ends. Details of the parameters are given in Table 1. The beam structure is shown in Figure 1. The BFRP rebar used in the tests was manufactured by ShangHai Russia & Gold Basalt Fiber CO., LTD.

TABLE I   DETAILS OF TEST SPECIMENS.

Beam Num.	Concrete strength	Upper rebar	Lower rebar	Steel stirrup	Reinforcement ratio(%)
M1-1,2	C20	2 Φ 12	S2 Φ 11	Φ 5@80	0.36
M2-1,2	C30	2 Φ 12	S2 Φ 16	Φ 5@80	0.77
M3-1,2	C40	2 Φ 12	S3 Φ 16	Φ 5@80	1.15
M4-1,2	C20	2 Φ 12	B2 Φ 11	Φ 5@80	0.36
M5-1,2	C30	2 Φ 12	B2 Φ 16	Φ 5@80	0.77
M6-1,2	C40	2 Φ 12	B3 Φ 16	Φ 5@80	1.15

Note: S-steel reinforcement B-BFRP reinforcement

A. Experimental Loading and Test

Loading device of the sample beam is shown in Figure 1.

The sample beam bears the three-point concentrated load transferred by the distributive beam over which a jack is set up. In order to avoid local stress concentration or even local concrete crush, pads are put on the beam bearing and three-point bend. Replacement meter monitors the settlement of middle span and abutment and strain gage measures strain on beam section at different heights. The test is conducted under step load, with unit load of 5KN before appearance of concrete crack and unit load of 3KN after that. The load holding duration is 2 min, meanwhile experimental data is recorded. When the crack grows to some extent, change the loading way from step loading to sustained loading until the concrete sample is crushed and test specimens damaged.

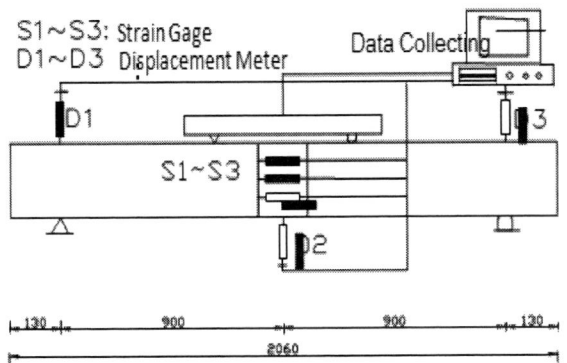

FIGURE I. LOADING TEST SET-UP(UNITS: MM).

III. EXPERIMENTAL RESULTS AND ANALYSIS

A. Crack Development in BFRP Reinforced Concrete Beam

a. Conduct test under step load until crack appears. During this period, since there is no crack in concrete beam, the total cross section is loaded. Concrete beam is characteristic of elastic deformation at low deformation rate, which means stress and strain are proportional. The load-deflection relationship is linear.

b. As the load increases, the tensile strain reaches its utmost limit on the concrete's edge. In this case, the tensile stress approaches tensile strength of the test specimen and cracks begin to appear on the weakest stress section of concrete beam. With cracks on the cross section, concrete slab under strain stop working and, correspondingly, strain on this area is transferred to BFRP rebar. Obviously, stress is redistributed. Tensile stress increases with the increasing of load. Because of low elastic modulus of BFRP rebar, cracks develop very quickly. The area of concrete sample under stress decreases, while the growth rate of deflection increases. Although curve slope is much smaller due to yield, the curve is approximately linear. With sustainedly increasing of load, the sample concrete under stress is crushed while the BFRP rebar being tested doesn't reach its nominal yield point. Therefore, failure in BFRP reinforced concrete beam occurs, when concrete under stress is crushed, much earlier than the BFRP in tensile area reaches its ultimate tensile strength, which tends to cause failure in high-reinforcement-ratio beam.

As load increases continuously, deflection value of BFRP reinforced concrete beam increases greatly, cross section is

deformed rapidly and reaches its ultimate deformation, which is one of the main causes of structure failure. Based on relevant theories and experimental analysis, failure in BFRP reinforced concrete beam occurs in three cases. The first, when the structure is designed to be high-reinforcement-ratio beam, concrete under stress is crushed, namely, failure in high-reinforcement-ratio BFRP reinforced concrete beam; the second, when the structure is designed to be middle-reinforcement-ratio beam, BFRP reinforced concrete beam is deformed greatly, causing structure failure, namely, failure in middle-reinforcement-ratio BFRP reinforced concrete beam; the third, when FPR rebar is fractured, brittle failure occurs, abruptly and with no warning signs, namely, failure in low-reinforcement-ratio concrete beam.

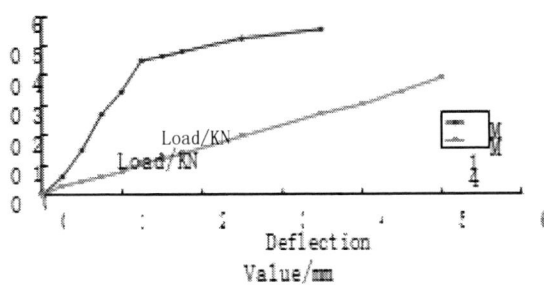

FIGURE II. M1,M4 BEAMS' LOAD-DEFLECTION CURVE.

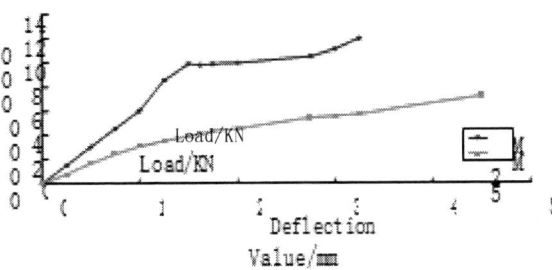

FIGURE III. M2,M5 BEAMS' LOAD-DEFLECTION CURVE.

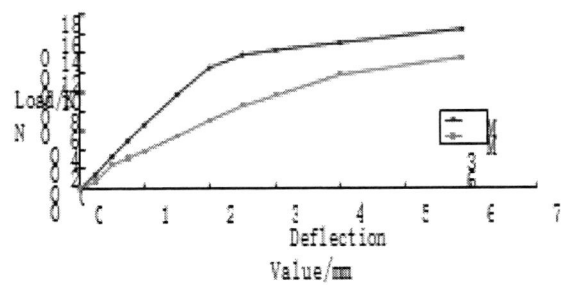

FIGURE IV. M3,M6 BEAMS' LOAD-DEFLECTION CURVE.

Specimens M1 & M4, specimens M2 &M5, specimens M3 & M6 are of the same enforcement ratio respectively, with same concrete strength, but different types of longitudinal bars inside. As shown in Figure 2 Figure 3 and Figure 4, reinforced concrete beam strengthened with BFRP bar, marked M1 M2 and M3, and ordinary reinforced concrete beam, marked M4

M5 and M6, are tested and compared. At the beginning of loading, stress on M1 M2 and M3 increase quickly but deflection of them increase slowly, while as for M4 M5 and M6, the opposite is true. The reason is analyzed as follows: bonding strength between BFRP bar and concrete was overestimated in designing. As a matter of fact, bonding strength doesn't come up to expectation. As load is being exerted, slip occurs, which results in cracks widening and, ultimately, beam failure. While considering BFRP bar, a new type of building material, as the replacement of reinforcing steel to bear longitudinal stress, the major concerns are how to enhance bonding strength between BFRP bar and concrete, how to improve production technology, and how to produce high-bonding-strength BFRP rebar.

B. Comparing the Theoretical Values and Experimental Values of Sample Beams' Deflection

The comparison is based on the calculating formula of FRP-strengthened-concrete beams' deflection, proposed by Cangarao. In the formula, assumed concrete slab under load cracks significantly, cracking equivalent moment of inertia I_{cr} is applied to central beam, and ACI formula is applied to calculate I_{cr} of top section, then the moment of inertia is calculated as follows:

$$I_e = \frac{23 I_c I_{cr}}{8 I_{cr} + 15 I_c} \qquad (1)$$

Apply the present formula to obtain value of the inertia moment, and insert it into the formula of mid-span deflection, the result is as follows:

$$\Delta_{\max imum}^{midspan} = \frac{p l_1}{E_c}\left(\frac{4 l_1 l_2 + l_2^{\,2}}{8 I_{cr}} + \frac{l_1^{\,2}}{3 l_e} \right) \qquad (2)$$

The result shows that experimental data of BFRP reinforced concrete beam's loading capacity agrees well with theoretical values.

IV. SUMMARY

If bonding between them is strong enough, BFRP rebar and concrete should co-work properly, and consequently the loading capacity of beam specimens should be improved greatly. But during the test, bonding between BFRP and concrete is not as strong as expected, therefore results of the test fail to embody BFRP bar's advantage in strength. BFRP rebar's debonding from concrete restricts its application severely, so the first concern should be improving the technology for debonding resistant BFRP rebar. Under the same loads, the deflections of the beam specimens are all large, mainly due to the fact that the elastic modulus of the BFRP rebar is small, while its tensile strength is high, but not fully utilized, so the emphasis in design should be put on the use of pre-stressing tendons. Based on the experimental results, recommendations on relevant design and applications are of great importance to engineering practice and help improve the design guidance for fiber composite concrete structure.

REFERENCES

[1] Baorong Huo. Theoretical and Experimental Study on BFRP Concrete Structure [D]. Doctor Paper of Liaoning Technical University, 2011:29-43.

[2] Kelley L, Michal L. Design philosophy for structural strengthening with FRP [J]. Concrete International, 2000, （2）:77-82.

[3] Xue Weichen, Wang Xiaohui, Zhang Shulu. Bond properties of high strength CFRP strands[J]. ACI Material Journal, 2008, 105(1):303-311.

[4] ACI Committee 440. State-of-the-Art-Report on Reinforced Plastic for Concrete Structures. Detroit, Michigan: American Concrete Institutes February, 1996.

[5] KWON I B, KIM C Y, CHOI M Y. Distributed strain and temperature measurement of a beam using fiber optic BOTDA sensor [J]. Proceeding of SPIE, 2003, 50(57):486-496.

[6] ALAHHAB I M, CHO Y T, NEW SON T P. Comparison of the methods for discriminating temperature and strain in spontaneous Brillion based distributed sensors [J]. Optics Letters, 2004, 29(1):26-28.

[7] Kelley L, Michal L. Design philosophy for structural strengthening with FRP [J]. Concrete International, 2000, （2）:77-82.

International Conference on Power Electronics and Energy Engineering (PEEE 2015)

Experimental Investigating Effect of Reprocessing on Properties of Composites based on Recycled Polypropylene

F. Gu, P. Hall, N.J. Miles

Faculty of Science and Engineering
Nottingham University
No. 199, East Taikang Rd, Ningbo, Zhejiang, China. P. R.

Abstract—This paper is aimed to provide knowledge and understanding to promote the use of recycled materials via experimental study. In this paper, for the first time, effect of reprocessing on properties of recycled PP/talc composites has been investigated via series of experimental tests, while both virgin and recycled PP were used as comparatives. The materials were reprocessed in two different routes, multiple extrusion cycles and multiple injection moulding cycles. Some materials were taken out for testing during each cycle. The tests include mechanical, rheological and thermal. The results were plotted and discussed, and for the first time, results from the two different reprocessing routes were compared. Some phenomena were observed and fitted in others' studies and prediction. Also, the complexity of composites made from recycled materials under reprocessing was detected, and in need of further research.

Keywords-reprocessing; recycled; polypropylene; talc-filled; extrusion; injection moulding; chain-scission

I. INTRODUCTION

Due to various reasons, such as pressure from legislations [1~3], saving natural resource (mainly crude oil which plastics are made from), reducing cost, landfill and emission, plastic recycling has attracted a broad interest since 1950s. Even after decades of practices and researches, it still poses a difficult challenge both for industry and for academia. Comparing with landfill, incineration for energy recovery and chemical recycling, mechanical recycling has been proved to be the most viable way to deal with the challenge, for it reduces the consumption of natural resources as well as a reduced landfill and material cost [4]. It was estimated that using recycled plastics could reduce greenhouse gas emissions by about 80% [5]. But when compared to virgin plastic, recycled plastic tend to show lesser performances due to degradation phenomena that occur during the product's first life and reprocessing [6~12]. Thus, the effect of reprocessing became a popular topic, and many studies have investigated the effect of reprocessing on their structure, thermal and mechanical properties [4, 13~17]. González-González, et al. [14] have identified the chain scissions during multiple extrusion which are linked to the melt temperature and resulted molecular weight losses. Brennan, et al. [18] recovered acrylonitrile-butadiene-styrene (ABS) and high-impact polystyrene (HIPS) from waste computer equipments, and found elongation at break and impact strengths were reduced

considerably during reprocessing. Su, et al. [15] investigated the influence of reprocessing on the mechanical properties and structure of polyamide 6, and reported an increment in the tensile yield stress, flexural strength and modulus but a decrement in Izod impact strength and the molecular weight. Scaffaro, et al. [4] found the tensile, flexural and impact properties of ABS deteriorated with reprocessing.

Polypropylene (PP), as one of the most common plastic materials, is usually being used in various applications while mixing with all sorts of fillers, such as minerals [19~23], clays [24] and fibres [25]. Among these fillers, talc is one of the most commonly used mineral filler which improves both the thermal and mechanical properties of PP [23], and a few studies have been conducted on reprocessing of talc-filled PP. Guerrica-Echevarría, et al. [13] studied talc-filled PP undergone multiple injection moulding cycles, and found that molecular weight decreased and break properties (such as Elongation at break) are related to the filler content and processing conditions. Bahlouli, et al. [26] studied the recycling effects on two high impact polypropylenes (HiPP), and found a better thermal and structural stability for talc filled HiPP. Wang, et al. [23] pointed out that during the repeated extrusion cycles, talc in PP slightly increased the Young's modulus and the yield strength.

Yet, most current studies only focused on reprocessing effect on virgin composites, such as talc-filled virgin PP, or recycling of waste plastics from different sources. To the best of our knowledge, the influence of the filler content and reprocessing cycle numbers on the properties of recycled PP-based composites has received little to no attention. Lack of research on recycled plastic based materials would certainly limit their use, and it is both uneconomic and environmental-unfriendly. Further, there is no research which focused on comparison of the effects of two major reprocessing routes (extrusion and injection moulding) on composites has been reported, and in industrial context, mechanical recycling usually involves both of the two procedures [27, 28]. In this paper, with the aim of promoting the understanding and use of composites based on recycled plastics, the effects from both extrusion and injection moulding cycles on the properties of talc-filled recycled PP composites were investigated via experimental methods. The results were discussed and correlated with previous researches,

© 2015. The authors - Published by Atlantis Press

177

which would provide a comprehensible knowledge base for reusing recycled plastics and their composites.

II. EXPERIMENTAL

A. Materials

The virgin PP (VPP) material used was a block co-polymer mainly being used for manufacturing automobile parts, with a trade-name of PPB-MO2-V and produced by Yangzi Sinopec. The material was used as received, and has an average particle size of 3.0 mm, density of 0.9 g cm^{-3}, as shown in Figure 1. The recycled PP (RPP) used was some grey pellet recovered from white post-customer storage boxes, has an average particle size of 3.0 mm, density of 1.0 g cm^{-3}, as shown in Figure 2, and was used as received.

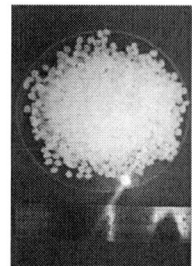

FIGURE I. PPB-MO2-V (VPP).

FIGURE II. GREY RECYCLED PP (RPP).

The talcum powder (talc) used in this work was bought from a local factory, has an average particle size of 12.5 um, a density of 2.7 g cm^{-3}, as shown in Figure 3, and was used as received.

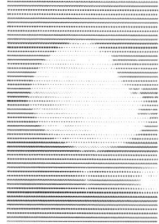

FIGURE III. TALCUM POWDER.

The coupling agent, maleic anhydride grafted polypropylene (MAPP) used in this work was bought from Nanjing Deba Chemical Co.,Ltd, has an average particle size

of 2.5 mm, a density of 0.9 g cm^{-3}, with the grafted rate of 0.8%, as shown in Figure 4, and was used as received.

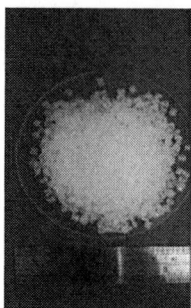

FIGURE IV. MAPP.

The compositions used in the paper were shown in Table 1.

TABLE I. THE COMPOSITIONS OF TESTING MATERIALS (WT.%).

Designation	PPB-MO-V	Grey RPP	Talc	MAPP
VPP	100	0	0	0
RPP	0	100	0	0
RPP-T20	0	75	20	5
RPP-T40	0	50	40	10

B. Sampling and Reprocessing

1) Extrusion cycles

All materials for multiple extrusion cycles were processed by a Kangrun KRSHJ-20 extruder; a co-rotating, intermeshing twin-screw extruder, with screw diameter of 22 mm and L/D=44. A single-screw feeder attached to the hopper was used for all the PP pellets. The processing temperatures were allowed to increase from 180°C to 200°C going from the hopper to the third barrel, and the temperatures of the last three barrels remained at 200°C, while the die temperature was set at 200°C. The screw rotation speed was 180 rpm, and the total mass flow rate was 5 kg h^{-1}. Blended strands were extruded into a water bath for cooling, and then pelletized by a cutter. The average extruded pellet size was 2.8 mm. This process was repeated 5 times under the same operating conditions, so the grinded material of each cycle was the starting material for the following reprocessing cycles.

Some pellets were taken in every extrusion cycle for making testing sample pieces. Those taken pellets were dried in a dry oven at 85°C for 12 h with constant air flow to keep the moisture content below 1% before being fed into the injection moulding machine.

Then these were injection moulded into ISO standard test specimens using a Haitian MA1200/370 injection moulding machine, and 2 (tensile pieces) or 4 (flexural or impact pieces) test specimens were produced per single injection moulding process, see Figure 5. The temperatures of five heating barrels were set at 190°C, 192°C, 195°C, 200°C, 200°C, with injection pressure of 50 MPa, injection speed of 50 g per second, packing pressure of 30 MPa for 10 s, cooling in moulds was allowed for 10 s. The mould was pre-heated to 50°C. The processing parameters were set in accordance with real manufacture [27].

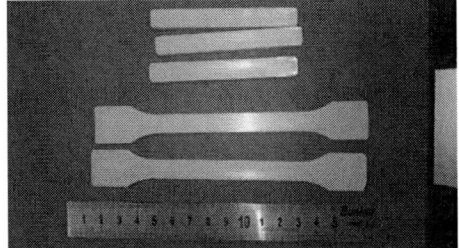

FIGURE V. TESTING SAMPLE PIECES.

2) Injection moulding cycles

All materials for multiple injection moulding cycles were processed by a Kangrun KRSHJ-20 extruder; a co-rotating, intermeshing twin-screw extruder, with screw diameter of 22 mm and L/D=44. A single-screw feeder attached to the hopper was used for all the PP pellets. The processing temperatures were allowed to increase from 180°C to 200°C going from the hopper to the third barrel, and the temperatures of the last three barrels remained at 200°C, while the die temperature was set at 200°C. The screw rotation speed was 180 rpm, and the total mass flow rate was 5 kg h^{-1}. Blended strands were extruded into a water bath for cooling, and then pelletized by a cutter. The average extruded pellet size was 2.8 mm.

The extruded pellets were dried in a dry oven at 85°C for 12 h with constant air flow to keep the moisture content below 1% before being fed into the injection moulding machine.

Then these were injection moulded into ISO standard test specimens using a Haitian MA1200/370 injection moulding machine, and 2 (tensile pieces) or 4 (flexural or impact pieces) test specimens were produced per single injection moulding process, the same as that shown in Figure 5. The temperatures of five heating barrels were set at 190°C, 192°C, 195°C, 200°C, 200°C, with injection pressure of 50 MPa, injection speed of 50 g per second, packing pressure of 30 MPa for 10 s, Cooling in moulds was allowed for 10 s. The mould was pre-heated to 50°C. The processing parameters were set in accordance with real manufacture [27].

This process was repeated 5 times under the same operating conditions. In each injection cycle, parts of the specimens were used for characterization purposes, while the remainders were grinded by a cutting mill (Retsch, Germany, model SM 200, as shown in Figure 6 left) in order to be reprocessed. The 4 mm sieve used was shown in Figure 6 right. The shredded plastic pellets were shown in Figure 7, with an average particle size of 3 mm.

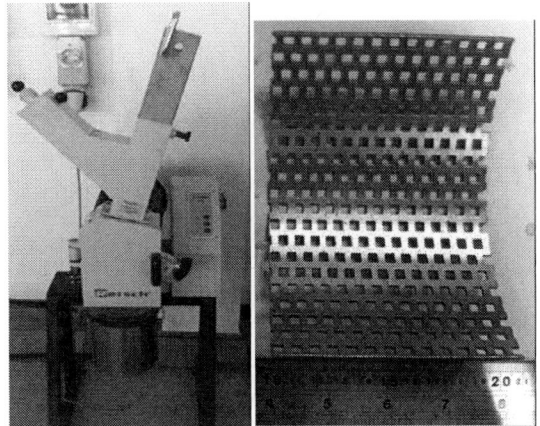

FIGURE VI. LEFT: CUTTING MILL, MODEL SM 200; RIGHT: THE 4 mm SIEVE.

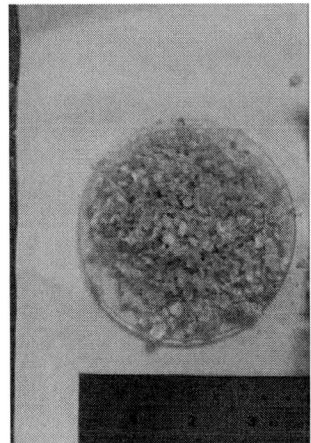

FIGURE VII. SHREDDED PLASTIC PELLETS.

C. Sampling and Reprocessing

1) Mechanical

The mechanical properties of composites were evaluated in terms of tensile, flexural, and impact properties, and all sample specimens were conditioned at 23°C and 50% R.H. for over 88 h before testing in accrodance with ISO291 specifications [30].

The tensile properties tested were tensile strength, yield strength and elongation at break, assessed in accordance with ISO527 specifications [31]. The gauge distance is 110 mm with fixers' moving speed of 50 mm min^{-1}, sampling rate at 200 pts s^{-1}, full-scale load range of 20 kN, performed on a Gotech Universal Testing Machine (model TCS-2000NE) at room temperature of 23°C and at 50% R.H. The tensile modulus was obtained by an extensometer (Epsilon Technology Co.Ltd) at tensile elongation of 1%.

The flexural properties tested were flexural modulus and flexural strength, assessed in accordance with ISO178 specifications [32]. The span was set to 64mm at a crosshead speed of 2 mm min^{-1}, sampling rate at 200 pts s^{-1}, full-scale load range of 20kN, performed on a Gotech Universal Testing Machine (model TCS-2000NE) at a room temperature of 23°C and at 50% R.H.

8 sample pieces were tested for each property tested and the average result taken if the coefficient of variance met the required limits (5% in accordance with ISO2602 specifications [33]).

2) Rheological

The flow behaviour of the materials was assessed using steady and dynamic shear rheology. The test was performed by utilising a dual-bore capillary rheometer (Rosand RH2200, Malvern Instruments) with two capillary dies with same radius of 1 mm but different length/radius ratios. Samples were pre-heated in the dual barrels at 190°C for 2 min, and measurements were carried out at 190°C under a shear rate ranging from 10 to 5000 s^{-1}, at room temperature of 23°C and at 50% R.H. Viscosity is plotted against shear rate, and the power law model was used to describe relationship between viscosity and shear rate as described in Eq.(1):

$$\eta = K\gamma^{n-1} \qquad (1)$$

where the consistency K corresponds to the viscosity value for a shear rate γ of 1 s^{-1} and the power-law index n characterizes the deviation of the Newtonian behaviour.

3) Thermal

The thermal properties of the materials were measured in form of the temperatures of deflection under load (TDL). The tests were conducted by using an HDT-VICAT test processor (CEAST model 6911.000) according to ISO75 [29], with constant heating rate of 50°C h-1 and a load of 0.45 MPa. The samples were immersed in silicon oil which filled the tank and preheated for 4 min at 40°C, therefore the tests were carried out at a room temperature of 23°C and at 50% R.H. An average 6 samples were prepared and tested in each group.

III. RESULTS AND DISCUSSION

In this session, the following equation was used to calculate the degradation rate (*DR*) of the composites' performance which was applied for all properties:

$$DR = \frac{P_O - P_A}{P_O} \times 100\% \qquad (2)$$

in which, the symbol P_O denotes the original performance obtained from initial tests once the specimens were made, the symbol P_A denotes he performance obtained after the ageing procedures were performed.

A. Effect of Reprocessing

Some of the experimental results from repeated extrusion cycles set were plotted in figures below.

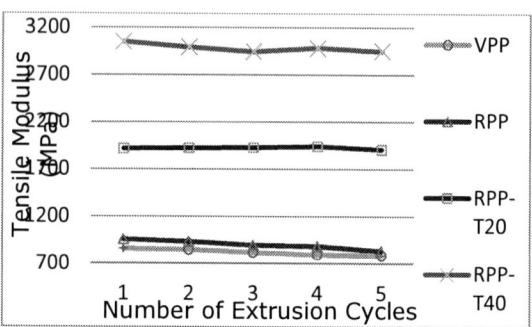

FIGURE VIII. PLOT OF TENSILE MODULUS (MPA) AGAINST NUMBER OF EXTRUSION CYCLES.

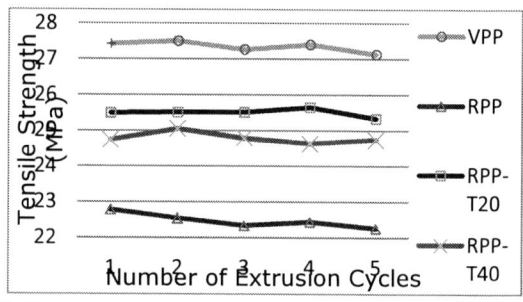

FIGURE IX. PLOT OF TENSILE STRENGTH (MPA) AGAINST NUMBER OF EXTRUSION CYCLES.

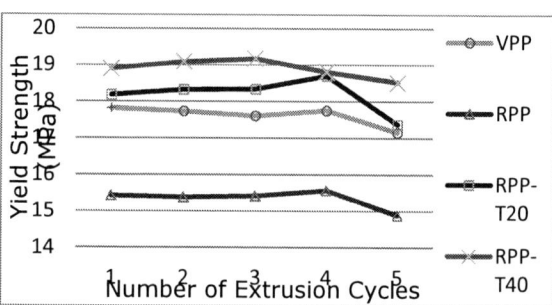

FIGURE X. PLOT OF YIELD STRENGTH (MPA) AGAINST NUMBER OF EXTRUSION CYCLES.

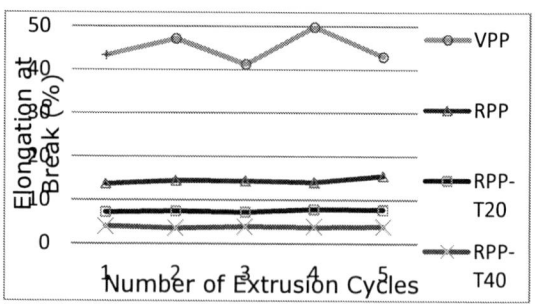

FIGURE XI. PLOT OF ELONGATION AT BREAK (%) AGAINST NUMBER OF EXTRUSION CYCLES.

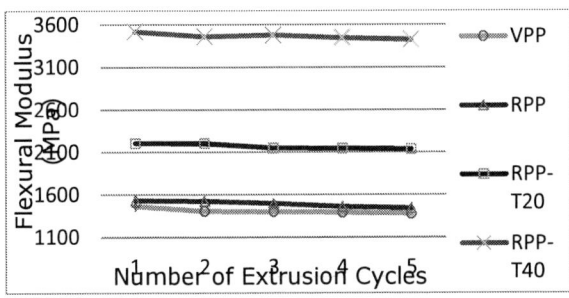

FIGURE XII. PLOT OF FLEXURAL MODULUS (MPA) AGAINST NUMBER OF EXTRUSION CYCLES.

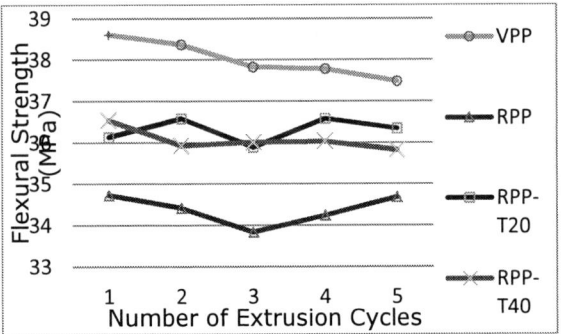

FIGURE XIII. PLOT OF FLEXURAL STRENGTH (MPA) AGAINST NUMBER OF EXTRUSION CYCLES.

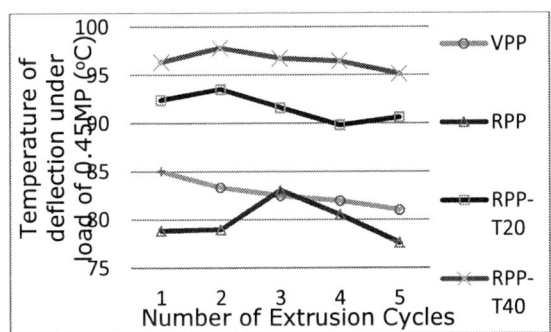

FIGURE XIV. PLOT OF TDL (°C) AGAINST NUMBER OF EXTRUSION CYCLES.

From the figures shown above, the performance of the virgin PP decreased slowly with the number of extrusion cycles, except for elongation at break. This probably resulted from chain-scissions in polymers occurred in elevated temperatures and rotating shear forces [8, 9, 14].

For recycled PP and talc-filled recycled composites, some properties exhibited a little improvement with reprocessing cycles initially and decreased with further extrusion, such as yield strength and TDL, while other properties are shown to be deteriorated shown to be slightly with reprocessing cycles. The elongation at break and flexural strength of recycled PP remained stable, and were slightly increased at the 5th cycle. For talc-filled recycled PP composites, their performances were comparatively more stable during repeated extrusions

except for elongation at break. The talc content improved the mechanical and thermal properties of recycled PP as predicted [20, 23].

*DR*s of some properties during extrusion cycles were summarized in Table.2 using Eq.(2).

TABLE II. *DR*s OF SOME PROPERTIES DURING EXTRUSION CYCLES (%, COMPARING WITH THE 1ST PROCESS).

	Number of Cycles	VPP	RPP	RPP-T20	RPP-T40
Tensile Modulus	2nd	1.42	2.49	-0.33	1.85
	3rd	4.89	6.50	-0.54	3.43
	4th	7.54	7.79	-1.23	2.14
	5th	8.53	12.86	0.48	3.31
Tensile Strength	2nd	-0.28	1.06	-0.06	-1.21
	3rd	0.58	1.94	-0.07	-0.15
	4th	0.06	1.51	-0.60	0.43
	5th	1.07	2.29	0.63	0.00
Yield Strength	2nd	0.50	0.30	-0.79	-0.95
	3rd	1.19	0.07	-0.87	-1.44
	4th	0.36	-0.91	-2.84	0.51
	5th	3.73	3.49	4.44	2.00
Flexural Modulus	2nd	3.95	0.50	0.12	1.54
	3rd	4.37	2.35	2.65	1.07
	4th	4.87	4.79	2.76	1.94
	5th	6.04	6.03	3.13	2.55
Flexural Strength	2nd	0.62	0.89	-1.24	1.68
	3rd	2.01	2.57	0.67	1.44
	4th	2.14	1.40	-1.23	1.39
	5th	2.92	0.14	-0.57	1.96
TDL	2nd	2.00	-0.13	-1.19	-1.56
	3rd	3.02	-5.28	0.87	-0.42
	4th	3.65	-2.07	2.81	-0.10
	5th	4.74	1.61	1.95	1.25
Average		2.91	2.17	0.37	0.94

In Table 2, virgin PP showed the largest *DR* values, because at low molecular weight (MW), the chain scission is random, but at higher MW it becomes MW-dependent, increasing with MW [35], and MW of virgin PP is larger than recycled PP [36].

In repeated heat-and-shear-involving cycles, the polymeric chain lengths become more unified than original lengths and MW, as shown in other's study [35], and the amorphous phase between the lamellae that has an increased mobility is increasing with the number of reprocesses [22], it could gave those polymers a better elongation at break. On the other hand, the presence of talc and contamination within PP matrix resulted in an increase in crystallinity [19, 23, 37], which would increase tensile properties, and reduce elongation at break. Thus, it is highly complicated in fully understand the pattern of recycled PP/talc composites, as shown in figures and Table.2, and needs further study. Still, the stabilizing effect of talc content was observed in recycled composites, as it did same effect on reprocessing of virgin PP/talc composites [23, 26, 37].

For rheological behaviours in the form of shear viscosity plot of the composites shown in Figure 15, and it quite obvious that reprocesses decreased the shear viscosity of virgin material, which might contribute to chain-scission actions took place in elevated temperatures during reprocessing. However, some talc-filled recycled composite

(RPP-T20) was comparatively more stable during multiple heat-involving procedures, as shown in Figure 15.

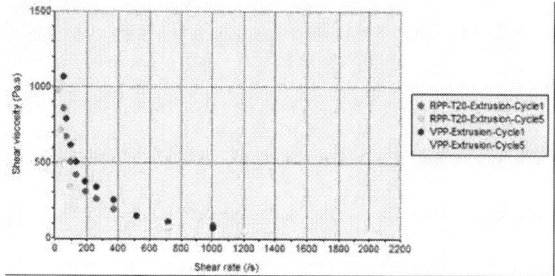

FIGURE XV. SHEAR VISCOSITY PLOT OF MULTIPLE EXTRUSIONS.

The results from repeated injection moulding cycles have shown the same pattern. It is supposed that the presence of talc does not cast a significant influence on the degradation mechanisms of the material matrix [13]), and the stability of recycled PP/talc composites compared might due to a gradual increase of delamination and dispersion of talc particles or agglomerates during the successive reprocessing, which resulted in an increased number of particles and a decreased particle size [23]. The consistency of flow properties of recycled PP/talc composites could lead to a more stable production rate, which would facilitate the use of such composites.

B. Comparison of Effect on Recycled Composites

Tensile results of talc-filled recycled composites from repeated injection moulding cycles set were plotted in Figure 16 to Figure 18, with comparison with multiple extrusion test set, since the tensile properties were found to be critical properties in reprocessing study [34].

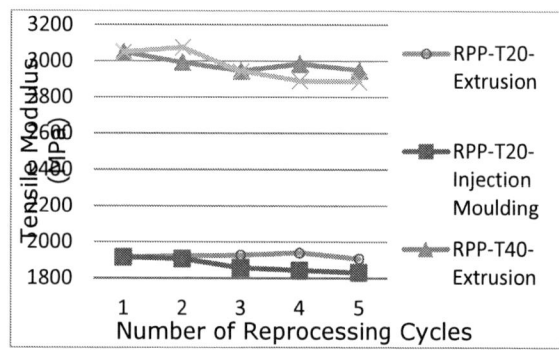

FIGURE XVI. PLOT OF TENSILE MODULUS (MPA) AGAINST NUMBER OF INJECTION MOULDING CYCLES COMPARING WITH EXTRUSION CYCLES.

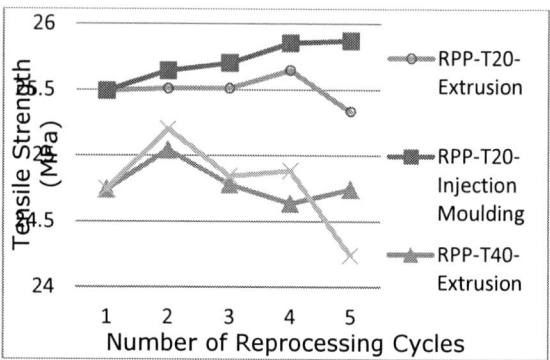

FIGURE XVII. PLOT OF TENSILE STRENGTH (MPA) AGAINST NUMBER OF INJECTION MOULDING CYCLES COMPARING WITH EXTRUSION CYCLES.

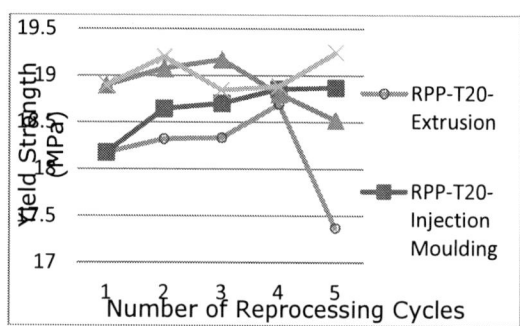

FIGURE XVIII. PLOT OF YIELD STRENGTH (MPA) AGAINST NUMBER OF INJECTION MOULDING CYCLES COMPARING WITH EXTRUSION CYCLES.

As shown in those figures, the materials processed by multiple injection moulding cycles are still preserving some good characteristics as the same as those being processed by multiple extrusion cycles. The tensile strength and yield strength are increased with number of injection moulding cycles, which would imply the increase of crystallinity as mentioned before. All tensile properties of RPP-T40 were increased initially, and then decreased with further re-injection moulding, for it is possible that crystallinity degree of RPP-T40 has reached certain limit, and the amorphous phase increased with further reprocessing.

For the injection moulding cycles involved shredding procedure, which also could be considered as an elevated temperature and high shear rate procedure, the materials endure twice such processes when compared to extrusion set. The chain-scission effect took the ruling place in material matrix, and it would explain the lower tensile modulus than extrusion set.

To sum up, the recycled PP/talc composites still maintain some good performance during reprocesses, and were more stable when comparing to both virgin and recycled PP in most properties, as shown in the figures Table.2. Thus, reusing of rejected parts which made from the talc filled recycled composites could be feasible.

IV. SUMMARY

The purpose of this paper is to investigate the effect of reprocessing on recycled material, for fulfilling the knowledge gap resulted from lack of such research and promoting the use of recycled plastic, and using recycled plastics has been already proved to be both economic and environmental-friendly.

For the first time, recycled PP was compounding with talc and coupling agent with different concentration for reprocessability study, where virgin and recycled PP were used as comparatives. The materials were subjected to 5 reprocessing cycles from two different routes, extrusion and injection moulding. The experimental part consists of mechanical, rheological and thermal.

The results showed the talc content has similar effect as it did in virgin PP, such as performance enhancing and stabilization. The recycled PP/talc composites were superior than virgin PP in most properties, and were more stable under multiple reprocessing cycles than both virgin and recycled PP. In both reprocessing routes, the performances of the recycled PP/talc composites were similar and constant, it showed the potential of using recycled materials.

There are still something unclear in this moment which displayed in complicated pattern of some properties. It requires further study to understand the competing mechanisms within recycled polymer composites under reprocessing, and how talc stabilize the recycled material.

ACKNOWLEDGEMENTS

This work was financially supported by the Innovation Team of Ningbo Science and Technology Bureau (2011B81006), International Technological Cooperation Project of the Ministry of Science and Technology (2012DFG91920) and Industrial Technology Innovation and Industrialization Project of Ningbo (2014A35001-2).

REFERENCES

[1] Directive 2000/53/EC of the European Parliament and of the Council of 18 September 2000 on end-of life vehicles - Commission Statements. Official Journal L 269, 21/10/2000, P.0034-0043.

[2] Directive 2002/96/EC of the European Parliament and of the Council of January 27th 2003 on waste electrical and electronic equipment (WEEE).

[3] [3] Directive 2012/19/EU of the European Parliament and of the Council of 4 July 2012 on waste electrical and electronic equipment (WEEE).

[4] Scaffaro, R., Botta, L., and Di Benedetto, G., Physical properties of virgin-recycled ABS blends: Effect of post-consumer content and of reprocessing cycles, Euro. Poly. 48 (2012) 637-648.

[5] Makuta, M., Moriguchi, Y., Yasuda, Y. and Sueno, S., Evaluation of the effect of automotive bumper recycling by life-cycle inventory analysis. J. Mater. Cycle. Waste. Manag., 2 (2000) 125-137.

[6] Valenza, A. and La Mantia, F.P., Recycling of polymer waste: Part II-Stress degraded polypropylene, Poly. Degrad. Stab., 20 (1988) 63-73.

[7] Ehrig, R.J., Plastics Recycling: Products and Process, Hanser, New York, 1992.

[8] La Mantia, F.P., Recycling of PVC and Mixed Plastic Waste, ChemTec Publisher, Toronto, 1996.

[9] Samperi, F., Puglisi, C., Alicata, R. and Montaudo, G, Thermal degradation of poly(butylene terephthalate) at the processing temperature, Poly. Degrad. Stab., 83 (2004) 11-17.

[10] Rex, I., Graham, B.A. and Thompson, M.R., Studying single-pass degradation of a high-density polyethylene in an injection molding process., Poly. Degrad. Stab., 90 (2005) 136-146.

[11] Zhou, Y., Rangari, V., Mahfuz, H., Jeelani, S., and Mallick, P.K., Experimental study on thermal and mechanical behavior of polypropylene, talc/polypropylene and polypropylene/clay nanocomposites, Mater. Sci. Eng. A., 402 (2005) 109-117.

[12] Scaffaro, R., La Mantia, F.P., and Dintcheva, N.T., Effect of the additive level and of the processing temperature on the re-building of post-consumer pipes from polyethylene blends, Euro. Poly., 43 (2007) 2947-2955.

[13] [13] Guerrica-Echevarría, G., Eguiazábal, J.I., and Nazábal, J., Effects of reprocessing conditions on the properties of unfilled and talc-filled polypropylene, Poly. Degrad. Stab., 53 (1996) 1-8.

[14] González-González, V.A., Neira-Velázquez, G., and Angulo-Sánchez, J.L., Polypropylene chain scissions and molecular weight changes in multiple extrusion, Poly. Degrad. Stab. 60 (1998) 33-42.

[15] Su, K.H., Lin, J.H., and Lin, C.C., Influence of reprocessing on the mechanical properties and structure of polyamide 6, J. Mater. Proc. Technol. 192-193 (2007) 532-538.

[16] Beg, M.D.H., and Pickering, K.L., Reprocessing of wood fibre reinforced polypropylene composites. Part I: Effects on physical and mechanical properties, Compos. A: Appl. Sci. Manuf., 39 (2008) 1091-1100.

[17] Rahimi, M., Esfahanian, M., and Moradi, M., Effect of reprocessing on shrinkage and mechanical properties of ABS and investigating the proper blend of virgin and recycled ABS in injection molding, J. Mater. Proc. Technol., 214 (2014) 2359-2365

[18] Brennan, L.B., Isaac, D.H. and Arnold, J.C., Recycling of acrylonitrile-butadiene-styrene and high-impact polystyrene from waste computer equipment, J. Appl. Poly. Sci., 86 (2002) 572-578.

[19] Díez-Gutiérrez, S., Rodríguez-Pérez, M.A., De Saja, J.A., and Velasco, J.I., Dynamic mechanical analysis of injection-moulded discs of polypropylene and untreated and silane-treated talc-filled polypropylene composites, Poly., 40 (1999) 5345-5353.

[20] Denac, M., Musil, V., Šmit, I., and Ranogajec, F., Effects of talc and gamma irradiation on mechanical properties and morphology of isotactic polypropylene/talc composites, Poly. Degrad. Stab., 82 (2003) 263-270.

[21] Gafur, A., Nasrin, R., Mina, F., Bhuiyan, A.H., Tamba, Y., and Asano, T., Structures and properties of the compression-molded isotactic-polypropylene/talc composites: Effect of cooling and rolling, Poly. Degrad. Stab., 95 (2010) 1818-1825.

[22] Mnif, N., Massardier, V., Kallel, T., and Elleuch, B., New (PP/EPR)/nano-CaCO3 based formulations in the perspective of polymer recycling. Effect of nanoparticles properties and compatibilizers, Poly. Adv. Technol., 21 (2010) 896-903.

[23] Wang, K., Bahlouli, N., Addiego, F., Ahzi, S., Rémond, Y., Ruch, D., and Muller, R., Effect of talc content on the degradation of re-extruded polypropylene/talc composites, Poly. Degrad. Stab., 98 (2013) 1275-1286.

[24] Boumbimba, R.M., Wang, K., Bahlouli, N., Ahzi, S., Rémond, Y., and Addiego, F., Experimental investigation and micromechanical modeling of high strain rate compressive yield stress of a melt mixing polypropylene organoclay nanocomposite, Mech. Mater., 52 (2012) 58-68.

[25] Bourmaud, A., and Baley, C. Investigations on the recycling of hemp and sisal fibre reinforced polypropylene composites, Poly. Degrad. Stab., 92 (2007) 1034-1045.

[26] Bahlouli, N., Pessey, D., Raveyre, C., Guillet, J., Ahzi, S., Dahoun, A., Hiver, J.M., Recycling effects on the rheological and thermomechanical properties of polypropylene-based composites. Mater. Des., 33 (2012) 451-458.

[27] [27] Personal communication with workshop manager of Ningbo Tokai Minth Automotive Parts Co., Ltd. on 21/05/2012.

[28] Lindahla, P., Robèrta, K.H., Nya, H. and Bromana, G., Strategic sustainability considerations in materials management, J. Clean. Prod. 64 (2014) 98-103.

[29] Plastics - Determination of temperature of deflection under load - Part 2: Plastics and ebonite, published by the International Organization for Standardization (ISO), 2004.

[30] Plastics - Standard atmospheres for conditioning and testing, published by the International Organization for Standardization (ISO), 2008.

[31] Plastics - Determination of tensile properties - Part 2: Test conditions for moulding and extrusion plastics, published by the International Organization for Standardization (ISO), 2012.

[32] Plastics - Determination of flexural properties, published by the International Organization for Standardization (ISO), 2010.

[33] [33] Statistical interpretation of test results - Estimation of the mean -- Confidence interval, published by the International Organization for Standardization (ISO), 1980.

[34] Bernardo, C.A., Cunha, A., and Oliveira, M.J., The recycling of thermoplastics. Prediction of the properties of mixtures of virgin and reprocessed polyolefins, Poly. Eng. Sci. 36 (1996) 511-519.

[35] Canevarolo, S.V., Chain scission distribution function for polypropylene degradation during multiple extrusions, Poly. Degrad. Stab., 70 (2000) 71-76.

[36] GU, F., Hall, P., Miles, N.J., Ding, Q., and Wu, T., Improvement of mechanical properties of recycled plastic blends via optimizing processing parameters using the Taguchi method and principal component analysis, Mater. Des., 62 (2014) 189-198

[37] Wang, K., Addiego, F., Bahlouli, N., Ahzi, S., Rémond, Y., and Toniazzo, V., Impact response of recycled polypropylene-based composites under a wide range of temperature: Effect of filler content and recycling, Compos. Sci. Technol., 95 (2014) 89-99.

Properties of SnAgCu Solders Bearing Al Nanoparticles

L. Zhang, L. Sun, Y.H. Guo, Y. Min
School of Mechanical & Electrical Engineering
Jiangsu Normal University
Xuzhou 221116, China

Abstract—**In order to enhance the properties of SnAgCu lead-free solders, the Al nanoparticles was selected as the additives. The effects of Al nanoparticles on wettability and mechanical properties of solder and solder joints were studied. The results showed that a small amount of Al can enhance the wettability of SnAgCu solders. With the N_2 atmosphere, the wettability of SnAgCu-xAl can be increased obviously, which can be attributed to the N_2 can resist the oxidation of molten solder. Combing different fluxes, the wettability of lead-free solder can represent variation, the suitable flux can improve the wettability of SnAgCu-xAl solders. Moreover, the mechanical property of solders can be improved obviously with the addition of Al nanoparticles.**

Keywords-lead-free solder; wettability; N2 atmosphere; mechanical property

I. INTRODUCTION

SnPb alloys have been used extensively in chip attachment and surface-mount processes in electronic packaging[1], the excellent properties showed by Pb-contained solders were not doubted until it was realized that lead is a hazardous element because of its toxicity[2]. Therefore, considering the health and environmental safety, the investigation of lead-free solders play an important role in electronic industry. Among series of lead-free alloys, SnAgCu solders have been regarded as the most promising solders that can replace traditional SnPb solders in electronic packaging[3,4]. Moreover, due to the requirement of high-density and high reliability in electronic device, the properties of SnAgCu solder should be improved to meet the development trend.

The addition of nanoparticles into solders was proposed as an effective way to improve the properties. Liu[5] found that the addition of grapheme nanosheets can enhance the wettability of SnAgCu solder, and low the CTE values. ZnO nanoparticles can reduce the β-Sn grain size and spacing between Ag_3Sn and Cu_6Sn_5 particles, this obviously improve the yield stress and ultimate tensile strength of SnAgCu solder[6]. Addition of SiC nanoparticles to SnAgBiIn lead-free solder refined its microstructure and improved its electromigration reliability under high current stress[7]. Al nanoparticles can improve the fatigue life of SnAgCu solder joints in QFP device under thermal cycling[8].

In this paper, the wetting balance method was used to analyze the wettability of SnAgCu solder bearing Al nanoparticles at different atmospheres and fluxes. In addition, the mechanical property and fatigue life of SnAgCu-Al solder joints were tested, the content of Al nanoparticles was optimized.

II. EXPERIMENTAL

SnAgCu-based alloys were prepared via mechanically incorporating different contents of Al nanoparticles for about 15 minutes to promote uniform particles distribution. A Rhesca SAT-5100 wetting tester was used in the wetting experiments following IPC-J-STD-003B (solderability tests for printed boards) and Japanese Industry Standard JIS Z 3198-4 (Test methods for lead-free solders-Part 4: Methods for solderability test by a wetting balance method and a contact angle method)[9]. Moreover, different atmospheres (air and N_2) and different fluxes (R and RMA) were selected to be used in the solderability testing. And the SnAgCu-xAl solders were used for soldering of R0805 resister, and STR-1000 tester was used to testing the shear force of SnAgCu-xAl solder joints in R0805 resister.

III. WETTABILITY

Wetting is crucial for soldering because it plays an essential role to ensure the good bonding between the solder materials and the substrate, wettability between the solder and substrate is also an important issue in reliability of electronic products[10]. The wetting time and maximum wetting force in the wetting balance testing can be used to evaluate the effect of Al nanoparticles on the wettability of SnAgCu solders. Figure 1~Figure 4 show the data of wettability with different Al content, temperatures, atmospheres and fluxes. It is found that when the content of Al nanoparticles is less than 0.1%, the wetting force increases and wetting time decreases as Al increases. When the content of Al nanoparticles is 0.1%, the wetting force of solders gives a maximum increase, and wetting time shows a maximum decrease. When the content of Al is over 0.1%, there is an opposite tendency for the wetting data to happen obviously.

© 2015. The authors - Published by Atlantis Press

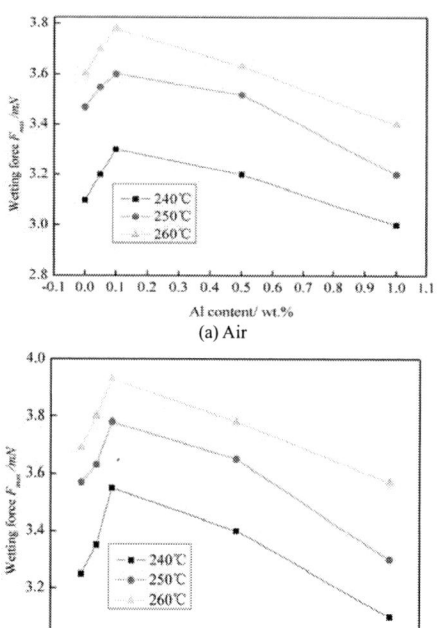

(a) Air

(b) N$_2$

FIGURE I. WETTING FORCE OF SnAgCu-xAl SOLDERS DURING DIFFERENT ATMOSPHERES WITH R FLUX.

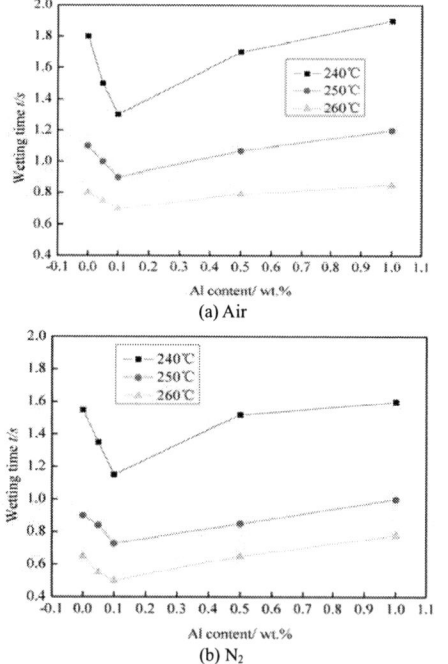

(a) Air

(b) N$_2$

FIGURE II. WETTING TIME OF SnAgCu-xAl SOLDERS DURING DIFFERENT ATMOSPHERES WITH R FLUX.

Moreover, comparing with the wettability of SnAgCu-xAl solders with air atmosphere, with N$_2$ atmosphere, the wetting force increases significantly, meanwhile the wetting time drops, which demonstrates that the N$_2$ atmosphere can improve the solderability of solders, which can be attributed to the N$_2$ can resist the oxidation of molten SnAgCu-xAl solders. And the fluxes transformation (R-RMA) can also improve the wettability of solders, however, the amplitude of enhancement is not higher than N$_2$ atmosphere. This is due to the minimal amount of oxygen in the N$_2$ atmosphere as compared to ambient, during heating of solder, although flux activation is present, some sort of oxides would still form. Prabhu[11] also demonstrated the advance effect of N$_2$ atmosphere in protecting the molten solders. Cheng[12] found the wettability was significantly improved by using the nitrogen protection, was more profound for the SnAgCu lead-free solders.

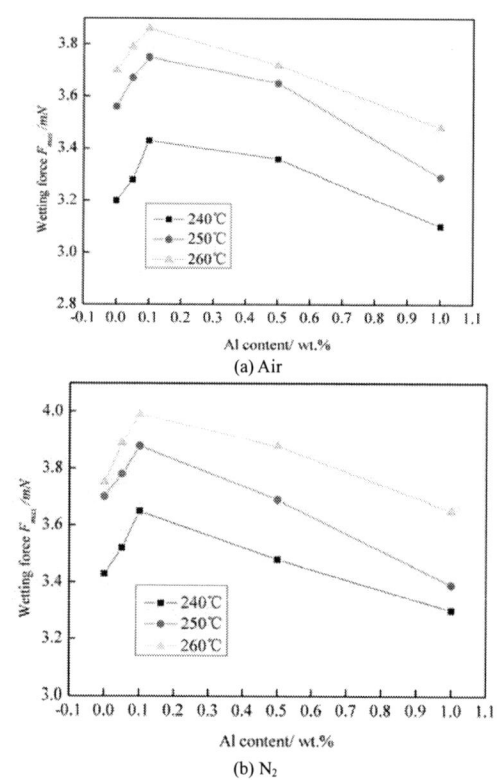

(a) Air

(b) N$_2$

FIGURE III. WETTING FORCE OF SnAgCu-xAl SOLDERS DURING DIFFERENT ATMOSPHERES WITH RMA FLUX.

186

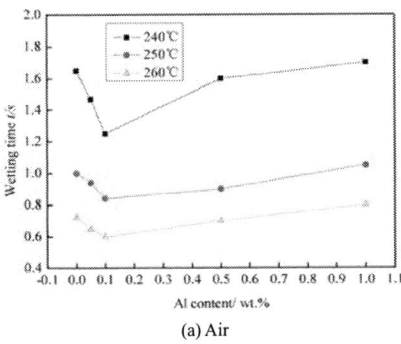

(a) Air

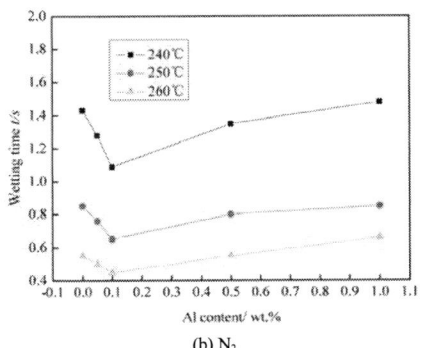

(b) N$_2$

FIGURE IV. WETTING TIME OF SnAgCu-xAl SOLDERS DURING DIFFERENT ATMOSPHERES WITH RMA FLUX.

IV. MECHANICAL PROPERTY

Shear force of SnAgCu-xAl solder joints in R0805 resister was tested, which can be used to evaluate the effect of Al nanoparticles on the mechanical properties of SnAgCu solder joints. Figure 5 show the shear force of SnAgCu solder joints bearing different Al content, the results revealed that with the addition of Al nanoparticles as reinforcement, shear force can be improved significantly. When the Al content is more than 0.1%, the improvement effect is little, this can be attributed to the strength effect of nanoparticles, hindering the movement of dislocations and refinement of microstructure matrix of solder joints. Then it is can be concluded that the optimal content of Al nanoparticles is 0.1%.

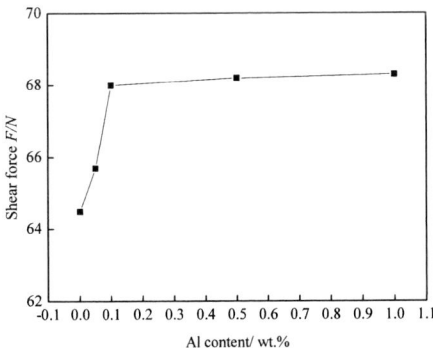

FIGURE V. SHEAR FORCE OF SnAgCu-xAl SOLDER JOINTS IN R0805 RESISTER.

V. CONCLUSIONS

With the addition of Al nanoparticles into SnAgCu alloys, the wettability of solders can be improved obviously, the optimum content of Al is 0.1 wt %, and with the N$_2$ atmosphere, the wettability of SnAgCu-xAl can be increased obviously, which can be attributed to the N$_2$ can resist the oxidation of molten solder. Combing different fluxes, the wettability of lead-free solder can represent variation, the suitable flux can improve the wettability of SnAgCu-xAl solders. And the shear force of SnAgCu-0.1Al is higher than that of SnAgCu solder.

ACKNOWLEDGMENTS

The present work was carried out with the support of the Natural Science Foundation of Jiangsu Province (BK2012144), the Natural Science Foundation of China (51475220) and the Natural Science Foundation of the Higher Education Institutions of Jiangsu Province(12KJB460005).

REFERENCES

[1] L Zhang, K N Tu. Structure and properties of lead-free solders bearing micro and nano particles[J]. Materials Science and Engineering: R: Reports, 2014, 82:1-32.

[2] S M A A Mohd, A M M A Bakri, H Kamarudin, M Bnhussain, H M H Zan, F Somidin. Solderability of Sn-0.7Cu/Si$_3$Ni$_4$ lead-free composite solder on Cu-substrate[J]. Physics Procedia, 2011, 22:299-304.

[3] A A EI-Daly, A E Hammad, G S AI-Ganainy, M Ragab. Properties enhancement of low Ag-content Sn-Ag-Cu lead-free solders containing small amount of Zn[J]. Journal of Alloys and Compounds, 2014, 614:20-28.

[4] L Zhang, J G Han, Y H Guo, L Sun. Properties and microstructures of SnAgCu-xEu alloys for concentrator silicon solar cells solder layer[J]. Solar Energy Materials and Solar Cells, 2014, 130:397-400.

[5] X D Liu, Y D Han, H Y Jing, J Wei, L Y Xu. Effect of graphene nanosheets reinforcement on the performance of Sn-Ag-Cu lead-free solder[J]. Materials Science & Engineering A, 2013, 562:25-32.

[6] A A EI-Daly, T A Elmosalami, W M Desoky, M G EI-Shaarawy, A M Abdraboh. Tensile deformation behavior and melting property of nano-sized ZnO particles reinforced Sn-3.0Ag-0.5Cu lead-free solder[J]. Materials Science & Engineering A, 2014, 618:389-397.

[7] Y Kim, S Nagao, T Sugahara, K Suganuma, M Ueshima, H J Albrecht, K Wilke, J Strogies. Refined of the microstructure of Sn-Ag-Bi-In solder, by addition of SiC nanoparticles, to reduce electromigration damage under high electric current[J]. Journal of Electronic Materials, 2014, Doi:10.1007/s11664-014-3377-x.

[8] L Zhang, X Y Fan, Y H Guo, C W He. Microstructures and fatigue life of SnAgCu solder joints bearing nano-Al particles in QFP devices[J]. Electronic Materials Letters, 2014, 10(3):645-647.

[9] H Wang, S B Xue, F Zhao, W X Chen. Effects of Ga, Al, Ag, and Ce multi-additions on the wetting characteristics of Sn-9Zn lead-free solder[J]. Rare Metals, 2009, 28(6):600-605.

[10] L Zhang, J H Cui, J G Han, Y H Guo, C W He. Microstructures and properties of SnZn-xEr lead-free solders[J]. Journal of Rare Earths, 2012,30(8):790-793.

[11] K N Prabhu, M Varun, Satyanarayan. Effect of purging gas on wetting behavior of Sn-3.5Ag lead-free solder on nickel-coated aluminum substrate[J]. Journal of Materials Engineering and Performance, 2013, 22:723-728.

[12] F J Cheng, F Gao, J Y Zhang, W S Jin, X Xiao. Tensile properties and wettability of SAC0307 and SAC105 low Ag lead-free solder alloys[J]. Journal of Materials Science, 2011, 46:3424-3429.

International Conference on Power Electronics and Energy Engineering (PEEE 2015)

A Review: The Wettability and Oxidation Resistance of Sn-Zn-X Lead-Free Solder Joints

L. Sun

School of Mechanical and Electrical Engineering
Jiangsu Normal University
Xuzhou, China

L. Zhang

School of Mechanical and Electrical Engineering
Jiangsu Normal University
Xuzhou, China

Abstract-With the development of lead-free solders, Sn-Zn solder was attracted increasing attention due to its low melting point, cost saving and excellent mechanical properties. However, there are many problems need to resolved, such as poor wettability and oxidation resistance. In order to overcome these shortcomings, more recent studies on Sn-Zn solder have proposed strategies aimed at obtaining a good wettability and solderability by adding a third or fourth elements, such as In, Ga, P, Bi, Ni, Cr, Ag, Cu, Al and rare earth. This work summarizes the effects of alloying elements to the wettability and oxidation resistance of Sn-Zn lead-free solders.

Keywords-lead-free solders; alloying elements; wettability; oxidation resistance

I.   INTRODUCTION

For a long time, the traditional Sn-Pb solder has been widely used in the electronic devices. Due to the increasing environment and toxic concerns, governments of many countries have established laws to prohibit the use of Pb from electronic packaging [1-3]. Today, studies on the lead-free solder have made breakthrough progress, Sn-Ag, Sn-Cu, Sn-Zn and Sn-Ag-Cu solders have been developed. Among series of lead-free solders, Sn-Zn solder was highly recommended as the promising lead-free solder of the next generation, owing to its favorable melt temperature, low cost and excellent mechanical properties [4,5]. However, because the high activity of Zn, the wettability and oxidation resistance of the Sn-Zn alloys are poor, which is a major obstacle to its application [6]. In order to overcome these drawbacks, two methods are taken. One is to develop a new kind of flux which is suited to lead-free solder and the other is to add alloying elements into the Sn-Zn solder.

In this paper, we summarize the development of Sn-Zn solders, and analyze the effects of alloying elements on wettability and oxidation resistance.

II.   WETTABILITY

Wettability of solder can be defined as the ability of the molten solder to spread over on a substrate during the reflow process [7]. The better wettability is to evaluate the performance of important indicator. For Sn-Pb solder, due to the existence of Pb, the solder alloy owns better wettability. But for lead-free solders, the wettability may be dropped obviously due to the replacement of Pb. Hence, adding alloying elements is an effective method to improve the wettability of Sn-Zn solder alloy.

The addition of alloying elements can be classified as follows:

(i) Add surface-active elements, which can easily accumulate at the solder interface in the molten state.

(ii) Add a large amount of low-active elements to reduce the activity of Zn.

(iii) Add more reactive than Zn elements, using oxide film that is preferentially formed on the surface of Zn, which can improve the wettability.

The addition of 5%~10% In into Sn-9Zn solder can improve the wettability [8]. The researcher has attributed to solder flux. Yu et al. [9] has also confirmed it. Fima et al. [10] has reported the effect of 0.5%~1.5% In on Sn-Zn solder under Ar protective atmosphere and found that the Ar atmosphere can better decrease the surface tension and improve the wettability. Ga was incorporated into Sn-9Zn solder, which can significantly change the wettability and optimal content was about 0.5% [11]. Due to the surface-active, Ga would accumulate at solder interface in the melting state then the surface tension of the liquid solder was decreased [12]. The similar phenomenon can be found for P addition into Sn-Zn solders [13]. Zhou et al. [14] confirmed that the addition of Bi to the Sn-Zn solder decreases surface tension of the liquid solder, and suppose it is effective to enhance the wettability. The effect of Ni addition to Sn-9Zn lead-free solder on the wettability of aluminum and copper base metal was studied by Huang et al. [15]. It is found that the wettability of Sn-9Zn-xNi solders on Al substrate was much better than on copper substrate. In addition, with increasing of Ni content, the wettability on copper substrate was slightly enhanced but became worse on Al substrate. Due to the high mutual solubility Al and Zn, Al atoms can dissolve from the substrate into Sn-Zn based solders during reflow, to form Al-Zn solid solution during solidification. Chen et al. [16] studied the effect of Cr on the wettability of Sn-9Zn solder compared with Sn-37Pb and Sn-3.5Ag-0.75Cu. It is indicated that the wettability of Sn-9Zn-xCr solder is poorer than both solders. However, adding Cr element can slightly improve the wettability of Sn-9Zn solder alloy. The addition of trace amount of Ag into Sn-9Zn solder can't obviously change the wettability [17]. With the 0.5% and 1.0% addition, the wetting angle is slightly lower. With the Ag addition concentration increasing, the wetting angle increases. The reason could attributed to Sn-Zn-xAg solder has better oxidation resistance

© 2015. The authors - Published by Atlantis Press

than Sn-Zn solder, but when the Ag content exceed a certain, the liquidus temperature increased [18], and the fluidity of the solder decreases with the same soldering temperature. So the wetting angle increases slightly. Yu et al. [19] found that the addition of Cu can alter the wettability of Sn-9Zn solder. Results show that the wetting angle of Sn-9Zn is 120° while that of Sn-9Zn-10Cu is about 54°. Fima et al. [20] has also investigated the effect of Cu on Sn-8.8Zn solder, the soldering temperature was 250°C and ALU33®flux was used. It is found that the wetting angle of Sn-8.8Zn is about 42°, with 0.5%Cu addition, the wetting angle was decreased to about 32°. Some researchers have proposed the ideas that through adding more active Al to inhibit oxidation of Zn then improve the wettability of Sn-Zn lead-free solder. But according to the existing literature, we find that researchers have different views on improvement the wettability of Sn-Zn solder with the addition of Al. Chen et al. [21] has found that adding Al may deteriorate the wettability of the solder. Huang et al. [22] proposed that adding Al can no signification change the wettability of Sn-Zn solder on copper substrate. However, Chen et al. studied the effect of Al under N2 protective atmosphere and ZnCl2 + NH4Cl flux as well as no-cleaning flux were used, the wettability of solder was extremely improved [23]. Comprehensive results of the study, we can find that only a small amount of Al can show this enhancement effect. When the addition is excessive, the negative effect can be found obviously.

Trace amount of RE elements (Ce and La), can reduce the surface tension of Sn-9Zn. When the RE content is 0.05 and 0.1%, the wettability is greatly improved with the RA flux [24]. The addition of Er into Sn-Zn was investigated by Zhang et al. [25]. When the addition of Er was 0.08%, the spreading area gave an 19.1% increase (Fig.1). Ce [26], La [27], Nd [28] and Pr [29] can also improve the wettability of Sn-Zn solders because of the lower surface tension caused by rare earth. Adding a proper amount of rare earth Pr and Nd to the Sn-9Zn-0.5Ga solders were studies by Xue et al. [30, 31]. With the addition of 0.08%Pr or 0.1%Nd can significantly improve the wettability of the solder. It is concluded that the rare earth elements can improve the wettability of Sn-Zn solder, only the optimal contents are different for different elements.

III. OXIDATION RESISTANCE

The poor oxidation resistance of the Sn-Zn solder is one of the most factors to hinder its development. Due to the existence of mass Zn element, the Sn-Zn solder is easily oxidized (Table 1), worsened the wettability of solder during soldering. Therefore, it is no doubt that the key issue is improving oxidation resistance of Sn-Zn solder.

TABLE I. THE PARTIAL PRESSURE'S DATA OF COMMON SOLDER'S OXIDE[32].

	Bi_2O_3	In_2O_3	SnO_2	PbO	ZnO	CuO
298K	9.0×10^{-39}	5.2×10^{-39}	6.9×10^{-92}	7.0×10^{-67}	2.3×10^{-112}	4.7×10^{-46}
400K	1.4×10^{-41}	4.4×10^{-50}	6.5×10^{-66}	2.2×10^{-47}	2.3×10^{-82}	6.5×10^{-32}
500K	1.5×10^{-31}	2.1×10^{-38}	8.9×10^{-57}	5.9×10^{-38}	4.6×10^{-63}	6.8×10^{-24}
600K	8.6×10^{-35}	1.4×10^{-35}	1.5×10^{-40}	1.8×10^{-28}	5.4×10^{-51}	2.0×10^{-18}

Oxidation resistance of Sn-9Zn-xGa solder was tested by means of thermal gravimetric analysis (TGA) method by Chen

et al. [11]. Fig. 2 shows that the oxidation resistance is enhanced with the addition of 0.5%Ga. The addition of Ag can improve the oxidation of the solder, and Sn-9Zn-0.3Ag solder exhibits much lower mass grain than Sn-9Zn in the liquid state at 245°C[33]. Lee et al. [34] reported the oxidation behavior of Sn-9Zn-xAg and Sn-9Zn-xCu solders during 85°C/85% relative humidity (RH) exposure. It is found that the addition of Ag or Cu can effectively improve the oxidation resistance, as compared with the Sn-Zn and Sn-Zn-Bi solders. Chang et al. [35] studied the effect of Ag and In addition on the oxidation resistance of Sn-9Zn solders and found that both Sn-9Zn-0.5Ag and Sn-9Zn-0.5Ag-1In solders have a higher oxidation resistance than that of Sn-9Zn solder. Chen et al. [16] demonstrated that the oxidation resistance of Sn-Zn solders can be improved with the addition of Cr, which can be attributed to segregation Cr in the sub-surface layer of Sn-Zn-xCr solder to prevent further oxidation. Adding Al element can obviously improve the oxidation resistance, however, an excessive amount of Al addition will form thick film and deteriorate the wettability of Sn-Zn solder [36].

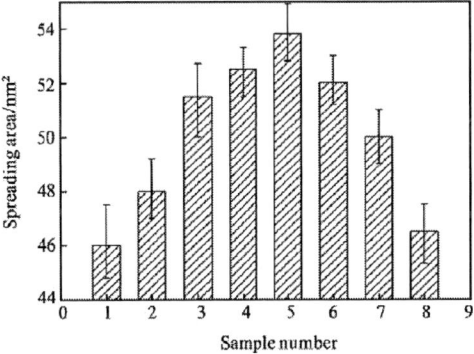

FIGURE I. SPREADING AREA OF SN-9ZN-XER [25].

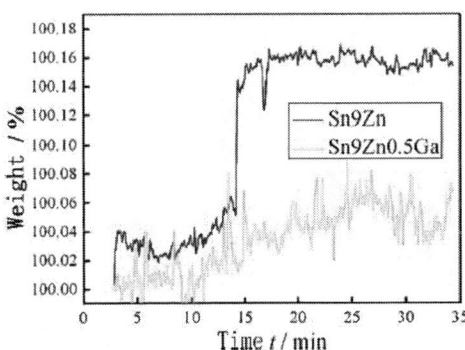

FIGURE II. TGA RESULTS OF SN-9ZN-0.5GA SOLDER COMPARED WITH THAT OF SN-9ZN SOLDER AT 235°C [11].

IV. CONCLUSIONS

To conclude this review, we can find that the addition of surface-active elements (Ga, P, Bi, Ni, and Cr) can improve the wettability of Sn-Zn solder. But when the additives are more than optimal content, the wettability of solder becomes stable, so it is difficult to improve the wettability greatly. The addition of In element can decrease the melting point of the Sn-Zn-In solder and enhances the wettability of the lead-free solder, but the In is very expensive, the addition of In can increase the cost of lead-free solders.Adding Ag can enhance the wettability. The addition of Al element can effectively enhance the oxidation resistance, but just only a small amount of Al addition can improve the wettability. Moreover, the addition of rare earth elements show the most obvious improvement of wettability, however, rare earth elements may worsen the oxidation resistance, because rare earth is more reactive than Zn.

In a word, adding alloy elements are difficult to significantly improve the wettability and oxidation resistance of Sn-Zn solders. However, we find a great enhancement effect on the wettability of the flux, so the development of flux for Sn-Zn solder may be the main aspect of future research.

ACKNOWLEDGEMENTS

This work was supported by the National Natural Science Foundation of China (No. 51475220), the Natural Science Foundation of Jiangsu Province (BK2012144) and the Natural Science Foundation of the Higher Education Institutions of Jiangsu Province (12KJB460005).

REFERENCES

[1] L. Zhang, K.N. Tu, Structure and properties of lead-free solders bearing micro and nanoparties, Mater. Sci. Eng., R. 82 (2014) 1-32.

[2] A. Fawzy, S.A. Fayek, M. Sobhy, E. Nassr, M.M. Mousa, G. Saad,Tensile creep characteristics of Sn-3.5Ag-0.5Cu (SAC355) solder reinforced with nano-metric ZnO particles,Mater. Sci. Eng., A. 603 (2014) 1-10.

[3] L. Zhang, J.G. Han, Y.H. Guo, L. Sun, Properties and microstructures of SnAgCu-xEu alloys for concentrator silicon solar cells solder layer, Sol.Energy Mater. Sol.Cells.130 (2014) 397-400.

[4] L.R. Garcia, L.C. Peixoto, W.R. Osório, A. Garcia, Globular-to-needle Zn-rich phase transition during transient solidification of a eutectic Sn-9%Zn solder alloy, Mater.Lett.63 (2009)1314-1316.

[5] W.R. Osório, L.C. Peixoto, L.R. Garcia, N. Mangelinck-Noël, A. Garcia, Microstructure and mechanical properties of Sn-Bi, Sn-Ag and Sn-Zn lead-free solder alloys, J. Alloys Compd.572 (2013) 97-106.

[6] J.X. Liang, T.B. Luo, A.M. Hu, M. Li, Formation and growth of interfacial intermetallic layers of Sn-8Zn-3Bi-0.3Cr on Cu, Ni and Ni-W substrates,Microelectron. Reliab.54 (2014) 245-251.

[7] S.K. Ghosh, A.S.M.A. Haseeb, A. Afifi, Effects of metallic nanoparticle doped flux on interfacial intermetallic compounds between Sn-3.0Ag-0.5Cu and copper substrate, 15thElectronics Packaging Technology Conference.(2013).

[8] M. McCormack, S. Jin, H.S. Chen, D.A. Machusak, New lead-free, Sn-Zn-In solder alloys, J. Electron.Mater.23 (1994) 687-690.

[9] S.P. Yu, C.L. Liao, M.H. Hon, The effects of flux on the wetting characteristics of near-eutectic Sn-Zn-In solder on Cu substrate, J. Mater. Sci. 35 (2000) 4217-4224.

[10] P. Fima, T. Gancarz, J.Pstruś, A. Sypień, Wetting of Sn-Zn-xIn (x = 0.5, 1.0, 1.5 wt%) Alloys on Cu and Ni Substrates, J. Mater.Eng. Perform.21 (2012) 595-598.

[11] W.X. Chen, S.B. Xue, H. Wang, Wetting properties and interfacial microstructures of Sn-Zn-xGa solders on Cu substrate, Mater. Des.31 (2010) 2196-2200.

[12] D.X. Luo, S.B. Xue, Z.Q. Li, Effects of Ga addition on microstructure and properties of Sn-0.5Ag-0.7Cu solder, J. Mater. Sci.-Mater.Electron.25 (2014) 3566-3571.

[13] H.Z. Huang, X.Q. Wei, D.Q. Tan, L. Zhou, Effects of phosphorus addition on the properties of Sn-9Zn lead-free solder alloy, Int. J. Min. MetMater.20 (2013) 563-567.

[14] J. Zhou, Y.S. Sun, F. Xue, Properties of low melting point Sn-Zn-Bi solders, J. Alloys Compd.397 (2005) 260-264.

[15] M.L. Huang, N. Kang, Q. Zhou, Y.Z. Huang, Effect of Ni Content on Mechanical Properties and Corrosion Behavior of Al/Sn-9Zn-xNi/Cu Joints, J. Mater. Sci. Technol.28 (2012) 844-852.

[16] X. Chen, A.M. Hu, M. Li, D.L. Mao, Study on the properties of Sn-9Zn-xCr lead-freesolder, J. AlloysCompd.460 (2008) 478-484.

[17] K. Berent, P. Fima, T. Ganacarz, J.Pstruś, Wetting and Microstructure Evolution of the Sn-Zn-Ag/Cu Interface, J. Mater.Eng. Perform. 23(2014) 1630-1633.

[18] K.L. Lin, C.L. Shih, Microstructure and Thermal Behavior of Sn-Zn-Ag Solders, J. Electron. Mater.32 (2003) 1496-1500.

[19] D.Q. Yu, H.P. Xie, L. Wang, Investigation of interfacial microstructure and wetting property of newly developed Sn-Zn-Cu solders with Cu substrate, J. AlloysCompd.385 (2004) 119-125.

[20] P. Fima, J.Pstruś, T. Gancarz, Wetting and Interfacial Chemistry of SnZnCu Alloys with Cu and Al Substrates, J. Mater.Eng. Perform.23 (2014) 1530-1535.

[21] X. Chen, M. Li, X.X. Ren, A.M. Hu, D.L. Mao, Effect of Small Additions of Alloying Elements on the Properties of Sn-Zn Eutectic Alloy, J. Electron. Mater.35 (2006) 1734-1739.

[22] M.L. Huang, X.L. Hou, N. Kang, Y.C. Yang, Microstructure and interfacial reaction of Sn-Zn-x(Al, Ag) near-eutectic solders on Al and Cu substrates, J. Mater.Sci. - Mater.Electron.25 (2014)2311-2319.

[23] L. Zhang, S.B. Xue, L.L. Gao, Z. Sheng, H. Ye, Z.X. Xiao, G. Zeng, Y. Chen, S.L. Yu, Development of Sn-Zn lead-free solders bearing alloying elements, J. Mater. Sci. - Mater.Electron.21 (2010) 1-15.

[24] C.M.L. Wu, C.M.T. Law, D.Q. Yu, L. Wang,The Wettability and Microstructure of Sn-Zn-RE Alloys, J. Electron.Mater.32 (2003) 63-69.

[25] L. Zhang, J.H. Cui, J.G. Han, Y.H. Guo, C.W. He, Microstructures and properties of SnZn-xEr lead-free solders, J. Rare.Earth.30 (2012) 790-793.

[26] Y.H. Hu, S.B. Xue,W.X. Chen, H. Wang, Analysis of microstructure and properties of Sn-9Zn-xCe solder, Trans. China Weld.Inst.31 (2010) 77-80.

[27] Z.A. Li, A. Mikula, Z.Y. Qiao, Surface Properties of the Sn-9Zn Alloy with the Trace Addition of Lanthanum,Monatsh. Chem.136 (2005) 1835-1840.

[28] Y.H. Hu, S.B. Xue, H. Wang, H. Ye, Z.X. Xiao, L.L. Gao, Effects of rare earth element Nd on the solderability and microstructure of Sn-Zn lead-free solder, J. Mater. Sci. - Mater.Electron.22 (2011) 481-487.

[29] Z.X. Xiao, S.B. Xue, Y.H. Hu, H. Ye, L.L.Gao, H. Wang, Properties and microstructure of Sn-9Zn lead-free solder alloy bearing Pr, J. Mater. Sci. - Mater.Electron.22 (2011)659-665.

[30] H. Ye, S.B. Xue, J.D. Luo, Y. Li, Properties and interfacial microstructure of Sn-Zn-Ga solder joint with rare earth Pr addition, Mater. Des.46 (2013) 816-823.

[31] D.X. Luo, P. Xue, S.B. Xue, Y.H. Hu, Microstructures and properties of Sn-Zn-Ga lead-free solder with rare earth Nd addition, Trans. China Weld. Inst.34 (2013) 57-60.

[32] M. Abtew, G. Selvaduray, Lead-free Solders in Microelectronics, Mater. Sci. Eng., R.27 (2000) 95-141.

[33] W.X. Chen, S.B. Xue, H. Wang, Y.H. Hu, Effects of Ag on Properties of Sn-9Zn Lead-Free Solder, Rare Metal. Mat. Eng.39 (2010) 1702-1706.

[34] J.E. Lee, K.S. Kim, M. Inoue, J.X. Jiang, K. Suganuma, Effects of Ag and Cu addition on microstructural properties and oxidation resistance of Sn-Zn eutectic alloy, J. AlloysCompd.454 (2008) 310-320.

[35] T.C. Chang, J.W. Wang, M.C. Wang, M.H. Hon,Solderability of Sn-9Zn-0.5Ag-1In lead-free solder on Cu substrate: Part 1. Thermal properties, microstructure, corrosion and oxidation resistance,J. AlloysCompd.422 (2006) 239-243.

[36] H. Wang, S.B. Xue, W.X. Chen, Effects of Al, Ga, and Ag on the surface properties and wetting reactions of Sn-9Zn-X solders, China Weld.18 (2009) 57-61.

International Conference on Power Electronics and Energy Engineering (PEEE 2015)

Tensile Behavior of High Temperature Cu-Cr-Zr Alloy

X.W. Zhang

School of Metallurgy and Engineering
Xi'an University of Architecture and Technology
Xi'an, China

Q.J. Wang

School of Metallurgy and Engineering
Xi'an University of Architecture and Technology
Xi'an, China

X. Zhou

School of Metallurgy and Engineering
Xi'an University of Architecture and Technology
Xi'an, China

B. Liang

School of Metallurgy and Engineering
Xi'an University of Architecture and Technology
Xi'an, China

Abstract-The tensile behavior of Cu-Cr-Zr alloy was investigated at different temperatures. The material investigated was heat treated to the following condition: solutionizing 980°C-40 min then water quenching, aging 460°C-4 h. The test temperature was ranged from room temperature to 600°C at intervals of 100°C. Curves describing the tensile properties were obtained. The Vickers hardness, microstructure and fracture surface of the tested alloys were also studied. The results reveal that the tensile strength of Cu-Cr-Zr alloys decreases with the increase of temperatures. Additionally, due to fully recrystallization, a combination of a high tensile strength with a fairly good plasticity can be achieved at 200°C. The suitable working temperature of Cu-Cr-Zr alloy ranges form room temperature to 300°C.The results also indicate that the mechanism of Cu-Cr-Zr alloy fracture is the nucleation, growth, extension and join of microvoid caused by second phase particle.

Keywords-Cu-Cr-Zr alloy; tensile properties; high temperature; fracture pattern

I. INTRODUCTION

Cu-Cr-Zr alloy, as one of the most promising functional materials, has been extensively researched in recent years [1-4]. With the combination of excellent electrical and thermal conductivity, good thermal stability at high temperature, relatively high strength, good fatigue resistance, outstanding resistance to corrosion and ease of fabrication [5-11] it has been widely used in the applications such as moulds for continuous caster, heat transfer materials, diverter target materials and other important industrial parts [12-15]. Under high temperature and alternating stress, fracture that caused by high temperature is the main factor leading to the final material failure. Therefore, to expand the application extension, there is an urgent need to study tensile fracture characteristics and mechanism of Cu-Cr-Zr alloy from room temperature to high temperature.

At home and abroad, the research of Cu-Cr-Zr alloy mainly concentrated in the preparation method, heat treatment process, and mechanical properties at room temperature, precipitates and microstructure during the last two decades [16-18]. However, there is still dearth of research on tensile behavior of

Cu-Cr-Zr alloy from room temperature to high temperature. The purpose of this study is to develop a better understanding of high temperature tensile properties by investigating the changes of microstructure and the observation of surface morphologies after tension to shed some light on the mechanism of Cu-Cr-Zr alloy fracture.

II. EXPERIMENTAL

The material used in the present investigations was a precipitation hardened Cu-Cr-Zr alloy. The alloy was supplied with a composition of Cu−0.84% Cr−0.15% Zr. The chemical composition of Cu-Cr-Zr alloy in wt% is listed in Table 1. The material received a solution annealing followed by an aging according to: 40 min at 980°C, water quenched, 4 hours at 460°C, water quenched. The microstructure of heat treated Cu-Cr-Zr alloy is shown in Fig. 1. Apparently, the grains are near equiaxial after solution and aging treatment. There still remains a large amount of Cr precipitates both at grain boundaries and inside grains.

Tensile specimens were machined with the gauge length of 20 mm and an oblong cross-section of 2×3mm2 as shown in Fig. 2, according to Standard GB/T228-2002 "metallic materials tensile testing at ambient temperature". The tensile tests were conducted on an Instron-8801 machine combined with a resistive furnace. The test temperature was ranged from room temperature to 600°C at intervals of 100°C, and the strain rate was of 4.3 × 10-4s -. The loading speed was 0.5 mm/min; the maximum load was 10 KN.

After the tension tests, the Vickers hardness of the tested specimens was determined, the microstructure was examined by an 'Olympus PMG3' optical metallographic microscope. By the help of SEM, surface morphologies after fatigue were investigated as well.

TABLE I. CHEMICAL COMPOSITIONS OF THE CU-CR-ZR ALLOY.

Chemical element	Cu	Cr	Zr	Fe
Mass fraction [%]	Balance	0.83-0.84	0.15-0.20	0.03

© 2015. The authors - Published by Atlantis Press

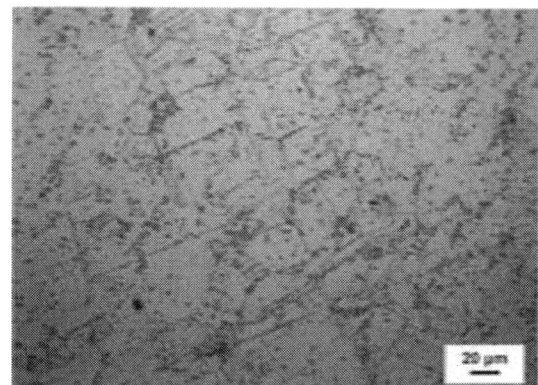

FIGURE I. INITIAL MICROSTRUCTURE OF CU-CR-ZR ALLOY.

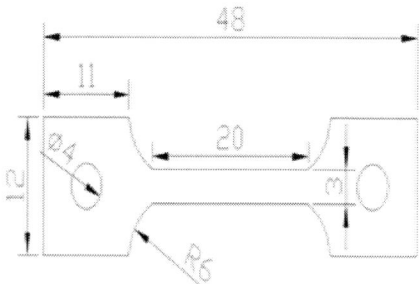

FIGURE II. DIMENSION OF TENSILE SPECIMEN.

III. RESULTS

A. Effect of Temperature on the Tensile Properties of Cu-Cr-Zr Alloy

Tensile tests at different temperatures were performed to figure out the effect of temperature in the mechanical performance of the Cu-Cr-Zr alloy. Fig. 3 shows the tensile stress–strain curves obtained from the tension tests at room temperature, 200°C, 300°C, 400°C, 500°C, and 600°C at a fixed strain rate. Data for ultimate tensile strength (UTS) and elongation to failure are summarized in Fig.4. As can be seen from Fig. 3 and 4, with the increasing of temperature, the tensile strength decreases gradually from room temperature to 600°C. Moreover, it can be found an interesting phenomenon about the evolution of elongation-to-failure: it first decreases from 18% to 14% continuously from 200°C to 400°C, but it dramatically increases with the reaching of 500°C. And it is up to 27% at 600°C. In addition, it is found in Fig. 3 at high temperature the uniform deformation stage represents more than half, almost 50% of the deformation. As a whole, with the increasing of temperature, uniform deformation also increases. It indicates that the plasticity of Cu-Cr-Zr alloy is good. The value of ultimate tensile strength is 308±15MPa at 300°C. It would ensure that Cu-Cr-Zr alloy has enough strength to be applied under the high temperature condition. Compared to the two curves in Fig. 4, the alloy has good strength and plasticity at the same time at room temperature to 300°C. So the suitable working temperature of Cu-Cr-Zr alloy ranges form room temperature to 300°C. All the results above reveal that temperature has remarkable effect on tensile strength and elongation.

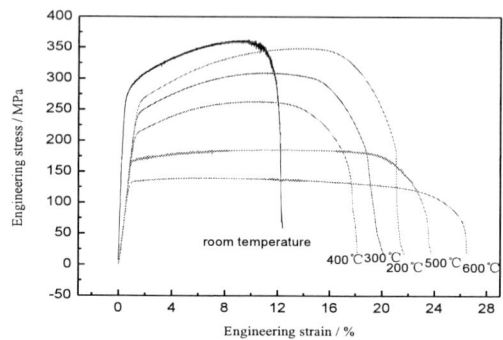

FIGURE III. ENGINEERING STRESS–STRAIN CURVES OF CU-CR-ZR ALLOY AT ROOM TEMPERATURE, 200°C, 300°C, 400°C, 500°C, AND 600°.

Hardness tests were performed to figure out of the effect of the high temperature in the mechanical performance of the Cu-Cr-Zr alloy. The results are shown in Fig. 5. As shown in Fig. 5, it can be seen that the temperature has great effect on the hardness of the alloy. The hardness of the alloy increases with the increase of temperature, and reaching the peak of 143.4HV at 200°C. Then with the rising of temperature, the hardness decreases by degrees. The hardness of the alloy at 300°C and 400°C are very close. The hardness decreases rapidly at 600°C. In summary, the temperature does have a big influence on the hardness of Cu-Cr-Zr alloys.

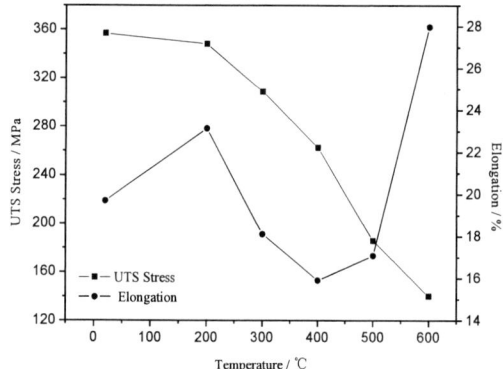

FIGURE IV. THE TENSILE STRENGTH AND THE ELONGATION OF CU-CR-ZR ALLOYS AT DIFFERENT TEMPERATURES.

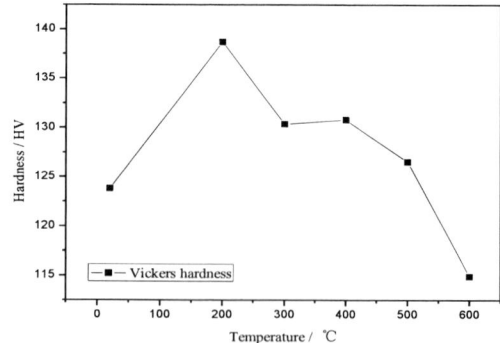

FIGURE V. THE HARDNESS OF CU-CR-ZR ALLOYS AT DIFFERENT TEMPERATURES.

B. Effect of Temperature on the Microstructure of Cu-Cr-Zr Alloy

The microstructures near tensile fracture of Cu-Cr-Zr alloys under different temperatures are shown in Fig. 6. It can be seen that at room temperature, the twin crystal occurred. Flake twin can be clearly observed in the matrix. When the temperature is 200°C, material is fully recrystallized and organization uniformity is good. The grain size is relatively small. So the tensile strength of Cu-Cr-Zr alloys is lower, while elongation is increased at 200°C. When the temperature reaches to 300°C, part of the material is recrystallized, the uniformity of the organization is poor. So the elongation is lower than 200°C. From Fig. 6(d) we can see, complete recrystallization happened with an equiaxed grain structure. The grain size is bigger than 200°C. Therefore, the tensile strength at 600°C is lower but the elongation is increased.

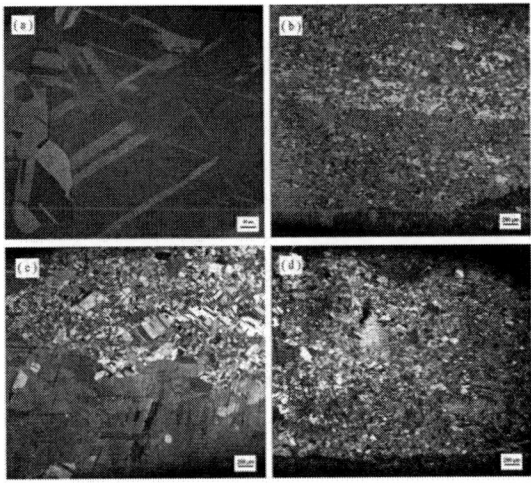

FIGURE VI. THE MICROSTRUCTURES NEAR TENSILE FRACTURE OF CU-CR-ZR ALLOYS UNDER DIFFERENT TEMPERATURES, (A) ROOM TEMPERATURE, (B) 200°C, (C) 300°C, (D) 600°C.

C. Effect of Temperature on the Fracture Appearance of Cu-Cr-Zr Alloy

The fracture surfaces of Cu-Cr-Zr alloys are shown in Fig. 7, respectively, as a function of temperature. Compared with the fracture appearance of Cu-Cr-Zr alloy at different temperatures, the mechanism of fracture can be figure out. The tensile is ductile fracture. The dimples appeared are the characteristic of the typical microvoid coalescence fracture. The fractures are mainly caused by broken particle, a large number of microvoids in different sizes and shapes, dimples and tearing ridges, which are taken as ductile fracture features. Under the condition of high temperature dimples vary in size and shape, accompanied by a small amount of cracks. While under the condition of low temperature, dimples are relatively uniform, showing Cu-Cr-Zr alloy has better plasticity. The mechanism of Cu-Cr-Zr alloy fracture is the nucleation, growth, extension and join of microvoid caused by second phase particle.

Can be seen from Fig. 7, the microstructures of fracture are mainly about dimples. With the increase of temperature, dimples grow up along the tensile direction, microvoids deepened, new small dimples occurred inside of the original dimples. From room temperature to 300°C, dimples are small and shallow. When the temperature is above 300°C, dimples become relatively bulky and deep, with a few of cracks.

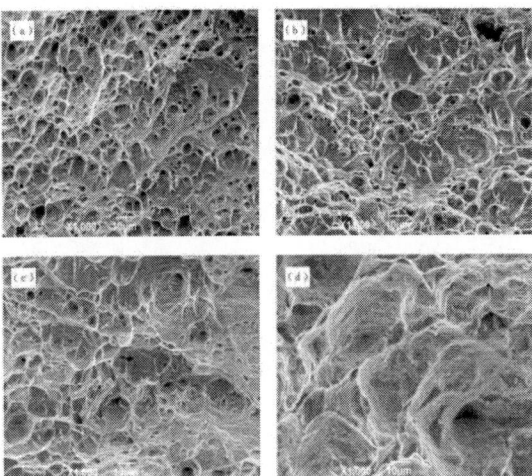

FIGURE VII. THE FRACTURE APPEARANCES OF CU-CR-ZR ALLOYS UNDER DIFFERENT TEMPERATURES, (A)ROOM TEMPERATURE, (B) 200°C, (C) 300°C, (D) 600°C.

IV. CONCLUSIONS

The effect of temperature on the tensile properties, mechanical performance, microstructures and fracture appearances of Cu-Cr-Zr alloys was investigated at ambient and high temperatures. From the analysis of the results, the following conclusions can be drawn:

1. Tensile strength of Cu-Cr-Zr alloys decreases with the increase of temperature. The alloy results in good material strength and hardness, while its ductility remains sufficient. So the suitable working temperature of Cu-Cr-Zr alloy ranges form room temperature to 300°C.

2. The hardness of the alloy increases from room temperature to 200°C. Then with the rising of temperature, the hardness decreases by degrees.

3. Recrystallization happens in Cu-Cr-Zr alloys with increase of temperatures, which leads to the tensile strength decreasing.

4. A typical ductile fracture mode with dimples is observed on the high temperature tensile fracture surface. The mechanism of Cu-Cr-Zr alloy fracture is the nucleation, growth, extension and join of microvoid caused by second phase particle.

ACKNOWLEDGEMENTS

This work was supported by the National Natural Science Foundation of China (Grant No. 51104113) and Research Program for Serving Local Funded by Shaanxi Provincial Education Department (Program No. 14JF013).

REFERENCES

[1] Liu J B, Zhang L, Dong A P, Effects of Cr and Zr additions on the microstructure and properties of Cu-6wt%Ag alloys, Mater. Sci. Eng. A 532 (2012) 331-338.

[2] N. Takata, Y. Ohtake, K. Kita, K. Kitagawa, N. Tsuji, Increasing the ductility of ultrafine-grained copper alloy by introducing fine precipitates, Scr. Mater. 60 (2009) 590-593.

[3] H. Fuxiang, M. Jusheng, N. Honglong, G. Zhiting, L. Chao, G. Shumei, Y. Xuetao, W. Tao, L. Hong, L. Huafen, Analysis of phases in a Cu-Cr-Zr alloy, Scr. Mater. 48 (2003) 97-102.

[4] H.T. Zhou, J.W. Zhong, X. Zhou, Z.K. Zhao, Q.B. Li, Microstructure and properties of Cu-1.0Cr-0.2Zr-0.03Fe alloy Mater. Sci. Eng. A 498 (2008) 225-230.

[5] J.W. Davis, G.M. Kalinin, Material properties and design requirements for copper alloys used in ITER, Nucl. Mater. 258-263 (Part 1) (1998) 323-328.

[6] S.A. Fabritsiev, S.J. Zinkle, B.N. Singh, Evaluation of copper alloys for fusion reactor divertor and first wall components, Nucl. Mater. 233-237 (Part 1) (1996) 127-137.

[7] M. Kulczyk, B. Zysk, M. Lewandowska, K.J. Kurzydłowski, Grain refinement in CuCrZr by SPD processing, Status Solidi. 207 (2010) 1136-1138.

[8] K.V. León, M.A. Munoz-Morris, Optimisation of strength and ductility of Cu-Cr-Zr by combining severe plastic deformation and precipitation, Mater. Sci. Eng. A 536 (2012) 181-189.

[9] X.F. Li, A.P. Dong, L.T. Wang, Z. Yu, L. Meng, Thermal stability of heavily drawn Cu-0.4wt.%Cr-0.12 wt.%Zr-0.02 wt.%Si-0.05 wt.%Mg, J. Alloys Comp. 509 (2011) 4092-4097.

[10] X.F. Li, A.P. Dong, L.T. Wang, Z. Yu, L. Meng, The stored energy in processed Cu-0.4wt.%Cr-0.12 wt.%Zr-0.02 wt.%Si-0.05 wt.%Mg, J. Alloys Comp. 509 (2011) 4670-4675.

[11] Z. Ding, S. Jia, P. Zhao, M. Deng, K. Song, Suitability of high-pressure xenon as scintillator for gamma ray spectroscopy, Mater. Sci. Eng. A 570 (2013) 87-91.

[12] I.J.C. Team, The impact of materials selection on the design of the International Thermonuclear Experimental Reactor (ITER), Nucl. Mater. 212-215 (Part 1) (1994) 3-10.

[13] I.S. Batra, G.K. Dey, U.D. Kulkarni, S. Banerjee, Microstructure and properties of a Cu-Cr-Zr alloy, J. Nucl. Mater. 299 (2001) 91-100.

[14] S. Higashijima, S. Sakurai, A. Sakasai, Heat treatment of CuCrZr cooling tubes for JT-60SA monoblock-type divertor targets, J. Nucl. Mater. 417 (2011) 912-915.

[15] V.R. Barabash, G.M. Kalinin, S.A. Fabritsiev, S.J. Zinkle, Specification of CuCrZr alloy properties after various thermo-mechanical treatments and design allowables including neutron irradiation effects, J. Nucl. Mater. 417 (2011) 904-907.

[16] Xia C, Jia Y, Zhang W, Study of deformation and aging behaviors of a hot rolled-quenched Cu-Cr-Zr-Mg-Si alloy during thermomechanical treatments, Materials & Design, 39 (2012) 404-409.

[17] Yu F X, Cheng J Y, Ao X W, Aging characteristic of Cu-0.6Cr-0.15Zr-0.05Mg-0.02Si alloy containing trace rare earth yttrium, Rare Metals, 30 (2011) 539-543.

[18] Tenwick M J, Davies H A, Enhanced strength in high conductivity copper alloys, Mater. Sci. Eng, 97 (1988) 543-546.

Polypyrrole/Sisal Fiber Composites for Energy Storage

H.D. Mo, C. Yang, C. Wei, F.A. Zhang, S.R. Lu, Z.Q. Wang, X.X. Huang

State Key Laboratory Breeding Base of Nonferrous Metals & Specific Materials Processing; Key Laboratory of New Processing Technology for Nonferrous Metals & Materials
Guilin University of Technology
Guilin, China

L.M. Zang

Department of Machine Intelligence and Systems Engineering, Faculty of Systems Engineering
Akita Prefectural University
Akita, Japan

Abstract—Polypyrrole-based electrodes were fabricated for energy storage applications using sisal fibers as a substrate material. It was found that the PPy nanoparticles deposited on the surface of sisal fiber cellulose (SFC) connected to form a continuous sheath by taking along the microcrystalline cellulose template. The as-prepared nanocomposites demonstrated a mass-specific capacitance of 367 F/g at 200 mA/g current density in supercapacitor application. Moreover, SFC/PPy electrode retained about 57.8% of Cs retained when the current density increased five times. This work provides a straightforward method to utilize renewable resource sisal microcrystalline cellulose for conducting composite, which could apply in sensors, flexible electrodes, and flexible displays. It also opens a new field of potential applications of micrometre-scale natural microcrystalline cellulose.

Keywords-cellulose; Conducting polymer; Supercapacitor; Renewable resource; Hemp

I. INTRODUCTION

Natural fibrous substances play a key role in worldwide modern industries [1]. The especial morphologies of the natural substances afforded gave superior properties in composites. Cellulose supported conducting polymer has received growing interest in recent years due to their better performance or new properties, which has largely potential applications such as batteries, sensors, electrical devices, etc. Polypyrrole (PPy) is one of the most widely investigated conducting polymers because of its excellent thermal and environmental stability and high electrical conductivity [2]. Unfortunately, the poor processability and inadequate mechanical properties of PPy limit its commercial application [2]. In order to overcome these problems, deposition of PPy on fiber surface of fabrics and yarn, such as cotton, silk, and cellulose derivatives have been widely investigated in the last few years. As a especial kind of cellulose, sisal microcrystalline cellulose is produced by sisal hemp. Sisal fiber is one of the most common nature fibers that is the most largely quantity and the most widely used [3-5]. Sisal is a renewable resource par excellence and can form part of the overall solution to climate change. Sisal fiber (SFC) exhibits an ultrafine fibrous network, highly crystalline structure, purity, low density and remarkable mechanical strength. Meanwhile, the raw material was easily obtained. In this view, electrically conducting composed of sisal fiber and polypyrrole that presents a successful combination of the inherent properties of each single component. These properties include high tensile strength, toughness, biocompatibility, high surface areas of the SFC and cytocompatibility, electronic and chemical properties of PPy.

In this paper, we prepared the SFC/PPy nanocomposite. The as-prepared nanocomposites were characterized by FTIR, SEM, and electrochemistry techniques. The SFC/PPy nanocomposites were further evaluated for their applicability for electrochemical energy storage application, and a high special capacitance was obtained.

II. EXPERIMENTAL SECTION

A. Materials

Sisal fibers were provided by Guangxi Sisal Group Co., Ltd., China. Pyrrole, ammonium persulfate, sodium p-toluene sulfonate, sodium tetraborate, acetic acid, and nitric acid were purchased from Nacalai Tesque, Inc., Japan. Pyrrole monomer dehydrated with calcium hydride for 24 h was distilled under reduced pressure before use. All other chemical reagents were in analytical grade. All solutions were prepared in deionized water.

B. Preparation

The sisal fiber cellulose (SFC) was produced from sisal fibers as reported [2]. The SFC/PPy nanocomposite was prepared via in situ oxidative polymerization-induced adsorption onto SFC micro-rods. A typical synthesis was as follows: STS (4.16 g) was dissolved in 100 ml of deionized water, and then a certain amount of SFC were added (50.9 mg, 107.4 mg, 241.8 mg, 414.4 mg, and 967.0 mg; named as: S1, S2, S3, S4, and S5, respectively). The mixture was then ultrasonically dispersed, and pyrrole (1 ml) was added into the mixture with vigorous stirring. Afterward, the mixture was mechanically stirred for 30 min at 0 °C. Then an aqueous solution (20 ml) of APS (0.90 g) was added drop by drop to the above mixture instantly to start the oxidative polymerization. The reaction was performed under mechanical stirring for 8 h. The resulting precipitates were washed with deionized water and ethanol several times. Finally, the product was dried in vacuum at 60 °C for 24 h to obtain of the product as a dark powder.

© 2015. The authors - Published by Atlantis Press

C. Measurements

The morphology and microstructure of the samples were investigated with a S-4800 field emission scanning electron microscope (SEM; HITACHI, Japan). The Fourier transform infrared spectroscopy (FTIR) measurements (Impact 400, Nicolet, Waltham, MA) were carried out with the KBr pellet method. Electrochemical experiments were performed on an electrochemical workstation (CHI660D, China) with a conventional three-electrode system. The working electrode was prepared by mixing the active material with 15 wt% acetylene black and 5 wt% Polyvinylidene fluoride (based on the total electrode mass) to form a slurry. Then the slurry was cast on Ni foam. A platinum foil and a saturated calomel electrode (SCE) were used as the counter and reference electrode, respectively. The strong electrolyte, 0.5 M Na_2SO_4 solution, was used to ensure high ionic strength. Cyclic voltammetry was performed in the voltage range of -0.6 V to 0.4 V at scan rates of 1, 5 or 10 mVs-1, respectively. Galvanostatic charge/discharge experiments were carried out in the potential range from -600 mV to 400 mV with an applied current density of 200, 400, 800 or 1000 mA/g, respectively.

III. RESULTS AND DISCUSSION

A. Morphology Analysis

Images showed that the SFC surface (Figure 1a) was the smooth. This observation clearly indicated that pretreatment could remove the hemicellulose and lignin coverings from the SFC surfaces. In SFC/PPy composites, the presence of PPy particles coated on SFC can be observed (Figure 1 c-d). The thickness layer of polypyrrole deposited on SFC increased and the PPy particles tend to agglomerate with increasing pyrrole concentration. Apparently, a uniform coating layer was achieved for SFC/PPy composites prepared with 30wt% and 50wt% SFC. As the SFC content increased, a more homogeneous coating was formed. From SEM observations, it is possible to state that the initial content of SFC determines the thickness of the PPy layer defines the final microstructure and coating quality.

B. FTIR Spectroscopy

The FTIR spectra of SFC and SFC/PPy composites are shown in Figure 2. The characteristic IR peaks of SFC includes a broad band at 3403 cm^{-1} attributed to O-H stretching vibration, a band at 2897 cm^{-1} assigned to aliphatic C-H stretching vibration. The SFC/PPy composites exhibited a band at 3429 cm^{-1} for N-H stretching vibration instead of 3403 cm^{-1} for O-H stretching, which results from the complete wrapping of PPy around SFC fibers. Similar quenching of aliphatic C-H stretching at 2897 cm^{-1} was also observed. Moreover, the characteristic peaks for PPy were identical to those for SFC/PPy composites, which include a band at 1542 cm^{-1} assigned to the C=C stretching vibration in the Py ring, band at 1459 cm^{-1} corresponding to the C=N stretching vibration in the Py ring [2]. Moreover, the absorbance of S=O stretching at 1180 cm^{-1} of sulfonate, indicated that PPy had been doped with sodium p-toluene sulfonate.

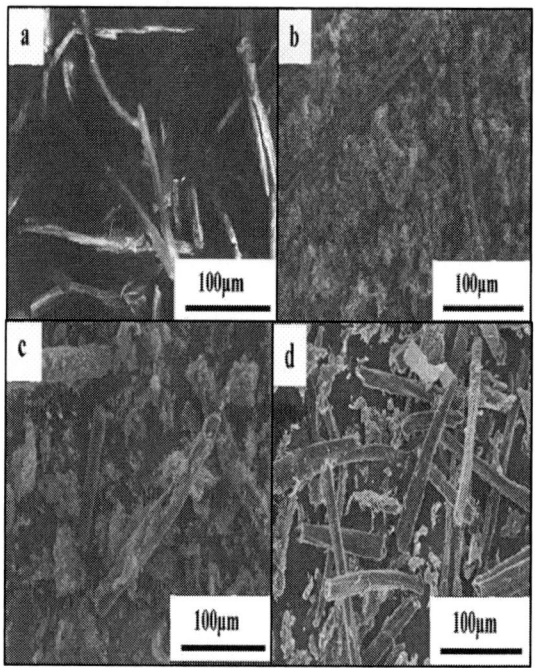

FIGURE I. SEM OF SFC AND SFC/ PPY COMPOSITES (A: SFC; B: S1; C: S3; D: S5).

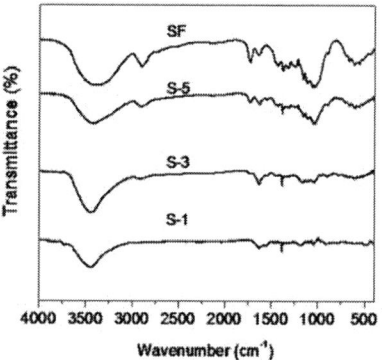

FIGURE II. FTIR SPECTRA OF SFC AND SFC/PPY COMPOSITES.

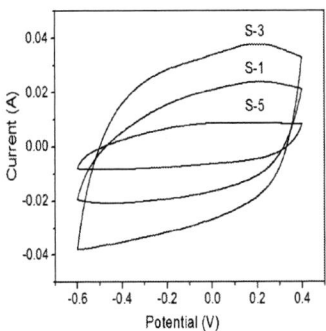

FIGURE III. CV CURVES OF SFC/PPY COMPOSITES.

C. *Electrochemical Performances*

The potential applications of as-prepared SFC/PPy nanocomposites were explored by fabricating the samples into supercapacitor electrodes and characterizing with cyclic voltammograms (CVs), and galvanostatic charge/discharge measurements. CVs response of SFC/PPy carried out at varied scan rates between -0.6 and +0.4 V is shown in Figure 3.

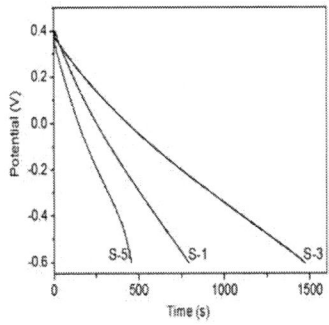

FIGURE IV. DISCHARGE CURVES OF SFC/PPY COMPOSITES(0.2A/G).

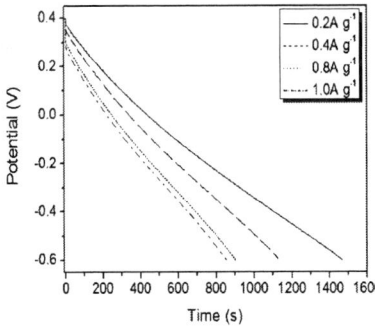

FIGURE V. DISCHARGE CURVES OF S3.

The composite showed a high degree of electroactivity, with the rectangular CV traces showing the transitions from reduced to oxidized forms, which demonstrates the retention of the important redox feature of conducting polymers in the as-synthesized SFC/PPy nanocomposites .

As illustrated in Figure 4, during the charge and discharge process, the charge curve of SFC/PPy is almost symmetric to its corresponding discharge counterpart with a slight curvature, indicating the pseudo capacitance behavior. The Cs values of SFC/PPy composites were calculated to be 195, 244, 367, 187, 114 F/g at current densities of 200 mA/g, respectively. According to the literature [7], the cellulose exhibited only negligible electroactivity. Galvanostatic discharge for S-3 sample was 367, 281, 227, 212 F/g performed at current densities of 200, 400, 800 and 1000 mA/g. About 57.8% of Cs was retained when the current density increased five times, which was attributed to the discrepant insertion-disinsertion behavior of alkali ion from the electrode to PPy.

IV. SUMMARY

It has been shown that it is possible to manufacture an electrical conducting composite material composed of natural cellulose and polypyrrole by direct chemical polymerization of pyrrole on sisal microcrystalline cellulose without the need for sophisticated techniques. The excellent performances of SFC/PPy composites pave the way toward promising applications in various electronic devices.

ACKNOWLEDGEMENTS

This work was supported by the National Natural Science Foundation of China (51303035, 21264005), Guangxi Natural Science Foundation (2013GXNSFBA019041), Training Programs of Innovation and Entrepreneurship for Undergraduates, the Guangxi Funds for a Specially-Appointed Expert, and the Guangxi Small Highland Innovation Team of Talents in Colleges and Universities. The correesponding authors are Chao Yang and Limin Zang.

REFERENCES

[1] M.M. Khin, A.S. Nair, V.J. Babu, R. Murugan, S. Ramakrishna, A review on nanomaterials for environmental remediation, Energy Environ. Sci. 5 (2012) 8075-8109.

[2] C. Yang, L.M. Zang, J.H. Qiu, E. Sakai, X.L. Wu, Y.K. Iwase, Nano-cladding of natural microcrystalline cellulose with conducting polymer: preparation, characterization, and application in energy storage, RSC Adv. 4 (2014) 40345-40351.

[3] H. Wang, L. Bian, P. Zhou, J. Tang, W. Tang, Core-sheath structured bacterial cellulose/polypyrrole nanocomposites with excellent conductivity as supercapacitors, J. Mater. Chem. A 1 (2013) 578-584.

[4] C. Yang, H.M. Mo, L.M. Zang, J.H. Qiu, H. You, X. Yang, Structural investigation of anionic functional poly(vinyl alcohol) doped polypyrrole nanospheres, Fiber. Polym. 15 (2014) 2019-2025.

[5] C. Yang, H. Mo, L. Zang, J. Qiu, E. Sakai, X. Wu, A facile method to synthesize polypyrrole nanoparticles in the presence of natural organic phosphate, Physica B 449 (2014)181-185.

[6] C. Yang, H.M. Mo, L.M. Zang, J.H. Qiu, X. Wang, H. You, Preparation and characterization of coaxial mullite/polypyrrole fibrous nanocomposites via self-assembling and in situ surface-initiated polymerization, Polym. Compos. 35 (2014) 892-899.

[7] L.M. Zang, J.H. Qiu, C. Yang,E. Sakai,Enhanced conductivity and electrochemical performance of electrode material based on multifunctional dye doped polypyrrole, J. Nanosci. Nanotechnol. In press (doi:10.1166/jnn.2015.10759).

Effect of Accumulated Strain on the Microstructure and Properties of TC4 Alloy Prepared by Continuous Various Cross-section Recycled Extrusion

X.L. You, Y.Y. Liu, J.S. Hua, K. Wang
Xi'an University of Architecture & Technology
Xi'an 710055, China

W. Yao
China Petroleum Pipeline Engineering Corporation Civil
Department
Langfang 065000, China

Abstract—The TC4 titanium alloy were deformed at 800℃, 2mm/s with various extrusion loops by a new sever plastic deformation method-Continuous Various Cross-section Recycled Extrusion, then the microstructure and microhardness of the specimen along the radial, the maximum deformation force in each extrusion procedure were also investigated. The results show that Continuous Variable Cross-section Recycled Extrusion is a better method for obtaining TC4 alloy with fine and equiaxed microstructure. With the increase of extrusion loops from 3 to 6, the grain size decreases and its size are about 2 to 3 um. As the extrusion loops increases from 6 to 8, the grain size is similar to that of the 6 loops. However, the grain size increases slightly due to multiple heating as the extrusion loops increases from 8 to 12. Moreover, the microhardness and maximum deformation force of the specimen in each extrusion procedure are higher while the extrusion loops is from 6 to 8. Therefore, the extrusion loop of 6 is the optimal, the microstructure is fine and equiaxed and it also distributes homogeneously.

Keywords-continuous various cross-section recycled extrusion; TC4 titanium alloy; accumulated strain; microstructure; properties

I. INTRODUCTION

Titanium alloys are widely used in the aerospace industry due to their low density, high strength and toughness as well as good high-temperature properties. Mechanical properties of titanium alloys are important criteria of material service capabilities both in aerospace and industrial applications. It is well known that the microstructure of the alloy is one of the important factors controlling its properties. If the titanium alloy with equiaxed, fine and homogeneous microstructure is obtained by some methods, its plasticity and strength will also improved significantly [1].

In recent years, in order to obtain the titanium alloys with equiaxed, fine and homogeneous microstructure, many methods are used, such as Equal Channel Angular Press(ECAP)[2-4], Multiple Forging(MF)[5], High Pressure Torsion(HPT)[6-7], Cyclic Extrusion Compression(CEC), Accumulative Roll Banding(ARB), High Ratio Extrusion(HRE), and Continuous Variable Cross-section Recycled Extrusion(CVCE)[8-10], etc.

Continuous Variable Cross-section Recycled Extrusion (CVCE) is a new method of sever plastic deformation(SPD), its advantages are as flows: (1) the large size block materials

with fine grain microstructure can be obtained; (2) the cost of die is low and the operation is simple; (3) the drum occurring in the traditional upsetting is eliminated, and the geometrical shape can be kept simultaneously; (4) the efficiency is high, and the quality is also good.

Therefore, in order to obtain the titanium alloy with fine and equiaxed microstructure, the TC4 titanium alloy is deformed with various extrusion loops by CVCE, and the effect of extrusion time on the microstructure, hardness and maximum deformation force have been investigated. The results will provide a new method with better process parameters and theoretical guidance for preparing the large size bulk titanium alloy with fine and equiaxed grain.

II. EXPERIMENTAL

The TC4 alloy used in the present work was a bar of Φ20mm in diameter. The beta transus temperature of the TC4 alloy is 989℃, and its chemical composition is shown in Table 1. The initial microstructure of the TC4 alloy is a typical equiaxed structure consisting of a large volume fraction of fine equiaxed α phase with a grain size of 12 um and a small amount of transformed β phase, as shown in Figure 1.

TABLE I. CHEMICAL COMPOSITION OF TC4 ALLOY (MASS FRACTION, %).

Al	V	O	N	C	Fe	H	Ti
6.5	4.3	0.2	0.05	0.10	0.30	0.02	Bal.

The cylinder specimens were electro-discharged machined from the bar with a length of 26 mm. In order to avoid the cracks formation during extrusion because of the stress concentration occurring in the sharp edge, the cut surfaces were filleted and shined with abrasive papers.

© 2015. The authors - Published by Atlantis Press

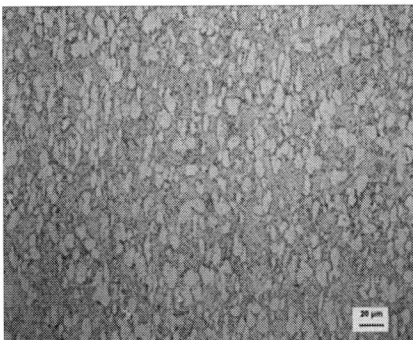

FIGURE I. INITIAL MICROSTRUCTURE OF TC4 TITANIUM ALLOY.

The flow diagram of CVCE is shown in Figure 2. It can be seen that a deformation loop is composed of four procedures, including two times extrusion from the cylinder to the frustum of cone in die cavity; two times upsetting from the frustum of cone to the cylinder. The deformation account is obtained by accumulated extrusion and upsetting. Finally, the TC4 alloy with fine and equiaxed microstructure will be obtained by accumulated strain.

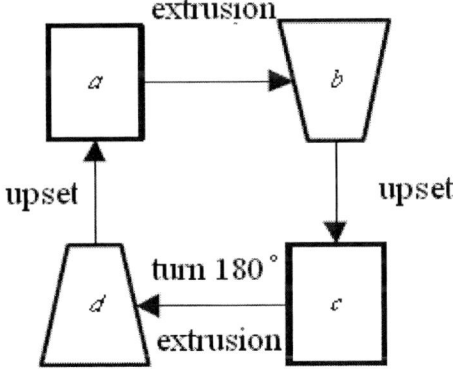

FIGURE II. EXTRUSION PROCESS FLOW OF CVCE.

The tests were carried out using a hydraulic press with controllable speed. The deformation temperature used was 800 °C [10], and the speed was 2mm/s. The extrusion loops were 3, 6, 8 and 12, respectively. The heating time was calculated by 0.6mm/ minute, and the specimen was heated by 10 minutes between two procedures for improving the plastic deformation ability of TC4 titanium alloy. In deformation, H13 tool steel was used as the die materials, and its temperature was kept at 550°C. After the total deformation, the specimen was cooled in water. Moreover, in heating and deformation, the surfaces of specimens were coated with special glass lubrication to resist oxidation. However, the surface of die cavity was coated with the mixing uniform graphite powder and oil.

The microstructure of TC4 alloy along the radial direction was observed by an optical microscopy (OLYMPUS PM-G3), and the microhardness was measured by 401MD microindenter with a load of 100gf applied for a dwell time of

15s. The specimen were etched with a solution containing 40vol.%HF+30vol.%HNO$_3$ in water.

III. RESULTS AND DISCUSSION

The microstructure of the TC4 alloy deformed at 800℃ with 2mm/s by various extrusion loops (various accumulated strain) is show in Fig.3. It can be seen that the grain size decreases as the extrusion loop increases from 3 in Fig.3a to 6 in Fig.3b, and the grain size of 6 loops is about 2 to 3um. As the extrusion loop increases from 6 to 8, the grain size is fine, this is similar to that of 6 loops in Figure3c. However, the grain size increases slightly as the extrusion loops increases from 8 to 12, and its grain size is about 3 to 4um. The reason is that with the increase of extrusion loop from 3 to 6 or 8, the accumulated strain increases, so the microstructure is broken sufficiently and show the fine and equiaxed shape. Moreover, although the accumulated strain increases with the increase of extrusion loop from 8 to 12, the microstructure is coarsen slightly due to multiple heating with the increase of extrusion loop.

From above mentioned, the better extrusion loop is from 6 to 8, the microstructure of grain size with about 2 to 3um can be obtained by CVCE. But considering the uniformity of microstructure distribution and deformation efficiency, the extrusion loop of 6 is the optimal.

FIGURE III. MICROSTRUCTURE OF THE TC4 ALLOY UNDERGONE VARIOUS DEFORMATION LOOPS(VARIOUS ACCUMULATED STRAIN): (a) 3; (b) 6; (c) 8; (d) 12.

The microhardness of the TC4 alloy deformed at 800°C with 2mms by various extrusion loops (various accumulated strain) is show in Figure4. It can be seen that the change law between microhardness and extrusion loop is in accordance with that of microstructure and extrusion loop. For example, while the extrusion loop is 3, 6, 8 and 12, the microhardness value is 278.7MPa, 298.7 MPa, 298.1MPa, and 287.9MPa, respectively. Based on the above microhardness value, the extrusion loop of 6 is also the optimal.

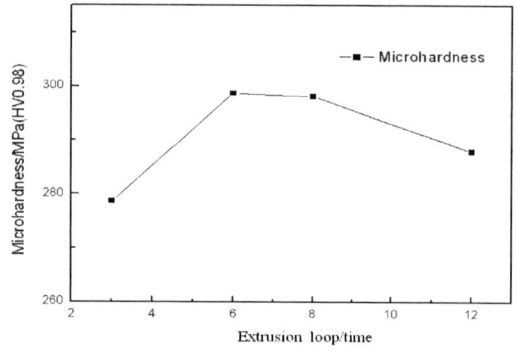

FIGURE IV. MICROHARDNESS OF TC4 ALLOY UNDERGONE VARIOUS DEFORMATION LOOPS AT 800°C WITH 2MM/S(VARIOUS ACCUMULATED STRAIN).

The relationship between maximum deformation force and extrusion loop from 4 to 10 in each extrusion procedure is shown in Table 2. It can be seen that as the extrusion loop is from 6 to 8, whatever in the extrusion procedure or the upset

one, the deformation force of the TC4 alloy is higher comparing to the other extrusion loop. The reason is that the grain size decreases and the homogeneity of distribution increases with increasing extrusion loop, which results in the increases of maximum deformation force. In addition, it is concluded that the deformation force in the extrusion procedure of each loop is smaller than that in the upsetting procedure. With the increase of extrusion loop from 9 to 10, the deformation force decreases obviously. For example, in the second procedure and fourth procedure of 8 and 10, the deformation force decreases from 400 KN to 309 KN and from 384 KN to 292 KN, respectively, this is related to the microstructure coarsening with the increase of deformation loop.

TABLE II. RELATIONSHIP BETWEEN MAXIMUM DEFORMATION FORCE AND EXTRUSION LOOP IN EACH EXTRUSION PROCEDURE (ACCUMULATED STRAIN).

Extrusion loop (loop)	Deformation force per procedure (KN)			
	1st	2st	3 st	4st
4	358	376	348	397
5	340	390	348	383
6	343	392	341	388
7	326	433	319	400
8	308	400	339	384
9	322	377	266	300
10	300	309	274	292

IV. CONCLUSIONS

1) The grain size decreases as the extrusion loop increases from 3 to 6, which is related to the increase of accumulated strain. When the extrusion loop increases to 8, the grain size is similar to that of the 6 loop. However, the grain size increases slightly due to multiple heating as the extrusion loop increases from 8 to 12.

2) The micro hardness and maximum deformation force of the specimen in each extrusion procedure are higher as the extrusion loop increases from 6 to 8, which are related to the fine microstructure and its homogeneous distribution.

3) Considering the uniformity of microstructure distribution and deformation efficiency, the extrusion loop of 6 is the optimal, and the microstructure of grain size with about 2 to 3um can be obtained by CVCE.

ACKNOWLEDGMENT

This work was financially supported by the Natural National Science Foundation of China (51101116).

REFERENCES

[1] S. M. Sastry, R N. Mahapatra, D. F. Hasson, Microstructural refinement of Ti-44Al-11Nb by severe plastic deformation, Scripta Materialia, 42(2000): 731-736.

[2] Alexander Korshunov, Tamara Kravchenko, Lev Polyakov, Andrey Smolyakov, Irina Vedernikova, Alexander Morozov, Effects of the number of equal-channel angular pressing passes on the anisotropy of ultra-fine titanium, Mater. Sci. and Eng. A , 493(2008): 160-163.

[3] X.C. Zhao, W.J. Fu, X.R. Yang, Terence G. Langdon, Microstructure and properties of pure titanium processed by equal-channel angular pressing at room temperature, Scripta Materialia, 59(2008): 542-545.

[4] X.C. Zhao, X.R. Yang, X.Y. Liu, X.Y. Wang, Terence G. Langdon, The processing of pure titanium through multiple passes of ECAP at room temperature, Mater. Sci. Eng. A, 527(2010): 6335–6339.

[5] G. A. Salishchev, R. M. Galeyev, S. P. Malysheva, M.M. Myshlyaev, Structure and density of submicrocrystalline titanium produced by severe plastic deformation, Nanostructured Mater., 11(1999): 407–414.

[6] S. Faghihi, A.P. Zhilyaev, J.A. Szpunar, F. Azari, H. Vali, M.Tabrizian. Nanostructuring of titanium material by high pressure torsion improves pre-osteoblast attachment, Adv.Mater., 19(2007): 1069-1073.

[7] F. Shahab, A. Fereshteh, Z.P. Alexander, A.S. Jerzy, V. Hojatollah, T. Maryam. Cellular and molecular interactions between MC3T3-E1 pre-osteoblasts and nanostructured titanium produced by high-pressure torsion Biomaterials, 28(2007): 3887-3895.

[8] C.R.Liu, H.X..Ren, Q.J.Wang, Microstructure and properties of AZ31 Magnesium alloy by Continuous Variable Cross-section Recycled Extrusion, Light Alloy Fabrication Technol., 37(6) (2009): 34-37 (In Chinese).

[9] J.Zhang, C.R.Liu, Q.J. Wang, K.S. Wang, Z.Q.Ma,Structural Evolution of Pure Aluminum1A85 During Continuous Variable Cross-section Recycled Extrusion, J. Mater. Sci. Eng., 28(6) (2010):930-933 (In Chinese).

[10] Y. Y. Liu, C.R. Liu, L. Wang, Y.Z. Zhang, Effects of processing parameters of Continuous Variable Cross-Section Recycled Extrusion on the microstructure and micro hardness of TC4 titanium alloy, Rare Met. Mater. Eng., 43(2) (2014):440-444 (In Chinese).

International Conference on Power Electronics and Energy Engineering (PEEE 2015)

Molecular Simulation Research on Domain Micro-structure of TSP-POSS/PU Hybrid Composites

R. Pan*, L.L. Wang
Chemistry and Material Science College
Sichuan Normal University
Chengdu City, PRC, 610041
*Corresponding author

Y. Liu
Key Laboratory of Special Waste Water Treatment
Sichuan Province Higher Education System
Chengdu City, PRC, 610041

Abstract—Trisilanolphenyl polyhedral oligomeric silsesquioxane (TSP-POSS) with three phenyl groups, was incorporated in concentrations of 6, 12 and 20wt% into 4, 4'-methylenebis(phenyl isocyanate) (MDI) and glycerol propoxylate to prepare TSP-POSS/PU hybrid composites models, respectively. The domain micro-structures of these hybrid composites models were characterized by mean square displacement and radial distribution function at molecular level. As the result shows: with TSP-POSS concentration increasing, the contacts of polymer chain is decreased and cage cores are inclined to be closer to each other due to the presence of humping cage structure of TSP-POSS unit. Meanwhile, high concentration of TSP-POSS apparently restricts the motion of polymer chains, which also demonstrates that TSP-POSS is as a rigid core linked to the backbones of hybrid composites.

Keywords-POSS; polyurethane; molecular dynamics; molecular mechanics

I. INTRODUCTION

As nano-particle, polyhedral oligomeric silsesquioxane (abbreviation POSS) has received increasing attention for its inorganic-organic hybrid nature. Most of POSS consists of a totally enclosed cage with eight silica atoms linking together via oxygen atoms.[1] Polymer properties, such as permeability, friction, mechanical and thermal properties, can be significantly altered by introducing different POSS molecules at various concentrations into polymer matrices. [2-5] As with other polymers, POSS can improve the properties of PU in applications ranging from elastomers, coatings to adhesives. However, there is seldom evidence in literature about the influence of TSP-POSS, as one of a few POSS with three phenyl groups, on polyurethane domain micro-structure with covalent incorporation at various concentrations. Since polyurethane has a large range of application and TSP-POSS is a good modification of filler, the mechanism about how the covalent inclusion of TSP-POSS into the PU backbone will tailor the domain micro-structure of hybrid composite is deserved to investigate.

II. SIMULATION

Accelrys Amorphous Cell module and COMPASS force field in Materials Studio software were adopted in all simulation process in our research, which has been used successfully for the simulation of polymer nanocomposites containing POSS [6-7]. As periodic boundary conditions imposed, an initial low density was used to construct the bulk

cubic structures of random hybrid copolymers. 10 initial configurations for each sample were optimized by molecular dynamics technique under the NPT (constant particle numbers, pressure and temperature) conditions at 4Gpa with a minimization involving 30000 steps to relax and equilibrate. After this minimization procedure, the density fluctuation of each system is less than 0.02 g/cm^3 under a given condition. Since these optimized configurations might not be in a local energy minimum state, an annealing procedure from 623 K to 273 K was applied on the above optimized configurations by conducting the velocity Verlet algorithm in NVT dynamics to reduce the possible potential energy. Finally, configurations with the highest energy were rejected and 10 configurations of each sample were selected for further analysis of composites' characteristics. The weight fractions of TSP-POSS in composites and sample codes are listed in Table 1. The molecular structures of TSPPOSS /PU hybrid composite are shown in Figure 1. [8]

TABLE I. CHARACTERISTICS OF TSP-POSS/PU HYBRID COMPOSITES.

Sample code	TSP-POSS: Voranol301 (mole ratio)	MDI: TSP-POSS (wt.%)	Initial density (g/cm^3)	Final density (g/cm^3)
6PU	1:5:4	6	1.163	1.177
12PU	1:3:2	12	1.174	1.187
20PU	1:3:1	20	1.202	1.204

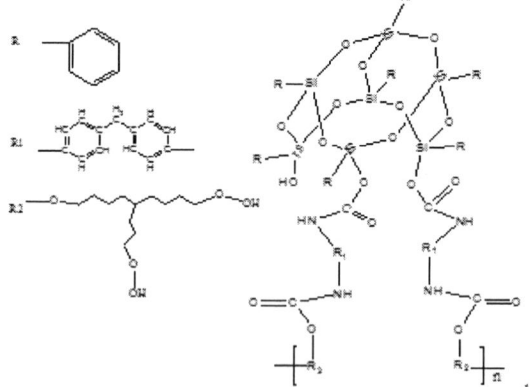

FIGURE I. CHEMICAL STRUCTURES OF TSP-POSS/PU HYBRID COMPOSITES.

© 2015. The authors - Published by Atlantis Press

III. RESULTS AND DISCUSSION

A. Validation of Simulation

The use of the COMPASS force field for our system was validated by the following method. Recently, it was also reported that the structure and energy of the POSS/PU hybrid composites were successfully simulated by use of the COMPASS force field. Model structure was generated through several cycles of molecular mechanics and molecular dynamics energy minimization. Table 1 lists the characteristic values of all hybrid composites. After the above minimization procedure, the density fluctuation of each system is less than 0.02 g/cm^3 under a given condition, indicating that the structure generated is fully relaxed and is in the equilibrium state which can be confirmed by energy optimization.

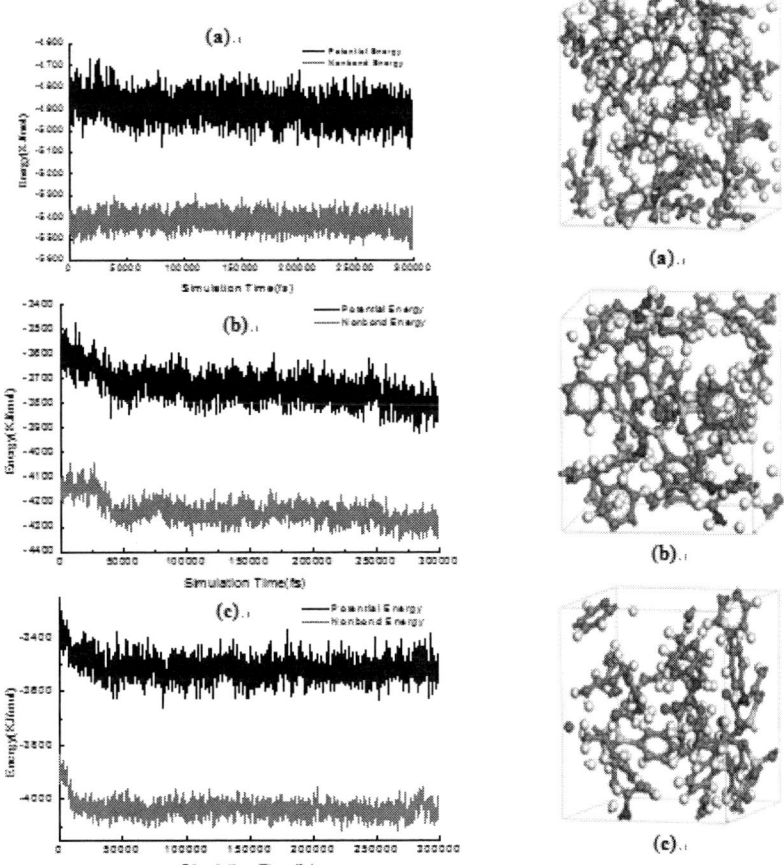

FIGURE II. ENERGY OPTIMIZATION AND 3D PERIODIC BOUNDARY CONDITIONS OF TSP-POSS/PU HYBRID COMPOSITES: (A). 6PU; (B). 12PU; (C) 20PU.

B. Radial Distribution Function Analysis

In statistical mechanics, the radial distribution function (pair correlation function) g(r) is a measure of the probability of finding a pair of atoms (α,β) that is separated by a radial distance rαβ, which is expected for a completely random distribution. Thus, details of polymer chain packing can be estimated by defining the Si–O pair in TSP-POSS core and atoms in polymer main chain as given reference particles in intermolecular pair correlation function, respectively.

Figure3(a) shows the intermolecular pair correlation function based on all backbone atoms in main chains of three samples. No significant diffuse peak can be observed, which indicates no obvious influence on the average spacing between neighboring chains. However, the value of g(r) decreases with increasing concentration of TSP-POSS, which means that at any given distance the number of contacts between neighboring chains is decreased due to the presence of TSP-POSS with humping cage structure [9-10].

Figure3(b) shows the intermolecular pair correlation function based on atoms (Si, O) of TSP-POSS cores in composite versus the atoms in the main chain. With the increase of TSP-POSS concentration, the number of contacts between TSP-POSS cores and main chains is slightly increased. It can be concluded that as the concentration increasing, TSP-POSS cores are inclined to be closer due to the inter-cage structure interactions.

203

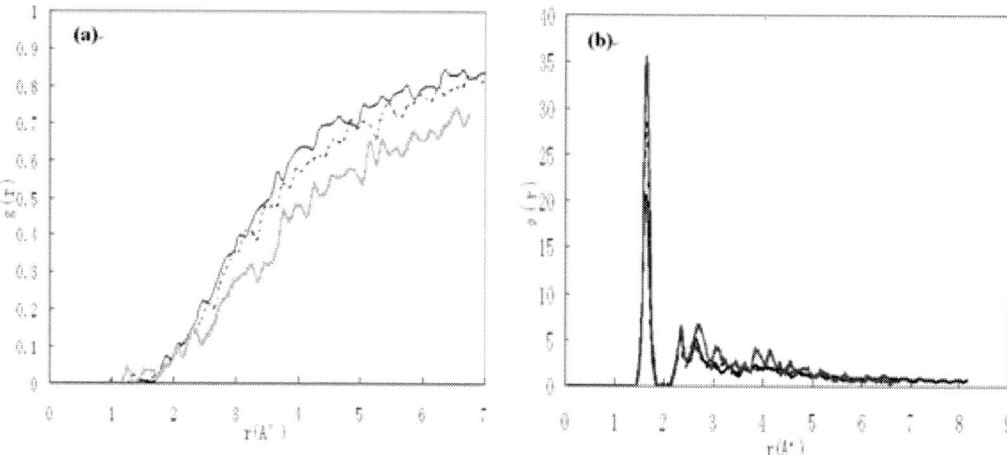

FIGURE III. PAIR CORRELATION FUNCTIONS G(R) FO 6PU (SOLID LINE), 12PU (DOTTED LINE) AND 20PU (DASHED LINE): (A) BASED ON ATOMS (SI, O) OF THE TSP-POSS CORES TO THE ATOMS IN THE MAIN CHAINS OF HYBRID COMPOSITES; （B）BASED ON ALL BACKBONE ATOMS OF HYBRID COMPOSITES.

C. Mobility Of Fractures In Polymer Chains In Tsp-Poss/Pu Hybrid Composites

The mobility of the atoms (molecules) in a simulation system is investigated by using mean square displacement (MSD), which can be computed from Eq.(1). [11]

$$MSD = [\ |\ (r_i(t) - r_i(0)\ |\ ^2].$$

Where r(0) is the initial positional coordinate of atom i (or molecules) and ri(t) is the coordinates of time.

Thus, the motion restrictions of polymer chains imposed by TSP-POSS structure can be evaluated by elucidating the mobility of atoms in polymer chains with and without the TSP-POSS incorporation. In Figure4a-c, the MSD of all atoms in the TSP-POSS/PU samples is compared to that of the Si-O pairs in TSP-POSS cores. Whatever concentration is, the MSD of TSP-POSS core is all lower than that of all atoms as a whole. This result confirms that the TSP-POSS cage linked to polymer backbones acts as a heavy core with less mobility. In Figure4d, it shows the comparison of enclosed cage cores mobility with different TSP-POSS concentration in hybrid composites. From the plot, the behavior of TSP-POSS is non-diffusive and stationary as MSD lower than 5 A^2 . Meanwhile, as the concentration of TSP-POSS increasing, the value of MSD decreases due to the motion restriction from humping cage reunion.

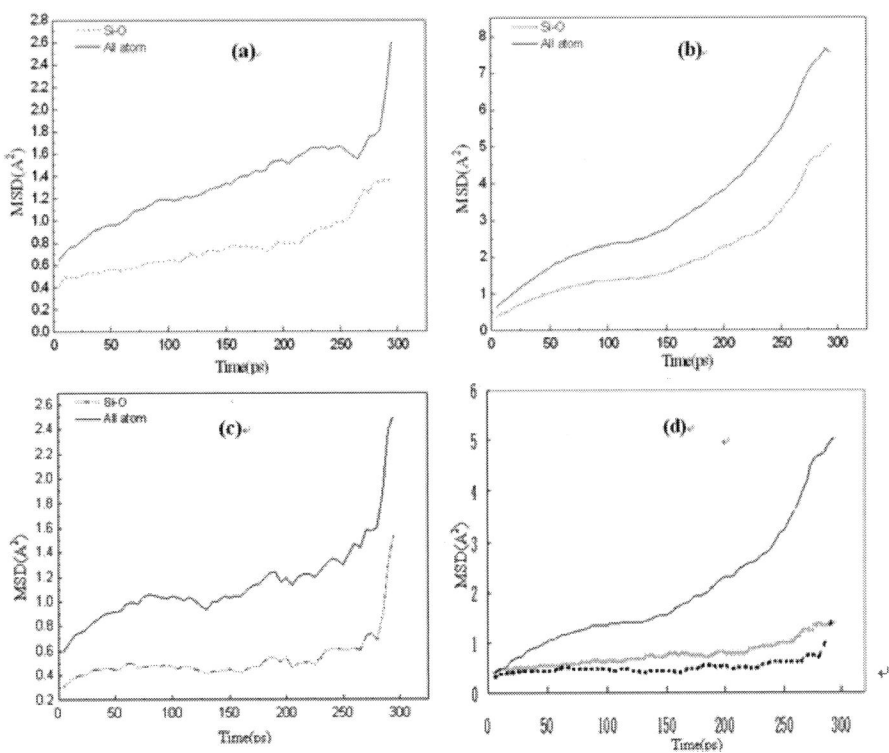

FIGURE IV. MEAN SQUARE DISPLACEMENT OF ALL ATOMS IN COMPOSITES (SOLID LINES) AND SI-O PAIRS IN TSP-POSS CORES (DOTTED LINES): (A).6PU; (B).12PU; (C).20PU. (D) MEAN SQUARE DISPLACEMENT OF SI-O PAIRS OF TSP-POSS CORES IN 6PU (DOTTED LINE), 12PU (SOLID LINE) AND 20PU (DASHED LINE).

IV. SUMMARY

In this study, molecular simulation was applied in study the effect of TSP-POSS cage structure on domain micro-structure of TSP-POSS/PU hybrid composites at molecular level. The fluctuation of density and optimized energy plots verified the accuracy of the models and the applied force field. As the result shows, with TSP-POSS concentration increasing in hybrid composites, polymer chains contacts is decreased due to the presence of humping cage structure in TSP-POSS fractures while cage cores are inclined to be closer to each other. Meanwhile, a significant decrease in the mobility of TSP-POSS cores and polymer backbones can be observed, which demonstrates that TSP-POSS cage, as a rigid core linked to the TSP-POSS/PU backbone, apparently restricts the motion of the whole polymer chain.

REFERENCES

[1] Information on http://www.hybridplastics.com/docs/user-v2.06.pdf

[2] Information on http://en.wikipedia.org/wiki/Polyurethane

[3] Jianqing Zhao, Yi Fu and Shumei Liu, Polyhedral oligomeric silsesquioxane (POSS)-Modified thermoplastic and thermosetting nanocomposites: A review.,Polym. Polym. Compos. 16(8)(2008) 483-500.

[4] Zheng L., Farris R.J. and Coughlin E.B., Novel Polyolefin Nanocomposites:Synthesis and Characterizations of Metallocene-Catalyzed Polyolefin Polyhedral Oligomeric Silsesquioxane Copolymers, Macromol.34(2001) 8034-8039.

[5] Waddon A.J., Zheng L., Farris R.J. and Coughlin E.B., Nanostructured Polyethylene-POSS Copolymers:Control of Crystallization and Aggregation., Nano Lett. 2(10)(2002) 1149-1155

[6] Zheng L., Waddon A.J., Farris R.J. and Coughlin E.B., X-ray Characterizations of Polyethylene Polyhedral Oligomeric Silsesquioxane Copolymers. ,Macromol.35(2001)2375-2379.

[7] Samy A. Madbouly, Joshua U. Otaigbe, Recent advances in synthesis, characterization and rheological properties of polyurethanes and POSS/polyurethane nanocomposites dispersions and films., Prog. Polym. Sci. 34(2009)1283-1332.

[8] Lingling Wang, Ming Zhang, Yong Liu and Rui Pan, Influence of trisilanolphenyl POSS on structure and thermal properties of polyurethane hybrid composites: a molecular simulation approach, Acta Polymerica Sinica. Doi: 10.11777/j.issn1000-3304.2015.14231

[9] Matthew Oaten and Namita Roy Choudhury, Silsesquioxane-urethane hybrid for thin film applications., Macromol. 38(2005) 6392-6401

[10] Stéphane Bizet, Jocelyne Galy Molecular dynamics simulation of organic-inorganic copolymers based on methacryl-POSS and methyl methacrylate., Polym. 47(2006) 8219-8227.

[11] Bruce X Fu, Benjamin S Hsiao, Joseph Schwab et al., Nanoscale reinforcement of polyhedral oligomeric silsesquioxane (POSS) in polyurethane elastomer., Polym. Int. 49(2000) 437-440

International Conference on Power Electronics and Energy Engineering (PEEE 2015)

Study on Fire Extinguishing Performance of Superfine Powder Fire Extinguishing Agent in a Cup Burner

T. Chen, X.C. Fu, J.J. Xia, L.S. Jing, C. Hu
Tianjin Fire Research Institute
MPS
Tianjin 300381, China

Abstract—**A modified cup-burner apparatus with a vibrating-spiral micro feeder was developed for testing the extinguishing performance of superfine powder fire extinguishing agent (SPEA). The suspension velocity of SPEA particles was calculated using the suspension-grade-model (SGD). The formation mechanism of SPEA-based aerosols was investigated theoretically, and the minimum air velocity for the SPEA particles was about 0.05m/s. Experiments were conducted on both ethanol and heptane fuels burning flames using $NH_4H_2PO_4$-based SPEA with air flow rate of 60 L/min. The results show that the minimum extinguishing concentrations (MEC) of $NH_4H_2PO_4$-based SPEA for ethanol and heptane fuel are 79.8g/m^3 and 98.8g/m^3, respectively. It is demonstrated that the $NH_4H_2PO_4$-based SPEA is a highly efficient fire-extinguishing agent, which is an average of 2~3 times more effective than CF_3Br, and an average of 5~8 times more effective than HFC-227ea.. This research is helpful for the engineering design of superfine powder fire extinguishing equipment, as well as has vital significance to the fire protection.**

Keywords-superfine powder fire extinguishing agent; extinguishing concentration; aerosol; cup-burner

I. INTRODUCTION

Superfine powder fire extinguishing agent (SFEA) with small size and large specific surface area can quickly spread and suspend in the air after being discharged, and then forms stable aerosols. Therefore, SFEA normally referred to as cold aerosol fire extinguishing agent[1-2]. Previous studies showed that SFEA has similar properties to gas fire extinguishing agent after being applied into the air. Superfine power fire extinguishing agent can bypass obstacles and suspend for some time when discharged by high pressure gas. Therefore, it can be used as a way of total flooding extinguishing in engineering application. Besides, SFEA can be used as the substitute of halon or hydrofluorocarbons (HFCs) fire extinguishing agent, such as Halon1301, HFC-227ea.and HFC-23, because it neither destroys the atmospheric ozone layer nor leads to global warming[3]. It will be widely employed in the future due to its low cost and cheap equipment. However, there are a lack of enough research on the performance and measurement of superfine powder total flooding as well as effective methods to evaluate the extinguishing performance of total flooding. Furtherly, the lack of key application parameters such as critical extinguishing concentration seriously limits the application of superfine power fire extinguishing technology.

Fire suppression model being closed to practical fire situation was used for assessing the extinguishing performance of superfine power fire extinguishing system. As described in GA 578-200<Superfine powder fire extinguishing agent>, fire extinguishing test was conducted in a 100m^3 space to evaluate performance of SFEA on standard fire. In every test, it will cost too much material, and it is difficult to determine critical fire extinguishing concentration. Fleming and Hmins[4-5] compared and analyzed cup-burner, Convection diffusion flame model and turbulent spray flame burner through studying the performance of large number of gas extinguishing agents. Results indicated that the characteristic of cup-burner flame structure was more close to real fire and reflected the interaction between fire extinguishing agent and fire rather than other models. Hirst[6] reported that fire in cup-burner was more difficult to being extinguished and required more 20% extinguishing concentration than fire in full size experiment test. Hirst thought that security coefficient had been considered essentially while extinguishing concentration was measured by cup-burner model. Many studies demonstrate that test methods based on cup-burner model is a scientific way to assess the total flooding performance of fire extinguishing agents, which has been accepted by the countries around the world and international organization for standardization.

Hamins[7] has measured the suppression effectiveness of $NaHCO_3$ by modified cup-burner. Whereas, the feeding rate is unstable, the calibration of feeding rate is too complicate and the object of study is too single. And the effect of suspension characteristics and air velocity on extinguishing concentration is also not considered. Given this, firstly, a modified cup-burner apparatus with a vibrating-spiral micro feeder was developed in this paper. Secondly, the formation mechanism of aerosols was investigated by theoretical calculation and test. Finally, this paper determined the extinguishing concentrations of $NH_4H_2PO_4$-based SPEA with different fuels. For comparison, experiments results using several common gaseous agents with the same test method were also reported.

II. EXPERIMENT

A. Fire Extinguishing Agent

As main material, $NH_4H_2PO_4$ powder was milled by Fluidized bed air flow crushing method before silicone oil was used to the surface modification of the SFEA. The particle size

© 2015. The authors - Published by Atlantis Press

distribution of $NH_4H_2PO_4$-based SPEA was measured by laser particle size analyzer (Mastersizer 2000), as shown in Figure 1. The results showed that D_{50} and D_{90} for $NH_4H_2PO_4$-based SPEA was 7.21 um and 13.29 um, respectively.

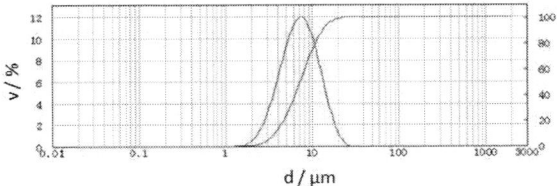

FIGURE I. PARTICLE SIZE DISTRIBUTION OF SPEA.

B. Experimental Apparatus

A modified cup-burner apparatus was developed for testing the extinguishing performance of SFEA with a vibrating-spiral micro feeder. As shown in Figure 2, circular cup was made of stainless steel with 25mm internal diameter and 1-2mm wall thickness.

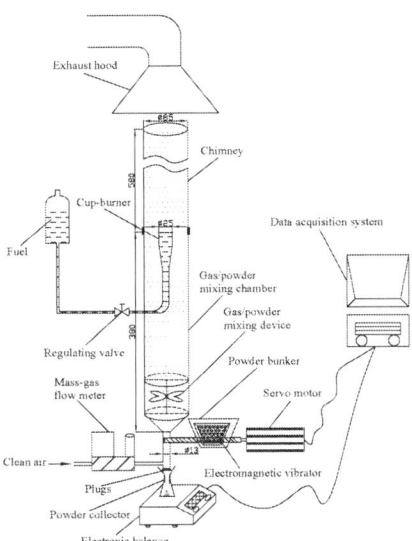

FIGURE II. SCHEMATIC DIAGRAM OF EXPERIMENTAL APPARATUS.

The chimney of burner with 85mm internal diameter was cylindrical. The feeding rate of superfine powder, which was calibrated through powder collector and powder quality collector, was regulated by controlling electromagnetic vibrator switch and adjusting the speed of servo motor. Air provided by compressor was dried through filter and then injected into the bottom of burner, flow rate of which was altered by gas mass flow meter (MFC D07-9E). Under electromagnetic vibration, superfine powder was driven by screw and injected into the bottom of cup-burner. Thereafter, superfine powder was driven into mixing chamber together with air and formed uniform aerosols under vigorously stirring through mixing apparatus. The mixing apparatus carried with blades rotating automatically under the effect of the airflow.

C. Experiment Procedure

Before the experiment, the powder feeding rate was calibrated. Firstly, 150g superfine powder was added into the powder bunker and air was stopped injecting into cup-burner, followed by removing the plug below air inlet. Secondly, the electromagnetic vibrator and servo motor were opened and then superfine powder was provided after setting up the speed of servo motor. Finally, the mass of agent in powder collector was recorded once a second by powder quality collector.

During the experiment, the plug was set up below air inlet and air was injected into cup-burner. Meanwhile, fuel was lit and burnt for 10s-30s. Then the electromagnetic vibrator and servo motor were opened after setting up the speed of servo motor. Superfine powder was driven into cup-burner with constant feeding rate and the time was recorded as agent contacted with fire. If the time exceeded 60s, while the fire still burned, the test continued after raising the speed of motor or decreasing airflow until the fire was put out. Parameters, such as attack time, motor speed and airflow, were recorded during this time. The powder delivery rate could be obtained from motor speed and extinguishing concentration was calculated by formula (1).

$$C = \frac{q}{V_{powder} + V_{air}} \times 1000 \qquad (1)$$

Where, C stands for extinguishing concentrations of SFEA by mass, g/m^3; q stands for SFEA powder delivery rate by mass, g/s; V_{air} stands for Air volume flow, m^3/s; V_{powder} stands for Volume flow of SFEA powder, m^3/s. In view of the fact that the powder density is far greater than the air, So this value(V_{powder}) can be ignored when the actual calculation.

III. RESULTS AND DISCUSSION

A. Relationship between Powder Delivery Rate and Motor Speed

In Figure 3, the mass of $NH_4H_2PO_4$ was recorded as motor speed changed. The feeding rate as a function of motor speed was shown in Figure 4.

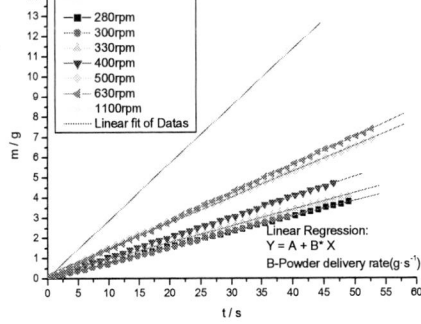

FIGURE III. CUMULATING MASS OF NH4H2PO4 POWDER CHANGING WITH TIME.

207

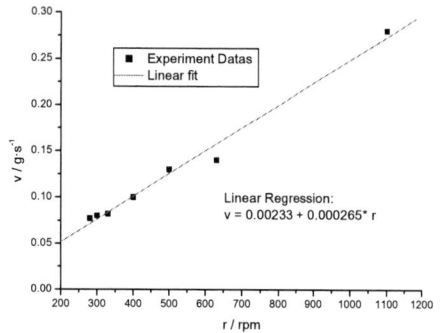

FIGURE IV. POWDER DELIVERY RATE AGAINST THE MOTOR SPEED.

Figure 3 indicates that the mass increases linearly with time at constant motor speed when Electromagnetic vibrator is open. Figure 4 shows that the feeding rate increases with motor speed and screw speed. According to different characteristics of superfine powder, feeding rate can be adjusted freely from 0.02g/s to 0.30g/s by changing motor speed. Therefore, modified apparatus with a vibrating-spiral micro feeder can supply continuous, stable and uniform powder which meets the requirement of fire extinguishing performance.

B. Aerosol Formation and Its Relationship with the Air Flow

In the system containing particle and airflow, airflow will move upward at certain speed. According to principle of gas-solid two phase fluid mechanics[8-9], the particle will fall when the speed of airflow is lower than the particle free settling velocity. The particle will rise when the speed is higher than the particle free settling velocity. If the speed of particle and airflow is equal, particle will swing in a horizontal plane and the speed of airflow is known as the free suspension velocity of solid particles or simply suspension velocity for short. Apparently, suspension velocity and the particle free settling velocity are equal in value, but opposite in direction. Suspension velocity depends on its resistance characteristics such as density, volume and shape. Besides, the density and dynamic viscosity of air also affects suspension velocity.

To obtain aerosols in cup-burner, it must be ensure that superfine powder move upward, that is to say, the speed of airflow must higher than suspension velocity. In general, the suspension velocity of powder particle is calculated using the suspension-grade-model (SGD) according to principle of gas-solid two phase fluid mechanics. Applicable particle size range formula for viscous resistance area is shown in formula (2).

$$d_s \leq 1.225 \left[\frac{\mu^2}{\rho(\rho_s - \rho)} \right]^{\frac{1}{3}}$$

(2)

Where, d_s stands for SFEA particle diameter, m; μ stands for dynamic viscosity of air, Pa•s, which is 0.0000179 Pa•s under standard conditions; ρ stands for air density, kg/m³, which is 1.205kg/m³ under standard conditions; ρ_s stands for powder

density, kg/m³, which is 1803kg/m³ of $NH_4H_2PO_4$-based SPEA.

The applicable particle size range of $NH_4H_2PO_4$-based powder in viscous resistance area can be calculated by formula (2): $d_s \leq 65\mu m$. The majority of $NH_4H_2PO_4$-based powder particle sizes are less than 30 microns, so can be determined in the viscous resistance area. The calculation formula of the suspension speed of viscous resistance area is shown in formula (3).

$$\upsilon_0 = \frac{(\rho_s - \rho)g}{18\mu} * d_s^2$$

(3)

Where, υ_0 stands for free suspension velocities, m•s⁻¹; g stands for acceleration of gravity, m•s⁻², which is 9.81 in here.

The suspension velocities for $NH_4H_2PO_4$-based powder particles with different sizes were calculated according to the formula (3), and the results is shown in figure 5.

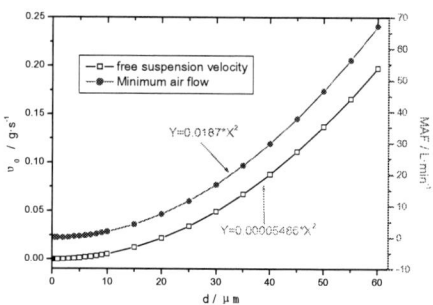

FIGURE V. SUSPENSION VELOCITIES FOR SUPERFINE NH4H2PO4 PARTICLES.

Figure 5 illustrates that the suspension velocity of 10um, 20um and 30um particle is 0.0055 m/s, 0.022 m/s and 0.049 m/s, respectively. The internal diameter of cup-burner is 85 mm, so airflow velocity is assure to be approximately above 17L/min (0.05m/s) as particle move upward during the experiment.

Considering the soft reunion and particle group effect of superfine powder, airflow velocity was chose to be 60L/min (0.18m/s) to make particle move at a certain speed and particle will move upward. The time is about 3s when particle move from cup-burner powder entrance to the burning mouth. It will take 3s at least when particle move to burning mouth after being discharging. Figure 6 shows the formation and fire extinguishing process of aerosols. As shown in Figure 6, after superfine powder is supplied for 3s, aerosols can be observed and begins interacting with flame, but the concentration of aerosols is too low to reach fire extinguishing concentration. When superfine powder is supplied for 5s, fire extinguishing concentration is achieved and fire was put out. Figure 6 illustrated that aerosols are homogenous in the cup-burner. In conclusion, the airflow velocity of 0.18m/s can assure that aerosols can be formed in cup-burner with the $NH_4H_2PO_4$-based SPEA.

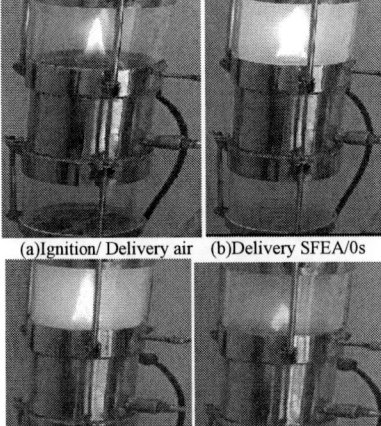

(a)Ignition/ Delivery air   (b)Delivery SFEA/0s

(c)3s   (d)5s

(e)flame extinguished

FIGURE VI. PROCESS OF SFEA AEROSOLS FORMATION AND FLAME SUPPRESSION.

C. Fire Extinguishing Performance for NH4H2PO4-based SPEA

Under the condition of that airflow velocity is 0.18m/s, the extinguishing concentration and time on ethanol and n-heptane fuel with $NH_4H_2PO_4$–based SPEA were determined, as shown in Table1.

TABLE I. EXPERIMENTAL RESULTS OF MEC WITH DIFFERENT FUELS.

Fuel	Air flow rate / $L\,min^{-1}$	Critical powder delivery rate /$mg\,s^{-1}$	Minimum extinguishing concentrations /$g\,m^{-3}$
Ethanol	60	79.8	79.8
n-heptane	60	98.8	98.8

Superfine powder moves upward under the carriage of airflow and then the mixture will interact with flame after getting into cup-burner. Superfine powder will spread into flame to extinguish fire. Although superfine powder has property similar to gas, diffusion of superfine powder is inferior. By contrast, the spreading time of superfine powder is longer. Especially, it will take much longer to achieve extinguishing concentration at the boot of flame. So, the powder concentration is referred to as fire extinguishing concentration after powder being injected into cup-burner for 30s when the fire is extinguished.

The experimental results show that the minimum extinguishing concentration of $NH_4H_2PO_4$–based SPEA on ethanol and n-heptane fuel were 79.8g/m³ and 98.8g/m³, respectively. The minimum extinguishing concentration of $NH_4H_2PO_4$–based SPEA is compared with other gaseous agents, as shown in Figure 7. It is found that the $NH_4H_2PO_4$-based SPEA is a highly efficient fire-extinguishing agent, which is an average of 2~3 times more effective than halon 1301, and an average of 5~8 times more effective than HFC-227ea.

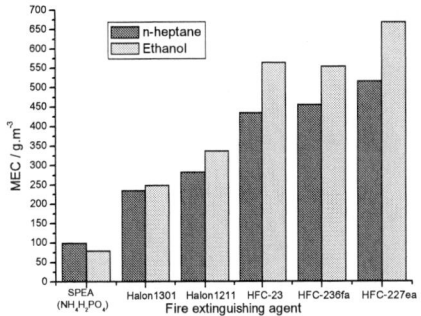

FIGURE VII. MINIMUM EXTINGUISHING CONCENTRATIONS OF DIFFERENT KINDS OF FIRE-EXTINGUISHING AGENTS.

It is indicated that the test methods based on the modified cup-burner apparatus can be used as a standard methods to evaluate suppression effectiveness and can provide a technique for comparing the performance of SPFE and gaseous agent. Also, it can provide a basis design parameters for the application of superfine powder fire extinguishing technology.

IV. CONCLUSIONS

This study involved an experimental investigation on the fire extinguishing performance of $NH_4H_2PO_4$-based SFEA in a modified cup burner apparatus, which was developed for testing the minimum extinguishing concentrations (MEC) of SFEA with different materials and particle sizes. Measurements were performed on flame burning both ethanol and n-heptane fuel using $NH_4H_2PO_4$-based powder as the SFEA. For comparison, experiments results using halon 1301 and CF_3CHFCF_3 with the same test method were also reported. The main conclusions of this study included the following:

(1) The $NH_4H_2PO_4$-based SPEA can be deliveried continuously and evenly in the range of 0.020 g/s to 0.30 g/s by the vibrating-spiral micro feeder of the modified cup-burner apparatus. Powder delivery rate was proportional to motor speed.

(2) According to the suspension-grade-model (SGD), those particles of the $NH_4H_2PO_4$-based SPEA prepared in the experiment were in the viscous resistance area by theoretical calculation. In order to ensure that all particles were upward movement, a minimum air velocity of 0.05m/s was required.

When the air flow reached 60 L/min, the $NH_4H_2PO_4$-based SPEA aerosols can be formed uniformly in the cup burner.

(3) It is demonstrated that the $NH_4H_2PO_4$-based SPEA is a highly efficient fire-extinguishing agent, which is an average of 2~3 times more effective than halon 1301, and an average of 5~8 times more effective than HFC-227ea.

(4) The test method and apparatus based on cup-burner can be used as a standard method to evalute fire extinguishing performance of SFEA. The method can provide a basis design parameters for the application of superfine powder fire extinguishing technology. Therefore, this research is helpful for the engineering design of superfine powder fire extinguishing equipment, as well as has vital significance to the fire protection.

ACKNOWLEDGEMENTS

The author acknowledgment the support of this work by the National Key Technology Research and Development Program (Project No. 2014BAK17B02),the National Natural Science Foundation of China (Grant Nos. 51176078), and the Application & Innovation Project of the MPS (Project No. 2012YYCXTJXF149)

REFERENCES

[1] Kranyansky M. Remote extinguishing of large fires with powder aerosols [J]. Fire and Materials,2006,30(5):371 -382.

[2] Zhou Xiaomeng, Jiang Lizhen, Chen Tao. Surface characteristics and fire extinguishing ability of superfine powder fire extinguishing agent[J].Journal of Combustion Science and Technology, 2009, 15(3):214-218.

[3] Du Lanping, Xie Delong, Dong Jingfei. Study on the Halon Alternative Technology[J]. Fire Science and Technonlogy, 2002,21(1):59-62.

[4] A. Hamins. Flame suppression effectiveness[R].NIST861,1994.

[5] Mikhail K. Studies of fundamental physical-chemical mechanisms and processes of flame extinguishing by powder aerosols [J]. Fire and Materials,2008,32(1) :27 -47.

[6] Hirst R. Measurement of flame-extinguishing concentration [J]. Fire technology, 1977(13):296-310

[7] A. Hamins. Flame extinction by sodium bicarbonate powder in a cup burner. Symposium (International) on Combustion,1998,27(2):2857–2864

[8] Xu Dayong, Dai Xiaoying, Hua Min. Numerical simulation of superfine powder extinguishing agent movement released in non-fire room[J]. Journal of Nanjing University of Technology, 2012,(6):130-135.

[9] Lu Zijian, Cao Wenzhong, Liu Jin. Research of Suspension of Particle in Reactor[J]. Chemical Engineering, 1997, 25(5):42-46.

International Conference on Power Electronics and Energy Engineering (PEEE 2015)

Lagrange Multiplying Method based Calculation of Material Flaw Thickness Exploiting Ultrasonic Multipath Detection

X.Z. Shen, S.J. Chen
Electrical and Automatic School
Shanghai Institute of Technology
Shanghai, China, 201418

Abstract—**A signal model based on the multipath higher order reflections of the flaw is reviewed to detect the sized flaw by ultrasound nondestructive testing, and the different paths are pre-defined and identified to form the corresponding equations. Three paths are studied, DRP, MP-C and MP-W. DRP and MP-C are utilized to measure the flaw from the top, and MP-W is from its bottom. Ultrasonic imaging is formed and synthesized by all the identified paths and all the tests to make the thickness of flaw be revealed. The thickness of the flaw is turned to two optimization problems based on different path profiles by only one pitch-catch measurement, and the problem is estimated based on Lagrange multiplying method. Simulations and experimentations demonstrate that the flaw sizing can be calculated utilizing the time-of-arrivals of the multipath signals, even with only one pitch-catch measurement.**

Keywords-ultrasonic nondestructive testing; multipath; sizing; flaw; lagrange multiplying method

I. INTRODUCTION

Sizing of material flaws is an important problem in ultrasonic nondestructive evaluation (NDE)[1]. One of the direct methods of flaw sizing is B-scan imaging from at least two different profiles, and transducer arrays are practical. However, when the space of installation of transducers is limited, it is hard to install enough sensors and the size is thus difficult to be observed. Therefore, it is necessary to develop a method to detect the Thickness with small number of transducers, or even with only one pair of transducers.

In this paper, we propose a method of sizing an isolated flaw using pitch-catch measurements.The multipahs are reviewed to show that the fusion of physical and virtual sensor data makes sizing the flaw possible[2][3]. Different from[2][3], an optimal model is given in this paper, which is solved by Lagrange multiplying method.

II. MULTIPATH SIGNAL MODEL AND SCENARIOS

A. Multipath Signal

Consider an ultrasonic pitch-catch measurement system. We model the measured ultrasound signal [2][3] at the receiving transducer,$r_{mn}(t)$ as follows,

$$r_{mn}(t) = \sum_i \alpha_{i,mn} s(t - \tau_{i,mn}) \tag{1}$$

where$s(t)$ is the ultrasonic pulse-echo wavelet, $\alpha_{i,mn}$ and $\tau_{i,mn}$ are respectively i-th path reflectivity and delay time of m-th transmitter and n-th receiver.See Figure 1.

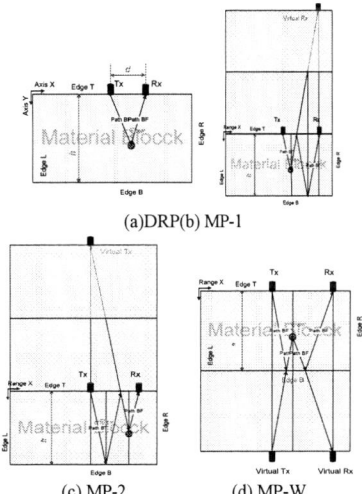

(a)DRP(b) MP-1

(c) MP-2   (d) MP-W

FIGURE I. DIRECT REFLECTION AND MULTIPATH SCENARIOS IN A PITCH-CATCH MEASUREMENT.

B. Multipath Delays and Equations

Consider a region of interest, which is a two-dimensional cross-section under the linear array, and a receiving mode backprojection beamforming algorithm is utilized to construct the image of the cross-section in interior material [3]. A sized flaw is assumed to be circle-like convex with its center at the position $p_f = (x_f, y_f, z_f)^T$ and its diameter d, and its top point at $p_{ft} = \left(x_{ft}, y_{ft} - \frac{d}{2}, z_{ft}\right)^T$, and its bottom point at $p_{fb} = \left(x_{fb}, y_{fb} + \frac{d}{2}, z_{fb}\right)^T$. The mth transmitting transducer is assumed to be located at $p_{t,m} = (x_{t,m}, y_{t,m}, z_{t,m})^T$, and the nth

© 2015. The authors - Published by Atlantis Press

receive transducer located at $p_{r,n} = (x_{r,n}, y_{r,n}, z_{r,n})^{\mathrm{T}}$. In general, we set $z_f = z_{ft} = z_{fb} = z_{t,m} = z_{r,n} = 0$.

The signal corresponding to DRP recorded at the nth receive transducer and m-th transmitter is given by

$$r_{0,mn}(t) = \alpha_{0,mn} s(t - \tau_{0,mn}) \tag{2}$$

where $\alpha_{0,mn}$ is the DRP reflectivity of the flaw, and $\tau_{0,mn}$ denotes the delay of the wavelet to travel for a round-like trip of DRP. Assuming a homogeneous material with constant ultrasonic propagation speed of v in the material, the time delay of DRP corresponding to p_{ft} can be calculated as

$$\tau_{0,mn} = (\|p_{t,m} - p_{ft}\| + \|p_{r,n} - p_{ft}\|)/v \tag{3}$$

Denote h as the thickness of the metallic object. The virtual transducer positions are respectively located at $p_{t1,m} = p_{t,m}, p_{r1,n} = (x_{r,n}, y_{r,n} - 2h, z_{r,n})$ for MP-1, $p_{t2,m} = (x_{t,m}, y_{t,m} - 2h, z_{t,m}), p_{r2,n} = p_{r,n}$ for MP-2, and $p_{tW,m} = (x_{t,m}, 2h - y_{t,m}, z_{t,m}), p_{rW,n} = (x_{r,n}, 2h - y_{r,n}, z_{r,n})$ for MP-W. The time delays of MP-1, MP-2 and MP-W corresponding to a sized target can be respectively calculated as

$$\text{MP-1: } \tau_{1,mn} = (\|p_{t1,m} - p_{ft}\| + \|p_{r1,n} - p_{ft}\|)/v \tag{4}$$

MP-2:
$$\tau_{2,mn} = (\|p_{t2,m} - p_{ft}\| + \|p_{r2,n} - p_{ft}\|)/v \tag{5}$$

MP-W:
$$\tau_{W,mn} = (\|p_{tW,m} - p_{fb}\| + \|p_{rW,n} - p_{fb}\|)/v \tag{6}$$

The ranges of multipaths, MP-1 and MP-2, are so close that they may be overlapped partly or completely. So we only can detect the combined overlapped version of MP-1 and MP-2, or the stronger one of the two if they are not overlapped, and it is denoted as MP-C. The TOA range trajectory of MP-C can be found as,

$$\tau_{C,mn} = \begin{cases} \dfrac{\|p_{t1,m} - p_{ft}\| + \|p_{r1,n} - p_{ft}\|}{v}, \\ \quad |x_{ft} - x_{tm}| < |x_{ft} - x_{rn}| \\ \dfrac{\|p_{t2,m} - p_{ft}\| + \|p_{r2,n} - p_{ft}\|}{v}, \\ \quad |x_{ft} - x_{tm}| \geq |x_{ft} - x_{rn}| \end{cases} \tag{7}$$

The top point, p_{ft}, also locates at the locus of DRP, and then the implicit function $\tau_{C,mn}$ can be obtained only with respect to x_{ft}.

C. Multipath Equations

If we have detected the above TOAs, $\tau_{0,mn}$, $\tau_{C,mn}$ and $\tau_{W,mn}$, the corresponding equations are respectively described as

DRP:
$$\sqrt{(x_{tm} - x_{ft})^2 + (y_{tm} - y_{ft})^2} + \sqrt{(x_{rn} - x_{ft})^2 + (y_{rn} - y_{ft})^2} = v\tau_{0,mn} \tag{8}$$

MP-C:
$$\begin{aligned} &\sqrt{(x_{tm} - x_{ft})^2 + (y_{tm} - y_{ft})^2} \\ &+ \sqrt{(x_{rn} - x_{ft})^2 + (y_{rn} - 2h - y_{ft})^2} \\ &= v\tau_{C,mn}, \quad |x_{ft} - x_{tm}| < |x_{ft} - x_{rn}| \\ &\sqrt{(x_{tm} - x_{ft})^2 + (y_{tm} - 2h - y_{ft})^2} \\ &+ \sqrt{(x_{rn} - x_{ft})^2 + (y_{rn} - y_{ft})^2} \\ &= v\tau_{C,mn}, \quad |x_{ft} - x_{tm}| \\ &\geq |x_{ft} - x_{rn}| \end{aligned} \tag{9}$$

MP-W:
$$\sqrt{(x_{tm} - x_{fb})^2 + (2h + y_{t,m} - y_{fb})^2} \\ + \sqrt{(x_{rn} - x_{fb})^2 + (2h + y_{rn} - y_{fb})^2} = v\tau_{W,mn} \tag{10}$$

These equations form 3 ellipses with focus points at the points of transducers. And the equation of DRP in (8) can be also expressed as

$$\frac{\left(x_{ft} - \frac{x_{tm}+x_{rn}}{2}\right)^2}{\frac{(x_{tm}-x_{rn})^2+(y_{tm}-y_{rn})^2}{4}} + \frac{\left(y_{ft} - \frac{y_{tm}+y_{rn}}{2}\right)^2}{\left(\frac{v\tau_{0,mn}}{2}\right)^2 - \frac{(x_{tm}-x_{rn})^2+(y_{tm}-y_{rn})^2}{4}} = 1 \tag{11}$$

III. Optimization Model and Its Solutions of Flaw Sizing

We redraw all the multipaths in Figure 2, where DRP with blue locus and ways, MP-C with orange locus and ways, MP-W with red locus and ways, and the flaw with red circle. The multipaths, DRP and MP-C, go through the top point of the flaw, and the multipath, MP-W, goes through the bottom point of flaw. The size of the flaw can thus be modeled as the following optimal problem,

$$d = \min\|p_{ft} - p_{fb}\|$$
$$= \min\sqrt{\left(x_{ft} - x_{fb}\right)^2 + \left(y_{ft} - y_{fb}\right)^2} \qquad (12)$$

Subject to

$$f_{0,mn} = \sqrt{\left(x_{tm} - x_{ft}\right)^2 + \left(y_{tm} - y_{ft}\right)^2}$$
$$+ \sqrt{\left(x_{rn} - x_{ft}\right)^2 + \left(y_{rn} - y_{ft}\right)^2}$$
$$- v\tau_{0,mn} = 0$$

$$f_{W,mn} = \sqrt{\left(x_{tm} - x_{fb}\right)^2 + \left(2h + y_{t,m} - y_{fb}\right)^2}$$
$$+ \sqrt{\left(x_{rn} - x_{fb}\right)^2 + \left(2h + y_{rn} - y_{fb}\right)^2}$$
$$- v\tau_{W,mn} = 0$$

$$f_{C1,mn} = \sqrt{\left(x_{tm} - x_{ft}\right)^2 + \left(y_{tm} - y_{ft}\right)^2}$$
$$+ \sqrt{\left(x_{rn} - x_{ft}\right)^2 + \left(y_{rn} - 2h - y_{ft}\right)^2}$$
$$- v\tau_{C,mn} = 0, |x_{ft} - x_{tm}| < |x_{ft} - x_{rn}|$$

$$f_{C2,mn} = \sqrt{\left(x_{tm} - x_{ft}\right)^2 + \left(y_{tm} - 2h - y_{ft}\right)^2}$$
$$+ \sqrt{\left(x_{rn} - x_{ft}\right)^2 + \left(y_{rn} - y_{ft}\right)^2} - v\tau_{C,mn}$$
$$= 0, \quad |x_{ft} - x_{tm}| \geq |x_{ft} - x_{rn}|$$

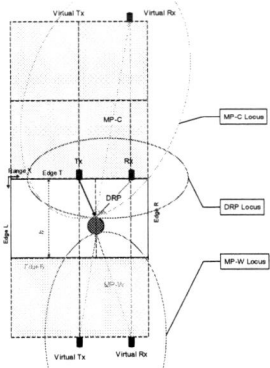

FIGURE II. DETECTION OF FLAW SIZE.

We here apply Lagrange multiplying method [6] to solve the optimization problem in(12). The segmented constrained equal equations can be seen as two problems. Thus, we construct two Lagrange functions as follows,

$$L_1 = \sqrt{\left(x_{ft} - x_{fb}\right)^2 + \left(y_{ft} - y_{fb}\right)^2}$$
$$+ \sum_{m,n}\left(\lambda_{0,mn}f_{0,mn} + \lambda_{W,mn}f_{W,mn}\right.$$
$$+ \lambda_{1,mn}f_{C1,mn}) \qquad (13)$$

and,

$$L_2 = \sqrt{\left(x_{ft} - x_{fb}\right)^2 + \left(y_{ft} - y_{fb}\right)^2}$$
$$+ \sum_{m,n}\left(\lambda_{0,mn}f_{0,mn} + \lambda_{W,mn}f_{W,mn}\right.$$
$$+ \lambda_{1,mn}f_{C2,mn}) \qquad (14)$$

As to the symmetry of MP-1 and MP-2 with respect to the center line of the transducers, the solution of the problem in (13) is theoretically the same as in (14). So we only consider one of them, such as the one in (13). However, we solve both problems and get two minimum values, and take the smaller value as the thickness of the flaw.

Now, we solve the problem in (13), and the procedure is also applied to(14). Set $x = \left[x_{ft}, y_{ft}, x_{fb}, y_{fb}, \lambda_{0,mn}, \lambda_{W,mn}, \lambda_{1,mn}\right]^{\mathrm{T}}$. The k-step update version of x is as follows,

$$x_{k+1} = x_k - \mu \frac{\mathrm{d}L}{\mathrm{d}x}\bigg|_{x=x_k} \qquad (15)$$

Here, μ is the step-size, which can be decided by the optimization step-size, $\mu_0 = \arg\min_\mu L(x_{k+1}) . \frac{\mathrm{d}L}{\mathrm{d}x}$ isthe gradient of Lagrange function, and here L can be either L_1 or L_2. When $\left|\frac{\mathrm{d}L}{\mathrm{d}x}\right|_{x=x_k}$ is less than one given terminal constants, such as 10^{-5} or the iterative number of steps is up to a given number, we terminate the updated iterative operation. After we get the minimum value of x, we can calculate the value of d in(12), and thus we get two values, d_1 and d_2 from the problems in (13) and . Then we take $d = \min(d_1, d_2)$.

IV. EXPERIMENTAL RESULTS

A. Experiment Setup

An aluminum block, alloy type 6061 is used to study the ultrasound flaw sizing. A side through hole(STH) of Φ2mm diameter is drilled into the material from side to simulate a sized specular flaw.Seven transducers are located on the top surface of the material and their x-positions related to the flaw center are aligned equally from -30 mm to 30mm with 10mm increment. 42 set of signals are sampled at 40MHz sampling rate. The DRP of the flaw appear around 16μs, whereas MP-W is located around 30.5μs. MP-C is located around 39.3μs.All the waveform envelopes show clear DRP, approximately between 16.0μs and 18.5μs, and 1st-order bottom reflection, approximately between 23.5μs and 25.1μs.

B. Ultrasonic Imaging

We have 42 set of data, and each set can be used as one path in (13) or to give overdetermined constraint equation. However, we here don't discuss how to solve the problem of so much constraint equations, but to discuss the ability of multipaths. We apply ultrasound imaging to show the possible solution of the flaw. Figure 3 shows the imaging results using different paths [3]. Figure 3a shows the image generated using DRP. The flaw top is identified. MP-W yields the bottom view of the flaw in Figure 3b. Figure 3c shows the top view obtained from MP-C. These images represent reflections at different profiles, and image fusion can be applied[3], shown in Figure 3d. The size of the hole is revealed by the fusion result, which is otherwise unavailable if only DRP observations are alone exploited. By observation, the thickness of the flaw is around 3-5mm.

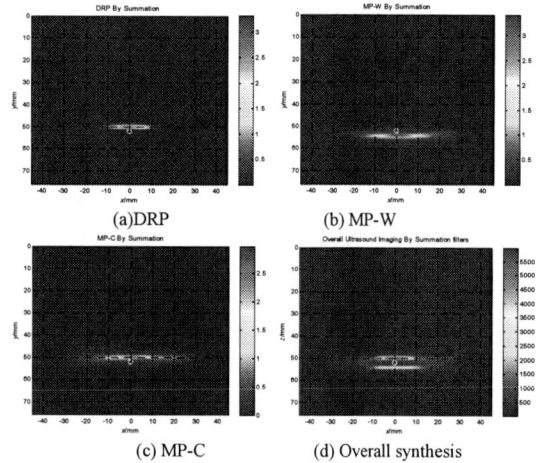

(a)DRP (b) MP-W

(c) MP-C (d) Overall synthesis

FIGURE III. ULTRASOUND IMAGING OF A HOLE VIA MULTIPATH EXPLOITATION.

C. Optimal Algorithm of Thickness

To show the ability of extended data due to the mulitpaths, we show here only one pitch-catch measurement can be applied to solve the size of the flaw. Figure4 shows the estimated thickness of the STH by the proposed method, where Figure4a shows the solutions of the thickness, the blue dotted line is solved by the optimal problem in (13) and the red dotted line is solved by the problem in . We can see that both solutions in problems (13) and are nearly the same in the sense that both the average thicknesss of the flaw is 3.3995mm. We take the minimum value of the solutions by (13) and as the estimated

thickness of the STH, and depicted in Figure4b, whose mean value is 3.19mm.

V. SUMMARY

Ultrasounds along different paths have different profiles of the flaw. We identifythe direct reflection path using conventional techniques, and then estimate the delay times of MP-C and MP-W in their active region. Thus we identify the multipaths, MP-C and MP-W. By fusing the identified ultrasonic multipaths, we can intuitively observe the size of the flaw. The flaw top point is formed by DRP and MP-C, and the flaw bottom point is exploited by MP-W. Thus the size from top to bottom of the flaw is observed by the ultrasound imaging of the flaw. At last, we turned it to an optimal problem, which is an equality constraint optimal problem. Lagrange multiplying method is applied to solve the problem, and the solution represent the thickness of flaw. Experiments shows the method is valid, although it is greater than the machined simulated hole.

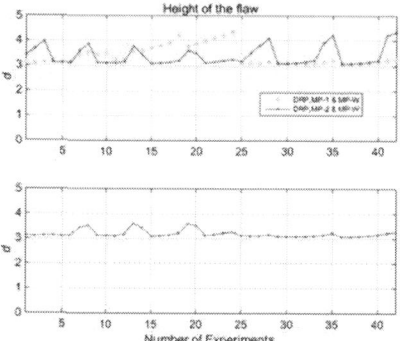

FIGURE IV. (a) ESTIMATED THICKNESS OF THE FLAW, THE BLUE DOTTED LINE BY THE OPTIMAL PROBLEM IN (17) AND THE RED DOTTED LINE BY THE PROBLEM IN (18); (b) THE MINIMUM VALUE OF THE SOLUTIONS.

ACKNOWLEDGEMENTS

This paper is supported by STCSM, No. 15ZR1440700.

REFERENCES

[1] L. W. Schmerr and S.-J. Song, Ultrasonic Nondestructive Evaluation Systems, Springer, 2007.

[2] Y. Zhang, X. Shen, R. Demirli and M. G. Amin, "Ultrasonic Flaw Imaging via Multipath Exploitation". Advances in Acoustics and Vibration. Vol. 2012 (2012), pp: 1-12.

[3] Xizhong Shen, Li Pan. Material Flaw Sizing By Ultrasonic Multipath Detection. Periodical of Advanced Materials Research Vols. 712-715, 2013, pp: 1067-1070.

[4] G. Alli and D. DiFlippo, "Beamforming for through-the-wall radar imaging," Chapter 3, in M. G. Amin (Ed.), Through the Wall Radar Imaging, CRC Press, 2010.

[5] Shen XZ, Yimin Z. et al. Ultrasound multipath background clutter mitigationBased on Subspace Analysis and projection.ICASSP 2012, March 25-30, 2012, Kyoto, Japan, pp: 2505-2508.

[6] Bertsekas D P, Nedic A, Ozdaglar A E. Convex Analysis and optimization. Athena Scientific Press, 2003.

Simple Synthesis for Hierarchical SiO$_2$ Tubes with Adjustable Mesoporous

Y.H. Zhang, Y. Deng, Y.H. Cai, W. Xiao
Research Institute for New Materials Technology
Chongqing University of Arts and Sciences
Chongqing 402160, China

L.L. Sun
School of Materials Science and Engineering
Chongqing University of Technology
Chongqing 400050, China

Abstract—With the aid of cotton template and precursor solution of mesoporous silica, hierarchical mesoporous SiO$_2$ tubes with size–adjustable mesoporous in the tube–wall were synthesized through sol–gel method. The products are tubular silica that inherit the hierarchical structure of initial cotton and contain mesopores in their wall. The size of mesoporous in the wall can be adjusted by change the mixture ratio of mesoporous silica precursor solution. Noble metal particles, silver and gold, were introduced into the surface of mesoporous silica tubes respectively to obtain a uniform distribution of noble metal nanoparticles on the tube–wall. This method can be applied to many nature substance templates with different structures to prepare various structural SiO$_2$ materials.

Keywords-natural substance template; SiO$_2$ tubes; size–adjustable; sol–gel chemistry

I. INTRODUCTION

One-dimensional silica-based nanostructural materials, such as nanotubes, have been vigorously investigated due to their specific mechanical, electrical, and chemical properties. These materials were generally synthesized by hydrolysis of TEOS on the surface of various templates, such as anodic aluminum oxide membranes,[1] crystalline fibers,[2] carbon nanotubes,[3] and silicon nanowires,[4] followed by removal of templates. Mesoporous silica materials, such as MCM-41/48, have found widespread application for industrial, technological and domestic purposes.[5,6] Combination of mesopores and tubular structure can obtain functional silica materials with complicated structures that can satisfy the demand for improved silica types with specific properties. Some researches indicated that tubular silica with a nanometer-sized hollow cavity have attracted increasing attention in natural science and materials science fields from the viewpoint of their potential applications and the continuing interest in fundamental phenomena specific to a confined nanospace.[7–9] Since the first report on the synthesis of silica tubes with mesopores in the wall by Mou's group,[10] a few mesostructured silica tubes have been prepared.[11–20] However, the simple morphology of these artificial template substrates mentioned above severely limits the variety of structures and morphologies of the products.

Here, we successfully synthesized mesoporous silica tubes through a simple and efficient method by using hierarchical cotton fibers as the hard template and cetyltrimethylammonium bromide (CTAB) micelles as the soft template This material inherits the .hierarchical morphology of cotton and the diameter of tubes range from tens nanometers to several microns. This hierarchical silica tubes with mesoporous wall was obtained by sol–gel process of tetraethyl silicate in the presence of cationic surfactant CTAB followed by the mixture covering on the cotton fibers. The size of mesopores in the wall was adjustable by varying the ratio of silica precursor. Then, we prepared Ag/silica and Au/silica composites through introduce silver and gold nanoparticles into the sample above, respectively.

II. EXPERIMENTAL

A. Materials

Cetyltrimethylammonium bromide (CTAB, 99.0%) and tetraethyl orthosilicate (TEOS, 28.4%) were obtained from Sinopharm Chemical Reagent Co., Ltd. Hydrogen tetrachloroaurate trihydrate (HAuCl$_4$·3H$_2$O, 99.99%), silver nitrate (AgNO$_3$, 99.85%) and sodium borohydride (NaBH$_4$, 99%) were purchased from Acros Organics. Milli-Q water (resistivity, 18.2 MΩ cm^{-1}) was used in all related cases. All the chemicals were used as-received without any further purification.

B. Synthesis of Mesoporous Silica tubes

In our experiment, two different kinds of mesoporous silica tubes have been prepared. Both silica samples were synthesized as follows. In a large narrow–mouth bottle, 1 g of CTAB was dissolved in a mixture of 3.5 ml of NaOH aqueous solution (2 M) and 480 ml of deionized water. After addition of 5 ml of TEOS with vigorous stirring for 10 min at 80 °C, the mixture was treated by two ways. One was directly filtered, washed with distilled water and dried at 90 °C overnight followed by calcined at 600 °C in air for 6 h. The obtained production was named MSP–1. The other method was adding the cotton in the mixture and then kept statically at the same temperature for another 10 min. Finally, the coated cotton was filtered, washed with distilled water and dried at 90 °C overnight followed by calcined at 600 °C in air for 6 h. The resulting silica product was typically in the form of tubes with mesopores in the wall, which was dominated as S–1. Sample MSP–2 and S–2 were made through the same process above. The only difference between two samples is the ratio of the initial silica precursors. The amounts of CTAB, NaOH (2 M), deionized water and TEOS were 2 g, 7 ml, 480 ml and 9 ml, respectively for S–2 sample that means the amounts of NaOH, CTAB and TEOS in the S–2 precursor were twice of those in S–1 precursor in the case of same amount of deionized water.

© 2015. The authors - Published by Atlantis Press

C. Preparation of Ag /S–1 and Au /S–1 Composites

The mesoporous silica tubes S–1 were immersed in 0.2 M AgNO$_3$ ethanol/water (1:1 v/v) solution for 2 h, followed by thorough washing with Milli–Q water and dried in air. Photoreduction of silver was achieved by subjecting the mesoporous silica tubes to UV irradiation (365 nm, 16 W) for 2 h. Then Ag /S–1 composite was obtained.

The mesoporous silica tubes S-1 were impregnating in 5 ml iced–cold 0.1 M NaBH$_4$ in ethanol/water (1:1 v/v) for 2 min, followed by washing thoroughly with deionized water and dried in air. Then, the sample was dipped in 5 ml iced-cold 2.5 × 10-4 M HAuCl$_4$ in ethanol/water (1:1 v/v) for another 2 min to obtain Au /S–1 composite

D. Characterizations

SEM observation was performed on a SIRON field emission scanning electron microscopy at an acceleration voltage of 25.0 kV. The TEM image were performed by a JEM-200CX transmission electron microscope (TEM) operating at an acceleration voltage of 160 kV, and a JEM-2010 transmission electron microscope operating at 200 kV. Adsorption and desorption isotherms of nitrogen were obtained at 77 K on a Micromeritics ASAP 2020 accelerated surface area and porosimetry analyzer. The linear part of the Brunauer–Emmett–Teller (BET) equation was used for the specific surface area determination.

III. Results and Discussion

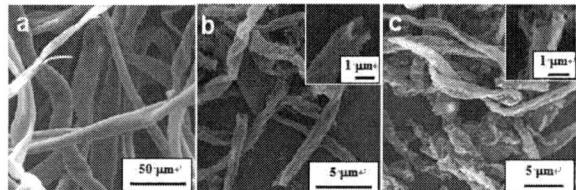

FIGURE I. FE-SEM IMAGE OF (a) COTTON FIBERS, (b) S–1 SAMPLE AND (c) S–2 SAMPLE. THE INSETS IN PANEL B AND C ARE ENLARGEMENT OF TOP PART OF SILICA TUBE.

A. Hierarchical Mesoporous Silica tubes

Cotton has a randomly crossed network of microfibers that are composed of nanofibers, and each individual nanofiber is an assemble of β-D-glucose polymer chains through multiple hydrogen bonding. The existence of surface hydroxyl groups provides a suitable platform for the deposition of a silica gel layer on the surface of cellulose nanofibers, and removal of cotton template results in the final hierarchical silica nanotubes. Figure 1a shows that bare cotton possess a macro to nanoscopic random morphological hierarchy and each fiber of cotton has a smooth surface. Figure 1b and 1c are SEM images of S–1 and S–2 samples after complete removal of the cotton fiber component, respectively, which indicate the resultant silica samples possess overall morphological characteristics of the initial cotton though the surface of each fiber became rough. The overview FE-SEM images show both sample are consisted of randomly interconnected nanotube assemblies with high aspect ratios. The outer diameters of the nanotubes range from nanometers to microns. The highly magnified SEM image of individual mesoporous silica tube

exhibits the obvious open ends nanotubular structures (Insets in Figure 1b and 1c). These observations demonstrate that the original morphology of the initial natural cellulose fibers is faithfully replicated by nanotubular silica film at nano-precision and the products can be regarded as negatively replication of the original structure of the initial cotton substance.

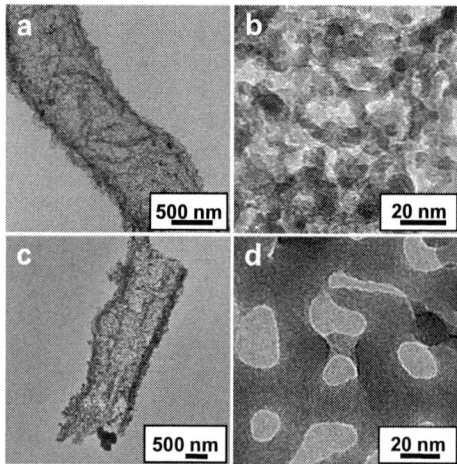

FIGURE II. (a) TEM AND (b) HRTEM IMAGES OF S–1 SAMPLE. (c) TEM AND (d) HRTEM IMAGES OF S–2 SAMPLE.

The TEM of individual tube of S–1 (Figure 2a) indicates that the tube wall of uniform thickness of 150–200 nm is clearly observed, which means the sample possesses tubular structure. HRTEM observation (Figure 2b) demonstrates more detail of the tube wall, which shows large number of mesopores with size of 10–20 nm randomly located in tube wall of the silica tube. TEM (Figure 2c) and HRTEM micrographs (Figure 2d) of S–2 indicate the S–2 is composed of silica tubes with a lot of irregular mesoporous of 15–30 nm in the wall. The difference on size of mesopores between S–1 and S–2 sample is most probably caused by the ratio variety of two initial MCM-41 precursor solutions. These results indicate that the whole samples prepared are hierarchical mesoporous silica tubes and the mesopores size can be altered by changing the initial precursor solution.

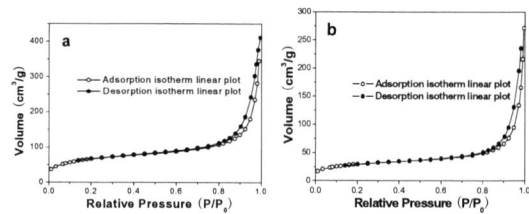

FIGURE III. N2 ADSORPTION/DESORPTION ISOTHERM OF (a) S–1 SAMPLE AND (b) S–2 SAMPLE.

Figure 3a and 3b shows the nitrogen adsorption-desorption isotherms for mesoporous silica nanotubes S–1 and S–2, respectively. The mesoporous silica nanotubes exhibit typical type IV adsorption isotherms with narrow hysteresis loops, which indicate the existence of mesopores in the structure of samples. The obvious adsorptions of N_2 at high relative pressure (e.g., $P/P_0 > 0.8$) in both isotherms indicate the formation of relative large additional mesoporosity. The pore sizes obtained from the nitrogen physisorption isotherms are 16 nm for S–1 and 23 nm for S–2 samples respectively, which agrees with those from observations of TEM images of S-1 and S-2 samples in Fig2. The large values of BET surface area (about 231.9m^2/g for S–1 sample and 102.4m^2/g for S–2 sample) and pore volume (0.63 cm^3/g for S–1 and 0.42 cm^3/g S–2) of our productions possibly attributed to the formation of larger holes due to the removal of CTAB micelles/silica composite particles at the starting of synthesis process.

B. Introduction of Ag or Au Nanoparticles in the Wall of Mesoporous Silica Tubes

Due to the large specific surface area, the mesoporous silica nanotubes provide a practical platform for noble metal Ag or Au particles. Here, the Ag nanoparticles were introduced by means of photocatalytic reduction of $AgNO_3$ to Ag. The incorporation of Au nanoparticles was through reduction of $HAuCl_4$ by $NaBH_4$. Figure 4a and 4c respectively show the FE-SEM images of individual Ag/S-1 and Au/S-1 tube, in which the cotton structure preserved perfectly that was as same as the S-1 sample (Figure 1b). It indicates that the introduction process of noble metal nanoparticles did not destroy the structure of sample. As presented in the TEM image (Figure 4c and d), a large number of individual Ag and Au nanoparticles with spherical shape are decorated on the surface of mesoporous silica nanotube. The size of these nanoparticles are quite uniform, which are measured about 7 nm and 3 nm for Ag/S-1 and Au/S-1, respectively.

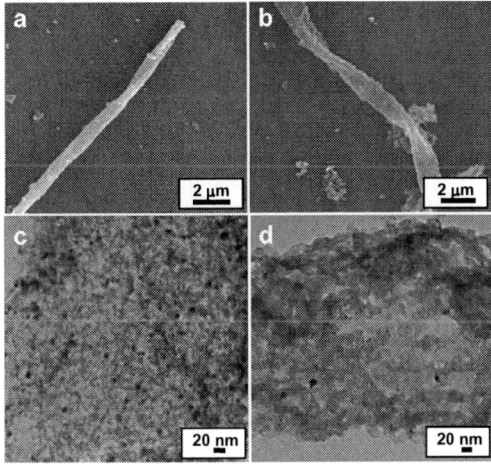

FIGURE IV. (a) SEM AND (b) TEM IMAGES OF AG/S–1 COMPOSITE. (c) SEM AND (d) TEM IMAGES OF AU/ S–1 COMPOSITE.

IV. CONCLUSIONS

Hierarchical mesoporous silica tubes have been fabricated through a simple route by employing cheap cotton cellulose fibers as hard template, which contributes to the formation of tubular structure, and mesoporous silica precursor as soft template that responsible for forming mesopores on the tube

wall. The final sample is a composite of a tubular structure and mesopores in its wall that can be adjustable by varying the solution ratio of silica precursor. The introduction of mesopores into the silica tubes can not only improve the structure but also enlarge the BET surface area of this composite, which is favor of the widely application in many fields. The currently developed method can be readily applied to prepare a variety of mesoporous silica materials with tailored structures, properties and functions for specific practical applications, such as catalysis, adsorption, microreactor, sensors, delivery and so on. Then noble metal particles, Ag and Au, have introduced the mesoporous silica tubes and made Ag (Au)/mesoporous silica tubes composite with uniform distribution Ag or Au particles.

ACKNOWLEDGMENTS

This work was financially supported by the National Natural Science Foundation of China (21101136), Project of International Cooperation and Exchanges NSFC (21310102011), the Key Project of Chinese Ministry of Education (212144), Natural Science Foundation Project of CQ CSTC (cstc2012jjA50037), Scientific and Technological Research Program of Chongqing Municipal Education Commission (No. KJ121203, 121220) and Chongqing University of Arts and Sciences (No. R2012CJ15, Y2011XC45).

REFERENCES

[1] D.T. Mitchell, S.B. Lee, L. Trofin, N. Li, T.K. Nevanen, H. Söderlund, C.R. Martin, J. Am. Chem. Soc. 124 (2002) 11864.

[2] (a) J. Zygmunt, F. Krumeich, R. Nesper, Adv. Mater. 15 (2003) 1538. (b) S. Naito, M. Ue, S. Sakai, T. Miyao, Chem. Commun. (2005) 1563.

[3] B.C. Satishkumar, A. Govindaraj, E.M. Vogl, L. Basumallick, C.N.R. Rao, J. Mater. Res. 12 (1997) 604.

[4] R. Fan, Y. Wu, D. Li, M. Yue, A. Majumdar, P. Yang, J. Am. Chem. Soc. 125 (2003) 5254.

[5] (a) J.S. Beck, J.C.Vartuli, W.J Roth, M.E.Leonowicz, C.T. Kresge, K.D. Schmitt, C.T.W. Chu, D.H. Olson, E.W. Sheppard, J. Am. Chem. Soc. 114 (1992) 10834. (b) C.T. Kresge, M.E. Leonowicz, W.J. Roth, J.C. Vartuli, J.S. Beck, Nature 359 (1992) 710.

[6] (a)A. Corma, Chem. Rev. 97 (1997) 2373. (b) N.K. Mal, M. Fujiwara, Y. Tanaka, T. Taguchi, M. Matsukata, Chem. Mater. 15 (2003) 3385. (c) S. Huh, H.T. Chen, J.W. Wiench, M. Pruski, V.S.Y. Lin, J. Am. Chem. Soc. 126 (2004) 1010. (d) M. Vallet-Regi, A. Ramila, R.P. del Real, J. Perez-Pariente, Chem. Mater. 13 (2001) 308. (e) C.Y. Lai, B.G. Trewyn, D.M. Jeftinija, K. Jeftinija, S. Xu, S. Jeftinija, V.S.Y. Lin, J. Am. Chem. Soc. 125 (2003) 4451. (f) D. Brunel, A.C. Blanc, A. Galarneau, F. Fajula, Catal. Today 73 (2002) 139.

[7] S. Iijima, Nature 354 (1991) 56.

[8] J.M. Schnur, B.R. Ratna, J.V. Selinger, A. Singh, G. Jyothi, K.R.K. Easwaran, Science 264 (1994) 945.

[9] P.G. Collins, A. Zettl, H. Bando, A. Thess, R.E. Smalley, Science 278 (1997) 100.

[10] H.P. Lin, C.Y. Mou, Science 273 (1996) 765.

[11] S.M. Yang, I. Sokolov, N. Coombs, C.T. Kresge, G.A. Ozin, Adv. Mater. 11 (1999) 1427.

[12] M. Harada, M. Adachi, Adv. Mater. 12 (2000) 839.

[13] M. Adachi, T. Harada, M. Harada, Langmuir 16 (2000) 2376.

[14] F. Kleitz, F. Marlow, G.D. Stucky, F. Schüth, Chem. Mater. 13 (2001) 3587.

[15] F. Kleitz, U. Wilczok, F. Schuth, F. Marlow, Phys. Chem. Chem. Phys. 3 (2001) 3486.

[16] F. Kleitz, F. Marlow, G.D. Stucky, F. Schüth, Chem. Mater. 13 (2001) 3587.

[17] Z.L. Yang, Z.W. Niu, X.Y. Cao, Z.Z. Yang, Y.F. Lu, Z.B. Hu, C.C. Han, Angew. Chem. Int. Ed. 42 (2003) 4201.

[18] Z.J. Liang, A.S. Susha, Chem. Eur. J. 10 (2004) 4910.

[19] Y.T. Yu, H.B. Qiu, X.W. Wu, H.C. Li, Y.S. Li, Y. Sakamoto, Y. Inoue, K. Sakamoto, O. Terasaki, S.N. Che, Adv. Funct. Mater. 18 (2008) 541.

[20] S.W. Bian, Z. Ma, L.S. Zhang, F. Niu, W.G. Song, Chem. Commun. (2009) 1261.

[21] Y. Sun, B. Wiley, Z.Y. Li, Y. Xia, J. Am. Chem. Soc. 126 (2004) 9399.

[22] S.J. Park, T.A. Taton, C.A. Mirkin, Science 295 (2002) 1503.

[23] N.L. Rosi, D.A. Giljohann, C.S. Thaxton, A.K.R. Lytton-Jean, M.S. Han, C.A. Mirkin, Science 312 (2006) 1027.

[24] M.S. Chen, D.W. Goodman, Science 306 (2004) 252.

[25] H.M. Chen, R.S. Liu, M.Y. Lo, S.C. Chang, L.D. Tsai, Y.M. Peng, J.F. Lee, J. Phys. Chem. C 112 (2008) 7522

[26] Y. Sun, Y. Xia, Science 298 (2002) 2176

International Conference on Power Electronics and Energy Engineering (PEEE 2015)

Effect of Synthesis Conditions on the Growth of ZnO Nanorods via the Solution Deposition Method

C.M. Zhang, T. Meng, S.Y. Yao, S. Huang, D.D. Wang

Beijing Institute of Graphic Communication
Beijing 102600, China

Abstract—**The ZnO nanorod films were synthesized using zinc acetate-sodium hydroxide aqueous solutions under different synthesis conditions. The effect of synthesis conditions on structural properties and morphology of ZnO films was investigated using field emission scanning electron microscopy and X-ray diffraction. The results demonstrated that the morphology of ZnO films was determined by the concentration of the precursors. The predominantant c-axis growth of hexagonal lattice was observed, which confirmed that high-quality ZnO nanorod films were obtained. At last, the mechanism of formation of varying morphologies was discussed.**

Keywords-ZnO nanorod film; solution deposition; morphology

I. INTRODUCTION

With the rapid developments in nanoscience and nanotechnology, there has been great interest in ZnO nanomaterials because ZnO is a Ⅱ-Ⅵsemiconductor with a wide and direct band gap(3.37eV) and an exaction-binding-energy of about 60meV. [1-2] As a result, it is useful for light-emitting devices, solar cells, gas sensors and laser diodes, etc.[3-6] ZnO is also environmentally friendly, stable, indefinitely and can be synthesized easily and inexpensively into different shapes, such as rods, wires, flowers, ribbons and pillars.[7] For the development of novel devices, the fabrication of ZnO nanorod films with highly oriented, aligned and ordered arrays is of critical importance, and the well-controlled synthetic procedures of ZnO nanorod films have been the focus of crystal synthesis.

Over the past few years, oriented arrays of ZnO nanorods or nanowires have been synthesized with various methods. The low-temperature aqueous solution deposition was frequently employed and is attracting considerable attention because it is easy to handle, low temperature (60-100 °C), less expensive, and environmentally amicable.[8] With this kind of method, well-aligned ZnO nanorod films have been successfully synthesized in zinc salt aqueous solutions using different chelating agents, such as hexamethylenetetramine (HMT), dimethylamineborane (DMAB), and ethane-1,2-diamine(EN). Recently, zinc nitrate and sodium hydroxide solution without any surfactants was reported to synthesize ZnO with different shapes such as stars, nultipods, spikes and nanorods, and the system of zinc acetate and sodium hydroxide solution was also used to fabricated bunch-shaped nanowires.[9-11] In the present work, we selected the zinc acetate and sodium hydroxide system, and employed the aqueous solution deposition method

to synthesize the nanorod films. The influence of the molar ratio of Zn^{2+} to NaOH in the morphologies and the structures of the films was studied.

II. EXPERIMENTAL DETAILS

Firstly, a textured ZnO seed layer was prepared on a glass substrate. The glass substrate was spin-coated with 0.0025M zinc acetate dehydrate-ethanol solution, and then was transferred in muffle for treatment at 400 °C for 30 min. Then the aqueous solution containing $Zn(CH_3COO)_2$ and NaOH with different molar ratios was prepared at ambient temperature by stirring for 20min. Subsequently, pretreated glass substrates were submerged in the aqueous solutions. Then the solution was heated to 65°C and the growth time was 6 hours. Finally, as-deposited substrates were taken out and washed with deionized water and ethanol, and then dried at room temperature.

The morphology of ZnO films were examined by using a field emission scanning electron microscope (FESEM; S4800, Hitachi). The crystalline phase of products was identified by X-ray diffraction (XRD; D8 ADVANCE, Bruker) using Cu Ka radiation (λ= 0.15406 nm).

III. RESULTS AND DISCUSSIONS

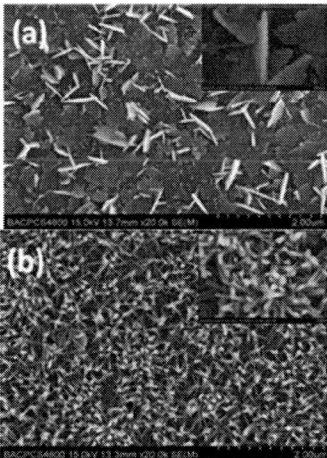

FIGURE I. FE-SEM IMAGES OF ZnO FILMS GROWN UNDER DIFFERENT MOLAR RATIO OF NaOH to Zn^{2+}: (a)6:1, (b)8:1.

© 2015. The authors - Published by Atlantis Press

FIGURE II. FE-SEM IMAGES OF ZnO FILMS GROWN UNDER DIFFERENT MOLAR RATIO OF NaOH to Zn^{2+}: (a)10:1,(b)12:1,(c)16:1,(d)20:1.

In this study, ZnO films were grown in aqueous solution of zinc acetate dehydrate and sodium hydroxide. NaOH provides the hydroxide ions (OH^-) to the solution. The initial concentration of OH^- in the solution had a great influence on the growth of ZnO films. So a set of samples were grown on the pre-treated substrates under the condition of different molar ratio between NaOH and $Zn(CH_3COO)_2 \cdot 2H_2O$, in order to investigate the effect of the concentration of NaOH on the structures of ZnO films. In the precursory solution, the initial zinc concentration was 0.05M, which was kept as a constant and the molar ratio of NaOH to $Zn(CH_3COO)_2 \cdot 2H_2O$ changed from 6:1 to 25:1.

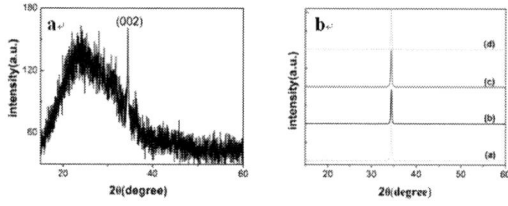

FIGURE III. (a) XRD PATTERN OF ZnO FILMS PREPARED UNDER THE CONDITION THAT THE MOLAR RATIO OF NaOH TO Zn^{2+} WAS 8:1. (B) XRD PATTERN OF ZnO FILMS GROWN UNDER DIFFERENT MOLAR RATIO OF NaOH TO Zn^{2+}: (a)10:1,(b)12:1, (c)16:1,(d)20:1.

Figure1-2 showed the morphological evolution of ZnO films as growth conditions vary with the molar ratio. When the molar ratio was 6:1, randomly distributed nanosheets were observed on the surface of the substrate (Figure1 (a)). The size of the nanosheet was about 300nm and the thickness was about 25nm. The magnified view (inset image) showed that the surface of the nanosheet was rough and was aggregated by nanoparticles. However, for samples under higher molar ratio (Figure1 (b)), the low density nanorod film was obtained and the rod diameter was about 35nm. Then with the further increasing of the molar ratio (Figure2), dense nanorod arrays formed and the diameter of the nanorod increased from 60nm to 200nm. When the molar ratio reached 25:1, there was no ZnO nanorod on substrate, and even the ZnO seed film on substrate

was dissolved by OH^-. The result displayed a broad scope of the molar ratio allowable for rod growth in this system and the diameter of nanorod can be adjusted by the molar ratio of NaOH to Zn^{2+}. Most top ends of ZnO nanorods was well faceted and flat hexagonal symmetry, illustrating the nanorods grow along c-axis direction.

Figure3 showed the XRD diffraction pattern of ZnO films prepared with different molar ratio. For all the samples, the significantly high intensity of (002) diffraction peak was observed, indicating that ZnO were preferentially oriented along the c-axis direction as well.

In this work, the possible reaction process for the growth of ZnO film can be described as the following equation[10]:

$$Zn(CH_3COO)_2 \cdot 2H_2O \rightarrow Zn^{2+} + 2CH_3COO^- + 2H_2O \quad (1)$$

$$Zn^{2+} + 2OH^- \rightarrow Zn(OH)_2 \quad (2)$$

$$Zn(OH)_2 + 2OH^- \rightarrow [Zn(OH)_4]^{2-} \quad (3)$$

$$[Zn(OH)_4]^{2-} \leftrightarrow ZnO + 2OH^- + H_2O \quad (4)$$

Zinc acetate dehydrate provides Zn^{2+} ions required for growing ZnO nanorod and NaOH produces OH^- in water solution. When the molar ratio of NaOH to Zn^{2+} changed from 10:1 to 20:1, the form of zinc ion in the solution should be predominantly $[Zn(OH)_4]^{2-}$ (Eq.2,3). $[Zn(OH)_4]^{2-}$ can be adsorbed on the positively charged (0001) plane of ZnO seeds on the substrate. Then the crystal structure of ZnO was gradually constructed by dehydration between OH^- present on the (0001) plane and the OH^- ligands of the hydroxyl complexes (Eq.4).[12] It is well known that that the crystal ZnO was the polar crystal, and polar faces ((0001) and (0001) faces) with surface dipoles are thermodynamically less stable than nonpolar faces, often undergo rearrangement to minimize their surface energy and also tend to grow more rapidly. So ZnO tends to grow as an one-dimensional nanostructure along the c-axis with (0001) and form the rod shape.During the growth of ZnO, the supersaturation was the key driving force. The Zn^{2+} ion and NaOH concentrations should be varied to create the degree of supersaturation.[11-12] If the molar ratio of NaOH to Zn^{2+} was higher than the optimal, ZnO cannot form on the substrate, and even the ZnO seeds were etched away (Eq.3); if too low, most of the zinc ions precipitated out of solution (Eq.2), and the low concentration of $[Zn(OH)_4]^{2-}$ was not benifical for the growth of nanorods. When the molar ratio was 6:1, a large amount of $Zn(OH)_2$ appeared in the solution, and $Zn(OH)_2$ can be tansformed to ZnO by dehydration and hence nucleation of ZnO occurred. Then the nuclei aggregated to nanosheet.[9] With the increasing of the molar ratio, the equilibrium shifted to left (Eq.3) and less ZnO precipitated out by homogeneous nucleation. As a result, the solution became favorable for rod growth and the nanorods were synthesized. With the further increasing of the molar ratio, the higher concentration of

$[Zn(OH)_4]^{2-}$ improved the growth rate of ZnO rod and prompted precursors to accumulate together, which result in high density of nanorod and the increase of the rod diameter.

IV. CONCLUSIONS

A system deriving from $Zn(CH_3COO)_2 \cdot 2H_2O/NaOH$ was demonstrated as suitable to prepare ZnO nanorod arrays on the ZnO-coated glass substrate. The preparing conditions such as precursor concentration, the molar ratio of NaOH to Zn^{2+} have a great influence on the morphology of ZnOnanorods. The predominant c-axis growth of hexagonal lattice was observed, which confirmed that high-quality ZnO nanorod films were obtained. At last, the mechanism of formation of varying morphologies was discussed in detail.

ACKNOWLEDGMENTS

This work was financially supported by Initial funding for the Doctoral Program of BIGC (09000114/129),Research Project of Beijing institute of graphic communication (23190114030) and the project of Beijing college students' scientific research (081501141056).

REFERENCES

[1] Taeseup Song, Jae WoongChoung, Won Il Park, John A.Rogers, Ungyu Paik, Advanced Materials.20 (2008) 4464.

[2] H.Chik, J.Liang, S.Cloutier, N.Kouklin, J.Xu, Applied Physics Letters.84 (2004) 3376.

[3] WaldoJ.E.Beek, Martijn M. Wienk, MartijnKemerink, Xiaoniu Yang, RenA.J.Janssen, J.Phys.Chem.B, 109 (2005) 9505.

[4] D.I.Suh, S.Y.Lee, T.H.Kim, J.M.Chun, E.K.Suh, O.B.Yang, S.K.Lee, Chemical Physics Letters. 442 (2007) 348.

[5] C.H.Ku, J.J.Wu, Applied Physics Letters. 91 (2007) 093117.

[6] R.Thitima, C.Patcharee, S. Takashi, Y. Susumu, Solid-State Electronics. 53 (2009) 176.

[7] R.C.Pawar, J.S.Shaikh, A.A.Babar, P.M.Dhere, P.S.Patil, Solar Energy. 85 (2011) 1119.

[8] C.M.Shin, J.H.Heo, J.H.Park, T.M.Lee, H.Ryu, B.C.Shin, W.J.Lee, H.K.Kim, Physica E. 43, (2010) 54.

[9] S.Navaladian, B.Viswanathan, J. Nancsci. Nanotechnol.11 (2011) 10219.

[10] Xiulan Hu, Yoshitake Masuda, Tatsuke Ohji, Nagahiro Saito, Kazumi Kato, J. Nancsci. Nanotechnol. 11 (2011) 1.

[11] Juan Zhao, Zhengguo Jin, Xiaoxin Liu, Zhifeng Liu, Journal of the European Ceramic Society. 26 (2006) 3745.

[12] Renee B. Peterson, Clark L. Fields, Brian A, Gregg. Langmuir. 20 (2004) 5114.

International Conference on Power Electronics and Energy Engineering (PEEE 2015)

Study of the Numerical Parameter Optimization Method based on Nano-indentation Process

J.S. Ding, G.Q. Shi, G.F. Shi
Changchun University of Science and Technology
Ji Lin, China, 130012

Abstract—In order to solve critical process parameters characterize aluminum grating is proposed to optimize the impact of stress calculation values are based on a nano-indentation experiment - Method strain relationship parameters. By establishing numerical approximation model to fit the experimental data of training, and the experimental data for numerical optimization. Experimental results show that: the method to calculate the yield strength aluminum material property parameters, namely strain hardening exponent, and obtained more satisfactory aluminum material properties parameter elastic modulus and yield strength relationship by comparing the value of the deviation can be found in the obvious effectiveness of this approach, but also proved a reliable orthogonal experiment. The research process parameters characterize the relationship between the aluminum grating characterization and constitutive model provides a reference and basis for, and also provide a basis for the development of intelligent raster scored.

Keywords-grating aluminum film; substrate; nano-indentation; grating ruling

I. INTRODUCTION

Mechanical process issues grating ruling is consistent is the key to troubled grating manufacturing personnel, and long, the film still has not mastered the mechanical properties of the grating and the strain constitutive relation, because the stress grating film scored-strain characterization of the relationship between a aspects dependent on the physical properties of the film material and additional material of the grating itself, more importantly, on the other hand depends on both the equipment and the engraving tool scribe physical geometry. This makes it difficult to extract relations or discriminating between direct and deformation in the experiment. Therefore, to explore suitable for grating fabrication process today, reveal the mechanism of the grating mechanical property characterization constitutive relationship is particularly important characterization methods[1] [2].

US Richardson Grating Laboratory (Richardson Grating Laboratory) after a long period of research and exploration, has been carved out of the world can be directly largest grating, the maximum area of up to 400mm×600mm [3]. Reed et al., The film separated from the substrate to carry out a conventional tensile test [4], but this test method is particularly difficult to implement, and is generally difficult to obtain the desired tensile curve, and therefore such a method is not easy to obtain a wide range of applications. Shut [5] and Bader [6] Similarly tried separately release film study, just peeling method (thermal stress method) which uses a different,

and they have a relatively large film temperature change during the study, so the use of thermal stress the method difficult to obtain mechanical strength of the film was difficult to apply to the film material at room temperature. Shuai Hua Shang et al [7] using the finite element method with the indentation method of combining layers of composite material W/Al bilayer metal film department to do some research to prove that the elastic modulus and hardness of the W film number the average of the corresponding bulk material parameters varied.

Based on the scoring process of the existing research on the grating is proposed based on indirect indentation experiments, a lot of data, and the data were fitted and gradually approaching to reflect grating constitutive relations approach. Aims to characterize the process of grating material stress - strain constitutive relationship provides the reference and basis, and also provides the basis for the development of intelligent raster scored.

II. THE MECHANISM CHARACTERIZATION OF THE GRATING

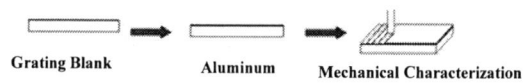

Grating Blank   **Aluminum**   **Mechanical Characterization**

FIGURE I. DEPICTS A SIMPLE FLOW CHART OF MANUFACTURING GRATINGS.

As shown in Figure 1, this paper is coated grating blank one or more layers of metal films on a glass substrate and then coated by a diamond knife to perform characterization, in scribing knife through extrusion coating process to make it happen plastic deformation, to form the desired "Groove", i.e., it is a simple process for preparing a ruled grating. Study the fundamental reason is not mastered the mechanical constitutive relation raster layers in the composite aluminum and aluminum grating groove formed under the laws graver understanding of the role is unclear, thus affecting the design and subsequent diamond knife scribe solving process parameters, and even affect the replication slotted grating production quality. Therefore, the mechanical properties of aluminum manufacturing clear grating is particularly important.

III. GRATING RULING ESTABLISHED THE MODEL FITTING APPROACH

Based on material mechanics, stress elastoplastic material base member-strain relations power hardening model,

© 2015. The authors - Published by Atlantis Press

222

$$\sigma = \begin{cases} E\varepsilon & \sigma \le \sigma_y \\ R\varepsilon^n & \sigma \ge \sigma_y \end{cases} \qquad (1)$$

When $\sigma = \sigma_y$, there

$$\sigma_y = E\varepsilon_y = R\varepsilon^n \qquad (2)$$

Wherein the stress, ε is the strain of the total, ε_y when the stress σ_y corresponding strain, i.e. the yield strain, R is the coefficient, and

$$\varepsilon = \varepsilon_y + \varepsilon_p \qquad (3)$$

Where, ε_p total strain to remove part of the linear part of the strain can be called plastic strain.

So the load P can be represented by the following parameters:

$$P = P(h, E, v, E_i, v_i, \sigma_y, n) \qquad (4)$$

Wherein, E, v, Ei and vi are compressed material and the modulus of elasticity and Poisson's ratio of the indenter; σ_y is the yield strength of the pressed material; n is pressed material strain hardening index. While the above formula can be expressed as:

$$P = P(h, E^*, \sigma_r, n) \qquad (5)$$

Take σ_r and h as the basic amount, according to the dimension of Π theorem, the above equation can be written style,

$$C = \frac{P}{h^2} = \sigma_r \Pi_1(\frac{E^*}{\sigma_r}, n) \qquad (6)$$

Thus, the constitutive relation to aluminum expressed as a quadratic curve fitting equation relationship, then it will use nano-indentation measured data for training and approximation, numerical model has been reached for grating ruling process.

IV. NANO-INDENTATION EXPERIMENTS

A. Sample Preparation

79g / mm rough structure of aluminum-chromium film - glass substrate layer composite structures were coated by vacuum evaporation method crossing. Aluminized chromium plating film with a thickness of $10\,\mu m$, $1\,\mu m$, respectively, a chrome film is the elastic modulus and Poisson's ratio were 240GPa, 70GPa, 0.3, a glass substrate elastic modulus and Poisson's ratio were 70GPa, 0.3.

B. Test Method

Swiss CSM's commercial nanoindentation testing research 79g / mm echelle rough, indenter Berkovich indenter for. Indentation depth coverage Echelle mechanical scribing depth of 0.5-$5\,\mu m$, depth interval of $0.5\,\mu m$, the depth of each test was repeated six groups, loaded (Time 10s) by controlling the displacement of the way, Paul load (time 10s), uninstall (time 10s), a total of 60 group test.

C. Test Results

Figure 2 is a resultant indentation load - displacement curve, Table 1 is a set of data indentation. From the figure it can be observed slight curve fluctuations, but the overall trend is in line with load - unload law. Show pushed the process more reliable.

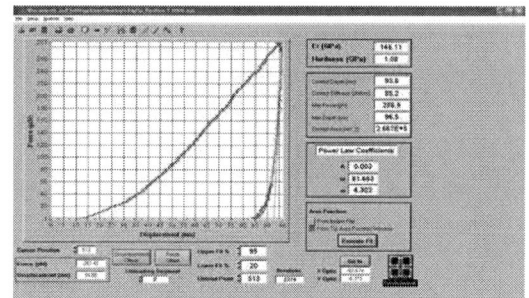

FIGURE II. ILLUSTRATES THE INDENTATION CURVES.

TABLE I. A GROUP OF INDENTATION DATA.

$h_{max}/\mu m$	K	x	H	E /GPa	K_{mean}	x_{mean}	H_{mean} /GPa	E_{mean} /GPa
0.5	8.138	1.091	0.8634	90.4119	13.359	1.053 2	0.856	87.668
	29.8	0.897	0.7909	85.9201				
	6.094	1.151	0.8596	87.05953				
	4.864	1.18	0.9156	81.93206				
	8.219	1.1	0.8815	94.21422				
	23.04	0.9	0.8263	86.47135				

V. ORTHOGONAL DATA APPROXIMATION AND OPTIMIZATION

During orthogonal optimization, load curve and select analog instrumentation indentation load curve of the same deep pressure 10 points (0.5 to $3\,\mu m$, interval $0.5\,\mu m$) corresponding to the load deviation Δi and $\Sigma\Delta i$ to optimize the target amount. Figure 3 shows the simulated load test load curve deviation between the curves.

The experimental data obtained in the test group before adding 5, all relative optimization target amount $\Sigma\Delta$ corresponding sum is about to take a first level of factor E was 605.0805, denoted Ⅰ. The data from the second set of experiments 5 obtained by adding all the corresponding optimization target amount relative $\Sigma\Delta$ adding factor E is about to take 2 level was 566.6331, denoted Ⅱ. The third set of numbers 11,12,13,14,15 five experimental data obtained by adding 543.7966, denoted Ⅲ. The fourth set of experimental data obtained by adding 518.7891, denoted Ⅳ. The fifth set of experimental data obtained by adding 498.5228, denoted Ⅴ. Characterized by poor quantity of various factors on the extent of the impact test objectives, that is, 5 groups and subtracting the minimum and maximum recorded as R. The larger the poor R factors affecting its corresponding test target and the greater the amount of $\Sigma\Delta$ analyze orthogonal table can be found in important order for each factor is., The curve shown in Figure 4 after optimization.

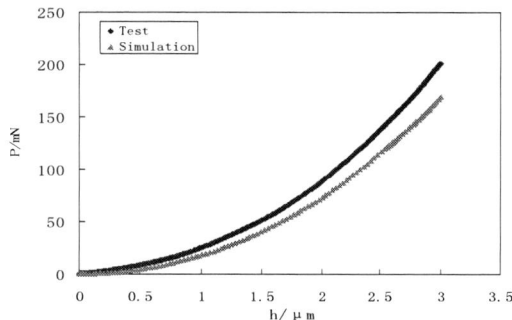

FIGURE III. THE SIMULATION OF LOAD DEVIATION BETWEEN THE CURVE AND THE TEST LOAD CURVE.

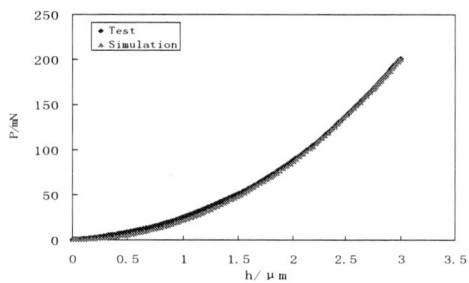

FIGURE IV. THE PARAMETRIC TEST LOAD CURVE OPTIMIZED LOADING CURVE WITH THE SAME ANALOG SCALES.

As shown, the effect is obvious, $\Sigma\Delta$ only 13.16mN, you can find it both desirable than either a set of orthogonal test to prove the reliability of the orthogonal experiment also proved the feasibility of the proposed method.

VI. CONCLUSION

By characterizing the constitutive relation of aluminum grating, nano-indentation experiments and numerical approximation research fitting method to obtain the following conclusions:

(1) Based on the analysis carved aluminum grating technology, the establishment of an aluminum grating dimensional characterizations approximation model;

(2) study the numerical approximation algorithm experimentally obtained by nano-indentation data, calculate the yield strength aluminum material property parameters, namely strain hardening exponent, and obtained more satisfactory aluminum material properties parameter elastic modulus and yield strength relationship by comparing the value of the deviation can be found in the obvious effectiveness of this approach, but also proved a reliable orthogonal experiments, provides a reference and basis for the characterization of the grating parameters characterize the relationship between aluminum and the constitutive model.

REFERENCE

[1] Fang T H, Jian SH R, Chu D S.Investigation of material properties of thin copper films[J].Applied Surface Science,2004,228(4):365.

[2] Ahn J H,Kwon D.Investigation of material properties through finite element modeling of microindentation test[J].Materials Science and Engineering A,2000,285(1):172.

[3] Erwin G. Loewen and Robert S. Wiley,Large diffraction grating ruling engine with nanometer digital control system,Proc. SPIE.[J].1987:88~95.

[4] Reed D T, Daully J W. A new method for measuring the strength and ductility of thin films [J]. J. Mater. Res., 1993, 8(7): 1542-1550.

[5] Shut C J , Cohen J B. Determination of yielding and debonding in Al-Cu thin films from residual stress measurements via diffracrtion [J]. J. Mater. Res, 1991, 6 (5): 950- 956.

[6] Bader S. Comparison of mechanical properties and microstructure of Al (1wt, %Si) and Al (1wt, %Si, 0.5wt, %Cu) thin films [J]. Thin Solid Films, 1995, 263(2): 175-181.

[7] SHANG Shuaihua, YANG Ping, LI Chun. Nano-indentation experiment of W/Al bilayer-film system and its finite element simulation[J]. ELECTRONIC COMPONENTS AND MATERIALS. 2009,28(11):60-63

Gefitinib, as a New Stent Coating Material, Specifically Inhibits Smooth Muscle Cells Proliferation Through Inhibition of EGFR/Akt Pathway Phosphorylation

F. Li, S.Y. Wang, J. Luo, Z.X. Wu, T. Xiao, O. Zeng, J. Yang*

The First Affiliated Hospital of University of South China
Hengyang, China, 421001

C. Chu

The Second Affiliated Hospital of University of South China
Hengyang, China, 421001
*Corresponding author

Abstract—The objective of this paper is to investigate the effects of gefitinib as a new stent coating material on proliferation of smooth muscle cells and the expression and phosphorylation of EGFR/Akt protein. Rat smooth muscle cells were cultured in medium with gefitinib (10^{-2}μmol/L-10μmol/L) for 24h-72h. MTT assay was used to test the inhibition of cell proliferation. Western-blot was used to detect the expression of EGFR, Akt, phosphorylated EGFR (p-EGFR) and phosphorylated Akt (p-Akt). MTT assay showed that the inhibitory effect of gefitinib on smooth muscle cells' proliferation was in a time and concentration dependent manner. Western-blot showed the expression of EGFR and Akt has no significant change between gefitinib group and paclitaxel group in smooth muscle cells, but gefitinib could significantly inhibit the phosphorylation of EGFR and Akt in smooth muscle cells compared with paclitaxel. It is concluded that Gefitinib could significantly suppress the proliferation of smooth muscle cells; the mechanism might be by inhibiting the phosphorylation of EGFR and Akt.

Keywords-coating material; cytotoxicity; stent; gefitinib; smooth muscle cells; phosphorylation

I. INTRODUCTION

Percutaneous Coronary Intervention(PCI)is the main method to treat the coronary heart disease at present. Drug-eluting stents have been widely used to reduce the occurrence of in-stent restenosis. However, as stent coating materials of drug-eluting stents, paclitaxel and rapamycin may cause endothelial damage and stent thrombosis, which is due to the non-selective cytotoxicity of the coating materials of the drug-eluting stent. To develop new stent coating materials, it is critical to selectively inhibit smooth muscle cells proliferation and makes the effect to endothelial repair decrease to the minimum extent.

A main mechanism of restenosis after Percutaneous Coronary Intervention (PCI) is that smooth muscle cells migrate to injured vascular intima and keep on proliferating after phenotypic transformation[1, 2] while the proliferation and migration of endothelial cells contribute to repair injured vascular intima and may prevent from thrombosis[3]. Currently, the main research direction of prevention and treatment on restenosis is to effectively inhibit the excessive proliferation of smooth muscle cells (SMCs) and reduce damage to endothelial cells (ECs). Epidermal Growth Factor Receptor (EGFR) is involved in the phenotypic transformation of smooth muscle cells and plays an important role in the regulation of signal pathway during the development of restenosis[4]. Gefitinib as an EGFR inhibitor can bind to intracellular region sites of EGFR tyrosine kinase together with ATP in a competitive manner, and obviously inhibits autophosphorylation of tyrosine kinase on surface receptor of EGFR transmembrane cell so as to inhibit cell proliferation[5]. EGFR is required for Akt activation. Blocking EGFR signalling amplifies the apoptotic response to TGF-beta1. Our previous research had shown that Gifitinib could inhibit the proliferation of SMCs without effecting ECs. This experiment is to investigate the effects and possible mechanisms of gefitinib on proliferation of smooth muscle cells in order to provide new ideas to clinically prevent and reduce restenosis after PCI.

II. MATERIALS AND METHODS

A. Cells and Main Reagents

Rat vascular smooth muscle cells were provided by the Experimental Animal Center, University of South China. Gefitinib and paclitaxel were purchased from AstraZeneca (England). Antibodies for EGFR, Akt, phospho-EGFR and phospho-Akt were purchased from Cell Signaling Technology (Beverly, MA).

B. Cell Culture

Rat vascular smooth muscle cells were cultured in medium, supplemented with 10% FBS at 37°C in a humidified incubator with 5% CO_2. When they grew to 80% area of the Petri dish, the cells were subcultured 2 to 3 times after digestion with 0.25% trypsin.

C. MTT Assay

About 2×10^4/ml cell suspension was made from logarithmic phase cells, 200 μl/hole of it was added to 96 well plates, then cultured in the incubator. Two hours later after cells were adhered, Gifitinib and paclitaxel with different concentrations (10^{-2}μmol/L, 10^{-1}μmol/L, 1μmol/L, 10μmol/L)

© 2015. The authors - Published by Atlantis Press

were added into the cells respectively and wait for 48 hours. Meanwhile, Gifitinib and paclitaxel(1μmol/L) were added to smooth muscle cells for 24h, 48h and 72h, respectively. After this, 20μl MTT was added into every hole, and 4 hours later, supplemented with 150μl 10% SDS. The absorbance at 490 nm was recorded using a 96-well microplate reader.

D. Western Blot Analysis

Cells were lysed in protein lysis buffer and then were quantified. Each sample was subjected to 10% SDS-PAGE and the separated proteins were transferred to PVDF membranes. The membranes were incubated with EGFR, Akt, phospho-EGFR and phospho-Akt antibody respectively. Then primary antibodies were detected with a secondary antibody and finally the membranes were subjected to chemiluminescence detection assay.

E. Statistical Analysis

Data were shown as means ± Standard Error (SE). The inhibition data of Gifitinib and paclitaxel on smooth muscle cells were analyzed by SPSS (V18.0) using one-way analysis of variance (ANOVA). P <0.05 was considered to be statistically significant.

III. RESULTS

A. Gefitinib Inhibits the Proliferation of Smooth Muscle Cells

Smooth muscle cells were exposed to Gefitinib with four different concentrations (from 10-2μmol/L to 10μmol/L) for 48h. It showed that Gefitinib inhibited the proliferation of smooth muscle cells in a concentration-dependent manner (Figure 1). Then smooth muscle cells were dealt with Gefitinib (1μmol/L) for 24h, 48h and 72h respectively. And the results showed that Gefitinib could inhibit the proliferation of smooth muscle cells in a time-dependent manner (Figure 1).

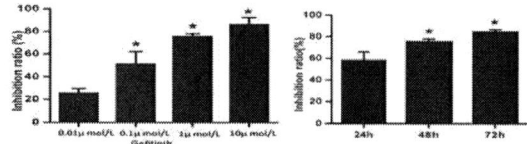

FIGURE I. THE INHIBITION EFFECT OF GEFITINIB ON PROLIFERATION OF SMOOTH MUSCLE CELLS WITH DIFFERENT CONCENTRATIONS AND TIMe.(*P<0.05, VS 0.01μmol/L GROUP OR 24h GROUP).

B. Gefitinib Affects the Phosphorylation of EGFR of Smooth Muscle Cells

Western-blot was used to test the expression level of EGFR and pEGFR after smooth muscle cells had been treated with Gefitinib for 48h. The expression of EGFR showed no statistical difference between the two groups (*P*>0.05), while expression of pEGFR in Gefitinib group was much less than the control group. Gefitinib inhibited EGFR protein phosphorylation of smooth muscle cells obviously (*P*<0.05) (Figure2).

C. Gefitinib Affects the Phosphorylation of Akt of Smooth Muscle Cells

Western-blot was used to test expression level of Akt and

pAkt after smooth muscle cells had been treated with Gefitinib for 48h. It showed that expression of pAkt in smooth muscle cells was much lower by treating with Gefitinib (*P*<0.05) while expression of Akt showed no statistical difference (*P*>0.05). Gefitinib could inhibit Akt protein phosphorylation of smooth muscle cells (Figure2).

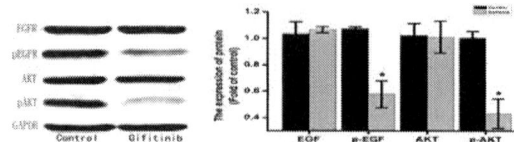

FIGURE II. THE EFFECTS OF GIFITINIB ON EXPRESSION AND PHOSPHORYLATION OF EFGR/AKT OF SMOOTH MUSCLE CELLS.(*P<0.05, VS CONTROL GROUP).

IV. DISCUSSION

Percutaneous Coronary Intervention (PCI) has become an important treatment method for vascular reconstruction. It can significantly reduce the incidence of acute cardiac events and related mortality. However, it brings a high rate of restenosis after PCI, in which its main mechanism is related with the migration of smooth muscle cells to intima and its proliferation after phenotypic transformation. Drug-eluting stents can inhibit proliferation of smooth muscle cells and then significantly reduce the rate of restenosis after PCI [6]. Paclitaxel and rapamycin, as a new stent coating material, have been proven to be effective in inhibiting VSMCs proliferation and now are widely used in drug-eluting stents to prevent restenosis after PCI. However, there is still something unsatisfactory. Though paclitaxel and rapamycin can inhibit proliferation of smooth muscle cells, it would either produce significant cytotoxicity on endothelial cells to inhibit endothelial repair and result in incomplete or delayed stent endothelialization. Therefore, in order to prevent stent restenosis and thrombosis, it is critical to choose a drug that can inhibit the proliferation and migration of smooth muscle cells to the greatest extent and meanwhile inhibit endothelial repair to the minimum extent. Our previous researches had shown that Gifitinib, an EGFR inhibitor, could effectively inhibit the proliferation of smooth muscle cells; meanwhile, its inhibition on endothelial cells is significantly small. Gefitinib may bind to intracellular region of EGFR to inhibit phosphorylation of EGFR and further proliferation of smooth muscle cells. EGFR is a very important outpost of information transfer [7]. Once the receptor binds to its ligand, it can promote cellular migration, differentiation and proliferation [8]. Therefore, it is possible that the AKT pathway plays a role in vascular smooth cell proliferation and apoptosis, and its abnormality leads to restenosis. In addition, recent studies have suggested that AKT activation is facilitated via EGFR ligand secretion, EGFR activation and subsequent c-Src phosphorylation. In this study, gefitinib could suppress the phosphorylation of EGFR protein on surface of smooth muscle cells, and it has a dose-time dependent manner. It is also observed that Akt protein expression of smooth muscle cells has been reduced by gefitinib, particularly obvious in down-regulation of the expression of phosphorylated Akt protein. The results suggest that gefitinib has a significant inhibitory effect on phosphorylation of EGFR/Akt pathway of the smooth muscle cells, which may be related to its relatively selective cytotoxic effect.

It can be concluded from the obtained results that gefitinib, as an EGFR inhibitor, can influence phosphorylation of its downstream Akt protein by inhibiting the auto-phosphorylation of EGFR, and then inhibit celluar proliferation. EGFR is one of the key signaling pathways in excessive proliferation of smooth muscle cells. It can be observed in this experiment that, either EGFR targeted drugs gefitinib or TS-oriented drug, their specificity may inhibit proliferation of smooth muscle cells. The cytotoxic effects of both drugs on endothelial cells are relatively small, while TS is a drug with more potential drug. These cytotoxic drugs particularly targeting the smooth muscle cells are expected to become the next generation of stent eluting material to displace with the old non-specific cytotoxic drugs, such as paclitaxel and rapamycin.

ACKNOWLEDGEMENTS

This work was financially supported by the National Natural Science Foundation of China (81270181) and National Natural Science Foundation of China (81202830).

REFERENCES

[1] Tsaousi A, Mill C, George SJ. The Wnt pathways in vascular disease: lessons from vascular development[J]. Current opinion in lipidology,2011,22:350-7.

[2] Orr AW, Hastings NE, Blackman BR, et al. Complex regulation and function of the inflammatory smooth muscle cell phenotype in atherosclerosis[J]. Journal of vascular research,2010,47:168-80.

[3] Carmeliet P, Moons L, Stassen JM, et al. Vascular wound healing and neointima formation induced by perivascular electric injury in mice[J]. The American journal of pathology,1997,150:761-76.

[4] Igura T, Kawata S, Miyagawa J, et al. Expression of heparin-binding epidermal growth factor-like growth factor in neointimal cells induced by balloon injury in rat carotid arteries[J]. Arteriosclerosis, thrombosis, and vascular biology, 1996,16:1524-31.

[5] Burton A. What went wrong with Iressa?[J]. The lancet oncology,2002,3:708.

[6] Landau C, Lange RA, Hillis LD. Percutaneous transluminal coronary angioplasty[J]. The New England journal of medicine,1994,330:981-93.

[7] Parise Junior O, Carvalho LV, Miguel RE, et al. Prognostic impact of p53, c-erbB-2 and epidermal growth factor receptor on head and neck carcinoma[J]. Sao Paulo medical journal = Revista paulista de medicina,2004,122:264-8.

[8] Cullen M. Second-line treatment options in advanced non-small cell lung cancer: current status[J]. Seminars in oncology,2006,33:S3-8.

International Conference on Power Electronics and Energy Engineering (PEEE 2015)

A Method for Composite Wing Box Optimization with Manufacturing Constraints

X.P. Zhong, P. Jin, Q. Han

School of Aeronautics
Northwestern Polytechnical University
Xi'an, China, 710072

Abstract—An optimization strategy is presented for composite wing box structure using genetic algorithm with manufacturing constraints. Firstly, on the basis of region division and key region defined, both fiber orientation angle and covered length of each ply are treated as design variables to optimize laminate thickness and stacking sequence simultaneously while maintaining laminate ply continuity. Secondly, a parallel genetic algorithm (GA) is adopted to accelerate optimization process. Finally, a composite wing box is used to demonstrate the effectiveness of the proposed method. The result shows that the optimization method can obtain optimum design with light weight which satisfies manufacturing constraints for large composite wing structure.

Keywords-composite wing box; manufacturing constraints; ply continuity; parallel genetic algorithm

I. INTRODUCTION

Advanced fiber-reinforced composites gain increasing interest and are growingly used in automotive, aerospace, and marine structures due to their high stiffness-to-weight and strength-to-weight ratios. Designs of laminate configurations of composite panels, including laminate thicknesses and stacking sequences, are of vital importance to achieve the required structure mechanical behaviors for specific sets of loading conditions.

For large composite wing structure, the panel is often divided into several smaller sub-regions to avoid overdesigning the wing structure due to different load level of each region. A common problem with the design of composite structure consist in that the laminate configurations between adjacent regions may vary from each other, i.e. different thickness and stacking sequence exist, causing incompatibilities in stacking sequence. The incompatibilities across panel design boundaries will bring manufacturing difficulties, increase costs and lead to structural integrity issues. Thus, stacking sequence blending design of laminate panels has received more and more attention.

Since the term blending was introduced by Kristinsdottir et al. to illustrate the ply continuity problem [1], many papers have contributed to the investigation of the composite structure blending models. Liu [2] and Toropov [3] developed and applied two measures of continuity in terms of material composition and stacking sequence between adjacent panels to obtain blended composite panel. Soremekun et al. [4] used DARWIN's sub-laminates and design variable zone features to design completely blended composite structures. Adams [5, 6]

achieved a blended design using a guide based genetic algorithm and developed two blending models, inner and outer blending, to maintain the ply continuity between adjacent laminates. Campen [7] proposed two new blending definitions, generalized blending and relaxed generalized blending.

In this paper, an optimization method based on parallel genetic algorithm that can make laminate thickness and stacking sequence be optimized simultaneously is presented to find the global optima. This method divides the panel to be optimized into individual regions for sufficient reduction in weight. An indicator that identifies the key region (thickest region) from which plies originates and can cover a number of adjacent regions is determined. All plies start from the key region and a ply can be stopped between adjacent regions if the required strengths and stiffness are satisfied. Once the ply is dropped, it is not allowed to be added back to the latter regions. For a single ply, two different indicators are incorporated in the optimization formulations for the design of composite panels. The first indicator uses an integer θ to determine its ply angle. $\theta=1,2,3$ and 4 corresponds to the fiber angle of -45, 0, 45 and 90 degrees. The second indicator uses an integer L to determine how many regions a ply occupies, e.g. $L=2$ represents the ply continues through the first two regions, $L=0$ implies that the ply does not exist.

II. OPTIMIZATION MODEL OF A WING BOX

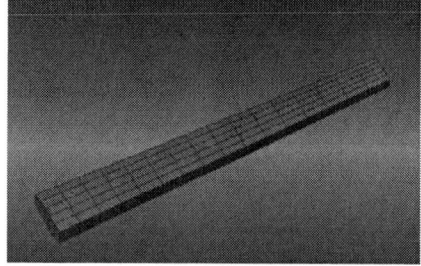

FIGURE I. FINITE ELEMENT MODEL OF WING BOX.

A composite wing box which is the main load-carrying component is to be optimized. Figure 1 shows the finite element model of the wing box. The elements can be grouped into several regions along the span wise direction. In each region, the properties of the elements are identical, i.e. the same thickness and lay-up. The optimization model for the composite wing is stated as follows.

© 2015. The authors - Published by Atlantis Press

A. Objective Function

The objective is the mass of the composite wing.

$$min \ W = W_0 + \sum_{i=1}^{n} \rho h S_i \tag{1}$$

Here W_0 is the weight that is kept constant during the optimization process. ρ and h are referred to as the material density and ply thickness, respectively. S_i stands for the total area of ply i, and n denotes the number of plies.

B. Design Variables

The design variables in the optimization formulation are the two different indicators of each ply, θ_i and L_i. θ_i is the ply angle, and L_i defines the regions covered by the ply. In equation (1), S_i is related to L_i.

C. Constraints

The following subsections describe the constraints imposed on the optimization of the composite wing.

1) Strength constraints: Strength constraints are introduced to limit the magnitude of strains in tension, compression and shear undertaken by the laminate. Design strain levels of 4000με in both tension and compression and 4000με in shear are imposed.

2) Stiffness Constraints: Wing tip displacement constraint and torsion angle constraint are taken in account to keep the bending and torsion stiffness of the wing to prevent the aero loads be deteriorated dramatically.

3) Manufacturing constraints:

a. The lay-up is balanced, i.e. the number of 45° plies and -45° plies is identical in each region.

b. Due to the damage tolerance requirements, the outer plies of the skin laminate should always contain at least one set of ±45° plies.

c. The number of plies in any one direction stacked consecutively is limited to 4.

4) Bucking constraints: Bucking constraints are considered as Reference 8.

III. OPTIMIZATION ALGORITHM

It has been recognized that optimal design of composite structures is a global optimization problem with multiple local optima and a complex design space, and the GA is an ideal tool for a discrete design problem and a multiple local optima problem.

The fiber orientation angles are often limited to 0°, ±45° and 90° to ease manufacturing effort and reduce cost. In this paper, each individual representing a single design is encoded into a digital string. The 0° ply is encoded into the digit 1, the 45° ply the digit 2, the −45° ply the digit 3, and the 90° ply the digit 4. For example, a design [0/45/0/−45/90] is coded as [1 2

1 3 4]. The population is composed of a set of such digit strings appended by the substring representing the length covered by each ply. The elitist selection scheme is incorporated to evolve population in the algorithm. Promising individuals from an extended population consisted of parent and offspring population are selected as the new parent population in next generation. And mating selection from the parent population to recombination is conducted by using a tournament strategy. Thus well balance can be reached between the survival pressure of each individual and diversity maintaining of the population during the optimization process.

The master-slave parallel GA [8] is adopted to decrease optimization time. The master-slave GA system is divided into one master process and several slave processes. The master process controls the whole population and carries on the operations of selection, crossover, and mutation. The slave processes receive the subpopulation sent by the master process and evaluate their fitness functions, and then send the results back to the master process. Information sending and receiving are the key points in the parallel GA. Functions MPI_Send and MPI_Recv in the MPI lib [9] are used to pass information between the master and slave processes. The frame for the master-slave parallel GA based on MPI is shown in Figure2.

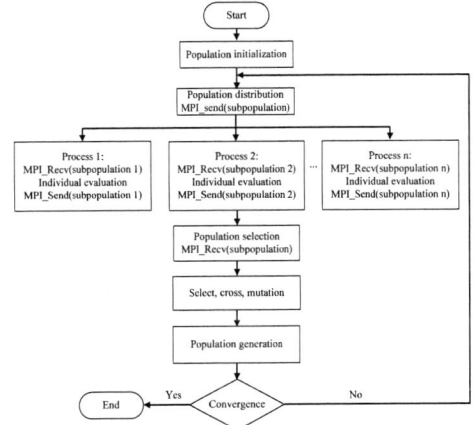

FIGURE II. MASTER-SLAVE PARALLEL GA BASED ON MPI.

IV. OPTIMIZATION RESULTS

The objective of this optimization is to find the lightest weight of the composite wing box in figure 1. The behavior constraints, global stiffness constraints, local strength constraints, buckling constraints, and composite manufacturing constraints are imposed. The top and the bottom skin layer are the targeted design panels. The wing box model is divided into 8 regions as in Figure 3, and Figure 4 details the wing section. For simplicity, the top and the bottom skin of the wing box in the same region have the same stacking sequence in this optimization problem. The possible ply orientations for the design panels are 0°, ±45° and 90°, while all other panels are fixed to the design of $[0_4/45_4/0_4/-45_4/90_4]_S$. Table 2 shows the material properties of T300/N5208, where ρ is the density and t is the thickness of a ply.

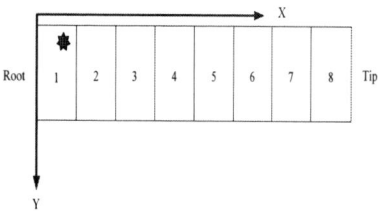

FIGURE III. REGION DIVISION ALONG SPAN WISE OF WING BOX.

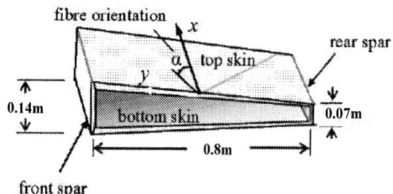

FIGURE IV. GEOMETRY OF WING CROSS SECTION.

TABLE I. MATERIAL PROPERTIES OF T300/N5208.

E11 [GPa]	E22 [GPa]	μ_{12}	G[GPa]	ρ[kg/m3]	t[mm]
127.56	13.03	0.3	6.41	1577.76	0.127

The master-slave parallel GA presented above is employed to solve the wing box optimization problem. It is supposed that the initial top and bottom skins are both comprised of 100 plies. The stacking sequence for each panel is symmetric about its mid-plane. The running parameters of GA code are as follows: the population size is 400, crossover probability is 0.9, and mutation probability is 0.05. The maximum generation is set to 400.

Table 2 details the optimal design of the composite skin obtained by this optimization method. Results show that the wing root (the key region) is the thickest region, and some plies are dropped off in a latter thinner region. The design meets the manufacturing requirements. Therefore, for such large composite wing box structure an optimal blended design is found using the method proposed in this paper.

TABLE II. STACKING SEQUENCE OF UPPER SKIN REGIONS.

Region	Stacking sequences	Number of plies
1	$[45/-45/0_2/45/90/45/0_4/45_2/0_2/-45/90/-45/0_3/45/0_4/-45_3/0]_s$	60
2	$[45/-45/0_2/45/90/45/0_4/45/0_2/-45/0_3/45/0_2/-45_2]_s$	48
3	$[45/-45/0_2/90/45/0_4/45/0_2/-45_2/0_4]_s$	38
4	$[45/-45/0_2/90/0_3/45/0_2/-45/0_3]_s$	30
5	$[45/-45/0/90/0_3/45/0_2/-45/0_3]_s$	28
6	$[45/-45/90/0_3/45/0_2/-45/0_3]_s$	26
7	$[45/-45/0_2/45/0_2/-45/0_3]_s$	22
8	$[45/-45/0_2/45/0/-45/0]_s$	16

V. CONCLUSIONS

Composite structure is often divided into multiple regions, each of which is optimized ignoring ply continuity in conventional multi-level optimization method. Mismatches between adjacent regions do inevitably occur in the obtained optimum. An optimization strategy is proposed for large composite structure using GA with manufacturing constraints by treating fiber orientation angle and covered length of each ply as design variables. Parallel master-slave GA based on MPI is used to improve the optimization efficiency. It is readily to get a blended design by considering manufacturability such as ply continuity in the proposed method. A composite wing box structure is optimized to demonstrate the effectiveness of this method. The optimal results show that a blended light weight design can be obtained using the method proposed in this paper.

ACKNOWLEDGEMENT

This study was supported by National Natural Science Foundation of China (No.11402204).

REFERENCES

[1] B.P. Kristinsdottir, Z.B. Zabinsky, M.E. Tuttle, S. Neogi, Optimal design of large composite panels with varying loads, Composite Structures. 51 (2001) 93-102.

[2] B. Liu, R.T. Haftka, Composite wing structural design optimization with continuity constraints. In: Proceedings of the 42nd AIAA/ASME/ASCE/AHA/ACS structures, structural dynamics and material conference (2001).

[3] Toropov, V.V., Jones, R., Willment, T., Weight and manufacturability optimization of composite aircraft components based on a genetic algorithm. In: Proceedings of 6th World Congress of SMO, Brazil, (2005).

[4] G.A. Soremekun, Z. Gurdal, C. Kassapoglou, Stacking sequence blending of multiple composite laminates using genetic algorithm, Composite Structures. 56 (2002) 53-62.

[5] D.B. Adams, L.T. Watson, Z. Gurdal, C.M Anderson-Cook, Genetic algorithm optimization and blending of composite laminates by locally reducing laminate thickness, Advances in Engineering Software. 35 (2007) 35-45.

[6] Seresta, O, Gurdal, Z., Adams D.B., Watson, L.T., Optimal design of composite wing structures with blended laminates, Composites: Part B. 38 (2007) 469-480.

[7] J.M. Van Campen, O. Seresta, M.M. Abdalla, Z. Gürdal, General blending definition for stacking sequence design of composite laminate structures. In: Proceedings of the 49th AIAA/ASME/ASCE/AHS/ASC Structures, Structural Dynamics and Materials Conference, Schaumburg, IL, USA,(2008).

[8] P. Jin, B. F. Song, and X. P. Zhong, Structure optimization of large composite wing with parallel genetic algorithm, Jornal of Aircraft. 48 (2011) 2145-2148.

[9] M.T. McMahon, L.T. Watson, A distributed genetic algorithm with migration for the design of composite laminate structures, Parallel Algorithms Appl. 14 (2000) 329-62.

International Conference on Power Electronics and Energy Engineering (PEEE 2015)

Preparation of HA/TiO$_2$ Biological Coating on Titanium Alloy

H.P. Shao, S.J. Wu, T. Lin, Z.M. Guo

Institute for Advanced Materials & Technology
University of Science and Technology Beijing
Beijing 100083, China

Abstract—In this paper, hydroxyapatite and TiO$_2$ biological coatings on TC4 titanium alloy substrate was prepared by electrophoretic deposition (EPD) method. The coatings were characterized by scanning electron microscopy (SEM). By subsequent different conditions of heat treatment process (vacuum and high purity argon), corrosion resistance of composite biological coatings were investigated by potentiodynamic polarization curves method. The results showed that the coatings were dense and can be tightly combined with substrate after sintering when TiO$_2$ content was 10 g·L^{-1}. Corrosion resistance of specimens treated by heat treatment in argon atmosphere was superior to specimens treated by heat treatment in vacuum.

Keywords-electrophoretic deposition; HA; suspension; corrosion resistance; heat treatment

I. INTRODUCTION

Hydroxyapatite (HA, Ca$_{10}$(PO$_4$)$_6$(OH)$_2$) is one of the most common types of bioactive ceramic materials, which has the same structure as inorganic of human body bone. It has non-toxic after implantation in the human body and no rejection in vitro owing to its excellent bioactivity and biocompatibility[1-3]. It can stimulate or induce bone tissue growth and can form ceramic materials containing phosphorus and calcium with bone tissue. However, the flexural strength and fracture toughness are lower than those of human compact bone, and the poor mechanical performance has restricted the use in load-bearing parts in the human body [4]. Titanium alloy has excellent properties such as good corrosion resistance, biocompatibility and so on. It can be used in the manufacture of medical apparatus and instruments, prosthesis and auxiliary treatment equipment etc., which is one of the widely used metal implant materials[5-6]. HA coated to titanium alloy surface not only has a good biological activity, but excellent mechanical properties of titanium alloy substrate and good biological properties of HA ceramics, which has become a hot topic of current biomedical materials research[7-8].

In this paper, the different content additives TiO$_2$ were mixed with HA to prepare composite coatings of different content additives and HA on titanium alloy substrate surface by the ultrasonic dispersion and EPD method. The influence of different content TiO$_2$ on HA biological coatings of titanium alloy substrate was studied. By subsequent different conditions heat treatment process, corrosion resistance

properties of HA/TiO$_2$ composite biological coatings on TC4 titanium alloy surface were studied.

II. EXPERIMENTAL PROCEDURE

TC4 titanium alloy substrate cut into 20 mm×10 mm×1 mm was chosen as samples. The samples were polished and smooth by mechanical grinding. They were putted into acetone for degreasing, rinsed with deionized water to remove surface residual liquid, and they were cleaned in deionized water by ultrasonic dispersion. Afterwards, the samples were treated by HF solution 100 ml/L + HNO$_3$ solution 300ml/L. Finally they were taken out immediately, putted into flowing deionized water and then cleaned in deionized water by ultrasonic dispersion to use for EPD.

The n-butyl alcohol was selected as solvent and triethanolamine was used to adjust pH value to 7~8. HA and TiO$_2$ were mixed fully by different mass ratio. Then a stable suspension can be obtained by ultrasonic stirring after 1 h. Suspension ages for 24 h can be used for EPD experiment. A stainless steel cylinder was as an anode, titanium alloy substrate as a cathode which was placed in the stainless steel cylinder axis. The distance between cathode and anode was 10 mm, deposition voltage was 30 V and deposition time was 120 s. After deposition, the samples were dried in air and then stored in desiccators. The dried samples were placed in a tube resistance furnace for heat treatment. This experiment adopted two kinds of environment of heat treatment-vacuum and high purity argon (purity ≥ 99.99%, mass fraction). Heat treatment temperature was 800 ~ 850 ℃, heating rate was 5 ℃·min^{-1}, hold for 1 h, then cooled to room temperature in the furnace. The structure characteristics of the coatings were observed and analyzed by SEM. Corrosion resistance of implanted materials was investigated by potentiodynamic polarization curves method.

III. RESULTS AND DISCUSSIONS

Figure1 shows the SEM photos of composite biological coatings of TC4 titanium alloy substrate which was made up of different content TiO$_2$ and HA. Shown as in Figure1(a), when TiO$_2$ was 2 g·L^{-1} in suspension, much holes existed in the coatings, the structure was loose. Furthermore the binding state between coating and substrate interface was not close and existed obvious boundary. When TiO$_2$ increases gradually, loose structure will be improved to some extent and had no obvious boundary as shown in Figure 1 (a) and (b). On the other hand, when TiO$_2$ content was high (10 g·L^{-1}), coatings

© 2015. The authors - Published by Atlantis Press

231

were uniform and dense relatively. And thickness of HA/TiO$_2$ composite biological coatings was about 25 μm. Between coatings and substrate interface combined closely, there was no obvious boundary. This showed that coatings were dense and can be tightly combined with substrate after sintering when TiO$_2$ content was 10 g·L^{-1}.

Figure2 showed electrochemical corrosion polarization curves of the coatings which were treated by different conditions of heat treatment process in simulated body fluid. Shown as in the Figure 2, it also experienced a stable passivation phenomenon and a stable and less corrosion current density in the anode region as the voltage increased. Then the corrosion current density increased rapidly due to the breakdown of the passive film. When corrosion potential became high, it re-entered a stable passivation stage.

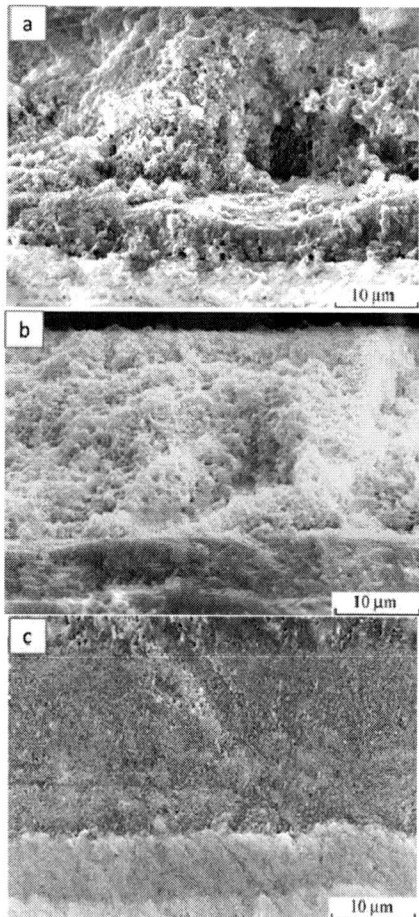

FIGURE I. SEM IMAGES OF CROSS-SECTIONAL OF COMPOSITE COATINGS WITH DIFFERENT CONTENT OF TiO2 AFTER HEAT TREATMENT AT 800 ℃ IN VACUUM(a)HA 10 g·L^{-1} + TiO$_2$ 2 g·L^{-1}; (b) HA 10 g·L^{-1} + TiO$_2$ 6 g·L^{-1}; (c) HA 10 g·L^{-1} + TiO$_2$ 10 g·L^{-1}.

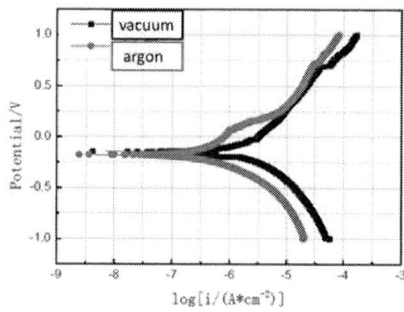

FIGURE II. ELECTROCHEMICAL CORROSION POLARIZATION CURVES OF THE COATINGS WHICH WERE TREATED BY DIFFERENT CONDITIONS OF HEAT TREATMENT ENVIRONMENT IN SIMULATED BODY FLUID.

TABLE I. CORROSION CURRENT AND CORROSION POTENTIAL CORRESPONDED TO DIFFERENT HEAT TREATMENT ENVIRONMENT.

Sample	E_{corr} vs SCE / V	I_{corr} /μA·cm^{-2}
vacuum	-0.147	2.92
argon	-0.178	0.845

Table 1 was the electrochemical parameters that were obtained by fitting polarization curve data. Shown as in the Table1, corrosion potential had little difference between samples treated by heat treatment in vacuum and samples treated by heat treatment t in argon atmosphere. But the corrosion current density in argon atmosphere was significantly less than the corrosion current density in vacuum, that's to say, the samples treated by heat treatment t in argon atmosphere had a better corrosion resistance

Figure 3 was AC impedance Nyquist diagram (the vertical axis was the imaginary part of the impedance value, the abscissa was the real part of the impedance value) of the coatings were treated by different heat treatment environment. As we can see, capacitive arc radius of samples treated in argon atmosphere was larger than other, and exhibited a better corrosion resistance that also confirms the results of the analysis of polarization curves.

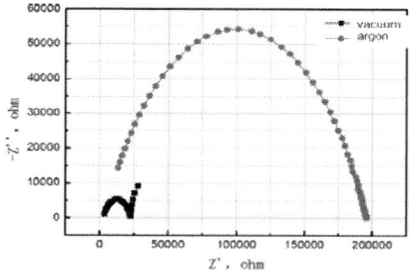

FIGURE III. AC IMPEDANCE NYQUIST DIAGRAM IN DIFFERENT HEAT TREATMENT ENVIRONMENT.

IV. CONCLUSIONS

HA /TiO$_2$ composite biological coatings were prepared successfully by adding additives TiO$_2$ to HA to form a stable suspension. Thickness of HA/TiO$_2$ composite biological coatings was about 25 μm. When TiO$_2$was 2 g·L^{-1}, there were some holes existed in the coatings and the structure was loose; when TiO$_2$ content was 10 g·L^{-1}, the coatings were uniform and dense relatively. Corrosion resistance of specimens were treated by heat treatment in argon atmosphere was superior to specimens were treated by heat treatment in vacuum.

ACKNOWLEDGEMENTS

This work was financially supported by the National Natural Science Foundation of China (No.51274039) and Production, Education & Research Project of Guangdong Province (No.2011A090200091).

REFERENCES

[1] Wang Z C, Ni Y J, Huang J C. Effects of suspension power content on the electrophoretic deposition of hydroxyapatite coatings [J]. *Journal of the Chinese Ceramic Society*, 2008, **36**(5): 626-630.

[2] Lídia Ágata de Sena, Mônica Calixto de Andrade, Alexandre Malta Rossi, Gloria de Almeida Soares. Hydroxyapatite deposition by electrophoresis on titanium sheets with different surface finishing [J]. *Journal of Biomedical Materials Research*, 2002, **60**(1): 1-7.

[3] Byung-Dong Hahn, Jung-Min Lee, Dong-Soo Park, Jong-Jin Choi, Jungho Ryu, Woon-Ha Yoon, Byoung-Kuk Lee, Du-Sik Shin, Hyoun-Ee Kim. Mechanical and in vitro biological performances of hydroxyapatite–carbon nanotube composite coatings deposited on Ti by aerosol deposition [J]. *Acta Biomaterialia*, 2009, **5**(8): 3205-3214.

[4] Chen F, Lin C J, Wang Z C. The electrophoretic deposition of nano hydroxyapatite coating on titanium surface [J]. *Electrochemistry*, 2005, **11**(1): 67-70.

[5] Shengjiang Wu, Huiping Shao, Ran Wei. Ito. Influences of different surface treatments on titanium substrate to HA/TiO$_2$ bioactive coatings [J]. *Advanced Materials Reaserch*, 2014, **893**: 508-511.

[6] Yang D H, Shao H P, Fan L P, Lin T, Guo Z M. Porosity and mechanical properties of porous Ti-7.5Mo alloy for medical applications by gelcasting [J]. *Journal of University of Science and Techno logy Beijing*, 2011, **33**(9): 1122-1126.

[7] Boon Sing Ng, Ingegerd Annergren, Andrew M Soutar, K.A. Khoret, Anders E.W. Jarfors. Characterization of a duplex TiO2/CaP coating on Ti6Al4V for hard tissue replacement [J]. *Biomaterials*, 2005, **26**: 1087-1095.

[8] Huiping Shao, Shengjiang Wu, Tao Lin, Ran Wei. Study of corrosion-resistance on biomedical coating by different surface treatments [J]. Applied Mechanics and Materials, 2014, 563: 391–395.

International Conference on Power Electronics and Energy Engineering (PEEE 2015)

Preparation of Sub-micro SiO$_2$ Particles by Sol-gel Method

T. Lin, H.P. Shao, H. Zheng, L. Zhang

Institute for Advanced Material & Technology
University of Science and Technology Beijing
Beijing 100083, China

Abstract—In this paper, the sub-micro silica particles were prepared successfully by sol-gel method. The ammonia was introduced as a catalyst and silane coupling agent as surfactant. The impact of reaction temperature, ethyl orthosilicate (TEOS) concentration, the amount of silane coupling agent and the order added to the solution on the size of particles were investigated. The sub-micro SiO$_2$ particles were determined by SEM. The result showed that the average size of particles changed from 200nm to 600nm while the reaction temperature was from 20℃ to 50℃, concentration of TEOS was 0.2mol/L to 0.3mol/L and 0.075 to 0.125ml dispersant was added after reaction. At room temperature, the silica particles grew up more easily on the condition that the concentration of TEOS was 0.3 mol/L and added 0.075 to 0.125ml dispersant after the reaction, the sub-micro silica particles with the average size of 600nm were prepared.

Keywords-silicon; sub-micro; sol-gel method; silane coupling agent; TEOS

I. INTRODUCTION

SiO$_2$ particles are widely used in the column packing, structural ceramic materials and cosmetics, paint, ink additives, etc. due to its mobility and high mechanical strength[1-2]. At the same time, it becomes functional because of no toxicity, high biological activity and it is suitable for its surface silanols as modified bridges [3].The system of sub-micro SiO$_2$ is 100nm to 1000nm, sub-micro SiO$_2$ can be applied to liquid body armor [4]. The main component of liquid armor is a special "shear dense liquid" (STF), the extremely small spherical particles (such as SiO$_2$ spherical particles) of this substance will be mixed into a liquid which is non-volatile, non-toxic and with good mobility forming a suspension or gel. This new liquid material usually very easily deformed, the extremely small hard particles is in suspension. However, once under the attract, the previous suspension collision point charged into the hard particles aggregating into clusters of particles, so that the shear thickening fluid instantly becomes very hard to stop the deadly strike from the human body[5].

There are many methods to prepare nano-silica microspheres, such as mechanical alloying method, hydrothermal synthesis method[6-7], microemulsion method[8], sol-gel method[9], precipitation, radiation synthesis preparation method[10]etc. The sol-gel method has become one of the preferred methods of preparation of spherical SiO$_2$

because it's simple process and low cost. Ultrasonic agitations were introduced based on the traditional sol-gel method in this article [11]. Sub-micro SiO$_2$ particles were prepared by TEOS hydrolysis reaction, then centrifuged, washed and dried at the end of reaction.

II. EXPERIMENTAL PROCEDURE

Preparation of A solution: mixing 28% concentrated ammonia, ethanol and deionized water in a flask, then carried on ultrasonic agitation; Preparation of B solution; mixing TEOS and ethanol uniform; Then the solution B was added to solution A rapidly, adding a certain amount of silane coupling agent KH-570, and reduced the strength of the ultrasound after a minute, the reaction was continued at room temperature for 3 to 6 hours. The resulting material was centrifuged, washed and spin-steamed grinding flour. The silica particles were analyze by scanning electron microscopy (SEM).

III. RESULTS AND DISCUSSION

A. The Effect of Order of Adding the Dispersing Agent KH-570 on the Particles

The volume of the reaction system (made up by ethanol, water, ammonia, and TEOS) was controlled to about 100ml. Solution A was 9ml 28% concentrated aqueous ammonia, 16.25 ml ethanol and 22.50ml water. Solution B was 6.75ml TEOS and 45.5ml ethanol. Solution B was mixed uniformly and then added into solution A rapidly. The ultrasound intensity was 35W at the moment then the ultrasound intensity was adjusted to 20W after a minute. Finally, the powder was prepared by rotary evaporation and grinding after 4h reaction at room temperature. The SEM images of the particles were showed as Figure 1. The SiO$_2$ particles were spherical particle and their size was about 600 nm, shown as in Figure 1 (a, c). The concentration of the TEOS was 0.3 mol/L. The dispersing agent KH-570 immediately was added after the solution B was added, shown as Figure 1 (b, d). The TEOS was not complete hydrolysis in Figure 1 b, however, the particle was liked as floc in Figure 1 d, which because that the SiO$_2$ nucleation was packaged by silane coupling agent KH-570. They prevented further growth and agglomeration, and the nanoparticles were easily formed. It can be seen that dispersing agent KH-570 was added to the solution suitably after reaction complete, namely, the agglomeration of SiO$_2$ particles can be improved after the nucleation.

© 2015. The authors - Published by Atlantis Press

234

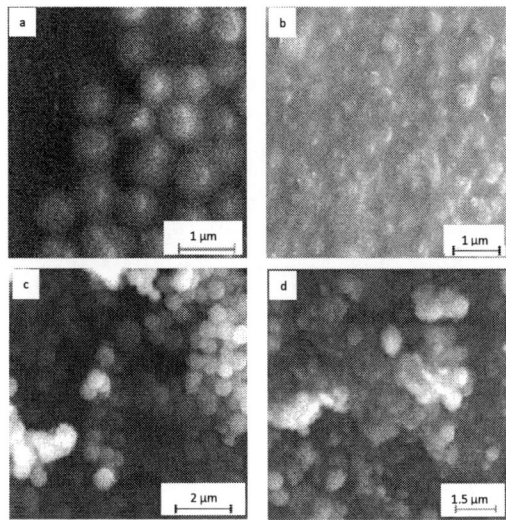

FIGURE I. SEM IMAGES OF SiO$_2$ PARTICLES (a) 0.075 ml KH-750 ADDED AFTER REACTION; (b) 0.075 ml KH-750 ADDED ONLY AFTER ADDING THE B SOLUTION; (c) 0.125 ml KH-750 ADDED AFTER REACTION; (d) 0.125 ml KH-750 ADDED ONLY AFTER ADDING THE B SOLUTION.

B. The Effect of the Amount of the Dispersing Agent KH-570 on the Particles

The volume of the reaction system (made up by ethanol, water, ammonia, and TEOS) was controlled to about 100ml. Solution A was 9ml 28% concentrated aqueous ammonia, 16.25 ml ethanol and 24.75ml water agitated at the 35W ultrasound intensity . Solution B was 4.5ml TEOS and 45.5ml ethanol. Solution B was mixed uniformly and then added into solution A rapidly. The ultrasound intensity was adjusted to 20W after a minute. Finally, the particles were prepared by rotary evaporation and grinding after 4h reaction at room temperature. Figure 2a showed the 300nm SiO$_2$ particles was prepared at the room temperature (20 ℃), the concentration of TEOS was 0.2 mol/L, and the silane coupling agent KH-570 was 0.15ml. Figure2b showed that SiO$_2$ were floc while adding 0.15ml silane coupling agent KH-570 immediately after the solution B was added. And the SiO$_2$ particles of 500 nm were prepared by adding 0.075 ml silane coupling agent KH-570 after reaction as Figure2c. Silica hydrolysis was not complete hydrolysis by adding 0.125ml silane coupling agent KH-570, shown as Figure2d. Because the hydroxyl (-OH) was informed by unmodified SiO$_2$ microsphere surface reaction with water molecules. The hydroxyl interacted in anhydrous ethanol and combined by chemically or hydrogen and the silane coupling agent occurred hydrolysis during preparing the SiO$_2$. The inorganic group of hydrolysis was adsorbed on the surface of SiO$_2$ particles which were joined with hydroxyl on the surface of SiO$_2$ network structure by dehydration synthesis, hindering the role of hydrogen bonding between the particles and preventing agglomeration. If the amount of coupling agent was too small, the surface of SiO$_2$ particles was incomplete and modified was ineffective; and if opposite, it will cause unnecessary waste, and excess coupling will cause reunion.

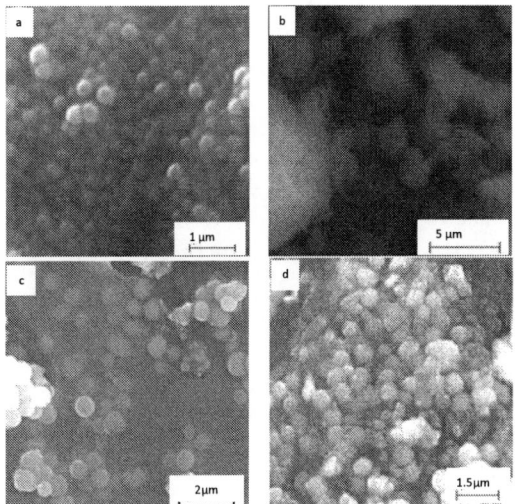

FIGURE II. SEM IMAGES OF SiO$_2$ PARTICLES (a) 0.15 ml KH-570; (b) 1.5 ml KH-570; (c) 0.075 ml KH-570; (d) 0.125 ml KH-570.

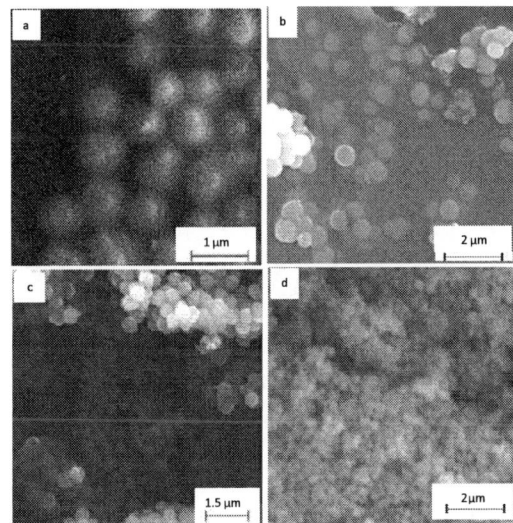

FIGURE III. SEM IMAGES OF SiO$_2$ PARTICLES (a) T=20℃, CTEOS: 0.3 mol/L, 0.075mlKH-570; (b) T=20℃CTEOS: 0.2 mol/L, 0.075mlKH-570; (c) T=50℃, CTEOS: 0.3 mol/L, 0.125mlKH-570; (d) T=50℃, CTEOS: 0.2 mol/L, 0.125mlKH-570.

C. The Effect of Concentration of TEOS on Synthetic Silica Particles

Figure 3 (a,b) showed the SiO$_2$ particles were prepared by adding 0.075ml dispersant after reaction, and the concentration of TEOS was 0.3 mol/L or 0.2 mol/L at room temperature (20 ℃). Figure 3 c and d showed the SiO$_2$ particles were prepared by adding the 0.125ml dispersant, and the concentration of TEOS was 0.3mol/L and 0.2 mol/L at 50 ℃. Figure 3 (a) shows the size of SiO$_2$ particle size about 600 nm was slightly larger than the particle size about 500 nm shown in Figure b. Figure 3 (c) showed the size of SiO$_2$ particle approximately 400 nm. It was because the main source of Si was TEOS; the concentration of TEOS affected the particle size. When the other reaction condition was same, the change of concentration of TEOS in the system caused the change of rate of hydrolysis and polymerization, and therefore affected the size of SiO_2 particles.

IV. CONCLUSIONS

(1)The average size with 200-600 nm of sub-micro SiO_2 particles was prepared successfully by sol-gel method. The SiO_2 particle was spherical while the silane coupling agent KH-570 was added after the reaction.

(2) The 200-600 nm SiO_2 particles were prepared with the concentration of TEOS was 0.2-0.3 mol/L, 0.075-0.125 ml dispersant, and the reaction temperature is from 20℃ to 50℃. The silica particles grew up more easily on the concentration of TEOS was 0.3mol/L, the dispersant was added after reaction at room temperature 20 ℃ and the particle size was about 600 nm.

ACKNOWLEDGEMENTS

This work was financially supported by the National Natural Science Foundation of China (No.51274039).

REFERENCES

[1] Qian Lihai, Luo Yuanfang, Jia Zhixin, Jia Demin. SBR-SiO2. Preparation and properties of super-hydrophobic coatings [J]. Functional Materials, 2013, 44 (5): 722-726

[2] Zhuo Xiaoyu, Zhang Qiuyu, Wang Xiaoqiang, et al. Preparation and modification of sub-micron SiO2 particles [J]. Chemical Engineering, 2010, 38(3): 72-75.

[3] Guangfa W Y Z L J, Zhong Z. Study of Spherical Silica in W/O Micro-emulsion Process with Orthogonal Experiment [J]. Lubrication Engineering, 2007, 4(1): 12-16.

[4] Liu Ling, Xu Qiang, Wang Fuchi, et al. Rare Metal Materials and Engineering [J]. 2009, 38(2): 780-782.

[5] Liu Wei, Yang Jinlong, Xiao Meng, et al. Functional Materials[J]. 2011, 42 (Suppl. 4): 632-634.

[6] Meseguer F,Blanco A,Miguez H,et al. Synthesis of inverse opals[J].Colloids and Surfaces A: Physicochemical and Engineering Aspects, 2002, 202(2): 281-290.

[7] Mi Gang, Chen Ping, Ren Nan. Chemical Journal[J]. 2008, 29(12): 2511-2515.

[8] Zhu Saifen, Wu Zhouan. Shanghai Chemical Industry [J]. 2001, 8(1): 22-25.

[9] Beck J S, Vartuli J C, Roth W J, et al. A new family of mesoporous molecular sieves prepared with liquid crystal templates [J]. Journal of the American Chemical Society, 1992, 114(27): 10834-10843.

[10] Park S K, Kim K D, Kim H T. Preparation of silica nanoparticles: determination of the optimal synthesis conditions for small and uniform particles [J]. Colloids and Surfaces A: Physicochemical and Engineering Aspects, 2002, 197(1): 7-17.

[11] Ding Guanjun,Zhu Mingwei,Qian Guodong et al. Rare Metal Materials and Engineering [J].2004, 33(3):15-18.

Impact of Computer Music Technology on the Effect of the Information Memory of Audiences

Y. Qin, D.J. Li
Business College
Nankai University
Tianjin, 300071, China

Y. Qin, J.T. Yang
School of Management
Tianjin University of Traditional Chinese Medicine
Tianjin, 300193, China

Abstract—**In the digital era to the core of computer technology, computer music came into being; it brings a series of new topics to disciplines to study. In the field of communication, the computer music technology is an element commonly used in the production of information, which can effectively achieve all kinds of information dissemination by the music computer synthesized. Through reviewing the recent research on computer music technology development and information dissemination effect, this paper analyzes the impact of computer music has on the audience memory in information dissemination.**

Keywords-computer information technology; information production; communication effect

I. INTRODUCTION

Computer music is the product of combining development of science and technology and music and arts. When promoting social and economic development, science and technology can also improve the art form, content and communication vector, promote the application and development of computer music technology in different fields. Computer music technology is the especial computer application technology ,the process of which is converting the musical sound information into machine-specific binary digital information and entering into the computer ,and such digital information will be processed via the specialized computer applications (music software),then the processing result is converted to voice information (computer music composition) and output to the viewers. Computer music technology includes the generation of digitized sound and digital real-time process of the sound two components, namely the so-called MIDI and digital audio technologies, which are the two core technologies of computer music. They both use the computer to process the information and produce the music, but they are two different technologies owning their special technical characteristics. With the development of these two types of technology, revolutionary changes have produced in various fields involving the use of music, and using computer technology to synthesize the song has become an important form commonly used in information production. Kilgour, Jakobson and Cuddy (2000) and other scholars are concerned about the effect of using computer music in advertising production, which confirmed that the memory effect is better to sing than to say expressing the same language content [1]. But some studies have shown that there is no difference of the effect whether using music technology or not, and even to sing has a worse memory effect than to say [2].

For the confusion of the study, this paper based on the development of computer music technology background, analysis of computer music technology to consumers remember information relevant literature to explore the impact of information and communication computer electronic music effects. Faced with the confusion of the above studies, this paper analyzes of relevant literature about the effect of computer music technology on consumers memory information, based on the development of computer music technology background, to explore the impact of computer electronic music on the effects of information communication.

II. HISTORY OF PRODUCTION AND DEVELOPMENT OF COMPUTER MUSIC TECHNOLOGY

Reviewing the entire development process of computer music technology, we have found that computer musical sound processing technology represented in the sequencer keeps the simultaneous development trend with computer musical sound producing technology and personal computer.

First, the generation of computer musical sound processing technology. Early in 1955, through special perforated roll tape, the first American-made synthesizer "RCA Electronic Music Synthesizer" can bring the sound with musical treatment in the manner of computing program. It was not until the 1980s, the computer music sound processing technology could be able to walk into the composer's studio and enter into the real practical stage with spread applications of the specially designed computer music Handler "Sequencer" on a Macintosh computer and PC Windows operating system.

Second, development of the computer music processing technology. With the advent of the low-cost personal computers in the early 1980s, in 1984,the American Passport company launched the first MIDI sequencer —Disings MIDI / 4 for personal computers in the history of computer musical development . The so-called sequencer is actually a music processing program for computer, which emerged as a landmark music creation tool with the development of the integrated circuit chip technology and the popularity of the personal computer. In addition to the general sensed sequencer software, in the last decade of the 20th century ,there appeared automatic orchestration software such as Band -in -a -Box, Jammer Pro dedicated to the creation of an accompaniment music, production of sheet music software (such as Sibelius, Finale, Encore, Overture) ,digital audio processing software (such as Cool Edit Pro, Sound Forge), multi-track audio

© 2015. The authors - Published by Atlantis Press

synthesis software (such as Samplitude Producer 2496, Vegas Audio), mastering software (such as T -Racks24), etc. The combination of the computer music applications software owning their different expertise with the traditional sequencer software has opened up a host of extremely broad application space for computer music sound processing technology.

III. THE APPLICATION OF COMPUTER MUSIC TECHNOLOGY IN INFORMATION DISSEMINATION

The advantage of computer technology is the ability to handle high level digital information, so the computer music is characterized by the digital processing of sound. Digitized sound and voice processing technology greatly improve the fidelity of sound, enriching the expressive power of music, which makes musical sound quality and music construct ability present unprecedented leap. Application and development of computer music is becoming increasingly popular in the business popularity world, which has a direct relationship with contemporary social cultural atmosphere in postmodern period. On the one hand, the professional becomes more and more detailed, and the degree of it becomes more and more high. While commercialization swept every corner of society, which makes the popular culture become the strongest voice of the era. On the other hand, since the electronic music successfully emerged in the business world and the popular music field in the 1960s, computer technology has developed further in the business application and popularization of music .its powerful functions and editing capabilities make it more like a duck to water in the cultural market .And many foreign music critics consider the computer music more as commercial music production.

When used in the dissemination of information, computer music technology mainly for the music production of decomposition and synthesis of the computer. Further audio processing will be done in this field, samly by means of computer, which seeks to break through the traditional forms of music and sound features, aimed to seek new sound designs, sound effects and musical expressions. From the perspective of music creation, it is a new way of composing, setting composer, playing, production at an organic whole, which can not only broaden the infinite possibilities of sound clips, also bring the dual audio experience for music. It can be said to be a special kind of music genres and types when using this approach to create music, owning its unique music style, language and aesthetic features. Differing from any other previous musical expressions ,this kind of music has its own characteristics and system, the reaction of the recipient to information dissemination caused by which is a revolutionary change especially in the melody, songs structure, harmony, orchestration, and other performances.

IV. IMPACT OF THE COMPUTER MUSIC TECHNOLOGY ON THE AUDIENCE MEMORY INFORMATION

One view is supported by some scholars that it helps the audience memory information in the form of computer music to present vocabulary, which cannot be explained by a widely accepted mechanism .And scholars attributed the promotion role of music to the following aspects:

First, the music has its own structural features that can

assist people to learn and memory the lyrics. Wallace (1994) believed that have been determining the length of the lyrics, setting the syllables whether be emphasized or not, and highlighting some of the specific elements of the lyrics, music as a framework made melody and lyrics closely work together to make every song as a coherent whole [3]. Other scholars believe that when the listeners are recalled the songs it is based on certain specific melody to fill into corresponding words. Therefore, the computer music is not just information, but a comprehensive and integrated framework, why it can assist people to learn and memory the lyrics.

Second, it has become slower to output information to the listener since the information can be presented in the form of computer music. In other words, it is the key factor in promoting memory that the pronunciation is slower to sing the language content than to say. Tests showed that it would do the same effect when the duration time in singing is controlled to be the same with speaking out the information content. This fully demonstrated that the key reason to learn and remember easily is the slower speed to sing than to say.

Third, most people like to listen to music or sing and repeatedly hum after hearing a song. It's just this fully exercises and repeat that play a significant promoting role on memory.

Another view is negative. Although it can help to remember information in the form of a song to present from a few aspects, many studies have shown that to say the language information is better than to sing to help listeners understand and remember information content [4].The reason is that to say has more advantages than to sing on information transmission.

First, the listener may probably treat the lyrics from voice perspective rather than from the perspective of understanding the semantic meaning of the lyrics when by means of music to transfer information .Some scholars have found an interesting phenomenon in the experiment that the listener may sometimes replace the original words with phonetically similar words when recalled, for example when test subjects were recalled to replace the original text posterity by prosperity, which two words are phonetically similar but have very different semantics. Other researchers also reached a similar conclusion that the subjects sometimes replace the original word by the same number of syllables word when recalled, or replace by a meaningless | la | syllable when the word is missing in order to ensure the integrity of the song structure. All the above results have proved that although the melody provides a framework for learning language content, singing is a relatively shallow learning, the listener may probably treat the lyrics from voice perspective rather than from the perspective of understanding the semantic meaning of the lyrics.

Second, although the computer music is more abstract to well express feelings and render atmosphere, it has limited capacity of displaying the information logic and helping listeners deeply understand the semantics. While the language is more conducive for the listeners to do deep processing of information logic and semantics in order to express specific and identified information. In the way to say to present computer music to the listeners is more conducive if subjects consciously try to learn and remember the lyrics for the purpose. Therefore, in many studies, to say is more conducive to listeners to remember the lyrics than to sing.

V. CONCLUSIONS AND IMPLICATIONS

The computer music technology has played an important role in the production and dissemination of information, being an integral and important form in information dissemination, which has promoted the computer music technology move toward diversification trend. For information producers, the computer music technology has already become an important means of information production and dissemination. It has greatly enhanced the quality and efficiency of information production and dissemination that computer music technology continues to improve and the software and hardware keep constantly updated ,which provide a scientific and effective music creation conditions for information dissemination workers so that the creation of music tend to be modern and digitized . But the computer music can only be as a technical means or compositional tools no matter how powerful it is in technology and how new the musical element is. Although the composer has mastered high tech hardware, been familiar with the computer music technology, the music still should only come from the heart rather than turning composing into a sound mind games relying on the way in technology. The production and dissemination of information do the same that the computer music technology can only be as a supplementary tool, with which the composer will become more efficient, and the information production and dissemination more colorful.

ACKNOWLEDGEMENT

Supported by the National Natural Science Foundation of China (Nos 71372099).

LITERATURE REFERENCES

[1] A. R. Kilgour, L. S. Jakobson and L. L. Cuddy, Music training and rate of presentation as mediators of text and song recall, J. Memory & Cognition, 28 (2000) 700-710.

[2] S. L. Calvert, Impact of televised songs on children's and young adults' memory of educational content, J. Media Psychology, 3 (2001) 325-342.

[3] W. T. Wallace, Memory for music: Effect of melody on recall of text, J. Journal of Experimental Psychology: Learning, Memory, and Cognition, 20 (1994) 1471-1485.

[4] S. L. Calvert, R. L. Billingsley, Young children's recitation and comprehension of information presented by Songs, J. Journal of Applied Developmental Psychology, 19 (1998) 97-108.

[5] X Lin, Advertising slogan and analysis, Central South University Press, Changsha, 2007.

International Conference on Power Electronics and Energy Engineering (PEEE 2015)

Design and Realization of Temperature Measurement System based on PROTEUS Software

C.L. Wang, P.Y. Chen, H.L. Hu

The School of Physics and Electronic Engineering
Xingtai University
Xingtai, 054001, China

Abstract—**Computer aided design is an important method in industrial design. The process of single chip micro-controller unit (MCU) based on Proteus simulation software is ocular, rapid, easy debugging, with low cost and high efficiency. DS18B20 temperature sensor has high accuracy, wide range and anti-interference properties. Hardware design, software editing, component welding and system debugging are involved in circuit design. Under the environment of Proteus, the hardware design and software debugging of single chip micro-controller and digital circuit can be completed easily. This can shorten the cycle of product development and enhance design efficiency. It can be applied to the t range of temperature occasions from -55°C to +125°C.**

Keywords-temperature; proteus software; simulation; microcontroller; design

I. INTRODUCTION

With the development of computer technology, all kinds of computer-aided design and production increase daily. Man-machine dialogue is a kind of work mode which is used in industrial design. The computer has interactive, rapid access and automatic processing and other functions. It can not only show the design effect and high efficiency, but also can do further simulation of the design content, show even 3D virtual component, device or product. The production of temperature measurement system is based on Proteus simulation software, combined with the microcontroller circuit principle. The circuit design process involves the hardware design, software programming, component and system for welding debugging. The design of hardware and software, the single-chip microcomputer and digital electronic circuit and convenient debugging can be completed in the Proteus environment. The product development cycle can be shortened and the design efficiency improved.

The purpose of this paper is to demonstrate the temperature measurement system which is designed under the *Proteus* software. First it shows the circuit, then programming, finally simulates the virtual system.

II. CIRCUIT OF PRINCIPLE

Temperature measurement system is mainly composed of single chip AT89C51 control chip, DS18B20 temperature sensor, liquid crystal monitor LCD1602 components.

AT89C51 is a kind of lower voltage, high performance CMOS 8 bit micro-processor with 4kbit flash

register(FPEROM-Flash Programmable and Erasable Read Only Memory).

DS18B20 Is a Tiny Temperature Sensor. Its range is -55°C~+125°C. In the range of -10~+85°C , the accuracy is ±0.5°C . The field temperature is transmitted directly by way of "first bus line" digit mode, thus the anti-interference of the system is greatly enhanced. This tiny temperature sensor is suitable for the measurement of field temperature in abominable environment. DS18B20 only need to be connected with micro-processor by way of one port line to achieve bi-directional communication. It communicates with the MCU of AT89C51 through single bus protocol. The AT89C51 port of P3.3 connects the DQ end of DS18B20 chip. LCD1602 links the SCM P0. P2 port of AT89C51 .The main function of MCU AT89C51 is to control and complete the collection of temperature and display information. The schematic diagram drawn by the Proteus software interface is shown in figure 1[1, 2].

Liquid crystal 1602LCD refers to liquid crystal module (displays characters and numbers) which displays the content of 16X2, i.e. display two lines and each line has 16 characters. Most character liquid crystal is based on 16X2 liquid crystal chip [3]. Liquid crystal LCD1602 monitor is connected with the P0 and P2 ports of single chip micro-controller. The single chip micro-controller controls and completes the temperature collection and information display.

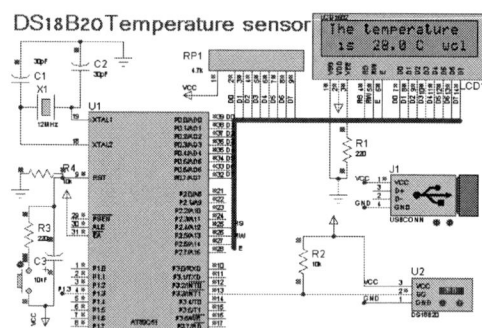

FIGURE I. PRINCIPLE CIRCUIT.

III. SOURCE PROGRAM DESIGN

The editing of source program mainly includes the realization of temperature acquisition, its display and control module. They are the temperature sensor DS18B20 reading and

© 2015. The authors - Published by Atlantis Press

writing program, the display program of LCD1602 and microcontroller control program. DS18B20 temperature reading programming is given as below [4,5]:

```
void tmpDelay(int num)
{
        while(num--) ;
}
void Init_DS18B20()
{
        unsigned char x=0;
        DQ = 1;
        tmpDelay(8);
        DQ = 0;
        tmpDelay(80);
        DQ = 1;
        tmpDelay(14);
        x=DQ;
        tmpDelay(20);
}
unsigned char ReadOneChar()
{
        unsigned char i=0;
        unsigned char dat = 0;
        for (i=8;i>0;i--)
        {
                DQ = 0;
                dat>>=1;
                DQ = 1;
                if(DQ)
                dat|=0x80;
                tmpDelay(4);
        }
        return(dat);
}

void WriteOneChar(unsigned char dat)
{
        unsigned char i=0;
        for (i=8; i>0; i--)
        {
                DQ = 0;
                DQ = dat&0x01;
                tmpDelay(5);
                DQ = 1;
                dat>>=1;
        }
}

unsigned int Readtemp()
{
        unsigned char a=0;
        unsigned char b=0;
        unsigned int t=0;
        float tt=0;
        Init_DS18B20();
        WriteOneChar(0xCC);
        WriteOneChar(0x44);
```

```
        Init_DS18B20();
        WriteOneChar(0xCC);
        WriteOneChar(0xBE);
        a=ReadOneChar();
        b=ReadOneChar();
        t=b;
        t<<=8;
        t=t|a;
        tt=t*0.0625;
        t= tt*10+0.5;
        return(t);
}
```

The program can be written or compiled directly in the Proteus software platform. Then copy the generated HEX files into AT89C51. One can also open the Keil and uVision3 software[1,6,7,8,9]. Click the new project in the menu. After pop-up the dialog box, create a project file and select the SCM model. Write the original program. When the program is compiled, click the compile connection and compile, which generates hex files. Check and see no error appears, create target (18B20.hex) file.

IV. CIRCUIT SIMULATION

The circuit debugging and analysis can be done under the Proteus software. Operate the loaded single chip controller target file and the simulation result will be given.

Open the already drawn schematic of design circuit on the platform of Proteus software and double click on the AT89C51 microcontroller. Show the edit element window, generates the source object code (18B20.hex) file and upload it into the microcontroller. This file is compiled as sixteen hexadecimal HEX file and it is shown as in figure 2. Then select suitable simulation method to analyze and debug design circuit. Third, Click "run" button to begin the simulation. System simulation result is shown in figure 1. Fourth, change the parameters and watch simulation results, then analyze the influence of parameters on the performance of the circuit [7, 8]. Fifth, Check the CPU registers, memory data changes. Figure 3, figure 4, Figure 5 show the CPU SFR Memory, CPU Registers and CPU Internal Memory data changes when the program are running respectively.

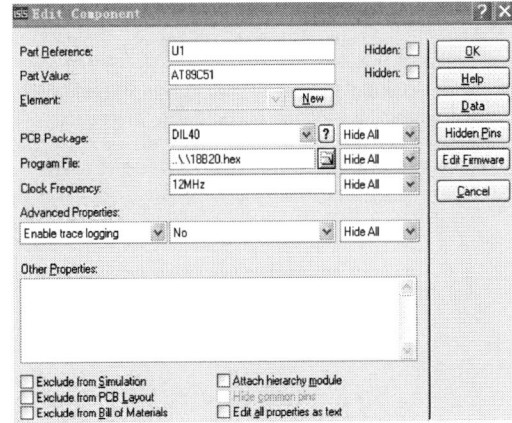

FIGURE II. EDIT WINDOW OF PARTS.

FIGURE III. 8051 CPU SFR MEMORY.

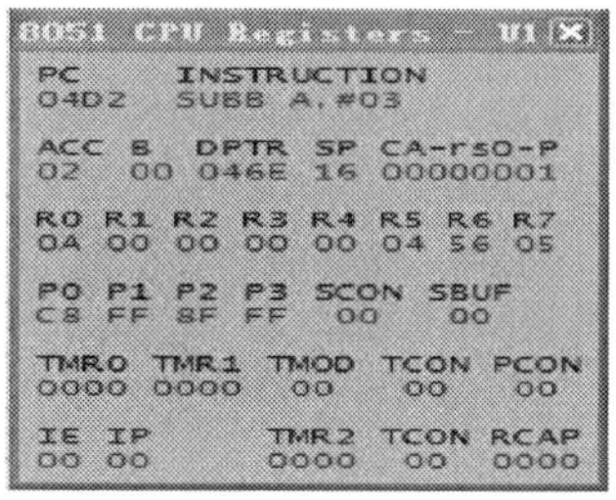

FIGURE VI. 3D SIMULATION.

VI. SUMMARY

Proteus circuit simulation fully reflects the ideas of circuit design, software design, circuit simulation, PCB circuit board design, the circuit welding and system debugging. The full understanding of the principle of the circuit is achieved from qualitative analysis to quantitative analysis; the achievement is realized from pure theory to the analysis of circuit design, circuit simulation experiment; the realization is achieved of the PCB circuit board design to the actual hardware installation and debugging. The simulation analysis and fabrication become the bridge from pure theory to practice. The design process is intuitive and quick to the comprehensive development of new products effectively. According to different applications, one can put the module package in other equipments, such as the realization of cable trench, water circulation system, air conditioning temperature control, agricultural greenhouse environment control temperature occasions.

FIGURE IV. 8051 CPU REGISTERS.

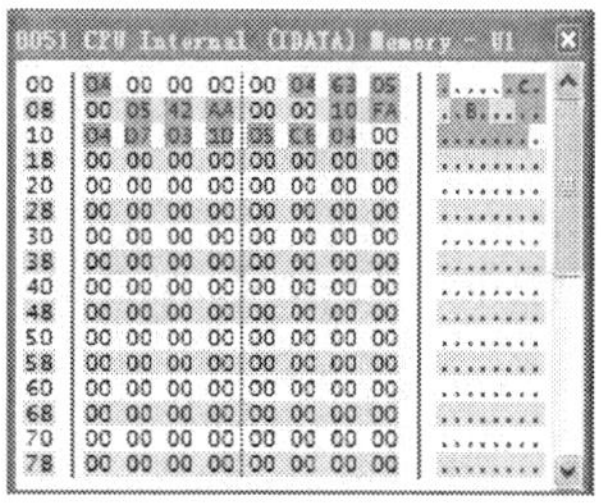

FIGURE V. INTERNAL MEMORY.

V. PCB CIRCUIT BOARD DESIGN

(1) Click the 'toolbox' of 'ARES' interface. Search 'component placement and editing' icon, select the elements in the list and click the left mouse button in a frame, put the selected element. (2) After placed element, click 'Tools'- 'Auto Router' from the menu. Dialog box, click the 'Begin Routing' button in the presence dialog box, it can be automatically complete the wiring. (3) Set level, set the rules, inspect wiring, check cyclic redundancy, check the design planning. (4) Click 'output' on the menu bar, and then click 'Gerber and Excellon output', open the dialog box, start to design. Generate Gerber file. Click 'output' on the menu bar again, then click 'Gerber view', pop up 'Gerber view', and check the selected optical drawing layer. Finish the final circuit PCB board. (5)When the wiring is completed, the 3D simulation effect is shown as figure 6.

ACKNOWLEDGMENT

This Project has been supported by Xingtai University.

REFERENCES

[1] He Jingkai. SCM System Design, Simulation and Application -- Based on the Keil and Proteus Simulation Platform [M]. Xi'an: Xi'an Electronic Science and Technology University Press, 2011.2 (in Chinese)

[2] Li Quanli. The SCM Theory and Interface Technology [M]. Beijing: Higher Education Press, 2013.12

[3] Yu chengbo. Sensor and Automatic Detection Technology [M]. Beijing: Higher Education Press, 2009.7 (in Chinese)

[4] Ding Xiangrong. C language Program Design and Keil C [M]. Guangdong: Guangdong Higher Education Press, 2013.9 (in Chinese)

[5] Geng Zhaoying.C, Application Programming Tutorial [M]. Beijing: People's Posts and Telecommunications Press, 2010.11 (in Chinese)

[6] Xu Aijun. Keil C51 Microcontroller High-level Language Application Programming and Practice [M]. Beijing: Publishing House of electronics industry, 2013.12

[7] Zhang Yigang. Principle and Application of Single Chip Microcomputer --C51 Programming +Proteus Simulation [M]. Beijing: Higher Education Press, 2012.11 (in Chinese)

[8] Du Shuqing. Based on Proteus and Keil C51 Microcontroller Design and Simulation [M]. Beijing: Publishing House of Electronics Industry, 2012.2

[9] Zhu Qinghui.Proteus Tutorial - electronic Circuit Design, Plate Making and Simulation [M]. Beijing: Tsinghua University Press, 2011.6

International Conference on Power Electronics and Energy Engineering (PEEE 2015)

Research on the Strategy of Group Vehicle Intelligent Perception and Traffic Route Guidance of Semantic Car Road Network

L.J. Tai, R.F. Hu, C.W. Chen

School of mechanical engineering
Ningbo University of Technology
Ningbo, China

Abstract—**Vehicle road network has become the development direction of the future transport system. In this paper, semantic modeling of car road network is built, semantic integration and sharing problems of heterogeneous networking information is solved. The Group car intelligent perception and effective cognitive method is explored. The dynamic traffic flow guidance strategies is proposed, to realize intelligent processing of massive car road network information.**

Keywords-car road network; traffic route guidance; intelligent transportation system

I. INTRODUCTION

With the high speed development of economy and the improvement of people's living standard, in recent years our country automobile industry showed leap type development. In the end of 2013, car ownership reach 137000000 vehicles, forecast it will reach 200000000 vehicles in 2020.But the attendant problem of traffic congestion, traffic accidents, environmental pollution and energy shortage has become a common problem the world facing.

Transportation, automobile manufacturing and other traditional separation industry is moving toward integrating with information communication, with the progress and application of the Internet, the Internet of things, cloud computing, data technology. At present, China's automobile industry developing at a surprising speed, with vigorously push forward of this emerging industry the Internet of things in our country the 12th five-year plan, make the intelligent transportation system of vehicle road network has become the development direction of the future transport system.

Vehicle road network is one of the most effective industry that promote the industrialization deply fuse information. Vehicle road network industry influence surface is very wide, value spread widely, led and radiation ability is very strong , from the automobile consumer services, automotive design and manufacturing, vehicle maintenance to finance and insurance, public transportation management, logistics and transport, urban public management etc. In addition, the development of vehicle road network will also improve the basic contradiction of social traffic, Including the contradiction between personal travel demand and traffic efficiency, the contradiction between travel safety and public traffic safety, the contradiction between energy saving and

carbon emission. Therefore, the vehicle road network technology is practical meaningful to study on the development of China's automobile industry.

This paper will carry on deeply research on semantic modeling, intelligent information perception and traffic route guidance on issues such as road of vehicle road network on the basis of previous studies. To provide guidance for the development of intelligent transportation system of China.

II. SEMANTIC MODELING OF CAR ROAD NETWORK

The heterogeneous of vehicle road network information must be solved firstly to complete the vehicle road network information integration. Ontology is an effective technology to realize semantic integration. If the ontology method is introduced in the knowledge expression of car road network, the recognized or default or implicit knowledge of car road network can be expressed by a display, and formal way, complete semantic integration of vehicle road network, realize knowledge semantic sharing of vehicle road network intelligent traffic control system .

Definition method of ontology in the field of car networking is described as follows.

A. Definition

Ontology of car network is a formal description of concept, relations, attributes connected in the field of car Network , defined as a five tuple: $O= (C, I, R, A, F)$.

Where: C is the concept set of car networking, is a collection of basic concept car in the car networking, is the result of the abstraction of example, its implication have clear description. I is the collection of instance, is specific instance of concept in concept set. R is a collection of relationship, including the relations between concept and concept, between association association, between concepts and association, association type can be the mapping that one to one, one to many, many to many. A is a set of attribute, is the description of the concept and feature of relation. F is a set of axiom, express the constraint rules between these elements.

III. RESEARCH SCOPE OF CAR NETWORKING ONTOLOGY

Vehicle road network relates to two fields the intelligent transportation and networking .Is the overlap that the two large-scale system in highway traffic application domain . So

© 2015. The authors - Published by Atlantis Press

243

in the analysis of vehicle road network knowledge field, It is necessary to consider the intelligence services that be provided in Intelligent Transportation and also consider the interconnection characteristics between car and car , between vehicles and infrastructure. Therefore, the vehicle road network research range can be determined as three sub field: service management field which provide intelligent traffic service, based equipment field which complete the basic functions of the Internet of things, knowledge management field that Implement effectively connecting with infrastructure and services. the field division of the relevant vehicle road network as shown in Figure 1.

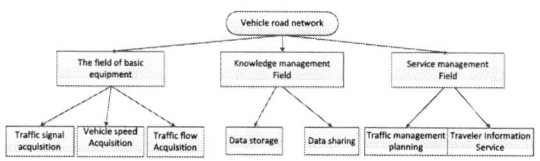

FIGURE I. HIERARCHICAL STRUCTURE OF CAR ROAD NETWORK FIELD.

IV. INTELLIGENT PERCEPTION OF GROUP OF VEHICLE

Figure 2 is the application hierarchical structure chart of perception layer. In the figure sensing layer of car road network is divided into sensing device layer, equipment network layer, equipment coordination layer. Where, sensing device layer can be abstracted as the sensing component, controlled components, the sensing component includes sensors, RFID, bar code, intelligent detection instrument, controlled components include valve switch, relay, some actuator device that work according to the logic relation , to realize the functions such as signal acquisition, processing, control . Equipment network layer includes various bus such as CAN bus, RS-485 bus, or wireless such as WSN, Bluetooth , WiFi etc, to realize communication connection between the sensing equipment and sensing equipment , between the sensing equipment and gateway of vehicle road network. Equipment coordination layer is gateway of car road network , to realize the unified management of the sensing device, make the sensing layer more real-time and reliable through the real-time scheduling.

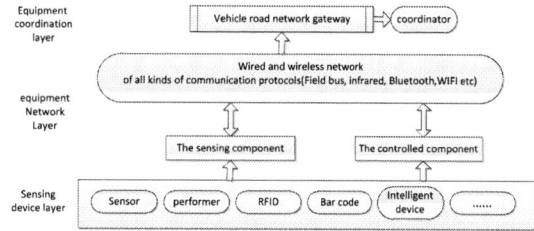

FIGURE II. STRUCTURE DIAGRAM OF CAR ROAD NETWORK SENSING LAYER.

V. TRAFFIC ROUTE GUIDANCE

The design of function of intelligent traffic route guidance system .

The system has function of the vehicle positioning, map browsing, optimal route guidance . The basic structure of the system is consists of five parts, as shown in Figure 3.

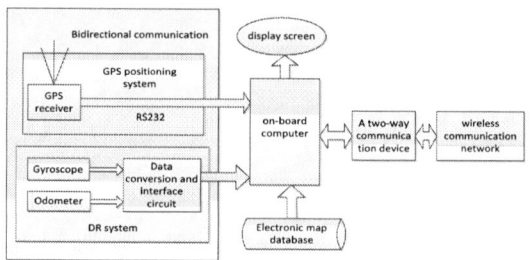

FIGURE III. INTELLIGENT TRAFFIC ROUTE GUIDANCE SYSTEM.

VI. CONCLUSION

This paper carry on deeply research on semantic modeling, intelligent information perception and traffic route guidance on issues such as road of vehicle road network on the basis of previous studies. To provide guidance for the development of intelligent transportation system of China.

ACKNOWLEDGEMENT

In this paper, the research was sponsored by the Nature Science Foundation of Zhejiang Province (Project No. Q14F010008) and the Nature Science Foundation of Ningbo City (Project No. 201401A6101039) and the national science and technology support program(Project No. 2012BAF12B11).

REFERENCES

[1] Saha A K, Johnson D B. Modeling Mobility for Vehicular Ad-Hoc Networks[C].Proc of the 1st ACM International Workshop on Vehicular Ad Hoc Networks,2004.91-92.

[2] B. Gallagher et al. Wireless Communications for Vehicle Saftey: Radio Link Performance and Wirelss Connectivity Methods[J]. IEEE Vehic. Tech. Mag,2006 1(4) 4-24.

[3] C.F. Mecklenbrauker et al. Vehicluar Channel Characterization and Its Implications for Wireless System Design and Performance[C]. Proc. IEEE, 2011.1189-1212.

[4] S.R. Dickey et al. Field Measurements of Vehicle to Roadside Communication Performance[C]. IEEE VTC-Fall, Baltimore(USA), 2007.2179-2183.

[5] J. Hourdakis, P. G. Michalopoulos, and J. Kottommannil. A practical procedure for calibrating microscopic traffic simulation models[J]. Transp. Res. Record, 2003 1852(3) 130–139 .

International Conference on Power Electronics and Energy Engineering (PEEE 2015)

Face Recognition Based on Gabor Wavelet Transform and Modular 2DPCA

H. Yan, P. Wang, W.D. Chen , J. Liu
College of Computer Science
Chongqing University of Technology
Chongqing, China

Abstract—Since the dimension of face features which is presented by Gabor wavelet is too high, there has large computation if using the feature by Gabor wavelet transform for recognition directly. A novel idea based on Gabor wavelet transform and modular Two-principal component analysis for face recognition is proposed. Firstly, face image feature is acquired by Gabor Wavelet transforming. Secondly, its dimension is reduced and eigenvectors are extracted by the method of modular 2DPCA. Finally, fusion with nearest neighbor classifier and support vector machine (SVM) is adopted to sort and distinguish. Experimental results on ORL and YALE show that the performance of proposed method is superior to other methods, such as modular 2DPCA and combination of Gabor wavelet transform and 2DPCA.

Keywords-Modular two-dimensional principal component analysis(modular 2DPCA); gabor wavelet; face recognition; Support vector machine(SVM)

I. INTRODUCTION

Face recognition has an active development in the past few decades due to its potential applications in access control, intelligent surveillance, automated video surveillance, law enforcement and identity authentication. With the same as other biometric technologies, There have been a lot of methods proposed for overcoming the difficulty of face recognition. Such as linear[1-5]methods have been widely used in face recognition. The principal component analysis(PCA)[1-3]and the linear discriminant analysis(LDA)[4,5] are two typical examples of linear transform methods.

The Gabor wavelet representation facilitates recognition without correspondence because it captures the local structure corresponding to spatial frequency, spatial localization, and orientation selectivity. As a result, the Gabor wavelet representation of face images should be robust to variations due to illumination and facial expression changes[6]. Although Gabor wavelets have a unique advantage in face recognition, but the dimension of face features which is presented by Gabor wavelet is too high. there has large computation if using the feature by Gabor wavelet transform for recognition directly. Cao et al.[7]came up with a idea that used PCA to reduce the Gabor feature dimension. However, before PCA transform two-dimensional images need to be converted into a one- dimensional vector matrix, which easily lead to the curse of dimensionality. Ma el at[8] used 2DPCA to reduce Gabor feature. Unlike Eigen face in which the

analysis and operation are based on 1D vector representation. 2DPCA is based on 2D matrix rather than 1D vector. Chen el at[9] proposed modular 2DPCA for face recognition. The method divides the face image into sub-images of the same dimension, which is helpful to highlight the local feature affected by posture, facial expression and so on. So it has been widely used in face recognition. The paper combines the Gabor wavelet transform and modular 2DPCA algorithm for face recognition and compares the recognition rate with PCA, LDA, Gabor wavelet transform combined with 2DPCA algorithm. The experimental results on ORL and YALE face database show that the algorithm has good recognition rate.

II. FACE FEATURE EXTRACTION BASED ON GABOR WAVELETS

Gabor wavelet transform has a good visual characteristics, and has a similar to mammalian visual system. It has the ability to time-frequency analysis combined with multi-scale and multi-directional characteristics. Gabor wavelet kernel function is defined as follows[10]:

$$G_{u,v}(z) = \frac{\left\| k_{u,v} \right\|}{\partial^2} e^{\left\| -\left\| k_{u,v} \right\|^2 \frac{\left\| z \right\|^2}{2\partial^2} \right\|} \left[e^{ik_{u,v}z} - e^{\frac{-\partial^2}{2}} \right] \quad (1)$$

Where $k_{u,v} = \begin{Bmatrix} K_v \cos\phi_u \\ K_v \sin\phi_u \end{Bmatrix}$, $K_v = \frac{k_{max}}{f^v}$, $\phi_u = \frac{u\pi}{N}$, $z = (x, y)$, μ and v defines the orientation and scale of the Gabor filters, k_{max} is the maximum frequency, and f is the spacing factor between kernels in the frequency domain. (x, y) represents the pixel coordinates. the spacing factor between kernels in the frequency domain. $\partial = 2\pi, k_{max} = \frac{\pi}{2}, f = \sqrt{2}$, $\frac{k_{u,v}}{\partial^2}$ determines the size of the Gauss window function. Facial feature extraction, usually choose five dimensions, eight directions.

Gabor feature of Face image is obtained by I(x,y) with a group of Gabor transform nuclear convolution. ts formula is as follows:

$$H(u, v, x, y) = G_{u,v}(z) \otimes I(x, y) \quad (2)$$

© 2015. The authors - Published by Atlantis Press

245

Where H(u,v,x,y) is the face of Gabor image, the face feature information in different frequencies and the different directions can be obtained through the changes of u and v parameter. However, its dimension are very high, In order to reduce its dimension, this paper adopts module 2DPCA to reduce the dimension.

III. MODULAR 2DPCA TRANSFORMATION

The modular 2DPCA is the extension of the 2DPCA,module 2DPCA divides each image into sub-image for each sub-image with the process of the 2DPCA respectively .

Assuming that there are A different persons in face samples library, the ith person has a training sample image matrixs: $I_{i1}, I_{i2}, I_{i3}, \cdots, I_{ia}$,($i = 1,2,3,..., A$). The size of training sample image I_{ij} ($i = 1,2,3,..., A$, $j = 1,2,3,..., a$) is $M \times N$, The $M \times N$ Block matrix is expressed as follows:

$$I_{ij} = \begin{Bmatrix} I_{ij(11)} & I_{ij(12)} & \cdots & I_{ij(1n)} \\ I_{ij(21)} & I_{ij(22)} & \cdots & I_{ij(2n)} \\ \cdots & \cdots & \cdots & \cdots \\ I_{ij(m1)} & I_{ij(m2)} & \cdots & I_{ij(mn)} \end{Bmatrix} \quad (3)$$

The size of each sub image is p x q ($p = M/m$, $q = N/n$). The mean of all the training sample :

$$B = \frac{1}{M} \sum_{i}^{A} \sum_{j}^{a} \sum_{k}^{m} \sum_{l}^{n} I_{ij}(kl) \quad (4)$$

$C = I_{ij} - B$ is the difference matrix between each sub image I_{ij} and B the average. The overall scattering matrix of the sub image:

$$S = \frac{1}{M} \sum_{i}^{A} \sum_{j}^{a} \sum_{k}^{m} \sum_{l}^{n} C_{ij(kl)} \times C_{ij(kl)}' \quad (5)$$

There $M = A \times a \times m \times n$ is the total number of sub image matrix of training samples. It is easy to prove that S is the nonnegative definite . Similar to 2DPCA, Needing to find a set of optimal projection vectors $V_1, V_2, ..., V_P$ .which meets the standarded orthogonality. Write it as $V = [V_1, V_2, ..., V_P]$, $P \in R^{(p \times q) \times P}$, which is named by the optimal projection matrix. The training samples are projected to the optimal projection matrix, the formula is:

$$W_i = V^T \times (I_{ij} - B)$$

IV. CLASSIFIER DESIGN

Design a good classifier is also a very important process in face recognition. K - nearest neighbor method（K—nearest neighbor, KNN for short), and Support vector machine (SVM) are common classifiers[11-12]. KNN is a simple and

nonparametric classification method in pattern recognition[11]. The main idea is comparing the Euclidean distance between test samples and all the training samples, the sample which is nearest to the test sample are of the same class. Supposing training sample B_i belonged to the class W_c , a test sample C_j, The definition of the distance formula:

$$d = \sum_{i=1}^{n} \left\| B_i - C_j \right\|_2 \quad (6)$$

B_i , C_j are the characteristic matrixs. If the distance of $d(B_1, C_1)$ is minimal, C_j belongs to w_c . SVM[12]is a supervised learning method, it has very good effects in dealing with high dimension data problems. Supposing a data set $\{(x_1, x_2 ... x_n)\}$, $x \in R^D$, Class Label $y \in \{1, -1\}$, Our purpose is to find a hyper plane $\omega \cdot x + b = 0$ making data separated. For the linear separable case, by solving a constrained extreme value problem, we can get the optimal classification plane. The optimal classification criterion function is:

$$f(x) = \text{sgn}(\sum_{i=1}^{n} a_i y_i (x_i \cdot x) + b) \quad (7)$$

In this formula: a_i, b is the optimal solution of the constrained extreme value problem. For the nonlinear separable, By introducing the kernel function $K(x_i, x_j) = \varphi(x_i) \cdot \varphi(x_j)$,The optimal classification criterion function (8) above turns into :

$$f(x) = \text{sgn}(\sum_{i=1}^{n} a_i y_i K(x_i, x) + b) \quad (8)$$

This paper based on KNN and SVM, proposes a new classifier fusion method - KNN+SVM method.

The idea of the algorithm: Firstly, extract features from face image using Gabor wavelet and 2DPCA, Secondly use KNN to filter feature vector, Finally using the SVM to classify and recognize. The algorithm flow chart is as follows:

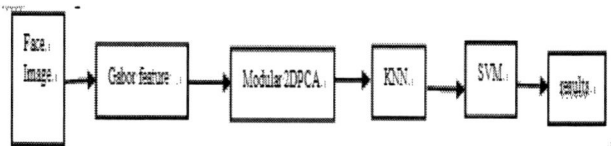

FIGURE I. KNN + SVM ALGORITHM FLOW CHART.

V. EXPERIMENT AND ANALYSIS

This experiment adopted the ORL and YALE face database respectively. ORL face database contains 40 people, everyone has 10 different image. Image contains certain illumination changes, expression change. The size of the image resolution is 112×92 pixels. YALE face database has 15 people, 165

246

images. Where illumination, expression and posture change is big. This article on the ORL database selects five images as training sample, five as testing sample, also selects five as training sample , six image as testing sample on the YALE database. Each sample is normalized to 128×128 .Figure 1 is one in ORL face database with 5 images and one in YALE with 5 images.

FIGURE II. THE FIRST ROW:5 IMAGES OF A MAN IN THE ORL DATABASE , THE SECOND ROW:5 IMAGES OF A MAN IN THE YALE DATABASE.

Figure 3 is the results of our methods and other methods. Each image is divided by four scheme as 2×2 , 4×2 , 4×4 , 8×4 ,The size of each sub-image is 64×64 , 32×64 , 32×32 , 16×32 . So their feature vector can be easily obtained.

The recognition rate of blocking is better than not block from the Figure 2. This is because after the image is partitioned, the face local feature information can be highlighted, and these information for the illumination, expression has a certain robustness. But the number of partitioned is not the more the better, when by 16 blocks, recognition rate is the highest, then the recognition rate did not change obviously when it is by 32 blocks. So this paper selects the best way of blocks of 4×4 .

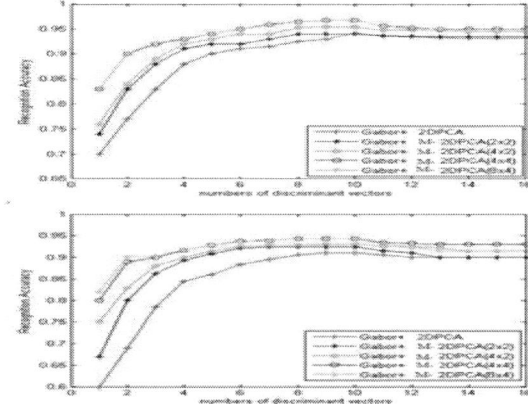

FIGURE III. THE RELATIONSHIP BETWEEN THE DIFFERENT BLOCK NUMBER RECOGNITION RATE AND THE FEATURE VECTORS IN THE 2 FACE DATABASES (%).

Table 1 displays that combining with Gabor wavelet and modular 2DPCA algorithm has the highest recognition rate on ORL and YALE. It demonstrates that the algorithm in this paper can give a better description of the local face information, and the dimensions of the local information is low, Table 1 also shows that the YALE recognition rate is lower than that of the ORL. It mainly due to the YALE face database changed large in illumination, expression than ORL face database.

Since the KNN algorithm is a simple classification, SVM has a very good effect on class classification, some researchers combine these 2 methods to classify [13]. This methods of combination have been widely used to face recognition. This paper uses the combination The experimental results is table 2, it shows that: the highest recognition rate was 96.5% using KNN+SVM in the ORL face database, 4.5% higher than using KNN alone and 2.4% higher than that of SVM. The highest recognition rate of KNN+SVM in the YALE database was 92.7%, 4.3% higher than KNN and 3.7% higher than that of SVM. Therefore, KNN+SVM has a better classification ability than KNN or SVM individually.

TABLE I THE IDENTIFICATION RESULTS OF EACH ALGORITHM(%).

method	ORL database			YALE database		
	optimal projection dimension	best feature dimension	recognition rate	optimal projection dimension	best feature dimension	recognition rate
Gabor+modular2DPCA	5	128x5	96.5	8	128x8	92.7
Gabor+2DPCA	10	128x10	94.5	12	128x12	91.5
modular 2DPCA	16	128x16	92.5	20	128x20	90
PCA	21	21	88	24	24	82
Fisher	18	18	92	24	24	89

TABLE II COMPARISON OF DIFFERENT CLASSIFIER (%).

method	ORL database	YALE database
KNN	92	88.4
SVM	94.1	89
KNN+SVM	96.5	92.7

VI. CONCLUSION

This algorithm makes full use of the property of the Gabor wavelet that it can effectively capture the face images of different spatial location, space, frequency and direction selectivity local structure characteristics, overcome the influence of illumination, scale, Angle , also modular 2DPCA can extract the feature whose dimensions are lower. Experimental results show that: the proposed algorithm has higher recognition rate than PCA, LDA and Gabor + 2DPCA.

ACKNOWLEDGEMENTS

This work is supported by National Natural Science Foundation of China (NO. 61173184) and the Graduate Innovation Foundation of Chongqing University of Technology (NO. YCX2013219).

REFERENCES

[1] Turk M, Pentland A. Eigenfaces for recognition [J]. Journal of Cognitive Neuroscience, 1991, 3(1): 71−86.

[2] Xiong Chengyi, Li Danting, Da Bangyou .Face recognition based on LBP and PCA feature extraction [J]. Journal of South—Central University for Nationalities,2011,(2):75-79.

[3] Yin Yong, Wen Juan. Face recognition based on modular PCA and SVD[J]. Journal of Chongqing University.2012,08:134-138.

[4] Zhang Jian, Xiao di. Face recognition method based on multi-scale adaptive LDA[J]. Computer Engineering and Design, 2012, 33(1): 332−335.

[5] LU Guifu, ZOU Jian, WANG Yong. Incremental complete LDA for face recognition[J]. Pattern Recognition, 2012, 45(7):2510-2512

[6] C. J. Liu et al.Gabor-Based Kernel PCA with Fractional Power Polynomial Models for Face Recognition[J], IEEE Trans. On PAMI, Vol. 26, No. 5,2004: 572-581.

[7] C. Liu, H Wechsler. Independent component analysis of Gabor features for face recognition[J], IEEE Trans. Neural Networks, Vol. 14, No. 4, 2003:919-928.

[8] Ma xiao yan et al.Face recognition based on Gabor wavelet and 2DPCA Computer Engineering and Applications[J].2006.10:55-57.

[9] CHEN Fu-bing, CHEN Xiu-hong, ZHANG Sheng-liang et al. A Human Face Recognition Method Based on Modular 2DPCA.Joumal of Image and Graphics[J].2006,11(4):580-585.

[10] T. S. Lee, Image Representation Using 2D Gabor Wavelets, IEEE Trans. On PAMI, Vol. 18, No. 10, October 1996,pp. 959-971.

[11] Cover T, Hart P. Nearest neighbor pattern classification[J].Information Theory, 1967, 13(1): 21−27.

[12] Schwenker F.Hierarchical Support Vector Machines for Multi-class Pattern Recognition[C]//Proceedings of the 4thIntenrational Conference on Knowledge—based Intelligent Engineering Systems& Allied Technologies. UK : University of Brighton，2000：561—565.

[13] CHEN Zhen zhou, LI Lei, YAO Zhengan. Feature-weighted K-Nearest neighbor algorithm with SVM[J]. Acta Scientiarum Naturalium Universitatis Sunyatseni, 2005, 44(1): 17−20.

International Conference on Power Electronics and Energy Engineering (PEEE 2015)

Study of Wuhan Metro Visual Communication Design under the Background of "Jiangcheng Culture"

X. Zhang
Wuhan Donghu University
Wuhan, China

Abstract—With the rapid urbanization in China, the metro construction of our country's first-tier cities has been under a basic model, and more and more second-tier cities join the metro construction tide. It is found that metro is a very good media platform according to the experiences of varied cities in different countries, all kinds of metro visual culture phenomenon have plentiful social-cultural meanings, metro culture becomes a special cultural landscape of the city. Nowadays, it's a important period for Wuhan, as a city famous for its history and culture in China, to promote its metro construction; how to dig out the essence of "Jiangcheng culture" to apply more cultural charm for Wuhan metro visual communication design and come into contact with the world gradually, has become a subject for us to think and study.

Keywords-metro visual communication design; visual culture; "Jiangcheng culture"

I. INTRODUCTION

With the rapid urbanization in China, the metro construction of our country's first-tier cities has been under a basic model, and more and more second-tier cities join the metro construction tide. It is found that metro is a very good media platform according to the experiences of varied cities in different countries, the media adventage of metro has been strengthened furtherlly with the continuous extension of metro lines and a sharp increase of urban metro, all kinds of metro visual culture phenomenon have been brought into public focus, these cultural shapes that based on figure, image and video have plentiful social-cultural meanings, metro culture becomes a special cultural landscape of the city.

For those cities with heavy historical and cultural foundation, metro visual culture can extend and improve their visual culture, even improve the contruction and development of urban culture, leading its trend and taste. It's a important period for Wuhan, as a city famous for its history and culture in China, to promote its metro construction; how to dig out the essence of "Jiangcheng culture" to apply more cultural charm for Wuhan metro visual communication design and come into contact with the world gradually, has become a subject for us to think and study.

II. CURRENT SITUATION OF WUHAN METRO VISUAL COMMUNICATION DESIGN

The gap of urban metro construction between our country and the developed countries is reduced, but our codes, systems and metro visual communication design that full of art and cultural charms still lag behind our modern metro facilities, the same applies to Wuhan metro visual communication design. Wuhan metro is trying hard to improve the public aesthetic taste and propagate "Jiangcheng culture" of Wuhan City, and has formed its own culture with Wuhan characteristics based on the study of urban metro culture development process at home and abroad. Wuhan metro cultural exhibition such as "Book Metro" and "Display the Future in Metro" is the first case in our country, which has fully embodied the humanistic ideals of Wuhan metro. However, for short history of Wuhan metro construction and imperfect culture penetration in existing metro system design, there are still the following problems:

Firstly, the logo of Wuhan metro consists of three equivoluminal color blocks, standing for this city's regional characteristics, namely "two river intersection and three town split", however, both the colors and figures of it are very similar to the logo of national inspection exemption and Motorola Corporation, the logo design is of low identification, and lack of culture.

Secondly, as the platform of mass culture, Wuhan metro provides multiple options of culture consumption for the citizens on the surface, but is totally used as a commercial publicity station in essence that based on the consideration of economic and commercial benefits, lack of cultural display obviously.

Thirdly, the content of cultural consumption provided by Wuhan metro management department is simple, mainly consists of commercials, artistic mural and underground sculpture, as well as limited distribution of Metro Daily, mainstream of playing game console, PSP and mobile game in crowded carriage, which affects metro cultural atmosphere badly.

III. SPECIFIC METHODS OF WUHAN METRO VISUAL COMMUNICATION DESIGN UNDER THE BACKGROUND OF "JIANGCHENG CULTURE"

A. Making the Best Use of Recognizable Figures and Integrating Underground Space Designs with Ground Space Culture

As a recognizable information carrier, figures have strong visuality and vision transmission. As a basic element of

© 2015. The authors - Published by Atlantis Press

creative expression, figures can transmit the design content and information to the audiences directly, vividly and effectively during visual communication, so as to evoke a strong response in the audiences.

While choosing figures, Jiangcheng characteristics shall be taken into account under the background of "Jiangcheng culture". The most visual chime and landmarks such as Yangtze River Bridge and Wuhan University, can be simplified as abstract figures, and then applied to logo design, advertisement design and public art design of Wuhan metro through varied creative design and expression, combined with different colors and words, so as to highlight the special cultural connotations of Wuhan City.

It is an important way to form the readability of each station that the design of visual works in the underground space shall echo the landmarks on the ground. In this way, designers can design a recognizable figure for each station according to local cultural characteristics, realize "one figure one station, knowing the history of Wuhan City through hundreds of stations", integrate the characteristics of "Jiangcheng culture" into metro space in visual form, so the style and subject of every station can be different from each other, and the passengers can have a clear impression of each station.

B. Refining Line Colors to Reflect the Cultural Features of Wuhan City

In 1933, the electronic engineer Henry C. Beck applied vivid colors to distinguish the metro lines, set up a new model of modern metro visual communication design. Wuhan metro has followed this model creatively.

Professor Fang Xing of Wuhan University of Technology, the CIS designer of Wuhan Metro Group, said that his inspiration of color selection came from local landscape and culture of each metro station, he led the team to take pictures of the most famous landscape around each line, such as East Lake, Guiyuan Temple and Yellow Crane Tower, found out the most outstanding color and the maximum color block through Mosaic amplification, so as to refined the colors of each metro line. In this way, realized "one color one line, displaying Wuhan features by ten lines" through the threads of urban landscape and cultural information, and the guidance of surrounding culture of each station, with the help of color elements to transmit the cultural connotation of each line color to the passengers. Moreover, how to combine the specific station color with station mark shall be the further study of Wuhan metro line-orientation identification design in the future.

C. Designing Distinctive Advertisement and Taking the Responsibility for Communicating "Jiangcheng Culture"

While the metro network construction becoming mature gradually and the arising of "cultural metro" concept, metro advertisement has become the most eye-catching mass media. In the communication process, metro advertisement can create more commercial value and promote the sustain growth for urban economy; meanwhile, it shall take the responsibility for public cultural communication and reveal high cultural quality.

For Wuhan is a famous historical city with more than 3,000 years of culture inheritance, we shall try to integrate the cultural elements of Wuhan city into metro advertisement in metro advertisement design to fully display the city's heavy cultural foundation. Chu opera, for instance, its relevant knowledge and the stage photo of classic characters can be implanted into metro advertisement, to attract the passengers to enjoy it through creative expression. In addition, the advertisements of introducing the city's famous historic and cultural sites, sightseeing, shopping, and recreation shall be set at each station, to make the metro as a "live map" of the city.

In Wuhan metro, "Theme Trains" including "Gourmet Food Train", "Prestigious School Train" and "Folk Train" can be opened. The metro advertisement shall conform to the subject of that "Theme Trains", for example, series of advertisements of special gourmet food in Wuhan City such as Hot-and-Dry Noodles, Instant Fish Soup Mix and fried Tofu skin can be designed for "Gourmet Food Train", and that of Hanchu Opera, Chu opera and Hubei Drum Art can be designed for "Folk Train", the out-of-town passengers can select different train according to their interests. In this way, the advertising can be more accurate, and the passengers can be more impressed by the internal cultural connotation.

D. Designing Tourism Products Derived from Metro Marks and Promoting "Jiangcheng Culture" through Visual Communication Design

The audiences of metro media consist of local citizens and tourists from other places. The special urban culture of metro space shall be displayed to these tourists through more designed products that can be taken away, other than monotonous public art design.

It is feasible to use remarkable Wuhan metro marks as tourism brand image, and then design series of tourism products such as stationery, T-shirt and toys full of publicity and characteristics, as well as small metro map and card that be carried easily, the application of Wuhan metro marks evolves from static symbols into visual symbols that can be moved in varied culture, cities and races all over the world, to promote the communication of "Jiangcheng culture" effectively.

IV. CONCLUSIONS

All the metro constructions in various countries can't do without national culture. The city's historic cultural tradition and humanistic spirit have been integrated into every urban metro, whether they realized it or not. The citizens have deep feelings for the metro in these cities, and travel by metro almost every day; people are interested in the design of metro station, foreign passengers and artists go there to find their inspiration, all people are looking forward to the new lines, which can give them sensation of new urban culture.

As a newcomer to the urban metro construction, Wuhan doesn't have heavy visual cultural accumulation compared to the foreign cities, so Wuhan metro construction shall focus on the establishment of personality and inherit its special "Jiangcheng culture" nowadays, namely the establishment of unique cultural charm. Following the development of Wuhan metro, its visual communication design shall enrich and

250

enlarge the connotation and denotation of "Jiangcheng culture" continuously, to make the citizens and passengers feel the features of Big Wuhan exactly, demonstrate the city's humanistic ideals and outstand its regional culture, so as to break down the space barriers of regional development and realize the optimal development of Wuhan city, and improve the cultural competence of Wuhan City in general.

ACKNOWLEDGEMENTS

School-level Youth Fund Project of Wuhan Donghu University (2013).

REFERENCES

[1] Fang Hua, Liu Jianjian. Environmental Visual Design of Hangzhou Metro from the Perspective of Urban Culture [J] Packaging World, May 2013

[2] Xi Xie. Comparative Study on Visual Conveying Design in the Subway of London and Nanjing [J] Hundred Schools in Arts, June 2008

[3] Feng Mingbing. On the Value of Metro Culture from the Perspective of Urban culture[J] Professors' Lecture, October 2012

[4] Wang Yun. The study of Regional culture in the recognition system of Wuhan metro[J] Home for You, October 2013

[5] Gao Huifang. Study on the Status and Countermeasures of Beijing Metro Culture Construction [J] Journal of Anhui Administration Institute, February 2013

International Conference on Power Electronics and Energy Engineering (PEEE 2015)

An Energy Consumption Model of Mobile Terminal Software Based on BP Neural Network

L.W. Liu, T.F. Zhan, Z.Y. Cai
College of Computer and Information technology
China Three Gorges University
Yichang, China

Abstract—**The energy consumption of software plays an important role in mobile application. Firstly, on the point of measurement, the characteristics of mobile terminal software are studied, and the functional relation between software energy and software characteristic is modeled as a nonlinear function. Secondly, based on BP neural network, an energy consumption model of mobile software is built. After the analysis of software characteristics in system architecture, the training and study mechanism of BP neural network in energy consumption measures is illustrated here. Lastly, BP training to fit the functional relation between software architecture and energy consumption is verified by experimental results.**

Keywords-mobile commerce; software energy; BP neural network; mobile terminal

I. INTRODUCTION

Along with the rapid advance of society and the development of economy, energy problem has become more and more serious, and the energy crisis has become the focus of social attention [1]. Now, the idea of energy saving has deeply rooted in the hearts of people. In energy saving, work and life are most closely related to our energy consumption with the increasing popularity of mobile terminals and all kinds of applications [2,3].

Reference [4] put forward a model of the mobile payment through the research in mobile agents and the existed mobile system. In mobile terminals, software architecture and mobile application are often designed in lightweight [5]. The rapid development of e-commerce is very popular in the world, and reference [6] considers the mobile terminal as an integral role in this respect. Then literature [7] presented a mobile terminal structure layered of a context-aware agent, and reference [8] proposed a model of mobile commerce value net. However, in these researches on mobile terminals and e-commerce, the impact of the software architecture itself on energy consumption is ignored.

This article puts forward a modeling method of the energy consumption of mobile terminal software. On the aspect of mobile architecture, a three level software architecture model-the component number, average complexity of component interface, average path complexity characteristics, is put forward. Assuming that there is a nonlinear function relationship between these characteristics and software energy consumption, the energy consumption of the mobile terminal software architecture model is established.

II. ENERGY CONSUMPTION MODELING OF MOBILE TERMINAL

Figure 1 shows a general system architecture of the mobile terminal software where the lower layers play an important role in the mobile terminal system. Among them, Graphical User Interface (GUI) provides an interface for users and runs on the Operating System Abstraction Layer (OSAL). The Protocol Stack Abstraction Layer (PSAL) refers to the sum of each layer protocol in network reflecting a network file transfer process. Device Abstraction Layer (DAL) is located at the interface between the operating system kernel and hardware circuit layer.

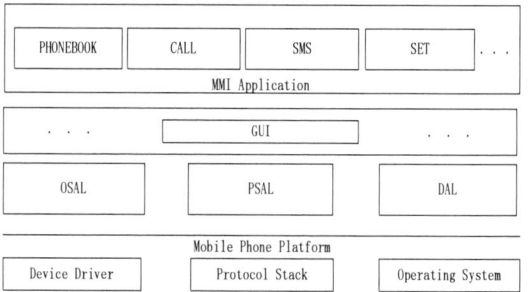

FIGURE I. SOFTWARE ARCHITECTURE MODEL.

Because of the layer model and the software architecture level characteristics, the relationship between the energy consumption and software is considered as a nonlinear function relation. Assuming E is the software energy consumption, a_1 is the initialization consumption of the software energy, b_n is the software characteristic value which is associated with energy consumption of software, and the value of a_n represents the parameter of the characteristics of various software, then the E can be expressed as

$$E = a_1 + a_2b_1 + a_3b_2 + ... + a_{n+1}b_n \tag{1}$$

A lot of mobile operating systems are used such as Androis, ios, Java, windows Mobile, symbian OS etc. There are a lot of mobile software that are related to these Operating systems; people have to download mobile software to their mobile terminals via WIFI from the Internet where there is a variety of Mobile software for people to use. Traditional architecture and energy consumption modeling use linear

© 2015. The authors - Published by Atlantis Press

regression method, which mainly consider the characteristics of software on architecture level, the n_{th} linear software function between energy consumption and the energy consumption model. Through the analysis of some software architecture level characteristics and relationship of energy consumption between different software, we can get a architecture model of energy consumption.

$$E_s = A \times T = f(B) \times T = f(L, C, R_c, R_w, DC) \times T \quad (2)$$

In the formula above, E_s is the consumption of the software and A is the average power consumption. T is the software running time and B is some software metrics associated with software power consumption characteristics, and f is the nonlinear function between software power consumption and software related measurement features. This article will use the BP neural network to carry on the fitting. L is the line number of effectively software code, C is component quantity, R_c is average complexity of component interface, the R_w is average path complexity, and DC is the average component coupling.

III. THE NONLINEAR STRUCTURE OF BP NEURAL NETWORK

Using BP neural network to realize nonlinear fitting of the specific process is as follows.

For large sample program of 4 level software architecture as shown in Figure 1, the characteristics including effective lines of code, member number, the average complexity of component interface, the average path complexity, and the average component coupling can be measured.

As GUI, OSAL, PSAL,DAL, etc., the 4 software metric values of the characteristics of preprocessing, after processing, can be taken as the input value of the BP neural network for study and update of neutral network. After running a sample program on power simulation experiment platform, the energy consumption values E of sample process can be calculated, where E will serve as the output value of the BP neural network.

Determining the specific structure of the BP neural network, and it includes the number of hidden layer, hidden layer node number and transfer function of hidden layer and output layer transfer function, etc.

Nowadays, many mobile terminals come with a lot of apps pre-installed, including weather apps, mail apps and games. By inputting sample application of software characteristics value and measured energy consumption values, the BP neural network can be trained and optimized for further determining the BP neural network weights and threshold of hidden layer by training.

The network in BP system refers to the inter–connections between different neurons in all layers in mobile system. The first layer with input neurons can send data via synapses to those in second layer, and then to the last layer via of output neurons more synapses. By feeding into the characteristics of target program to the trained BP neural network together with the predictive value of energy consumption, the actual energy consumption of target program can be compared to verify the validity of the modeling method.

The most important thing in neural networks is its possibility of learning. After determining the various parameters, the network can be tested. Through a large number of sample data input, if the minimum mean square error precision is lower than a pre-defined error, we will get the weights of BP neural network w_1, w_2 and threshold b_1, b_2. Lastly, the identification functions of each layer should be defined.

The task of training a neural network model is usually to select one model from a set of remains. Generally speaking, as the single hidden layer in BP neural network, the transfer function of input layer and hidden layer is tansig function, the transfer function of the output layer is purelin function. As Trainlm function, its convergence speed and the approximation error can achieve satisfactory results when the number of hidden layer nodes is 10, which can be found through many experiments. Therefore, the training function of BP neural network is defined as Trainglm, and the hidden layer node number of training function is 10.

IV. ENERGY OPTIMIZATION ALGORITHM

According to characteristics of GUI, OSAL, PSAL,DAL, etc., energy consumption modeling of software architecture level can be divided into the following 4 steps.

Step 1, Initialization.

Based on Figure 1, we assume there is a nonlinear function relationship between energy consumption and software characteristics of architecture level.

Five aspects of system structure and the energy consumption of software are isolated from each other and separated from software characteristics namely effective lines of code, number of components, the average complexity of component interface, the average path complexity, and the average component coupling.

Step 2, Measurement

By choosing reasonable measurement methods, line number of code, components, the average complexity of component interface, the average path complexity, average measurement of the characteristic components can be coupled in architecture level.

As the number of hidden layer, it is important to determine how much is in hidden layer to determine the network error. However, too many hidden layers will result in complicate network, and the network training time and the fitting time will increase. A three-layer BP network can be completed from any n to m dimensional mapping. Therefore, the number of hidden layer is determined as 1.

Step 3, iteration

Through the simulation experiment platform of power consumption and a set of benchmark programs, the value of *TS* and *ES* can be measured according to the formula (1). The average power consumption during the software running value can be concluded.

$$P = E_s / T \qquad (3)$$

After getting effective architecture characteristic measurements, it is necessary to further determine the structure of the BP neural network. Then it is to determine the number of hidden layer nodes. In the BP neural network, the number of hidden layer nodes has a huge impact on BP network performance.

Step 4, Training and Terminate

By training the BP neural network to fit the nonlinear correlation functions *f* of the input values for each software architecture level, characteristics measurement and average power output of the software can be gotten. It is typical to determine the number of nodes in hidden layer by the following empirical formula n :

$$n = \sqrt{a + b} + i \qquad (4)$$

where a is input layer node number, b is the output layer node number, and i is the constant between 1 and 10.

V.   EXPERIMENT AND ANALYSIS

In order to verify the effectiveness of the energy consumption model based on software characteristics, this paper adopts five target programs to verify this model; the program contains a common flow file and more complex MP3 file decoding program. In Win 7, 32bit and Matlab 7.0 environment, through specific environment simulation experiment, assuming the energy level is in the area [0,1], the verification results are shown in Figure 2.

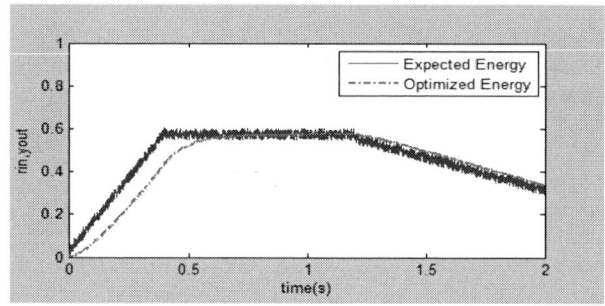

FIGURE II. EXPERIMENT RESULTS.

As we can see from Figure 2, target application data model of expected energy consumption value and the actual energy consumption value of maximum error is 33.4%, the minimum error is 3.1%, and the average error is 17.5%. The optimized energy consumption takes apparent advantages over that of expected energy, especially in step up period of software. It is not hard to see that there is a certain correlation between each characteristic and energy consumption values, and the most obvious characteristic is L which is on behalf of the effective lines of code.

VI.   CONCLUSIONS

This paper establishes a model of energy consumption and applies BP neural network as a function fitting method into energy management. The nonlinear function relationship between software characteristic quantities and software energy consumption is verified by an experiment. The future research is to extend these software characteristic quantities on more complex mobile terminal platform and software environment, and further research on optimization of energy consumption, etc.

ACKNOWLEDGEMENTS

This research was supported by the National Natural Science Foundation of China (No. 71471102), and Key Projects of Science and Technology Research Program of Hubei Provincial Department of Education in China (Grant No. D20101203).

REFERENCES

[1]   Zhu, Yuhao; Reddi, Vijay Janapa . High-Performance and Energy-Efficient Mobile Web Browsing on Big/Little Systems[C]. 19th IEEE International Symposium on High Performance Computer Architecture (HPCA). (Shenzhen, PEOPLES R CHINA . FEB 23-27, 2013), 2013: 13-24.

[2]   Song, Jaeki; Kim, Junghwan; Jones, Donald R.;   Application discoverability and user satisfaction in mobile application stores: An environmental psychology perspective[J].   DECISION SUPPORT SYSTEMS, 2014, 59:37-51.

[3]   Li, Gangmin; Shen, Zhun. A Middleware Architecture for Price Comparison Service on Mobile Phones [C]. st International Conference on Information Technology and Quantitative Management (ITQM).(Suzhou, PEOPLES R CHINA, MAY 16-18, 2013), 2013,17: 545-553.

[4]   Li, Yan; Hu, Xiaoqiang; Zeng, Liang. The Application of Mobile Agent in Mobile Payment[C]. International Conference on Computer Science and Network Technology (ICCSNT).(Harbin Normal Univ, Harbin, PEOPLES R CHINA, DEC 24-26, 2011), 2012,1-4:1612-1616.

[5]   Ruiz-Martinez, A.; Inmaculada Marin-Lopez, C; Sanchez-Martinez,D; SIPmsign:a lightweight mobile signature service based on the Session Initiation   Protocol[J].SOFTWARE-PRACTICE   & EXPERIENCE,2014,44(5):511-535.

[6]   Wang, Jiangjing; Jiang, Jiulei; He, Feng. Problem and Countermeasure on the Development of Mobile Electronic Commerce in China[C]. International Conference on Instrumentation, Measurement, Circuits and Systems (ICIMCS 2011).(Hong Kong, PEOPLES R CHINA, DEC 12-13, 2011),2012,127:861-866.

[7]   Peng Min-jing; Li Bo; Liu Meng. A Layered Context-aware Agent for Mobile Applications based on Users Needs Hierarchy [C]. International Conference on Mechatronics and Semiconductor Materials (ICMSCM 2013).(Xian, PEOPLES R CHINA, SEP 28-29, 2013), 2014,846-847: 1689-1692 .

[8]   Sun, Jie; Ren, Wei'anResearch on Mobile Business Value Network and Model Construction [C]. International Conference on Mechatronics and Information Technology (ICMIT 2013).(Guilin, PEOPLES R CHINA, OCT 19-20, 2013), 2014,462-463: 849-855.

International Conference on Power Electronics and Energy Engineering (PEEE 2015)

The Cultivation Mode about the Ability of Scientific Research of Undergraduate Based on the Dual-Tutorial System

X.D. Yuan

Materials Science and Engineering Institute
Shandong Jianzhu University
Jinan ,Shandong

J. Li

Yantai Lubao Steelpipe Co., Ltd
Yantai, Shandon

G.L. Yuan

Jinan Engineering Quality and Work Safety Supervision Station
Jinan, Shandong

Abstract—The dual-tutor system is a talent training mode under the Education Reform. Recent years, Institute of Materials Science and Engineering in Shandong Jianzhu University, has began to implement the dual-tutor system to improve comprehensively the quality of students. The cultivation mode about the capacity for undergraduate scientific research based on the dual-tutor system was discussed in detail through students' second class, open trials and the interest group of research. It is proved that students' capacity for scientific research has improved significantly.

Keywords-the dual-tutor system; open trials; undergraduate; capacity for scientific research

I. INTRODUCTION

The undergraduate tutor system originated from Oxford of England. Its originator was William·Wickham, the bishop of Winchester, who founded the "New School" at the beginning of the 14th century. It refers to a teaching system under the two-way selection between teachers and students. The teacher, who has a high level of expertise and possesses both political integrity and ability, can conduct an individual guidance aiming at students' learning, morality, life and psychology. The tutor system of Oxford brought a new revelation to talent training of Chinese universities. Peking University was regarded as an experimental unit to begin trying the tutor system in 2002. At present, with the education reform of high school progressing, there have been a variety of tutor system: masters tutor system, doctors tutor system, the head teachers system, the whole people tutor system, the whole journey tutor system, the scientific research tutor system, etc. [1]. Under a variety of tutorial system，the work of counselors and professional teachers is not clear, division of labor unclear and work often repeated, which directly affect the development of students' school work and interest of research. This paper mainly takes the Institute of Materials Science and Engineering of Shandong Jianzhu University as example, to discuss the cultivation mode on the capacity for undergraduate scientific research based on the dual-tutor system.

II. THE CONNOTATION AND THE IMPLEMENTATION PROCESS OF DUAL-TUTOR SYSTEM

Since 2006, the Institute of Materials Science and Engineering of Shandong Jianzhu University has began to try to implement dual-tutor system, whose main core thought was that the freshman and sophomore took ideological and political mentor – counselor as the principle thing; juniors and seniors regarded professional instructors - youth teachers as the prime thing. Explorations in this area have also been reported about other brother universities [2]. Students should be allowed to actively participate in teachers' research projects. Counselors' ideological guidance and professional teachers' guidance could be combined together organically through scientific research projects. This is conducive to managing students' daily life and improving the level of scientific research. The specific pattern is shown as Figure1. The significance of encouraging students of colleges and universities to actively participate in teachers' research projects is obvious [3-8].

Counselor is the main tutor of undergraduates during the freshman and sophomore year, who mainly takes charger of students' life, learning, work-study, the Party construction, campus culture and so on. During the freshman year, counselor divides students into different interest groups of scientific research based on the classification and their interest in the direction of their major, inviting different experts and professors with higher level of scientific research to give lectures according to different interest groups of scientific research. During the sophomore year, counselor should carry out a series of Design Competitions about students' career planning and Design Competitions about students' study arrangements according to different interest groups of scientific research, to help students identify good scientific research interest and build self-confidence.

At the first semester of the junior, the principle of "two-way selection" is adopted to carry out mutually selecting between students and tutors. Students select their tutors based

© 2015. The authors - Published by Atlantis Press

on major interests of their own and the research direction of teachers; teachers select students according to requirement of their issues and students' performance. The tutors of the students during the juniors and seniors year, are mainly the youth teachers, who mainly take charge of guiding students' major, cultivating their professional interests, improving the level of scientific research, etc. During the junior year, the tutors should actively encourage and guide students to participate in the nationwide undergraduates "Challenge Cup" Competition, the Casting Technology Design Contest, Concrete Design Competition etc, to encourage students to write academic papers. During the summer vacation, tutors should lead students to take part in " Substituted Post Internships" of social practice in summer and academic "Open Experiment" activities, to make them penetrate deep into enterprise and laboratory, to help them understand the expertise needs of the future work, to know the precautions and the process of operating the experimental apparatus, etc, to further enhance students' interest in their major and to improve students' capabilities in scientific research.

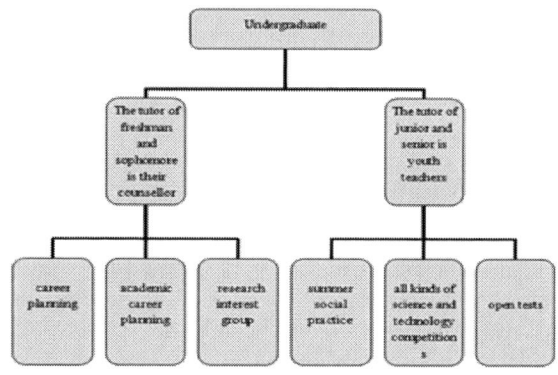

FIGURE I. THE CULTIVATION MODE OF UNDERGRADUATES' RESEARCH CAPACITY UNDER DUAL-TUTOR SYSTEM.

III. THE CULTIVATION MECHANISM OF STUDENTS' ABILITY ON SCIENTIFIC RESEARCH BASED ON THE DUAL-TUTOR SYSTEM

A. Guide Students to Do Well on Study Planning

On the condition of knowledge economy, the inexhaustible motive force for national economic development is that the innovation ability on scientific research of the successor under Socialist Modernization improves continually. For undergraduates to be as successor, it's imperative to improve themselves innovation capability on scientific research. In order to improve undergraduates' innovation ability on scientific research under the dual-tutor system, firstly tutors should do well on the students' academic guidance and guiding light. The Institute of Materials Science and Engineering of Shandong Jianzhu University, takes the dual-tutor system as guarantee, constantly pursuing improving students' research capabilities. Tutors establish different research interest groups according to different professional direction, such as: surface engineering group, casting group, mold design group, future welding group, metal materials group, green building team, CAD design teams, etc. And 30 research interest groups are included and each group has about 20 students. Then tutors

invite specialists and scholars on different research directions at home and abroad, to make professional reports depending on different research interest groups in order to enhance students' professional self-confidence. Every year about 40 specialists and scholars at home and abroad are invited to give different professional research reports. Students respond enthusiastically and the degree of participation is high. Finally, tutors establish evaluation form on students' profession comprehensive quality, according to students' performance in learning and life, the student leaders' feedback, as well as in-depth investigation on various professions. And reframe Design Competitions about students' career planning and Design Competitions about students' study arrangements to adjust students' profession direction, to enhance students' professional self-confidence, and to lay a solid foundation for improving students' research capacity .

B. Encourage Students to Participate in the Second Class

The second class is the best place where students put their professional knowledge into practice. Under the circumstances that Shandong Province is a big Province on manufacturing, the Institute of Materials Science and Engineering establishes the second class based on "research, practice, internship, teaching, and employment" relying on 20 enterprises, such as The Casting Association, The Surface Engineering Association, The Machinery Industry Association, The Academy of Sciences, The Limited Company of the King Kintai Casting in Shandong Province. The ideograph of the second class in the Institute Materials Science and Engineering is shown in Figure 2.

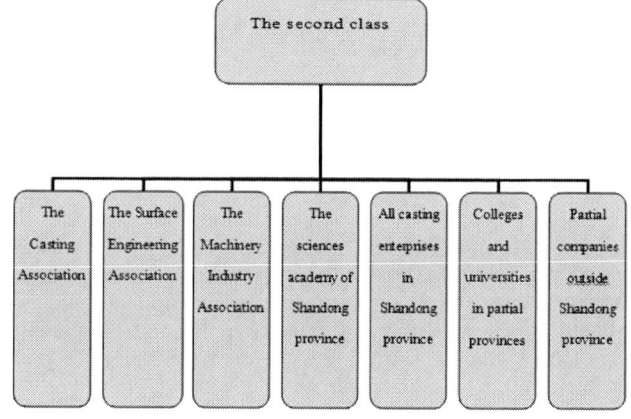

FIGURE II. THE SERVING MODE ON UNDERGRADUATES' SECOND CLASS.

The youth teachers are utilized as professional tutors to organize students to construct the second class, based on the form of "substituted post internship" organizing students to penetrate into enterprises, to understand their requirements, to know the relation between company and profession in order to further enhance students' professional ability; taking a series of undergraduate competitions such as "Challenge Cup" and the casting design competition like "The Perpetual Championship Cup" as the leader to encourage students to take part in research projects to improve the capacity for scientific research; regarding "Open Experiment" as possy, to organize students to know well the experimental procedure, the instruments' operation, combining the experimental knowledge with theoretical knowledge organically.

IV. THE RESEARCH CAPACITY OF STUDENTS IS SIGNIFICANTLY IMPROVED UNDER THE DUAL-TUTOR SYSTEM

A. Students' Employability is Significantly Enhanced

Based on the dual-tutor system and two-way managements and services for students, students' employability is also enhanced with their research capacity significantly improved. Recent three years, the graduates' first signing rate is more than 95% and the overall signing rate is 100% of the two majors in the Institute of Materials Science and Engineering. The employment companies are mainly National Nuclear Power, North Heavy Industries, China Railway Engineering Corporation, China Construction Group, Baosteel, Jinan Steel, Shandong Electric Power Group and other large enterprises. The mainly postgraduate universities are Tsinghua University, Harbin Institute of Technology, Beijing University of Aeronautics and Astronautics, Chinese Academy of Sciences, Shandong University and other renowned institutions. Especially in 2010, the number of the graduates who are admitted to the universities graduated as 211 and 985 universities, were 59 in all, accounting for about 20%. The number of the students who get a job from domestic famous enterprises is 52, accounting for about 17.6%.

B. Students' Research Achievement is Brilliant

Student's research capacity in dual-tutor system has been significantly improved, and their capacities on operating instruments, designing experiments, and writing papers have been improved significantly. Recent three years, the enthusiasm of students participating in social practice and various competitions is very high. There is a phenomenon that one teacher leads more than one team. Recent three years, students have achieved remarkable achievements in the social practice, technology innovation and other activities. 4 people have obtained the name of advanced individual in social practice of Shandong Province. 128 people have won the title of advanced individual in social practice of Shandong Jianzhu University. The number of awards with National Level in Science and Technology Competition like "Challenge Cup" is 7, in addition, 38 awards with provincial level, 4 patents for invention authorized by nation, 18 new type patents, and 14 academic papers first published (EI indexed).

V. CONCLUSIONS

Through different guidance of counselors and youth teachers, the talent cultivation mode under the dual-tutor system explores the cultivation mechanism on undergraduates' capacity for scientific research, to establish a more comprehensive research cultivation mode and effectively to improve the students' research capabilities, from students' research interest groups, career planning, academic career planning, social practice, science and technology competition, open-label trial, etc.

ACKNOWLEDGEMENT

Fund for the College science and technology plan of Shandong Province (Project No. J12LA11).

REFERENCES

[1] Wenqiang Zheng, Qian Ding, The first exploration on how the tutor system of research type impacting undergraduates' research level, The academic journal on Henan Staff medical college. 23 (2011) 762-763.

[2] Weizhong Liang, Zhenting Wang, etc. exploration for the training mode under undergraduate dual-tutor system about the major of Material Molding, Casting equipment and technology. 4 (2011) 33-34.

[3] Li Zhen,Yun Xue. Research on approaches of cultivating students' abilities based on undergraduate research projects, Talents training. 216 (2011) 27-29.

[4] Qinmin Xu, Tao Zhang. Quality training and practice of innovative and entrepreneurial undergraduates learning Engineering Majors, Laboratory Science. 14 (2011) 4-6.

[5] Baoliang Bi, Ying Chen,etc. approaches and suggestions on undergraduates participating in research, The academic journal of Yunnan Agricultural University. 5 (2011) 87-91.

[6] Shengyi Jiang, Empirical Study on training undergraduates' research ability, The academic journal of Guangdong University of Foreign Studies. 22 (2011) 96-99.

[7] Wanrong Gu, Ji Sun, etc. reflection on improving cultivation mechanism of undergraduates' research innovationton ability, Education and Teaching Management. (2011) 24-27.

[8] Haiping Lu. Analyze the factors affecting undergraduates' research and innovation abilities, The academic journal of Chongqing University of Science and Technology. 20 (2011) 163-164.

International Conference on Power Electronics and Energy Engineering (PEEE 2015)

Training Mode on Graduation Design of Engineering Students Based on the Dual-Tutor System

X.D. Yuan

Materials Science and Engineering Institute
Shandong Jianzhu University
Jinan, Shandong

X. Liu

Shandong Product Quality Inspection Research Institute
Jinan, Shandong

G.L. Yuan

Jinan Engineering Quality and Work Safety Supervision Station
Jinan, Shandong

Abstract—The dual-tutor system is a training model. The Institute of Materials Science and Engineering of Shandong Jianzhu University began to carry out the dual-tutor system in 2006 to improve the quality of cultivating students and to strengthen the construction of connotation. The training mode on graduation design of engineering students based on the dual-tutor system was mainly discussed in this paper. The importance of the reform of graduation design was investigated through the innovation of traditional graduation design, the graduation design mode of university-industry cooperation and the effect analysis on graduation design. It is proved that the training mode based on the dual-tutor system has remarkable effects on graduate design, which provides a new way of thinking to the construction of academy connotation.

Keywords-dual-tutor system; graduation design; school-enterprise cooperation; engineering

I. INTRODUCTION

The dual-tutor system originated from Oxford of England, which is a new talents training model. The dual-tutor system of Oxford brought new enlightenment to talents training of Chinese universities. In 2002, Peking University began to try the dual-tutor system as an experimental unit. Afterwards, various "similar tutor systems" are produced, for example, Master tutor system, Doctor tutor system, Head teachers system, Whole tutor system, Whole course tutor system, Research-type tutor system, etc [1]. Graduate design is the last course in the process of achieving the training objectives, which is also a test for the knowledge that undergraduates have learned in four years. In a sense, the quality of graduation design directly proves whether the training objective in the past four years is successful or not. Currently, the managements of graduation design under a variety of tutorial systems are not really the same. There is insufficiency in different Sponsored by the Research degree selecting a topic aimlessly for students, the design structure is unclear, lacking of theoretical knowledge, processing data is unscientific, experiment workload is too small, lacking of standardization in writing papers, etc. [2-6]. Given the status of traditional graduate design, this paper takes Materials Science and Engineering of Shandong Jianzhu University as an example to explore and analyze the training mode of graduation design based on the

dual-tutor system.

II. TRAINING SYSTEM OF DUAL-TUTOR SYSTEM

A. Implementation Process of the Dual-Tutor System

The implementation process of the dual-tutor system is shown in Figure 1. It can be seen in Figure 1 that the tutors of freshman and sophomore are mostly counselors, who mainly take charge of students' life, learning, taking a part-time job for their study, party building, and campus culture and so on. During the freshman year, counselor divides students into different interest groups of scientific research based on the classification and their interest in their major direction, inviting different experts and professors with higher level of scientific research to give lectures according to different interest groups of scientific research; during the sophomore year, counselor carries out a series of Design Competitions about students' career planning and Design Competitions about students' study arrangements according to different interest groups of scientific research, helping students to identify good scientific research interest and build self-confidence.

The students' tutors are mainly young teachers in junior and senior years, who mainly take charge of guiding students' major, cultivating their professional interests, improving the level of scientific research and instructing graduation design. In the first semester of junior year, the principle of "two-way selection" is adopted to carry out mutually selection between students and tutors. Students select their tutors based on major interests of their own and the research direction of teachers; teachers select students according to requirement of their issues and students' performance. During the junior year, the tutors should actively encourage and guide students to participate in the nationwide undergraduates "Challenge Cup" Competition, leading students to take part in "substituted post exercitation" in summer social practice and "open tests" activities of academy in order to lay a solid foundation for students' graduation design.

© 2015. The authors - Published by Atlantis Press

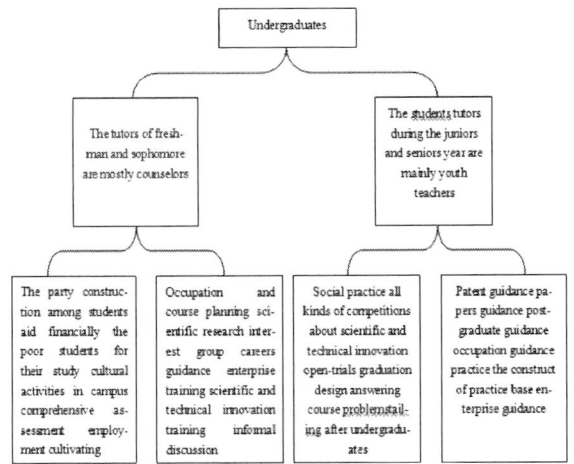

FIGURE I. IMPLEMENTATION PROCESS OF THE DUAL-TUTOR SYSTEM.

B. The Evaluation System on the Dual-Tutor System

To ensure the smooth implementation of dual-tutor system, and to keep it developing healthily, academy has established a set of scientific evaluation systems and has improved the dual-tutor system by methods of self-evaluation, students' assessment and comprehensive evaluation. And each of these three methods accounts for different proportions, whose results are generally classified into excellent, good, qualified and disqualified. On one hand, final evaluation results are regarded as a teacher's work performance. At the end of year, the teachers are given certain advantages and set the examples. On the other hand, final evaluation results are served as the content of assessing teachers titles work performance. And the teachers are given certain superiority. The principles of dynamic management, directional management and scientific management are adopted to enrich the dual-tutor system, to help the teachers by qualitative and quantitative guidance, to inspire teachers to broaden their horizons, to help students to focus on scientific facts, and to guide students to explore the peaks of science. There are certain reports in this area at home [7].

III. THE TRAINING MODE ON GRADUATION DESIGN

A. The Conventional Situation o Traditional Graduate Design

The graduation design is an important part in the undergraduate teaching system. The quality of graduation design directly affects the quality of training students. There are some problems existing in traditional graduate design, for example, the content of topics and the content of scientific research projects are inconsistent, which results in less teachers' input, unseriousness, and blindness. The required equipment in the process of carrying out the graduation design is insufficient, which results in students' poor operation ability, the experimental program being not fully implemented, and students' scientific design greatly restricted. The management system of graduate design is unsound, leading to many consequences like tutors are inattentive, and can't work actively, students are not serious and the workload is low. Tutors' workload is heavy and tutors have fewer opportunities to communicate

with students. Students' valid resources are fewer. The quality of graduation design is poor. The traditional graduate design is not in accordance with schools' standards, resulting in non-standard graduation design writing, unscientific data processing, and unprofessional research habits and so on.

B. The Innovation Mode of Graduation Design

Given the shortcomings of traditional graduation design, the Institute of Materials Science and Engineering of Shandong Jianzhu University has established a more scientific training and management mode for graduate design in the work requirements of dual-tutors system. The specific content is shown in Figure 2 and Figure 3. As we can see from the chart, graduation design time directed by tutors on campus is lengthened from half a year to one year. The students and tutors determine the topic and content of graduation design together based on students' research interests and tutors' research project with tutors' standard management, rigid attendance, serious guidance and students' serious attitude to study, deep thought and open innovation ability. In addition, in order to improve students' work ability and graduation design quality, the Institute of Materials Science and Engineering tries to adopt the university-enterprise cooperation mode to carry out the graduation design, to help communication between intramural tutors and enterprise tutors, and to help students to accomplish graduation design requirements together. There are also related graduation designs reports on school-enterprise cooperation [8].

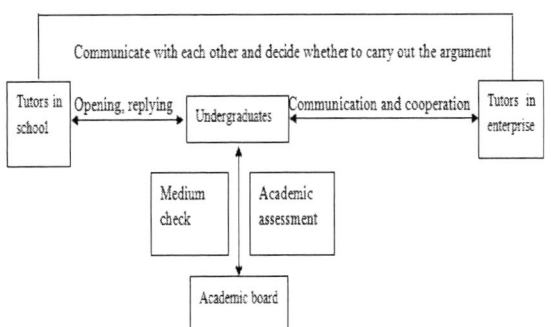

FIGURE II. THE MANAGEMENT MODE OF GRADUATION DESIGN.

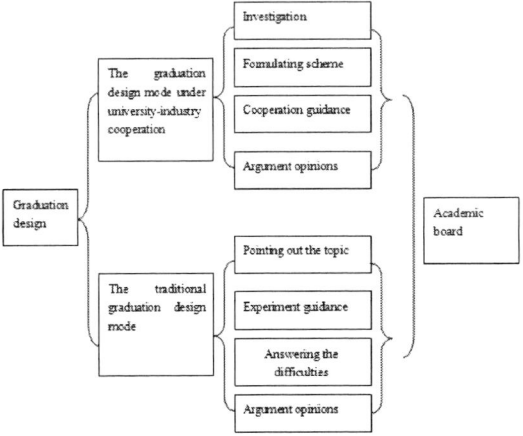

FIGURE III. THE TRAINING MODE OF GRADUATION DESIGN.

IV. THE EFFECT ANALYSIS OF GRADUATION DESIGN BASED ON THE DUAL- TUTOR SYSTEM

A. Students' Confidence at Work is Improved

Compared with the traditional graduation design, the effects on the innovation of graduate design is remarkable because of the dual-tutor system. The reform and innovation of intramural graduation design give students enough time to carry out graduation design. Students are able to think deeply, to perfect scheme, to study fully, and to strengthen theory. Students' practice abilities at work and their confidence to work after graduation are enhanced while students' graduation design quality is improved. The school-enterprise cooperation model improves and trains most of students' abilities at least three aspects, namely the ability of linking theory with practice, the ability of engineering drawings, and the ability of self-management. The improvement of these capabilities is bound to play an important role in students practice work.

B. The Structure of Graduation Design is Improved

Due to the influence of traditional graduation design, tutors choose the topic for graduation design based primarily on their professional advantages. Because of this choosing method, tutors can only study what they are already skilled at. However, the new training mode, especially the implementation of school-enterprise cooperation in training mode, improves the structure of graduation design, and broadens the horizons of professional instructors by obtaining external practice information of scientific research.

C. The Level of Teachers' Scientific Research is Improved

The development of graduation design of school-enterprise cooperation improves the structure of graduation design, and avoids the flaw of traditional graduation design. At the same time, tutors' research perspective, research ideas and research attitude are also changed. Tutors can correctly judge the practicality of scientific research project, adjust research ideas and increase available research content, thereby increasing the level of research work.

V. CONCLUSIONS

The training mode of the graduation design based on the dual-tutor system was experimented by Shandong Machinery Industry Association, Shandong Province Surface Engineering Association, and the province's major machinery manufacturing. The quality of graduation design and students' ability to get a job are improved. Teachers' research horizons are also broadened. The quality of teachers' research work is enhanced. Practice has proved that the feasibility of the training mode on graduation design based on the dual-tutor system is powerful. The effect of the training mode is significant. This kind of training mode should be widely implemented.

ACKNOWLEDGEMENT

Fund for the College science and technology plan of Shandong Province (Project No. J12LA11).

REFERENCES

[1] Wenqiang Zheng, Qian Ding. First exploration on how the research-tutors affect the scientific research level of undergraduates, The academic journal of Henan Staff Medical College. 23 (2011) 762-763.

[2] Fulin Teng, Jianhua Wang. How the tutor system influencing the quality of the graduation design of undergraduates, Chinese modern education and equipment. 17 (2011)140-141.

[3] Yanghua Ou, Yunqin Shen. The exploration and thought on how to improve graduation design quality of undergraduates, The innovation on technology and education. 15 (2011)201-203.

[4] Zhuojuan Yang, Xiaodong Yang. Differentiate and analyze the existence or abolishment of graduation design, Research on Higher Education in Heilongjiang. 9 (2011)160-162.

[5] Zhenyu Wang. The discussion on how improve the Improve the graduation design quality of engineering undergraduate, The innovation on science and technology. 30 (2011)176-177.

[6] Shuhong Xu, Shouzan Liu. The construct on evaluation index system of tutor system about undergraduate in high schools, Education in Heilongjiang province. 8 (2010)19-21.

[7] Xiaokai Xing, Changchun Wu. Practice and exploration on the graduation design directed by school-enterprise cooperation of research university, Educational reform. 5 (2011)84-86.

[8] Lingfeng Zhang, Jiuba Wen,etc. The training mode on undergraduates graduation design who are in the universities of school-enterprise cooperation based on the enterprise, Chinese modern education and equipment. 19 (2011)106-108.

International Conference on Power Electronics and Energy Engineering (PEEE 2015)

The Connectivity of Faulty Folded Hypercube Networks

D. Yuan, H.M. Liu, M.Z. Tang
Science College of China
Three Gorges University
China

Abstract—**The connectivity of a topology of a given network is one of the most important issues in determining whether a network can simulate routings of various lengths. In this paper, it shows that any two distinct nodes** x, y **are connected by paths of every length from** $d(x, y)$ **to** $2^n - 1$ **in folded hypercube networks** $FQ_n - F_e$ **with** $|F_e| \leq n - 1$ **if each node is incident with at least** m ($2 \leq m \leq n$) **fault-free links, where** $d(x, y)$ **is the distance between** x, y**, and** F_e **is the set of fault links.**

Keywords-network connectivity; folded hypercube networks

I. INTRODUCTION

The n-dimensional hypercube Q_n (or n-cube) is one of the most important topology of networks duo to its excellent properties such as regularity, recursive structure, small diameter, vertex and edge transitive and relatively short mean distance (Xu, J.M. 2003). In order to improve the performance of hypercube, the folded hypercube FQ_n has been proposed (Saad, Y. & Schultz, M.H. 1988).

In a large-scale hypercube network, should any component fail, it's desirable that the rest of the network continue to operate in spite of the failure. This leads to the graph-embedding problem with faulty edges and/or vertices. This problem has received much attention (Fu, J.S. 2006)-(Xu, J.M. & Ma, M.J. 2006).

A graph G with at least three vertices is pancyclic if there exist cycles of all lengths from 3 to $|V(G)|$. A graph is edge-pancyclicity (vertex-pancyclicity) if each edge (vertex) is contained in cycles of every length from 3 to $|V(G)|$. Further, a graph G is bipancyclicity if G is a bipartite graph whose cycles are necessarily of even length. A bipartite graph G is bipanconnected if G contains a path of length l connecting any distinct vertices x, y with $d(x, y) \leq l \leq |V(G)| - 1$ such that $2 \mid (l - d(x, y))$, where $d(x, y)$ denotes the shortest path between x and y. In this paper, we primarily explore the bipanconnection of a conditional faulty folded hypercube.

The problem of embedding paths in an n-dimensional hypercube and folded hypercube has been well studied. (Fu,

J.S. 2006) showed that $Q_n (n \geq 3)$ with faulty vertices $|F_v| \leq n - 2$ contains a path joining any two different vertices x and y in $Q_n - F_v$ with length of at least $2^n - 2|F_v| - 1$ (or $2^n - 2|F_v| - 2$) when $d(x, y)$ is odd (or even). (Hsieh, S.Y. et al. 2009) proved that if the folded hypercube FQ_n has just only one fault node, then FQ_n contains cycles of every even length from 4 to $2^n - 2$ when n is even, and cycles of every odd length from $n + 1$ to $2^n - 1$ when n is odd. (Liu, M. & Liu, H.M. 2012) addressed that there exists a cycle passing through all nodes in $FQ_n - F_e$ with the number of faulty edges $|F_e| \leq n - 1$ when n is odd. (Ma, M.J. et al. 2006) further demonstrated that $FQ_n - F_e (n \geq 3)$ with $|F_e| \leq 2n - 3$ contains a fault-free cycle passing through all nodes if each vertex is incident with at least two fault-free edges. (Kuo, C.N. & Hsieh, S.Y. 2010) improved the conclusion of (Ma, M.J. et al. 2006) and proved that $FQ_n - F_e$ with $|F_e| = 2n - 3$ contains a fault-free cycle of every even length from 4 to 2^n. (Kuo, C.N. et al. 2013) discussed the fault tolerance properties of FQ_n with both fault vertices and fault edges occurring and obtained that (i) FQ_n contains a fault-free path of length at least $2^n - 2|F_v| - 1$ (resp. $2^n - 2|F_v| - 2$) between any two fault-free nodes of odd (resp. even) distance if $|F_v| + |F_e| \leq n - 1$ when n is odd, (ii) FQ_n contains a fault-free path of length at least $2^n - 2|F_v| - 1$ between any two fault-free nodes if $|F_v| + |F_e| \leq n - 2$ when n is even.

In this paper, under the conditional fault model, we show that every pair of vertices x and y with distance $d(x, y)$ in $FQ_n - F_e (n \geq 3)$ with $|F_e| \leq n - 1$ are connected by paths of every length from $d(x, y)$ to $2^n - 1$ if each vertex is linked with at least m ($2 \leq m \leq n$) fault-free edges.

II. PRELIMINARIES

We follow (Xu, J.M. 2003) for graph-theoretical terminology and notation not defined here. A network is usually modelled by a simple connected graph $G = (V, E)$, where

© 2015. The authors - Published by Atlantis Press

$V = V(G)$ (or $E = E(G)$) is the set of vertices(or edges) of G. We define the vertex x to be a neighbor of y if $xy \in E(G)$. A graph G is bipartite if X, Y are two disjoint subsets of $V(G)$ such that $E(G) = \{xy \mid x \in X, y \in Y\}$. A graph $P = (u_1, u_2, \cdots, u_k)$ is called a path if the vertices $u_1, u_2, \cdots, u_k$ are distinct and any two consecutive vertices u_i and u_{i+1} are adjacent. u_1 and u_k are called the end-vertices of P. If $u_1 = u_k$, the path $P(u_1, u_k)$ is called a cycle (denoted by C). The length of a path P (a cycle C), denoted by $l(P)$ (or $l(C)$), is the number of edges in P (or C). In general, the distance of two vertices x, y is the length of the shortest (x, y)-path.

The n-dimensional hypercube Q_n (or, n-cube) can be represented as an undirected graph with 2^n vertices. Every vertex $x \in Q_n$ is labeled as a binary string $x_1 x_2 \cdots x_n$ of length n from $00 \cdots 0$ to $11 \cdots 1$. Two vertices u and v are adjacent if their binary strings differ in exactly one bit. For convenience, we call $e \in E(Q_n)$ an edge of dimension i if its end-vertices' strings differ in i th-bit. In the rest of this paper, denote $x^i = x_1 x_2 \cdots \overline{x_i} \cdots x_n$, $\overline{x_i} = 1 - x_i$, $x_i \in \{0,1\}$. The Hamming distance of two vertices $x = x_1 x_2 \cdots x_n$ and $y = y_1 y_2 \cdots y_n$ is as

$$H(x, y) = \sum_{i=1}^{n} |x_i - y_i|$$

The number of different bits between them. Let $d(x, y)$ be the shortest distance of x and y. Note that Q_n is a bipartite graph, and for any two distinct vertices x, y of Q_n, $d(x, y) = H(x, y)$.

As a variant of hypercube, the n-dimensional folded hypercube FQ_n is obtained by adding more edges between its vertices.

Definition 1. The n-dimensional folded hypercube FQ_n is a graph with $V(FQ_n) = V(Q_n)$. Two vertices $x = x_1 x_2 \cdots x_n$ and y is connected by an edge if and only if

(i) $y = x_1 x_2 \cdots \overline{x_i} \cdots x_n$ (denoted by x^i), or

(ii) $y = \overline{x_1} \overline{x_2} \cdots \overline{x_i} \cdots \overline{x_n}$ (denoted by $\overline{x}$).

Therefore, the hypercube Q_n is a spanning subgraph of the folded hypercube FQ_n obtained by removing the second type of edges $x\overline{x}$ ($x \in V(FQ_n)$), called complementary edges of FQ_n and denoted by $E_c = \{x\overline{x} \mid x \in V(FQ_n)\}$.

The first type of edges are defined to be the hypercube edges, and denoted by $E_i = \{xx^i\}, i = 1, 2, \cdots, n$.

The following results are useful in the proof of our method.

Lemma 1. $FQ_n - E_i$ is isomorphic to Q_n for any $i \in \{1, 2, \cdots, n, c\}$

Lemma 2 (Sun, C.M. 2012). Each edge of $Q_n - F_e$ with $|F_e| \leq n - 1$ is contained in cycles of every even length from 4 to 2^n.

Lemma 3 (Fu, J.S. 2006). Each edge of FQ_n is contained in every even cycle of length from 4 to 2^n if n is odd.

Lemma 4 (Fu, J.S. 2002). For any two vertices $u, v \in Q_n$, if $d(u, v) = d$, then there exists a path connecting u, v of length as $d, d + 2, \cdots, c$, where $c = 2^n - 1$ (or $2^n - 2$) for d is odd (or even).

Lemma 5 (Saad, Y. & Schultz, M.H. 1988). For any two vertices $u, v \in Q_n$, if $d(u, v) = k$, then there are n internal disjoint paths from u and v such that there are k paths of length k and $n - k$ paths of length $k + 2$.

Lemma 6 (Sun, C.M. 2012). For any two distinct vertices x and y of Q_n with at most $n - 2$ faulty edges, there exists a non-faulty (x, y)-path in Q_n of length l such that

if $|F_e| < d(x, y)$, then

$d(x, y) \leq l \leq 2^n - 1$ and $2 | (l - d(x, y))$;

if $|F_e| \geq d(x, y)$, then

$d(x, y) + 2 \leq l \leq 2^n - 1$ and $2 | (l - d(x, y))$.

Lemma 7 (Sun, C.M. 2012). There exist exactly $n - 1$ disjoint cycles in Q_n of length 4 that contain an edge xy in common.

By Lemma 7, we can immediately induct the following conclusion.

Lemma 8. There exist n disjoint cycles in an n-dimensional folded hypercube FQ_n of length 4 that contain an edge in common.

Lemma 9(Xu, J.M. & Ma, M.J. 2006). Let x, y be any two vertices of n-dimensional folded hypercube. If $H(x, y) = k$, then there exist $n + 1$ internal vertex-disjoint paths between x and y, among which there are k paths of length k and $n - k + 1$ paths of length $k + 2$ when

$1 \le k \le \lceil n/2 \rceil$, or k paths of length $n-k+3$ and $n-k+1$ paths of length $n-k+1$ when $\lceil n/2 \rceil \le k \le n$.

Lemma 10. n-dimensional folded hypercube FQ_n can be portioned into two subgraphs, denoted by FQ_n^0 and FQ_n^1, which are $(n-1)$-cubes along any dimension i such that for any $x = x_1 x_2 \cdots x_i \cdots x_n \in FQ_n^0$ satisfying $x_i = 0$. $x = x_1 x_2 \cdots x_i \cdots x_n \in FQ_n^1$ satisfying $x_i = 1$.

III. THE BIPANCONNECTIVITY OF FOLDED HYPERCUBE

Theorem. For any two distinct vertices x, y in FQ_n ($n \ge 3$) with $|F_e| \le n-1$, when n is an odd integer, there exists a fault-free (x,y)-path of length l such that

if $|F_e| < d(x,y)$, then

$d(x,y) \le l \le 2^n - 1$ and $2|(l - d(x,y))$;

if $|F_e| \ge d(x,y)$, then

$d(x,y) + 2 \le l \le 2^n - 1$ and $2|(l - d(x,y))$.

Proof. Notice that for any two distinct vertices x, y in FQ_n, if $d(x,y) \le \lceil n/2 \rceil$, then $d(x,y) = H(x,y)$; if $\lceil n/2 \rceil < H(x,y) \le n$, $d(x,y) = n - H(x,y) + 1$. Then we consider the cases when $d(x,y) = H(x,y)$ and $d(x,y) = n - H(x,y) + 1$.

Case 1. $d(x,y) = H(x,y)$

By Lemma 10, FQ_n can be partitioned into two $(n-1)$-cubes along some dimension i such that the vertices x, y are in the same $(n-1)$-cube and $|E_i \cap F_e| + |E_c \cap F_e| \ge 1$. Without loss of generality, assume that $x, y \in FQ_n^0$.

Case1.1.

$f_0 = |F_e \cap E(FQ_n^0)| = n-2$, $f_1 = |F_e \cap E(FQ_n^1)| = 0$,

$$|(E_i \cup E_c) \cap F_e| = 1$$

Select a faulty edge e and regard e as fault-free, then $f_0 - 1 = n-3$. With $d(x,y) = H(x,y)$, in FQ_n^0, there exists a (x,y)-path P of length l from $d(x,y) + 2$ to $2^{n-1} - 1$ and $2|(l - d(x,y))$. Next, we need to consider the case whether the faulty edge e is in any fault-free (x,y)-path or not.

(1) If this faulty edge $e = uv$ is in some (x,y)-path $P \subseteq FQ_n^0$ of length l.

Since $f_1 = 0$ and $|(E_i \cup E_c) \cap F_e| = 1$, then either the edges $\overline{uu, vv, uv}$ or $uu^i, vv^i, u^i v^i$ are fault-free. Without loss of generality, say, $\overline{uu, vv, uv}$ are fault-free. Lemma 4 and $d(\overline{u}, \overline{v}) = 1$ guarantees fault-free $(\overline{u}, \overline{v})$-path $P_1 \subseteq FQ_n^1$ of every odd length $l_1 \in \{1, 3, \cdots, 2^{n-1} - 1\}$. $P^* = (P - uv) \cup u\overline{u} \cup P_1 \cup \overline{v}v$ is a fault-free path connecting x, y with length of $l^* = (l-1) + 1 + l_1 + 1 = l + l_1 + 1$, $d(x,y) + 4 \le l^* \le 2^n - 1$, $2|(l^* - d(x,y))$.

Now, we consider a (x,y)-path of length $d(x,y) + 2$. With Lemma 5, there are $n-1$ internal disjoint paths between x and y in FQ_n^0, if every one of these $n-1$ paths contains at least one faulty edge, then there are at least $n-1$ fault edges in FQ_n^0, but $f_0 = n-2$. This means that there exists a fault-free path of length $d(x,y) + 2$.

(2) If this faulty edge $e = uv$ is not in any (x,y)-path.

That is, there exists a fault-free (x,y)-path $P \subseteq FQ_n^0$ of length l, where $d(x,y) \le l \le 2^{n-1} - 1$ and $2|(l - d(x,y))$. Then we only need to find those fault-free (x,y)-paths of length l^* which has the same parity as $d(x,y)$ and ranges from 2^{n-1} to $2^n - 1$. Choose an edge ab in one of the shortest fault-free (x,y)-path $P \subseteq FQ_n^0$ of length $2^{n-1} - 1$ when $d(x,y)$ is odd (or $2^{n-1} - 2$ when $d(x,y)$ is even) such that the edges aa^i, bb^i are non-faulty. Because $f_1 = 0$, the edge $a^i b^i$ is fault-free in FQ_n^1, and by Lemma 4, we can also construct a fault-free (a^i, b^i)-path $P_2 \subseteq FQ_n^1$ of every odd length l_1, $l_1 \in \{1, 3, \cdots, 2^{n-1} - 1\}$. Thus the desired (x,y)-path P^* of length l^* in FQ_n is constructed as $P^* = (P_1 - ab) \cup aa^i \cup P_2 \cup b^i b$ with length of l^*.

If $d(x,y)$ is odd, $l^* = 2^{n-1} + l_1$, and $2^{n-1} + 1 \le l^* \le 2^n - 1$, where $2|(l^* - d(x,y))$.

If $d(x,y)$ is even, $l^* = 2^{n-1} + l_1 - 1$, and $2^{n-1} \le l^* \le 2^n - 2$, where $2|(l^* - d(x,y))$.

Case1.2.

$f_0 = |F_e \cap E(FQ_n^0)| \le n-3$, $f_1 = |F_e \cap E(FQ_n^1)| \le n-3$.

Without loss of generality, suppose that $f_0 \geq f_1$. If $f_0 < d(x,y) = H(x,y)$, by Lemma 6, there exists a fault-free path (x,y)-path P_1 of length l_1, where $d(x,y) \leq l_1 \leq 2^{n-1} - 1$ and $2|(l_1 - d(x,y))$. Choose an edge ab in some fault-free (x,y)-path of length $2^n - 1$. By Lemma 8, the edges aa^i, bb^i are fault-free. $H(a^i, b^i) = 1$ and Lemma 6 guarantee that there being a fault-free (a^i, b^i)-path P_2 of length l_2, where $H(a^i, b^i) \leq l_2 \leq 2^{n-1} - 1, 2|(l_2 - H(a^i, b^i))$.

Thus $(P_1 - ab) \cup aa^i \cup P_2 \cup b^i b$ is a fault-free (x,y)-path of length $l_1 - 1 + 1 + l_2 + 1 = l_1 + l_2 + 1$.

On the other hand, if $f_0 \geq d(x,y)$, by Lemma 6, there are fault-free (x,y)-paths of length from $d(x,y) + 2$ to $2^{n-1} - 1$. The method of constructing such a fault-free (x,y)-path is similar to the case of $f_0 < d(x,y)$.

In particular, when $f_0 = f_1 = 0$, that is, all faulty edges are in $E_i \cap E_c$, by Lemma 1 and Lemma 6, the theorem is true.

Case 2. $d(x,y) \neq H(x,y)$.

By definition of FQ_n, $d(x,y) \neq H(x,y)$ implies that $[n/2] < H(x,y) \leq n$ and $d(x,y) = n - H(x,y) + 1$. The shortest path connecting x and y contains a complementary edge. Now, we can partition FQ_n into two subcubes such that the nodes x, y are not in the same subcube and $|E_i \cap F_e| + |E_c \cap F_e| \geq 1$. Suppose that $x \in FQ_n^0$ and $y \in FQ_n^1$. There are several cases need to be discussed.

Case 2.1. $f_0 = n-2, f_1 = 0$.

Choose a non-faulty edge $ax \in FQ_n^0$ such that the edge $\overline{aa}$ is fault-free (if $\overline{xx}$ is a non-fault link, $x = a$ is feasible), $d(\overline{a}, y) = d(x,y) - 2$ if ax is on a shortest (x,y)-path, and $d(\overline{a}, y) = d(x,y)$ if ax is not on a shortest (x,y)-path. Note that $f_1 = 0$ implies that FQ_n^1 is fault-free. By Lemma 4, we can find a fault-free $(\overline{a}, y)$-path P_1 of length l_1, $d(x,y) + 2 \leq l_1 \leq 2^{n-1} - 1$ (or $d(x,y) \leq l_1 \leq 2^{n-1} - 1$), l_1 and $d(x,y)$ have the same parity. Select a fault-free (x,y)-path $P = xa \cup a\overline{a} \cup P_1$ with length of $l = l_1 + 2$. Since $d(x,y) = n - H(x,y) + 1$, then $d(x,y) \leq l_1 \leq 2^{n-1} - 1$ (or $d(x,y) + 2 \leq l_1 \leq 2^{n-1} - 1$) and $2|(l - d(x,y))$. Similar to Case 1.1, those desired fault-free

(x,y)-paths of length from $2^{n-1} + 3$ to $2^n - 1$ can be constructed.

Case 2.2. $f_0 \leq n-3, f_1 \leq n-3$.

Because $f_0 \leq n-3$, choose a neighbor a of x in FQ_n^0 such that $ax, a\overline{a}$ are fault-free. Let $xa \cup a\overline{a} \cup P[\overline{a}, y]$ be a fault-free (x,y)-path, where $P[\overline{a}, y] \in FQ_n^1$ is a path connecting $\overline{a}$ and y.

If $|F_e| < d(x,y)$, we have $f_1 < H(\overline{a}, y)$ because $|E_i \cap F_e| + |E_c \cap F_e| \geq 1$ and $d(x,y) = 2 + H(\overline{a}, y)$. By Lemma 6, there exists a fault-free $(\overline{a}, y)$-path of length from $H(\overline{a}, y)$ to $2^{n-1} - 1$. Thus the desired paths, whose lengths are from $d(x,y)$ to $2^{n-1} - 1$, is constructed as $xa \cup a\overline{a} \cup P[\overline{a}, y]$.

If $|F_e| \geq d(x,y)$, by Lemma 8, there exists a shortest path P of length $d(x,y) + 2$ between x and y in FQ_n. Therefore, we only need to find fault-free paths of length from $2^{n-1} + 3$ to $2^n - 1$. By Lemma 6 and $d(x,y) = H(x,a) = 1$, there are fault-free (x,a)-paths $P[x,a]$ of length l' from $d(x,y) + 2$ to $2^{n-1} - 1$. Choose a fault-free $(\overline{a}, y)$-path $P[\overline{a}, y]$ of length $2^{n-1} - 1$. Then set $P^* = P[x,a] \cup a\overline{a} \cup P[\overline{a}, y]$ to be a new fault-free path of length l^* from $2^{n-1} + 3$ to $2^n - 1$, where $2|(l^* - d(x,y))$. The proof is finished.

ACKNOWLEDGMENT

This project is supported by NSFC (11371162) and NSFC(11171129) and HuBei (T201103).

REFERENCES

[1] Xu, J.M. 2003. Graph and Application of Graphs, Kluwer Academic Publishers, Dordrecht/Boston/London.

[2] Saad, Y. & Schultz, M.H. 1988. Topological properties of hypercube, IEEE Transaction Computers, 37(7): 867-872.

[3] Fu, J.S. 2006. Longest fault-free paths in hypercubes with vertex Faults, Information Sciences, 176: 759-771.

[4] Hsien, S.Y. & Kuo, C.N. 2009. H.L.Huang, 1-vertex-fault-tolerant cycles embedding on folded hypercube, Discrete Applied Mathematics, 157: 3094-3098.

[5] Liu, M. & Liu, H.M. 2012. Cycles in Conditional Faulty Enhanced Hypercube Networks, JOURNAL OF COMMUNICATIONS AND NETWORKS, 14(2): 213-221.

[6] Ma, M.J. & Xu, J.M. & Du, Z.Z. 2006. Edge-fault-tolerant hamiltonicity of folded hypercube, Journal of University of Science and Technology of China, 36(3):244-248.

[7] Kuo,C.N. & Hsieh, S.Y. 2010. Pancyclicity and bipancyclicity of conditional faulty folded hypercubes, Information Sciences, 180: 2904-2914.

[8] Kuo, C.N. & Chou, H.H. & Chang, N.W. & Hsieh, S.Y. 2013. Fault-tolerant path embedding in folded hypercubes with both node and edge faults, Theoretical Computer Science, 475: 82-91.

[9] Sun, C.M. 2012. Bipanconnectivity of faulty hypercubes with minimum Degree, Applied Mathematics and Computation, 218: 5518-5523.

[10] Liu, M. & Liu, H.M. 2013. Paths and cycles embedding on faulty enhanced hypercube networks, Acta Mathematica Scientia, 33(B): 227-246.

[11] Fu, J.S. & Chen, G.H. 2002. Hamiltonicity of the hierarchical cubicnetwork, Theory of Computing Systems, 35(1): 59-79.

[12] Xu, J.M. & Ma, M.J. 2006. Cycle in folded hypercube, Applied Mathematics Letters, 19: 140-145.

International Conference on Power Electronics and Energy Engineering (PEEE 2015)

Research and Application of Fuzzy Control with Multiple Weighted Factors by Genetic Algorithm

L.J. Dong

Department of Electric and Electronic
Wenzhou Vocational & Technical College
Wenzhou, Zhejiang, China

Abstract—**It is difficult to control the complex object with lagging uncertainty and nonlinearity effectively. To solve this kind of control problem, this paper presents a self-correction fuzzy controller with multiple weighted factors based on genetic algorithm. According to information achieved on line, it finds the global optimum weighted factors with a high speed by the improved genetic algorithm so that to amend and perfect the control rules. It also has done some simulation experiments in the tobacco-redrying control process. The simulation results demonstrate that this kind of control method can achieve good performance.**

Keywords-self-correction fuzzy controller genetic algorithm

I. INTRODUCTION

Tobacco-redrying is a process with nonlinearity uncertainty and varying parameters. It is difficult to control the whole process effectively only by a group of PID parameters. To meet the performance target of the tobacco-redrying process we should adopt an intelligent control strategy.

Fuzzy control is a technology to imitate the logical thinking of human beings without depending on the controlled object. It executes approximate reasoning based on the prior experience and knowledge of the field experts and has achieved excellent effect [1]. For the complex system with nonlinearity uncertainty, if wanting to obtain good performance then it needs perfect control rules. But for the factors such as nonlinearity, high order, time-varying random disturbance and so on, the fuzzy control rules become rough and inadequate perfect so that the control performance becomes deteriorated. To overcome the shortcoming of the fuzzy controller, we naturally consider improving the self-adaptive self-organizing and self-learning ability of the fuzzy controller. Genetic algorithm (GA) is based on the natural selection principle and genetic mechanism. It overcomes the local optimum defect of the common used BP algorithm and enhances the optimization capability. Now the genetic algorithm combining with fuzzy control technology possesses important status in the intelligence control field [2].

II. PROBLEM DESCRIPTION

Tobacco-redrying process is to refabricate the pre-redrying tobacco. It's a key plant to ensure the quality of the tobacco. During the whole redrying process the tobacco exchange heat with the heat transfer medium having a uniform temperature, stress and flow rate, which results all kinds of processes, for

example drying, cooling and dampening. The variation of the temperature and stress in each zone will affect the following zones [3]. It is a process with nonlinearity uncertainty and varying parameters.

The critical task of tobacco redrying is to control the intercoupling temperature and humidity. In order to control effectively, we should decouple them. we can use compensation method decouple, namely, use temperature variation to compensate for the humidity variation [3]. After that the temperature and humidity can be controlled respectively and the controlled plant can be treated as an object with concentrated parameters. So we use a first-order inertial object with pure delay to describe the whole process, given by

$$G(s) = \frac{K e^{-\tau s}}{TS + 1} \tag{1}$$

Where the static gain K, the pure delay τ and the constant times T of the object vary during different stages.

III. CONTROL SYSTEM DESIGN

A. Control Schematic Design

This paper designs a fuzzy self-correction control system shown in Figure 1. It consists of the two-dimension fuzzy controller, which has multiple weighted factors, and the genetic algorithm (GA). Without knowing the mathematical model and the input and output data, it uses the GA to find the optimal value of the weighted factors only by the information of the error e and the error variance ratio EC obtained online, so that the control rules can be self-corrected and perfect and the outputs of the controller can be varied according to different inputs E and EC.

B. Fuzzy Controller Design

1) Fuzzy controller structure design

This paper designs a two-dimension fuzzy controller, which has two inputs and one output. They are error e, variance ratio ec and u respectively, shown as follows:

© 2015. The authors - Published by Atlantis Press

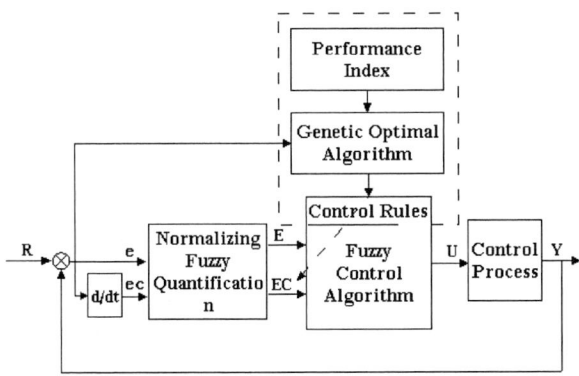

FIGURE I. FUZZY CONTROL SYSTEM BASED ON GENETIC ALGORITHM.

$$e(n) = y(n) - r(n) \qquad (2)$$

$$ec(n) = e(n) - e(n-1) \qquad (3)$$

$$u(n) = u(n-1) + \Delta u \qquad (4)$$

Where n is the sample time.

2) *Fuzzy quantification of the inputs*

Commonly, it utilizes quantification factor Ke and Kc to convert the factual basic interval of e and ec to the interval of the interrelated fuzzy set. For the two quantification factors have interrelationship so it is difficult to determine a suitable group of parameters for the control system. In addition, only by fix quantification factors, it is unable to keep the system under an optimal condition and having robust ability, which has large inertia and long process. The selection of the quantification factor has much to do with the control performance of the system. So some people have introduced an array of quantification factors or amended them according to different system conditions [4].

This paper adopts normalizing fuzzy quantification method, which not only can avoid selecting quantification factor but also can easily convert the precise value into fuzzy value within the setting fuzzy interval. It mainly includes normalizing and stepping fuzzification. Firstly, it executes normalizing process of e and ec by e/R and ec/R respectively, where R is the setting value of the system. Secondly, it divides them into several grades within the interval [0,1]. Based on that the fuzzy value of e and ec can be determined, written as E and EC.

C. *Fuzzy Rules Design*

The determination of fuzzy control rules is the critical step of the control table making. To gain good performance, we should perfect fuzzy rules. There have been many improvements in the establishment of language variable, fuzzy quantification grades, fuzzy membership and decision making of the fuzzy control output [4]. In fact they are all based on a kind of fuzzy controller, which describes their control rules in table. There still exists a common problem for the fuzzy controller that is the establishment of fuzzy rules having a lot of important parameters to determine. To overcome the above defection and to be easy for the fuzzy algorithm programming, many people have used an analytic expression to describe the fuzzy rule [2], which is convenient for the system control information expression and computer calculation. In brief, the fuzzy rule design is simplified and is handy to be amended.

On the basis of the predecessor, this paper presents a new type of analytic expression for synthesis reasoning and uses it to design fuzzy rule, given by

$$\Delta U = s \cdot E + q \cdot EC \qquad (5)$$

Where s and q are weighted factors, determined by E and EC, the normalizing fuzzy quantification value of e and ec. q and s given by

$$q = f(EC) = \begin{cases} t_1, & EC = -2 \\ t_2, & EC = -1 \\ t_3, & EC = 0 \\ t_4, & EC = 1 \\ t_5, & EC = 2 \end{cases} \qquad (6)$$

$$s = f(E) = \begin{cases} s_1, & E = -3 \\ s_2, & E = -2 \\ s_3, & E = -1 \\ s_4, & E = 0 \\ s_5, & E = 1 \\ s_6, & E = 2 \\ s_7, & E = 3 \end{cases} \qquad (7)$$

Where $s_1, s_2, s_3, s_4, s_5, s_6, s_7, t_1, t_2, t_3, t_4, t_5$ are the undetermined parameters.

IV. OPTIMIZATION ALGORITHM

Real-coded takes the advantages of naturally expressing, increasing the level of possible exploration of the search space without affecting the convergence characteristics. And it is more suitable for the continuous variable optimization than binary-coded. For this reason, this paper adopts real coding method to code these undetermined parameters. Firstly, the optimal seeking scope of the undetermined parameters waiting for optimization is confined to [0,1], and then they are arranged

267

according to the sequence $s_1, s_2, s_3, s_4, s_5, s_6, s_7, t_1, t_2, t_3, t_4, t_5$, thus to form an individual. The total number of initial population is 50 and the maximum generation is 100. This paper utilizes IAE performance index as the fitness function, defined by

$$J(IAE) = \int_0^\infty |e(t)| dt \tag{8}$$

To speed up the global optimization process, this paper has modified the genetic algorithm in the following aspects:

1) Fitness function scaling   This paper presents a new method for scaling. It assigns the fitness value to the individual according to the position arranged by descending order. The optimal problem in this paper is to find the minimum value. The scaling processes mainly involve two steps.

a. Assuming the scale of the population is N, we arrange the individuals with a descending order. In other word, the least fit individual, having the highest fitness value, is put in the first place while the fittest individual, having the lowest fitness level, is in the last place.

b. Assign fitness level to each individual according to its position. The scaling formula is written as

$$f(i) = \frac{2 \times (i-1)}{N}, \tag{9}$$

Where i means the position of the individual, $1 \le i \le N$.

After the fitness function are scaled by equation (9), if the individual is fitter then its fitness value is higher. Similarly, if the individual is less fit then its fitness value is smaller.

2) Selection   This paper presents an improved optimal selection method. In the first step, it arranges the population in terms of fitness value. Then it keeps the individuals with higher value and let them retain in the next generation directly. In the following step it selects the other individuals based on the arranging- selection mechanism [5]. The mainly reason for this is to extend the searching space without spoiling the best individual, so that the evolution process is heading in an optimal direction.

3) Recombination this paper uses recombination in place of crossover. The difference between them is that the recombination of the two parent individuals only generates one individual while by crossover it generates two. Assuming that the two parent generation individuals and the offspring individual are $x(x_1, x_2, \cdots, x_n)$, $y(y_1, y_2, \cdots, y_n)$ and $z(z_1, z_2, \cdots, z_n)$ respectively, the recombination mode given by

$$z_i = \alpha_i (x_i - y_i) + x_i \tag{10}$$

Where, α_i is a scaling factor chosen uniformly at random in the interval [-0.25,1.25].

4) Mutate   This paper introduces an self-adaptive mutation strategy. With the assumption that $x(x_1, x_2, \cdots, x_n)$ is the parent individual which is to be mutated, $y(y_1, y_2, \cdots, y_n)$ is the offspring individual and P_m is the mutation probability, if one component x_i of the parent individual is mutated, $x_i \in [a_i, b_i]$, then the corresponding component of the offspring is produced according to the self-adaptive mutation rule, given as

$$y_i = x_i + mut \times (b_i - a_i) \times delta \tag{11}$$

Where mut is 0,1 or -1. It has the same probability, namely $P_m/2$, to be 1 and -1. $delta = \sum_{i=0}^{m-1} \partial_i 2^{-i}$, where m=20 and ∂_i is 0 or 1. The probability for $\partial_i = 1$ is $1/m$.

Mutation probability should be self-adjusted based on the generation number [6-8], thus to maintain the diversity of the generation population, higher evolution rates and avoid premature convergence problem.

V.   SIMULATION RESULT

The mathematical model of the whole tobacco-redrying process is given by

$$G(s) = \frac{K e^{-\tau s}}{TS + 1}$$

From the mathematical model analysis we can know that K, T, τ of the model vary during different stages. So in the simulation process we let them vary according to $k(t) = \frac{a}{(b+t)}$, $T(t) = c + dt$ and $\tau(t) = e + ft$ correspondingly and adopt the unit step signal as the input.

The optimized parameter values of the fuzzy controller $s_1, s_2, s_3, s_4, s_5, s_6, s_7, t_1, t_2, t_3, t_4, t_5$ are shown in Table 1.

TABLE I .OPTIMAL VALUES OF THE UNDETERMINED.

s1	s2	s3	s4
0.645	0.435	0.427	0.419
s5	s6	s7	t1
0.391	0.09	0.496	0.693
t2	t3	t4	t5
0.654	0.519	0.337	0.761

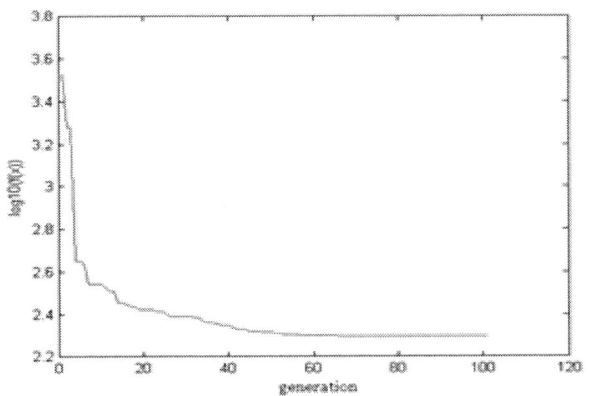

FIGURE II. THE VARIATION OF THE FITNESS FUNCTION VALUE IN THE OPTIMIZATION PROCESS.

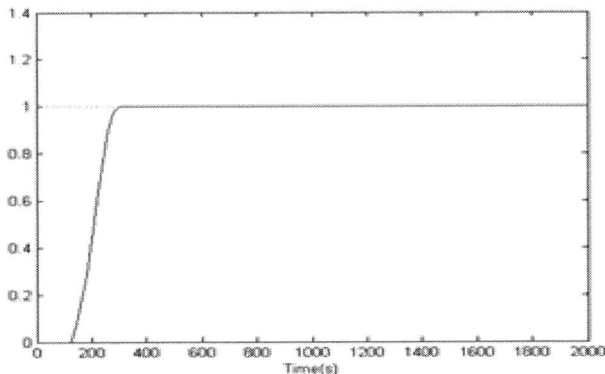

FIGURE III. THE UNIT STEP RESPONSE OF THE SYSTEM.

This is obtained by the improved genetic algorithm presented in this paper. Use the two-dimension fuzzy controller with multiple weighted factors based on genetic algorithm to control tobacco-redrying process the variety of the generation population fitness value during the optimizing process is shown in Figure 2. And the system response to the unit step signal is shown in Figure 3.Viewing Figure 3, we can know that the two dimension fuzzy controller with multiple weighted factors based on improved genetic algorithm can control effectively. Both the dynamic characteristic and the static characteristic are good.

VI. CONCLUSION

Using the error and error variance information obtained online this paper has designed a two dimension fuzzy controller with multiple weighted factors, which is optimized by the improved genetic algorithm, without knowing the mathematical model and the input and output data. The simulation result has demonstrated that this kind of controller can achieve good performance.

REFERENCES

[1] Feng Dongqing, Xie Songhe etc.Electronic Industry Press. 2003, pp. 35–46.

[2] Cai Zhixing, Intelligence Control Theory and Application [M]. the Publishing of QingHua University, 2013,pp.111-127, 287-294

[3] Chen Qi etc. Tobacco-redrying Intelligence Control System. Kun Ming University of Science and Technology, 2012.6

[4] Liu Xiangjie, Zhou Xiaoxin, Cai Tianyou. Present state and perspectives of the fuzzy control research. Control and Decision.2011.4.

[5] Chiu-Hung Chen. A Novel Crowding Genetic Algorithm and Its Applications to Manufacturing Robots, Industrial Informatics, IEEE Transactions 2014, pp 1705-1716.

[6] Pradhan, S.K. ; Subudhi, B.Fuzzy learning based adaptive control for a two-link flexible manipulator, Control Applications (CCA), 2013 IEEE International Conference, 2013,282-287

[7] Rong-Jong Wai, Muthusamy, R. Fuzzy-Neural-Network Inherited Sliding-Mode Control for Robot Manipulator Including Actuator Dynamics, Neural Networks and Learning Systems, IEEE Transactions, 2013:274-287

[8] Li Yang, Qing-Lan Jia.The application of improved evolutionary strategy algorithm in optimization. Machine Learning and Cybernetics, 2012 International Conference, 2012, pp1212-1217.

International Conference on Power Electronics and Energy Engineering (PEEE 2015)

Routing Problems in Emergency Logistics Based on Improved Data Envelopment Analysis

X.X. Zhu

School of Management, Harbin Institute of Technology
P.R. China

S.Y. Wang

School of Management, Harbin Institute of Technology
P.R. China
School of Architecture, Harbin Institute of Technology
P.R. China

Abstract—**Emergency decision making is still an important issue for unconventional emergency management. For non-standardized weight distribution in the data envelopment analysis model, an improved data envelopment analysis model by constructing two decision making units, which have the highest and the lowest efficiency to expand the scope of the efficiency value and to avoid the influence of artificial factors on the decision is proposed. Aiming at choices of routing optimization in emergency logistics, this paper also uses multi-objective decision-making to solve problems, i.e., time, cost and transportation distance of materials. Finally, the validity of the approach and the applied process of the proposed model are tested by a numerical example with the case of the earthquake emergency in Gansu Province, China.**

Keywords-the improved data envelopment analysis; multi-objective decision-making; route optimization; emergency logistics

I. INTRODUCTION

Natural disasters, accidents, public health, public security and other unexpected incidents frequently occur, and in areas with high population densities, human casualties can be great[1]. Therefore, how to efficiently transport materials in emergency disasters to the demand point becomes a key problem in the disaster relief process. Emergency logistics are special logistics activities to ensure that the demand of materials, personnel, and money are allocated to public health emergencies, major accidents and other emergencies which are sudden, uncertain, unconventional, and with urgency based on time constraints that benefit from a maximum of time management with an objective of minimum disaster losses [2-6].

Data envelopment analysis (DEA) has been applied in various areas such as hospital services, education (schools and universities), manufacturing, computers, peripheral firms, information technology investment, banks, and hotels [7-14].

II. DESCRIPTION OF MATHEMATICAL MODEL

A. Data envelopment Analysis (DEA)

Suppose there are n DMUs, with k input factors and m output factors, and let i $(1 \leq i \leq n)$ denote one of the n DMUs. The efficiency Ei of the ith DMU, with outputs $Y_i = (y_{1i}, y_{2i}, ..., y_{mi})^T > 0$ and inputs

$X_i = (x_{1i}, x_{2i}, ..., x_{ki})^T > 0$, i=1,2,...,n, where x_{ki} is the kth input of the ith DMU, y_{mi} is the mth output of ith DMU, is calculated by the following C2R model:

$$
\begin{cases}
\max(\dfrac{\mu^T Y_i}{v^T X_i}) \\
s.t. \quad \dfrac{u^T Y_j}{v^T X_j} \leq 1, j = 1,2,...,n \\
u \geq 0, v \geq 0
\end{cases} \quad (1)
$$

The above constraints restrict the efficiencies of all the DMUs to have an upper bound of 1. The ith DMU is efficient when Ei is equal to 1 and inefficient if Ei is less than 1. The variables $v = (v_1, v_2, ..., v_k)^T$ and $u = (u_1, u_2, ..., u_m)^T$ are the derived weights for the corresponding output and input factors. Model (1) is a fractional programming in which v and u have infinite solutions. Subjected to $v^T X_j = 1$ which has efficient solutions, then model (1) is transformed into an equivalent multiplier form of the linear programming model (2):

$$
\begin{cases}
\max \mu^{'} Y_j \\
s.t. \quad \dfrac{u^{'} Y_j}{v^{'} X_j} \leq 1, j = 1,2,...,n \\
v^{'} X_j = 1 \\
u \geq 0, v \geq 0
\end{cases} \quad (2)
$$

$$
\begin{cases}
\min(\theta) \\
s.t. \quad -Y_i + Y\lambda \geq 0 \\
\theta X_i - X\lambda \geq 0 \\
\lambda \geq 0
\end{cases} \quad (3)
$$

Linear programming model (2) is transformed into an equivalent binary form, i.e., the basic C2R model of DEA method (3).

© 2015. The authors - Published by Atlantis Press

θ is a scalar which represents the efficiency score of DMUi where $0 \le \theta \le 1$. λ is a constant vector. The binary form of the linear programming has fewer constraints [18] than the multiplier form which creates problems for the uneven weight distribution of the input and output factors.

B. Improved Data Envelopment Analysis Model

The specific steps of the improved DEA model are as follows:

Step 1: Construct DMUn+1 and DMUn+2 which have the highest and lowest efficiency,

Step 2: Determine the index weight of the input and output factors.

Evaluate the last n DMUs and DMU which has the highest efficiency by using the improved DEA model. DMUn+1 is the optimal DMU, so when i=n+1, $\theta = 1$. Then the optimal weights u^{*T} and v^{*T} are obtained, which satisfies $\dfrac{u^* y_j}{v^* x_j} = 1$.

Step 3: Solve the common weight vector of the input and output factors.

C. Establishing the Improved Data Envelopment Model

1)   Illustration

After a natural disaster, the National Center for Emergency Management informs technical institutions to make route optimization plans for the transportation of goods and services. Suppose there are n plans (n DMUs) to be evaluated as DMUj, j=1,2,3,4. In order to make sure the rescue goods and services arrive quickly after major disasters, "shortest time" is often treated as the objective to choose a plan.

Based on the case of the earthquake in the border region between Min County and Zhang County in Gansu Province, this paper gives an improved solution. We choose the total time, cost and transportation distance of materials as main reference indexes to determine the hierarchical structure of route optimization plans in emergency logistics. Suppose there are n lines to be chosen. Line i (i=1,2,...,n) has two types of indicators which are the positive index xki and negative index ymi. xki representing the rescue time, transportation cost and distance, and the smaller value the better. ymi represents the rescue ability of each plan, and the larger the better. Continue the efficiency evaluation for n lines according to the improved DEA model above. Construct line G (DMU7) which has the highest efficiency and line H (DMU8) which has the lowest efficiency. The input and output factors of line G are:

$$X_7 = (x_{1,7}, x_{2,7}, x_{3,7})^T, \; Y_7 = (y_{1,7})^T$$

Set input and output factors of DMU7 which has the highest efficiency the minimum value and the maximum value of the last 6 DMUs respectively, i.e.:

$$x_{2,7} = \min(x_{2,1}, x_{2,2}, ..., x_{2,6})$$
$$x_{3,7} = \min(x_{3,1}, x_{3,2}, ..., x_{3,6})$$

$$y_{1,7} = \max(y_{1,1}, y_{1,2}, ..., y_{1,6})$$

The input and output factors of DMU8 which have the lowest efficiency are:

$$X_8 = (x_{1,8}, x_{2,8}, \cdots x_{3,8})^T \; Y_8 = (y_{1,8})^T$$

Set input and output factors of DMU8 which have the lowest efficiency as the maximum value and the minimum value of the last 6 DMUs respectively, i.e.:

$$x_{1,8} = \max(x_{1,1}, x_{1,2}, ..., x_{1,6})$$
$$x_{2,8} = \max(x_{2,1}, x_{2,2}, ..., x_{2,6})$$
$$x_{3,8} = \max(x_{3,1}, x_{3,2}, ..., x_{3,6})$$
$$y_{1,8} = \min(y_{1,1}, y_{1,2}, ..., y_{1,6})$$

Evaluate the original 6 DMUs and the DMU which has the highest efficiency. DMU7 is the optimal decision-making DMU, so $\theta = 1$. We get the optimal weights of $u^{*'}$ and $v^{*'}$ which satisfies $u^{*'} y_j \big/ v^{*'} x_j = 1$.

For DMUs,

$$\begin{cases} \min_8(\theta) \to 0 \\ s.t. \quad -y_i + Y\lambda \ge 0 \quad i = 1,2,\cdots,8 \\ \qquad \theta x_i - X\lambda \ge 0 \\ \qquad u^{*'} y_j \big/ v^{*'} x_j = 1 \\ \qquad (u^{*''}, v^{*''}) \in (u^{*'}, v^{*'}) \\ \qquad \lambda \ge 0 \end{cases} \tag{4}$$

III.   EXAMPLE ANALYSIS

A 6.6-magnitude earthquake struck Northwest China's Gansu province at 07:45 am on Monday, 22nd of July, 2013. In the middle and late stages of the emergency, some materials in the emergency disaster relief needed to be transported from the emergency supply center to the emergency demand point and the road traffic network diagram is as shown in Figure 1. S is the supply center and D is the demand point. Ni (i=1,2,3,4,5) are the network nodes.

In the middle and later stages of emergency, the best route needs to be determined. There are six lines (from O to D), namely
A(line1):O→N1→N2→N5→D;B(line2):O→N1→N2→N5→D;C(line3):O→N1→N5→D;D (line4):O→N1→N5→D;E(line5):O→N1→N4→D; and F(lint6):O→N6→D. On the basis of the investigated information, the network in the middle and late stage of emergency and the main path optimization parameters according to the needs of the model are shown as Figure 1 and Table 1, respectively.

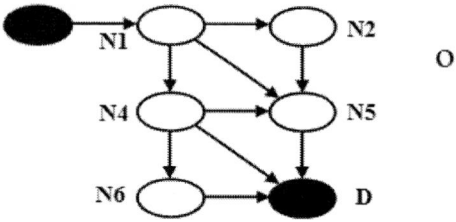

FIGURE I. NETWORK IN THE MIDDLE AND LATE STAGE OF EMERGENCY (ALIGNMENT).

TABLE I . INPUT AND OUTPUT FACTORS OF EACH PLAN (DMU).

DMU	time	cost	distance
A (DMU1)	24	60000	750
B (DMU2)	26	40000	890
C (DMU3)	18	48000	480
D (DMU4)	20	52000	810
E (DMU5)	17	70000	490
F (DMU6)	15	84000	540

TABLE II . RANKS OBTAINED BY DMUS BY USING DEA AND IMPROVED DEA.

DMUs	DEA	rank	Improved DEA	rank
A	0.830	4	0.667	6
B	1.000	1	0.833	2
C	1.000	1	0.800	3
D	0.901	2	0.722	5
E	1.000	1	1.000	1
F	0.881	3	0.778	4

From Table 2, the efficiency value using the data envelopment analysis (DEA) model is between 0.830 and 1 which has a difference of 0.170; and the efficiency value using the improved data envelopment analysis model is between 0.667 and 1 which has a difference of 0.333. The greater the efficiency difference expands the scope of the efficiency values which enhances the comparability of DMUs to achieve the goal of effectively distinguishing and sorting DMUs.

From the determined index weight of the two models, we can see there are broader weight constraints in using the DEA model. However, the improved DEA model allows the DMU which has the lowest efficiency to be the least and at the same time guarantees the efficiency value of DMU which has the highest efficiency to be 1. The different ways that the two models to determine index weight led to different efficiency estimation. In comparison, the index weight of the improved data envelopment analysis model is more canonical and the results of the evaluation are more reasonable and objective.

IV. CONCLUSION

With the benefits and simplicity of DEA, it has been applied in various areas. However, the DEA model has the problem of non-standardized weight distribution. Route problems of the transportation of materials after an emergency for disaster relief has multiple attributes as the central problem. Decision makers need to develop a balance between time, cost and distance in the middle and later stages of the emergency.

To solve these problems, this paper first points out a defect of DEA which has non-standard weight distribution. Then considering that problem two DMUs which strengthen the weight constraints to achieve the goal of the improved model is constructed. Secondly, the model considers time, cost and transportation distance of materials as the main objectives to evaluate route optimization plans. Finally, the case of the earthquake emergency in Gansu Province shows that the improved data envelopment analysis model is effective, which shows that the method is effective and feasible.

ACKNOWLEDGMENT

Support for this work was provided by the National Natural Science Foundation (No.71372091)

REFERENCES

[1] Mooney D D,Swift R J.A course in mathematical modeling[M].The Mathematical Association of America, 1996.234-245.

[2] Xianglong Sun,Jian Lu. A method of emergency logistics route choice based on fuzzy theory[J].Logistics for sustained economic development.ICLEM2010:2816-2822.

[3] Thomas K,Panayiotis M,Katerina P. The impact of replenishment parameters and information sharing on the bullwhip effect: a computational study[J]. Computers & Operations Research,2008,35(11):3657-3670.

[4] Thanassoulis E.A comparison of regression analysis and data envelopment analysis as alternative methods for performance assessment[J].OplRes.Soc,1993,44(11),1129-1144.

[5] Halme M,Joro T,korhonen P,et al. Value efficiency analysis for incorporating preference information in DEA[J].Management science,2000,45(2):103-115.

[6] Qinjie XIAO, Ruifang MOU.The Application of DEA/AHP in the Natural Disaster Emergency Logistics System[C].Logistics for Sustained Economic Development: ICLEM 2010, 2010:3946-3952.

[7] Jyoti,D.K.Banwet,S.G.Deshmukh. Evaluating performance of national R &D organizations using integrated DEA-AHP technique[J].International Journal of Proactivity and Performance Management.2008,57(5):370-388.

[8] M.Asmild,J.C.Parad,J.T.Pastor. Centralized resource allocation bbc models[J].The international journal of Productivity analysis,2004,22(3):143-161.

[9] Fare R, Grosskopf S. Network DEA[J].Socio-economic planning science,2000,34:35-49.

[10] Yu Dengke, Deng Qunzhao. Several thinking of the DEA method[J].Modern management science, 2012(10):20-24.

[11] Fang Lei.Resource allocation of emergency system based on the DEA model with preference information[J]. Systems engineering theory and practice, 2008,5(5):98-103.

[12] Wang Ting,Yi Shuping,Yang Yuanzhao. Performance evaluation method for business process of machinery manufacturer based on DEA/AHP hybrid model[J].Chinese journal of mechanical engineering,2007, 20(3):91-97.

[13] Yang Min, Wang Wei, Chen Xuewu.Decision-making method for mass rapid transit model selection based on DEA/AHP[J].Journal of highway and transportation research and development.2006,23(7):111-115.

[14] Liu Tianyu, Wang Meiqiang. Evaluation and selection of B2C enterprises to third-part logistics based on B2C enterprises based on DEA-Delphi method[J].Journal of Guizhou University,2013,30(4):136-140.

The Research of Numerical Simulation about the Improved Dense Medium Cyclone

X.B. Li, Q.Q. Huang
School of Mechanical Engineering
Anhui University of Science and Technology
Huainan, Anhui, PR China

X. Li
College of Chemistry and Materials Science
South-Central University for Nationalities
Wuhan, Hubei, PR China

Abstract—**The structure model, the CFD computing model and algorithm of flow field about the improved dense medium cyclone were analyzed in this paper. Then the simulation of flow field was done by using of CFD inside the improved dense medium cyclone. According to the simulation results, this paper analyzed the characteristics of the velocity field, the pressure field, the density field and the air column. The conclusion shows that the effect of medium separation and the distribution of flow field in the improved dense medium cyclone were improved obviously.**

Keywords- dense medium cyclone; optimization of the structure; simulation of the flow field

I. INTRODUCTION

Dense medium cyclone has the advantages of simple structure and easy to design, but the flow field internal is a three-dimensional flow field of strong turbulence, the characteristics of the fluid motion is very complex. In order to study the process of dense medium cyclone coal preparation from the view of fluid mechanics, to explore the impact of structure and parameters on heavy medium separation process, all the countries especially the researchers coal power is studied on the separation process, and made some progress. The research results are as follows:

(1) The sorting process experiment. Analysis of the flow characteristics and separation process on heavy medium cyclone in theory, including the velocity of the fluid field, the centrifugal force field and the density field distribution and particles, and the effect of various parameters on the separation, by using the high speed camera and particle tracking display technology, and puts forward some new arguments[1].

(2) mathematical model. A mathematical model of dense medium cyclone was bulit though mass test data and theoretical analysis. The design parameters of dense medium cyclone much of them are obtained through the experiment of the dense medium cyclone , to determine the relationship between the corresponding condition of geometric parameters, operating parameters and performance based on the empirical formula of production practice, but the application of each experience model range is too small, so we get the correct parameters must rely on a large number of trials . L. R. Plitt, АЙЛораорв, D·A·Daimler, Xue-shi Pang, Liang-yin Chu and so on were systematically studied on cyclone and put forward different mathematic models[2].

(3)the numerical simulation of the flow field. Numerical simulation was conducted of local flow field in heavy medium cyclone by J.Bassman and N.Lournes. The paper presents an improved structure of cyclone Through the simulation results, unfortunately, the results of improvement is unknown in domestic[3]. The simulations with different degrees and the density distribution of heavy medium cyclone were did by M.S.Breanan [4].

Flow field of dense medium cyclone for multiphase flow numerical simulation by computer,

which make the research visualization , provides a foundation for optimization of parameters of dense medium cyclone and CFD could achieve these possible , computational fluid mechanics has been gradually show attractive prospects.

II. STRUCTURE IMPROVEMENT AND MESH CYCLONE

Life and the separation effect was effected by different inlet forms of dense medium cyclone. Tangent to the material is the most way of dense medium cyclone. Figure1 (a) the inside streamlines and the outer flow field of tangent entrance almost perpendicular intersection therefore, when the Medium into, the cyclone will produce a flow disturbance area where is located between the cyclone cylinder and the feed pipe to the overflow pipe junction. The pressure feeding of cyclone will increase the feeding short-circuit enters the overflow and affect the separation efficiency by. And the direction of motion changed rapidly when the outer streamline undercut, wall of tangent point area the large coal is intense impact, especially of waste rock with the density of hardness and high. The entrance wall will appear sunken, resulting in flow more chaos and complex flow and make the destruction process intensifies, resulting in a decline in the performance of cyclone by using this kind of cyclone for a long time.

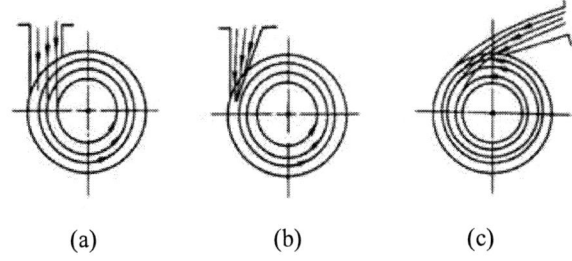

(a)   (b)   (c)

FIGURE I. DIFFERENT SKETCH OF INLET TUBE.

As Figure 1 (c), the number of intersecting angle of streamline reduced by cyclone spiral entrance or involute. This

© 2015. The authors - Published by Atlantis Press

can reduce the impact of media on the wall and improve the disturbance of flow. Another benefit is that the feeding in the bent pipe of different density of the component will be a small part of the separation, density gradually moves toward the wall outside, equivalent to the pre separation of small scale, which on this can improve the separation efficiency of the cyclone, pro-long its service life. In the feeding pipe and anumber of areas, from the test [4] Liang-yin Chu concluded that the separation efficiency of two-way inlet cyclone is beter than that of the single inlet cyclone, and it also has low energy consumption.

The changes of structural parameters based on classical models of dense medium cyclone which is one of the widely used, remodeling, a model of cyclone center symmetric involute form of the double inlet is established in this paper.

First, find out the equation of involute, by driving the equation of a curve, drawing into a double involute material pipe in the Solidworks, and then save the parasolid.x_t* format, completed all the geometry modeling and grid in the Gambit.

In order to avoid continuity problems affect the analysis result, where the use of split mesh generating method. First to the 4 part of the whole into the cyclone, the upper half part of an overflow pipe, cylinder (a circular cylinder),barrel body,a conical section combination body and a feeding tube,4 parts are used Hex/Wedge, hexahedral / wedge grid, the method of generation grid selection Cooper, the barrel type.

Figure2 is the new cyclone model in the grid map (looking down).

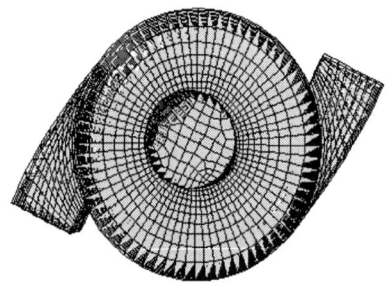

FIGURE II. MESH GIRD DISPLAY.

The geometric parameters of the new model as shown in Table 1.

TABLE I .IMPROVED DENSE MEDIUM CYCLONE.

Structure parameters	unit	Numerical
Cyclone diameter D	[mm]	800
The length of cylinder H	[mm]	600
The diameter of a feeding inlet D_e	[mm]	210
The diameter of overflow pipe D_o	[mm]	180
The diameter of underflow port D_u	[mm]	200
Cone angle	[°]	22
The depth of overflow tube insertion h	[mm]	300
Inlet pressure P	[MPa]	0.13
The inlet flow Q_e	[m³/h]	400
Overflow rate Q_d	[m³/h]	——

III. SWIRLER SETTING BOUNDARY CONDITIONS

A. Inlet Conditions

Inlet conditions are the choice of speed entrance, speed entrance can input many numerical information which were required for, associated with the cyclone model as follows:

① The relevant physical parameters of the medium components (Table 2);

②The medium into the cyclone: Including the components of the velocity ratio, volume ratio and the direction of the velocity;

③The turbulent situation of medium into the flow domain.

TABLE II MIXTURE PARAMETERS OF VARIOUS KINDS OF MEDIUM.

Material	Density /[kg·m⁻³]	Viscosity /[Pa·m⁻³]	Particle size/[mm]	Velocity of the entrance /[m·s⁻¹]	The volume ratio of the entrance[%]
suspension	1440	0.005	——	10	82
coal	1250		0.2~1	10	17
air	1.02	0.00001	0.001	0	1

The option settings Turbulence of submenu Momentum in the inlet speed. Choose the turbulent kinetic energy and turbulent dissipation in k-ε model and also can choose intensity and the hydraulic diameter.

B. Exit Conditions

Dense medium cyclone export to overflow and underflow port, two of exports are set to the boundary condition of Outflow, in outflow-top (overflow), the flow rate weighting column input 0.4, then this parameter underflow port outflow-b is input 0.6.

C. Wall Conditions

According to the requirements for simulation of cyclone, wall conditions do not have a separate settings, press the no slip boundary conditions, the default is a standard without heat exchange of the solid wall[5].

IV. SIMULATION RESULTS AND ANALYSIS

4 section selection of cyclone for section after treatment: 3 Z axes (cyclone axial) section , a X axis (parallel to the axis of the inlet velocity) section; There are the axial cross section of Improved dense medium cyclone, x=0mm (perpendicular to the direction of the inlet velocity the central shaft section);z=2000mm (the level surface beside the feed pipe).

A. Velocity Distribution

The following is the predicted velocity distribution. Figure 3 (a) is a mixture of the x=0mm surface velocity distribution. Figure 3 (b) is the axial velocity distribution nephogram of shaft face.

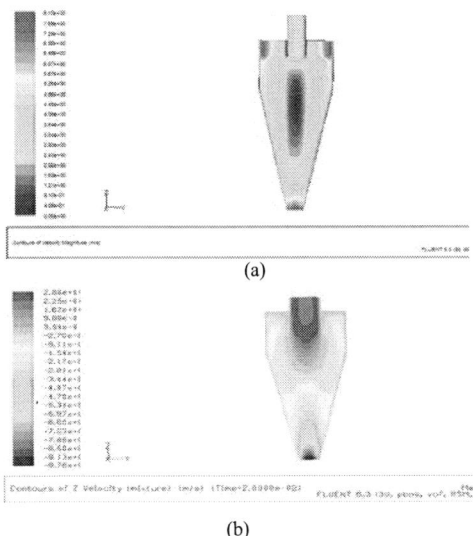

(a)

(b)

FIGURE III. THE VELOCITY DISTRIBUTION NEPHOGRAM OF SHAFT FACE.

B. Pressure Distribution

The total press of surface X=0mm and Z=2000mm are shown in Figure 4 (a), (b).

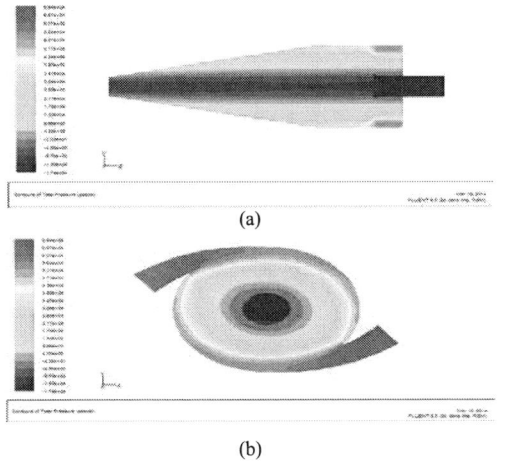

(a)

(b)

FIGURE IV. THE TOTAL PRESS ON THE SURFACES OF X=0MM AND Z=2000.

C. Density Distribution

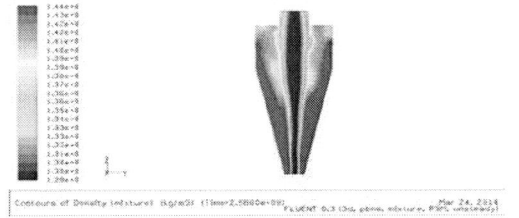

FIGURE V. IMPROVED DENSE MEDIUM CYCLONE AXIS SECTION CLOUD DENSITY DISTRIBUTION.

D. The Air Column

Air in the improved dense medium cyclone volume fraction distribution in shaft section is shown in Figure 6 (a) and air in the new cyclone volume fraction distribution z=2000mm is shown in Figure 6 (b).

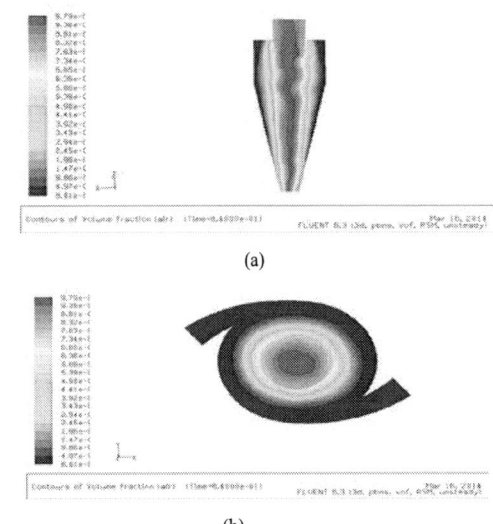

(a)

(b)

FIGURE VI. THE AIR VOLUME FRACTION DISTRIBUTION IN THE IMPROVED DENSE MEDIUM CYCLONE.

V. CONCLUSION

We get conclusion that the Improved dense medium cyclone inner velocity field, pressure field, density field and air column, Though the computer simulation of new model, analysis the simulation results of new model that conclusions are as follows:

(1)According to the simulation results on the velocity distribution of double inlet cyclone , the velocity distribution shows axial symmetry along the cyclone, the flow field distribution is more symmetrical balance, but also can obtain the larger axial velocity.

(2)Based on the simulation results on pressure of double inlet cyclone model , the pressure distribution is very uniform, almost central symmetric perfect, stable pressure field helps to improve separation efficiency.

Seen from Figure5, the simulation results of the double inlet cyclone show that the density distribution inside the cyclone is basically symmetrical, suspension density is big can with high density of coarse coal to underflow, overflow pipe of coal flow.

(3)The double inlet involute channel, its effect is very obvious, the flow field is stable after the medium entering the cyclone in a very short period of time, it terminate the iteration in the Time=8.4000e-01, the medium formed the air column through the underflow and the overflow pipe

when the medium into the cyclone after 0.84 seconds.

REFERENCES

[1] Feng Liu, Ai-jun Qian. CFD simulation of flow field in dense medium cyclone [J]. Coal Preparation Technology, 2004(5):10~15.

[2] Ai-jun Qian. Numerical simulation of flow field of heavy medium cyclone [D].Tangshan: Tangshan branch of Coal Science Research Institute, 2004.

[3] Shao-lei Zhou, Hong Wang. High efficient coal preparation of new technology and new equipment[M], China University of Mining and Technology press,2008.

[4] Liang-yin Chu, Wen-mei Chen. Cyclone[M]. Chemical Industry Press,1998.

[5] Anderson D A, Tannehill J C, Pletcher R H. Computational fluid mechanical and heat transfer[M].Washington DC:Hemisphere,1984:1999~2001.

International Conference on Power Electronics and Energy Engineering (PEEE 2015)

A Study on Solving the Nonlinear Seepage Flow Model of Three-Region Composite Reservoir

Y. Wang, X.X. Dong, S.C. Li, H.E. Li
School of Science, Xihua University
Chengdu, China

D.D. Gui
Beijing Dongrunke Petroleum Technology Co., Ltd.
Beijing, China

Abstract—**This paper builds a seepage model of three-region composite reservoir, in which quadratic-gradient effect, well-bore storage, effective radius and three kinds of outer boundary conditions (constant pressure boundary, closed boundary and infinity boundary) are considered. With Laplace transformation, the seepage flow model is transformed into a boundary value problem of three-region composite zero-order modified Bessel equation by the substitution of variables and introduction of dimensionless variables. Based on the similar structure of the solution of the boundary value problem of differential equation, this paper obtains solutions of dimensionless reservoir pressure and dimensionless bottom-hole pressure of three-region composite reservoir in Laplace space. This research not only contributes to further analyse the inherent law of the solution and compile corresponding well test analysis software, but also solves corresponding mathematical model of reservoir and supplements the study on composite reservoir.**

Keywords-composite reservoir; non-linear seepage; quadratic-gradient term; well-bore storage; effective radius; similar structure of solution; similar kernel function

I. INTRODUCTION

Composite reservoir is the reservoir that involves two regions with different properties (the rock-oriented one and the fluid-oriented one), namely, the inner region and the outer one which are separated by a discontinuous interface. There have been many foreign and domestic researches in the area of composite reservoir.

Reference 1 built seepage flow models of composite reservoir. Also, it studied pressure drop, transient behavior and transient pressure analysis of composite reservoir. Reference 2 proposed a composite reservoir model of rock with fractal characteristics. Numerical Solutions of the dynamic pressure of formation pressure is obtained by using the Laplace Numerical Inversion. Some other literatures [3-5] had studied the well test analysis models of the composite reservoirs, fractal composite reservoirs and multi-bed composite reservoirs which the well-bore storage effect and the skin effect are taken into consideration.

At the beginning of this century, the thought of similar structure of the solution of the boundary value problem of the differential equation began to form in Reference 6. Some gratifying results have been achieved. References [7-9] proposed mathematical model for well test analysis which perfectly described the fractal composite reservoir with the well-bore storage and the skin factor under three kinds of outer

boundary conditions (infinite boundary, constant pressure boundary and closed boundary). Solutions of the reservoir pressure and bottom-hole pressure in Laplace space were obtained by using the Laplace transform. A similar structure of the solution was discovered by analyzing the percolation characteristics of fractal composite reservoir under three kinds of outer boundary conditions. Reference s [10, 11] have established the seepage flow model of the nonlinear percolation model of two-region composite reservoir with binary pressure gradient and nonlinear seepage flow. Seepage flow equations were linearized by variable substitution. Solutions in Laplace space were obtained by Laplace transformation and real space solutions obtained by the Stehfest numerical inversion.

However, the above researches mainly studied the mathematical model of the two-region composite reservoir. Base on the above researches, this paper studied the mathematical model of the three-region composite reservoir. With Laplace transformation, the seepage flow model is transformed into a boundary value problem of three-region composite modified zero-order Bessel equation by the substitution of variables and introduction of dimensionless variables. Based on the theory of the similar structure of the solution of the boundary value problem of differential equation, this paper obtains solutions of dimensionless reservoir pressure and dimensionless bottom-hole pressure of three-region composite reservoir in Laplace space. The study simplifies solution procedure and provides a clear idea for compiling corresponding well test analysis software. It can be widely used for well test analysis.

II. THE NONLINEAR SEEPAGE FLOW MODEL OF THREE-REGION COMPOSITE RESERVOIR

To formulate the percolation model, the main assumptions are as follows:

1) The fluid is single-phase, low compressible and follows Darcy's law in two regions.

2) Reservoir has equal thickness, each direction is horizontal and has the same nature; tere exists impermeability around three regions respectively.

3) Neglect the capillary force and gravity effect;

4) The well production quantity is constant.

5) There is no additional pressure drop at the interface of the three seepage regions;

© 2015. The authors - Published by Atlantis Press

6) Formation pressure is initial reservoir pressure p_0 before producing.

Nonlinear seepage flow basic equations of three-region composite reservoir which consider the influence of well-bore storage and the effective radius $r_{we} = r_w e^{-S}$ are as follows:

Inter region:

$$\frac{\partial^2 p_1}{\partial r^2} + \frac{1}{r}\frac{\partial p_1}{\partial r} + C_{1L}\left(\frac{\partial p_1}{\partial r}\right)^2 = \frac{\phi_1 \mu_1 C_{1t}}{3.6k_1}\frac{\partial p_1}{\partial t},$$
$$r_{we} \leq r < \alpha r_{we}, t > 0, \tag{1}$$

Middle region:

$$\frac{\partial^2 p_2}{\partial r^2} + \frac{1}{r}\frac{\partial p_2}{\partial r} + C_{2L}\left(\frac{\partial p_2}{\partial r}\right)^2 = \frac{\phi_2 \mu_2 C_{2t}}{3.6k_2}\frac{\partial p_2}{\partial t},$$
$$\alpha r_{we} \leq r \leq \beta r_{we}, t > 0, \tag{2}$$

Outer region:

$$\frac{\partial^2 p_3}{\partial r^2} + \frac{1}{r}\frac{\partial p_3}{\partial r} + C_{3L}\left(\frac{\partial p_3}{\partial r}\right)^2 = \frac{\phi_3 \mu_3 C_{3t}}{3.6k_3}\frac{\partial p_3}{\partial t},$$
$$\beta r_{we} \leq r \leq R, t > 0, \tag{3}$$

Initial condition: $p_1(r,0) = p_2(r,0) = p_3(r,0) = p_0,$ (4)

Inter condition:

$$\begin{cases} p_w(t) = p_1|_{r=r_{we}} \\ \left(r\frac{\partial p_1}{\partial r}\right)\bigg|_{r=r_{we}} = \frac{\mu_1}{2\pi k_1 h}\left[1.842\times 10^{-3} Bq + 4.421\times 10^{-6} C\frac{dp_w}{dt}\right] \end{cases}, \tag{5}$$

Convergence condition:

(αr_{we} is the radius of initial region) ,

$$\begin{cases} p_1(\alpha r_\omega, t) = p_2(\alpha r_\omega, t) \\ \dfrac{k_1}{\mu_1}\dfrac{\partial p_1}{\partial r}\bigg|_{r=\alpha r_\omega} = \dfrac{k_2}{\mu_2}\dfrac{\partial p_2}{\partial r}\bigg|_{r=\alpha r_\omega} \\ p_2(\beta r_\omega, t) = p_3(\beta r_\omega, t) \\ \dfrac{k_2}{\mu_2}\dfrac{\partial p_2}{\partial r}\bigg|_{r=\beta r_\omega} = \dfrac{k_3}{\mu_3}\dfrac{\partial p_3}{\partial r}\bigg|_{r=\beta r_\omega} \end{cases} \tag{6}$$

Outer condition:

$$p_3(R,t) = p_0, \text{ or } \frac{\partial p_3}{\partial r}\bigg|_{r=R} = 0, \text{ or } p_3(r,t)|_{r=R\to\infty} = p_0 \tag{7}$$

Firstly, in order to facilitate the research and description, dimensionless variables are introduced:

$$p_{jD}(r_D, T_D) = \frac{542.867 k_1 h}{Bq\mu_1}[p_0 - p_j(r,t)],$$

$$C_{jLD} = C_{jL}\frac{Bq\mu_j}{542.867 k_j h}, (j=1,2,3), r_D = \frac{r}{r_w e^{-S}}, R_D = \frac{R}{r_w e^{-S}},$$

$$\lambda_1 = \frac{k_2\mu_1}{k_1\mu_2},$$

$$\lambda_2 = \frac{k_3\mu_2}{k_2\mu_3}, C_D = \frac{C}{6.283\times 10^6 \phi_1 C_{1t} h r_w^2}, T_D = \frac{3.600 k_1 t}{C_D \phi_1 \mu_1 C_{1t} r_w^2},$$

$$\sigma_1 = \frac{\eta_1}{\eta_2} = \frac{\phi_2\mu_2 C_{t_2}}{\phi_1\mu_1 C_{t_1}}\cdot\frac{k_1}{k_2}, \sigma_2 = \frac{\eta_2}{\eta_3} = \frac{\phi_3\mu_3 C_{t_3}}{\phi_2\mu_2 C_{t_2}}\cdot\frac{k_2}{k_3}.$$

Secondly, substituting variable are as follows:

$$p_{jD}(r_D, T_D) = -C_{jLD}\ln\left[u_j(r_D, T_D)+1\right],$$
$$p_{wD}(T_D) = -C_{1LD}\ln\left[u_w(T_D)+1\right].$$

Thirdly, the Laplace transform is taken to the seepage flow model of three-region composite reservoir with dimensionless variable t_D, i.e.

$$\bar{P}_{iD}(r_D, z) = \int_0^\infty e^{-zt_D} P_{iD}(r_D, t_D)dt_D \quad (i=1,2,3),$$
$$\bar{P}_{\omega D}(z) = \int_0^\infty e^{-zt_D} P_{\omega D}(t_D)dt_D$$

Finally, the boundary value problem of three-region composite modified Bessel equation with parameter z (where z is Laplace space variable) is obtained as below:

$$\begin{cases} \dfrac{d^2\bar{P}_1}{dr_D^2} + \dfrac{1}{r_D}\dfrac{d\bar{P}_1}{dr_D} = \dfrac{z}{C_D e^{2S}}\bar{P}_1, 1 \leq r_D < \alpha, \\ \dfrac{d^2\bar{P}_2}{dr_D^2} + \dfrac{1}{r_D}\dfrac{d\bar{P}_2}{dr_D} = \dfrac{\sigma_1 z}{C_D e^{2S}}\bar{P}_2, \alpha \leq r_D \leq \beta, \\ \dfrac{d^2\bar{P}_3}{dr_D^2} + \dfrac{1}{r_D}\dfrac{d\bar{P}_3}{dr_D} = \dfrac{\sigma_2 z}{C_D e^{2S}}\bar{P}_3, r_D \geq \beta, \\ \bar{P}_w(z) = \bar{P}_1(1,z), \left[(C_{1LD}+z)\bar{P}_w - \dfrac{d\bar{P}_1}{dr_D}\right]\bigg|_{r_D=1} = \dfrac{C_{1LD}}{z} \\ \bar{P}_{1D}(\alpha, z) = \bar{P}_{2D}(\alpha, z), \dfrac{d\bar{P}_{1D}}{dr_D}\bigg|_{r_D=\alpha} = \lambda_1 \dfrac{d\bar{P}_{2D}}{dr_D}\bigg|_{r_D=\alpha} \\ \bar{P}_{2D}(\beta, z) = \bar{P}_{3D}(\beta, z), \dfrac{d\bar{P}_{2D}}{dr_D}\bigg|_{r_D=\beta} = \lambda_2 \dfrac{d\bar{P}_{3D}}{dr_D}\bigg|_{r_D=\beta} \\ \bar{P}_{3D}(\infty, z) = 0 \text{ or } \bar{P}_{3D}(R_D, z) = 0 \text{ or } \dfrac{d\bar{P}_{3D}}{dr_D}\bigg|_{r_D=R_D} = 0 \end{cases} \tag{8}$$

III. Solving the Nonlinear Seepage Flow Model of Three-Region Composite Reservoir

Solutions of three regions of the BVP (8) which have the form of the product of continued fraction (the similar structure) are obtained as follows:

$$\overline{P}_1(r_D,z)=\frac{C_{1LD}}{z}\cdot\frac{1}{(C_{1LD}+z)+\frac{1}{\Phi_1(1,z)-1}}\cdot\frac{1}{\Phi_1(1,z)-1}\cdot\Phi_1(r_D,z),\quad(1\le r_D<\alpha)\tag{9}$$

$$\overline{P}_2(r_D,z)=\frac{C_{1LD}}{z}\cdot\frac{1}{(C_{1LD}+z)+\frac{1}{\Phi_1(1,z)-1}}\cdot\frac{1}{\Phi_1(1,z)-1}\cdot\frac{\Psi_{0,1}(\alpha,\alpha,\sqrt{z})}{\Phi_2(\alpha,z)\sqrt{z}\Psi_{1,1}(1,\alpha,\sqrt{z})+\lambda_1\Psi_{1,0}(1,\alpha,\sqrt{z})}\cdot\Phi_2(r_D,z),\quad(\alpha\le r_D\le R_D)\tag{10}$$

$$\overline{P}_3(r_D,z)=\frac{C_{1LD}}{z}\cdot\frac{1}{(C_{1LD}+z)+\frac{1}{\Phi_1(1,z)-1}}\cdot\frac{1}{\Phi_1(1,z)-1}\cdot\frac{\Psi_{0,1}(\alpha,\alpha,\sqrt{z})}{\left[\Phi_2(\alpha,z)\sqrt{z}\Psi_{1,1}(1,\alpha,\sqrt{z})+\lambda_1\Psi_{1,0}(1,\alpha,\sqrt{z})\right]}$$
$$\cdot\frac{\Psi_{0,1}(\beta,\beta,\sqrt{\sigma_1}z)}{\left[\Phi_3(\beta,z)\sqrt{\sigma_1}z\Psi_{1,1}(\alpha,\beta,\sqrt{\sigma_1}z)+\lambda_2\Psi_{1,0}(\alpha,\beta,\sqrt{\sigma_1}z)\right]}\cdot\Phi_3(r_D,z),\quad(\beta\le r_D\le R_D)\tag{11}$$

where the similar kernel function of outer region is delimited as below:

$$\Phi_3\left(r_D,z\right)=\Phi_{3j}\left(r_D,z\right)\quad\left(j=1,2,3\right).\tag{12}$$

The similar kernel function of middle region is delimited as below:

$$\Phi_2\left(r_D,z\right)=\Phi_{2j}\left(r_D,z\right)\quad\left(j=1,2,3\right).\tag{13}$$

The similar kernel function of inter region is delimited as below:

$$\Phi_1\left(r_D,z\right)=\Phi_{1j}\left(r_D,z\right)\quad\left(j=1,2,3\right).\tag{14}$$

where $j=1$ denotes that the outer boundary condition is infinite, $j=2$ denotes that the outer boundary condition is constant pressure, and $j=3$ denotes that the outer boundary condition is closed.

Case 1. When the outer boundary condition is infinite $\overline{P}_{3D}(\infty,z)=0$ (i.e. $j=1$),

$$\Phi_{31}\left(r_D,z\right)=-\frac{K_0\left(\sqrt{\sigma_2}z r_D\right)}{\sqrt{\sigma_2}z K_1\left(\beta\sqrt{\sigma_2}z\right)}\quad(\beta\le r_D\le\infty)\tag{15}$$

$$\Phi_{21}(r_D,z)=\frac{\lambda_2\Psi_{0,0}\left(r_D,\beta,\sqrt{\sigma_1}z\right)+\sqrt{\sigma_1}z\Psi_{0,1}\left(r_D,\beta,\sqrt{\sigma_1}z\right)\Phi_{31}(\beta,z)}{\lambda_2\sqrt{\sigma_1}z\Psi_{1,0}\left(\alpha,\beta,\sqrt{\sigma_1}z\right)+\sigma_1 z\Psi_{1,1}\left(\alpha,\beta,\sqrt{\sigma_1}z\right)\Phi_{31}(\beta,z)}\quad(\alpha\le r_D\le\beta)\tag{16}$$

$$\Phi_{11}(r_D,z)=\frac{\lambda_1\Psi_{0,0}\left(r_D,\alpha,\sqrt{z}\right)+\sqrt{z}\Psi_{0,1}\left(r_D,\alpha,\sqrt{z}\right)\Phi_{21}(\alpha,z)}{\lambda_1\sqrt{z}\Psi_{1,0}\left(1,\alpha,\sqrt{z}\right)+z\Psi_{1,1}\left(1,\alpha,\sqrt{z}\right)\Phi_{21}(\alpha,z)}\quad(1\le r_D\le\alpha)\tag{17}$$

Case 2. When the outer boundary condition is constant pressure $\overline{P}_{3D}(R_D,z)=0$ (i.e. $j=2$),

$$\Phi_{32}\left(r_D,z\right)=\frac{\Psi_{0,0}\left(r_D,R_D,\sqrt{\sigma_2}z\right)}{\sqrt{\sigma_2}z\Psi_{1,0}\left(\beta,R_D,\sqrt{\sigma_2}z\right)}\quad(\beta\le r_D\le R_D)\tag{18}$$

$$\Phi_{22}(r_D,z)=\frac{\lambda_2\Psi_{0,0}\left(r_D,\beta,\sqrt{\sigma_1}z\right)+\sqrt{\sigma_1}z\Psi_{0,1}\left(r_D,\beta,\sqrt{\sigma_1}z\right)\Phi_{32}(\beta,z)}{\lambda_2\sqrt{\sigma_1}z\Psi_{1,0}\left(\alpha,\beta,\sqrt{\sigma_1}z\right)+\sigma_1 z\Psi_{1,1}\left(\alpha,\beta,\sqrt{\sigma_1}z\right)\Phi_{32}(\beta,z)}\quad(\alpha\le r_D\le\beta)\tag{19}$$

$$\Phi_{12}(r_D,z)=\frac{\lambda_1\Psi_{0,0}\left(r_D,\alpha,\sqrt{z}\right)+\sqrt{z}\Psi_{0,1}\left(r_D,\alpha,\sqrt{z}\right)\Phi_{22}(\alpha,z)}{\lambda_1\sqrt{z}\Psi_{1,0}\left(1,\alpha,\sqrt{z}\right)+z\Psi_{1,1}\left(1,\alpha,\sqrt{z}\right)\Phi_{22}(\alpha,z)}\quad(1\le r_D\le\alpha)\tag{20}$$

Case 3. When the outer boundary condition is closed $\left.\dfrac{d\overline{P}_{3D}}{dr_D}\right|_{r_D=R_D}=0$ (i.e. $j=3$),

$$\Phi_{33}\left(r_D,z\right)=\frac{\Psi_{0,1}\left(r_D,R_D,\sqrt{\sigma_2}z\right)}{\sqrt{\sigma_2}z\Psi_{1,1}\left(\beta,R_D,\sqrt{\sigma_2}z\right)}\quad(\beta\le r_D\le R_D)\tag{21}$$

$$\Phi_{23}(r_D,z)=\frac{\lambda_2\Psi_{0,0}\left(r_D,\beta,\sqrt{\sigma_1}z\right)+\sqrt{\sigma_1}z\Psi_{0,1}\left(r_D,\beta,\sqrt{\sigma_1}z\right)\Phi_{33}(\beta,z)}{\lambda_2\sqrt{\sigma_1}z\Psi_{1,0}\left(\alpha,\beta,\sqrt{\sigma_1}z\right)+\sigma_1 z\Psi_{1,1}\left(\alpha,\beta,\sqrt{\sigma_1}z\right)\Phi_{33}(\beta,z)}\quad(\alpha\le r_D\le\beta)\tag{22}$$

$$\Phi_{13}(r_D,z)=\frac{\lambda_1\Psi_{0,0}\left(r_D,\alpha,\sqrt{z}\right)+\sqrt{z}\Psi_{0,1}\left(r_D,\alpha,\sqrt{z}\right)\Phi_{23}(\alpha,z)}{\lambda_1\sqrt{z}\Psi_{1,0}\left(1,\alpha,\sqrt{z}\right)+z\Psi_{1,1}\left(1,\alpha,\sqrt{z}\right)\Phi_{23}(\alpha,z)}\quad(1\le r_D\le\alpha)\tag{23}$$

where

$\Psi_{m,n}\left(\alpha,\beta,y\right)=I_m\left(\alpha y\right)K_n\left(\beta y\right)+\left(-1\right)^{m-n+1}K_m\left(\alpha y\right)I_n\left(\beta y\right)$ and $I_n\left(\cdot\right)$, $K_n\left(\cdot\right)$ are respectively the first and the second class of modified Bessel functions of order n [12].

IV. Symbol Description

The symbol meanings are listed below.

p — Reservoir pressure (MPa);

p_0 — Initial pressure(MPa);

p_w —Bottom-hole pressure (MPa);

q — Well yield (m^3/d);

t —Time (h);

r — The distance from any point in the reservoir to the center of well (m);

R — The outer boundary radius (m);

h —Reservoir thickness(m);

B —Oil volume factor, dimensionless;

k — Reservoir permeability(μm^2);

μ — The viscosity of fluid in reservoir ($mPa \cdot s$);

S —Skin factor, dimensionless;

C — Wellbore storage coefficient (m^3/MPa);

C_t — Total compressibility of reservoir, ($1/MPa$);

C_L — Fluid compressibility, ($1/MPa$);

r_w — Wellbore radius(m);

r_{we} — Effective wellbore radius,(m);

ϕ — Porosity, dimensionless;

σ — Elastic storativity ratio, dimensionless;

λ — Interporosity flow coefficient, dimensionless;

w —Well;

D —Dimensionless.

V. CONCLUSIONS

1) Using variable substitution and Laplace transformation, the seepage flow model of three-region composite reservoir is transformed into a boundary value problem of three-region composite zero-order modified Bessel equation. Solutions of dimensionless reservoir pressure and dimensionless bottom-hole pressure can be constructed by two linearly independent solutions of the basic equation and coefficients of boundary conditions. And these solutions have the form of the product of continued fraction, hence the similar structure.

2) The seepage flow model of three-area composite reservoir has the similar structure of solutions under three different outer boundary conditions. The difference is that similar kernel functions are different under three different outer boundary conditions.

ACKNOWLEDGMENT

This work is supported the by the Scientific Research Fund of the Sichuan Provincial Education Department of China (Grant No. 12ZA164).

REFERENCES

[1] D. Peon, AOSTRA, H. Chhina, AEC. Society of Petroleum Engineers. 1989, 5: 28-31.

[2] Poon D. International Meeting of the Pet. Soc. of. CIM / SPE9-534, 1995.

[3] K.L. Xiang, Y. Li, T.J. Li. Petroleum Exploration and Development,2001,28(5):49-52.

[4] Z.G. Yang, Y.X. Chen, X. Wang. Petrochemical Industry Application, 2011,30(3):16-19.

[5] J.X. Luo, L.H. Zhang, Y.L. Zhao, et al. Journal of Yangtze University (Nature Science Edition), 2011,8(2):65-67.

[6] S.C. Li, M.H. Jia. Journal of UEST of China, 2004, 33(Supp.):95-98.

[7] S.C. Li, T.P. Zheng. Journal of Jilin University (Engineering and Technology Edition), 2004, 34:104-107.

[8] X.R. Deng, S.C. Li. Journal of Xihua University (Natural Science Edition), 2005, 24(2):4-7.

[9] C.X. Xu, S.C. Li, W.B. Zhu. Drilling & Production Technology, 2006, 29(5):39-42.

[10] F.X. Zhang, B.C. Wang, Y.T. Fei, et al. Journal of Southwest Petroleum University (Science & Technology Edition), 2010,32(4):99-102.

[11] L. Zhang, Y.L. Jia, F.X. Zhang, et al. Journal of Daqing Petroleum Institute, 2011,35(5):54-59.

[12] Liu Shishi, Liu Shida. Special Function. Beijing: China Meteorological Press, 2002.

International Conference on Power Electronics and Energy Engineering (PEEE 2015)

A Study on Solving the Linear Seepage Flow Model with r_{we} of Composite Reservoir

Y. Wang, X.X. Dong, S.C. Li, H.E. Li
School of Science, Xihua University
Chengdu, China

D.D. Gui
Beijing Dongrunke Petroleum Technology Co., Ltd
Beijing, China

Abstract—**This paper studies the linear seepage flow model of composite reservoir under three different kinds of outer boundary conditions (infinite boundary, constant pressure boundary and closed boundary) which consider the well-bore storage and introducing the effective radius. On the basic of the theory of similar structure of solution of the BVP of differential equation, this paper obtains solutions of dimensionless reservoir pressure and dimensionless bottom-hole pressure of composite reservoir in Laplace space. Solutions of dimensionless reservoir pressure and dimensionless bottom-hole pressure have similar structure. Similar structures of solutions are convenient for analyzing the influence of reservoir parameters on reservoir pressure and bottom-hole pressure. The study supplements the other studies on composite reservoir.**

Keywords-composite reservoir; boundary value problem (BVP); seepage flow; similar structure of the solution; similar kernel function

I. INTRODUCTION

Composite reservoir is the reservoir that involves two regions with different properties (the rock-oriented one and the fluid-oriented one), namely, the inner region and the outer one which are separated by a discontinuous interface. There have been many foreign and domestic researches in the area of composite reservoir.

Reference1 built seepage flow models of composite reservoir. Also it studied pressure drop, transient behavior and transient pressure analysis of composite reservoir. Reference 2 studied the equation of pressure diffusion in radically composite reservoirs by the application of Green function, Fourier transform and Laplace transform and has given a general method to solve the class of equations. Since seepage flow models of composite reservoir usually can be transformed into BVPs of composite modified Bessel equation, so Reference 3 studied the BVP of second-order composite linear homogeneous composite modified Bessel equation. It studied the similar structure of the solution and put forward a new method for solving this class boundary value problem. References [4-6] studied the fractal composite reservoir under three kinds of outer boundary conditions (infinite boundary, constant pressure boundary and closed boundary) which consider the well-bore storage and the skin factor. Solutions of the reservoir pressure and bottom-hole pressure in Laplace space were obtained by using the Laplace transform. A similar structure of the solution was discovered by analyzing the percolation characteristics of fractal composite reservoir under

three different kinds of outer boundary conditions. References [7,8] obtained the similar structure of the solution of dimensionless reservoir pressure and dimensionless bottom-hole pressure of the composite reservoir under the three different kinds of outer boundary conditions (infinite boundary, constant pressure boundary and closed boundary). Furthermore, they made the theoretical graph and analyzed the impact of the well-bore storage and skin factor on dimensionless reservoir pressure and dimensionless bottom-hole pressure by using the numerical inversion. References [9-11] had studied the well test analysis models of the composite reservoirs, fractal composite reservoir and multi-bed composite reservoirs in which the well-bore storage effect and the skin effect are taken into consideration.

Based on the above researches, this paper studied the linear seepage flow model of composite reservoir which considers the well-bore storage and introduces the effective radius. By using Laplace transformation and introducing dimensionless variables, the linear seepage flow model is transformed into a boundary value problem of composite modified zero-order Bessel equation. Based on the theory of the similar structure of the solution of the boundary value problem of differential equation, this paper obtains solutions of dimensionless reservoir pressure and dimensionless bottom-hole pressure of composite reservoir in Laplace space. The study lays the foundation for studying the seepage flow of composite reservoir and supplements the study on composite reservoir.

II. THE LINEAR SEEPAGE FLOW MODEL WITH r_{we} OF COMPOSITE RESERVOIR

Here we studied the linear seepage flow model of composite reservoir which the well-bore storage and the effective radius were taken into consideration. To formulate the percolation model, the main assumptions are as follows:

1) The single-phase micro compressible fluid obeys Darcy's law;

2) The reservoir has equal thickness, each direction is horizontal and has a same nature;

3) The capillary single-phase horizontal flow without gravity effect;

4) The well production quantity is constant;

5) There is no additional pressure drop existing on the interface between the two seepage regions;

© 2015. The authors - Published by Atlantis Press

6) Formation pressure is initial reservoir pressure p_0 before producing.

The linear seepage flow basic equations of composite reservoir which consider the well-bore storage and the effective radius are as follows.

Inter region:

$$\frac{\partial^2 p_1}{\partial r^2} + \frac{1}{r}\frac{\partial p_1}{\partial r} = \frac{1}{\eta_1}\frac{\partial p_1}{\partial t} \quad r_\omega < r < \alpha r_\omega, t > 0 \quad (1)$$

Outer region:

$$\frac{\partial^2 p_2}{\partial r^2} + \frac{1}{r}\frac{\partial p_2}{\partial r} = \frac{1}{\eta_2}\frac{\partial p_2}{\partial t} \quad \alpha r_\omega \leq r \leq \beta r_\omega, t > 0 \quad (2)$$

Initial condition:

$$p_1(r,0) = p_2(r,0) = p_0 \quad (3)$$

Inner boundary condition:

$$\begin{cases} p_\omega(t) = p_1(r_{\omega e}, t) \\ \left(r\frac{\partial p_1}{\partial r} \right)\bigg|_{r=r_{\omega e}} = \frac{\mu_1}{2\pi k_1 h}\left(Bq + C\frac{dp_\omega}{dt} \right) \end{cases} \quad (4)$$

Convergence condition:

$$\begin{cases} p_1(\alpha r_\omega, t) = p_2(\alpha r_\omega, t) \\ \frac{k_1}{\mu_1}\frac{\partial p_1}{\partial r}\bigg|_{r=\alpha r_\omega} = \frac{k_2}{\mu_2}\frac{\partial p_2}{\partial r}\bigg|_{r=\alpha r_\omega} \end{cases} \quad (5)$$

Outer condition:

$$p_2(R,t) = p_0, or \quad \frac{\partial p_2}{\partial r}\bigg|_{r=R} = 0, \; or \; p_2(r,t)\big|_{r=R\to\infty} = p_0 \quad (6)$$

III. THE LINEAR SEEPAGE FLOW MODEL OF COMPOSITE RESERVOIR IN LAPLACE SPACE

In order to facilitate the research and description, firstly, dimensionless variables are introduced:

$$P_{iD}(r_D, t_D) = \frac{2\pi k_1 h}{Bq\mu_1}\left[p_0 - p_i(r,t) \right] \quad (i=1,2)$$

$$t_D = \frac{k_1 t}{\phi_1 \mu_1 C_{t_1} r_\omega^2}, \quad T_D = \frac{t_D}{C_D}, \quad r_D = \frac{r}{r_\omega e^{-S}}, \quad R_D = \frac{R}{r_\omega e^{-S}},$$

$$C_D = \frac{C}{2\pi\phi_1 C_{t_1} h r_\omega^2}, \quad \sigma = \frac{\eta_1}{\eta_2} = \frac{\phi_2 \mu_2 C_{t_2}}{\phi_1 \mu_1 C_{t_1}} \cdot \frac{k_1}{k_2}, \quad \lambda = \left(\frac{k_2}{\mu_2}\right)\bigg/\left(\frac{k_1}{\mu_1}\right)$$

Secondly, the Laplace transform is taken to the linear seepage flow model of composite reservoir with dimensionless variable t_D, i.e.

$$\bar{P}_{iD}(r_D, z) = \int_0^\infty e^{-z t_D} P_{iD}(r_D, t_D)\,dt_D \quad (i=1,2),$$

$$\bar{P}_{\omega D}(z) = \int_0^\infty e^{-z t_D} P_{\omega D}(t_D)\,dt_D.$$

Finally, the BVP of the ODE with parameter z (where z is Laplace space variable) is obtained as below:

$$\begin{cases} \dfrac{d^2 \bar{P}_{1D}}{dr_D^2} + \dfrac{1}{r_D} \cdot \dfrac{d\bar{P}_{1D}}{dr_D} = \dfrac{z}{C_D e^{-2S}}\bar{P}_{1D}, \quad 1 \leq r_D \leq \alpha \\[2mm] \dfrac{d^2 \bar{P}_{2D}}{dr_D^2} + \dfrac{1}{r_D} \cdot \dfrac{d\bar{P}_{2D}}{dr_D} = \dfrac{\sigma z}{C_D e^{-2S}}\bar{P}_{2D}, \quad r_D > \alpha \\[2mm] \bar{P}_{\omega D}(z) = \bar{P}_{1D}(1,z) \\[2mm] \left(r_D \dfrac{d\bar{P}_{1D}}{dr_D} \right)\bigg|_{r_D=1} = -\bar{q}_D(z) + z\bar{P}_{\omega D} \\[2mm] \bar{P}_{1D}(\alpha,z) = \bar{P}_{2D}(\alpha,z), \quad \dfrac{d\bar{P}_{1D}}{dr_D}\bigg|_{r_D=\alpha} = \lambda\dfrac{d\bar{P}_{2D}}{dr_D}\bigg|_{r_D=\alpha} \\[2mm] \bar{P}_{2D}(\infty,z) = 0 \; or \; \bar{P}_{2D}(R_D,z) = 0 \; or \; \dfrac{d\bar{P}_{2D}}{dr_D}\bigg|_{r_D=R_D} = 0 \end{cases} \quad (7)$$

IV. SOLVING THE LINEAR SEEPAGE FLOW MODEL OF COMPOSITE RESERVOIR

Solutions of two regions of the BVP (7) which have the form of the product of continued fraction (the similar structure) are obtained as follows.

$$\bar{P}_{1D}(r_D, z) = \bar{q}_D(z) \cdot \frac{1}{z + \cfrac{1}{-\frac{2}{z} + \Phi_1(1,z)}} \cdot \frac{1}{-\frac{2}{z} + \Phi_1(1,z)} \cdot \Phi_1(r_D, z) \quad (1 \leq r_D \leq \alpha) \quad (9)$$

$$\bar{P}_{2D}(r_D, z) = \bar{q}_D(z) \cdot \frac{1}{z + \frac{1}{-\frac{2}{z} + \Phi_1(1,z)}} \cdot \frac{1}{-\frac{2}{z} + \Phi_1(1,z)}$$

$$\cdot \frac{\Psi_{0,1}\left(\alpha, \alpha, \sqrt{\frac{z}{C_D e^{-2S}}}\right)}{\Phi_2(\alpha, z)\sqrt{\frac{z}{C_D e^{-2S}}}\Psi_{1,1}\left(1, \alpha, \sqrt{\frac{z}{C_D e^{-2S}}}\right) + \lambda\Psi_{1,0}\left(1, \alpha, \sqrt{\frac{z}{C_D e^{-2S}}}\right)} \cdot \Phi_2(r_D, z) \ (r_D > \beta)$$

(10)

where the similar kernel function of outer region is delimited as below:

$$\Phi_2(r_D, z) = \begin{cases} -\dfrac{K_0\left(\sqrt{\frac{\sigma z}{C_D e^{-2S}}}r_D\right)}{\sqrt{\frac{\sigma z}{C_D e^{-2S}}}K_1\left(\beta\sqrt{\frac{\sigma z}{C_D e^{-2S}}}\right)} & \text{Outer boundary condition is infinite} \\[2em] \dfrac{\Psi_{0,0}\left(r_D, R_D, \sqrt{\frac{\sigma z}{C_D e^{-2S}}}\right)}{\sqrt{\frac{\sigma z}{C_D e^{-2S}}}\Psi_{1,0}\left(\beta, R_D, \sqrt{\frac{\sigma z}{C_D e^{-2S}}}\right)} & \text{Outer boundary condition is constant pressure} \\[2em] \dfrac{\Psi_{0,1}\left(r_D, R_D, \sqrt{\frac{\sigma z}{C_D e^{-2S}}}\right)}{\sqrt{\frac{\sigma z}{C_D e^{-2S}}}\Psi_{1,1}\left(\beta, R_D, \sqrt{\frac{\sigma z}{C_D e^{-2S}}}\right)} & \text{Outer boundary condition is closed} \end{cases}$$

$(\alpha \leq r_D \leq \infty)$

(11)

The similar kernel function of inter region is delimited as below.

$$\Phi_1(r_D, z) = \begin{cases} \dfrac{\lambda\Psi_{0,0}\left(r_D, \alpha, \sqrt{\frac{z}{C_D e^{-2S}}}\right) + \sqrt{\frac{z}{C_D e^{-2S}}}\Psi_{0,1}\left(r_D, \alpha, \sqrt{\frac{z}{C_D e^{-2S}}}\right)\Phi_{21}(\alpha, z)}{\lambda\sqrt{\frac{z}{C_D e^{-2S}}}\Psi_{1,0}\left(1, \alpha, \sqrt{z}\right) + \frac{z}{C_D e^{-2S}}\Psi_{1,1}\left(1, \alpha, \sqrt{\frac{z}{C_D e^{-2S}}}\right)\Phi_{21}(\alpha, z)} \\ \qquad - \text{Outer boundary condition is infinite} \\[2em] \dfrac{\lambda\Psi_{0,0}\left(r_D, \alpha, \sqrt{\frac{z}{C_D e^{-2S}}}\right) + \sqrt{\frac{z}{C_D e^{-2S}}}\Psi_{0,1}\left(r_D, \alpha, \sqrt{\frac{z}{C_D e^{-2S}}}\right)\Phi_{22}(\alpha, z)}{\lambda\sqrt{\frac{z}{C_D e^{-2S}}}\Psi_{1,0}\left(1, \alpha, \sqrt{\frac{z}{C_D e^{-2S}}}\right) + \frac{z}{C_D e^{-2S}}\Psi_{1,1}\left(1, \alpha, \sqrt{\frac{z}{C_D e^{-2S}}}\right)\Phi_{22}(\alpha, z)} \\ \qquad - \text{Outer boundary condition is constant pressure} \\[2em] \dfrac{\lambda\Psi_{0,0}\left(r_D, \alpha, \sqrt{\frac{z}{C_D e^{-2S}}}\right) + \sqrt{\frac{z}{C_D e^{-2S}}}\Psi_{0,1}\left(r_D, \alpha, \sqrt{\frac{z}{C_D e^{-2S}}}\right)\Phi_{23}(\alpha, z)}{\lambda\sqrt{\frac{z}{C_D e^{-2S}}}\Psi_{1,0}\left(1, \alpha, \sqrt{\frac{z}{C_D e^{-2S}}}\right) + \frac{z}{C_D e^{-2S}}\Psi_{1,1}\left(1, \alpha, \sqrt{\frac{z}{C_D e^{-2S}}}\right)\Phi_{23}(\alpha, z)} \\ \qquad - \text{Outer boundary condition is closed} \end{cases}$$

$(1 \leq r_D \leq \alpha)$

(12)

where

$$\Psi_{m,n}(\alpha, \beta, y) = I_m(\alpha y)K_n(\beta y) + (-1)^{m-n+1}K_m(\alpha y)I_n(\beta y)$$

and $I_n(\cdot)$, $K_n(\cdot)$ are respectively the first and the second class of modified Bessel functions of order n [12].

V.   SYMBOL DESCRIPTION

The symbol meanings are listed below.

B —Formation volume factor (dimension);

C_t —Compressibility (MPa^{-1});

C —Well-bore storage coefficient (m^3 / MPa);

k —Permeability (μm^2);

p —Reservoir pressure (MPa);

p_ω —Well-bore pressure (MPa);

q —Production rate or injection rate (m^3 / d);

ϕ —Porosity (dimensionless);

R —The outer boundary radius (m);

r —Well-bore radius (m);

r_{we} —Effective well-bore radius,(m);

S —Skin factor (dimensionless);

μ —Viscosity ($MPa \cdot s$);

p_0 —Initial reservoir pressure (MPa);

h —Reservoir thickness (m);

z —Laplace space variable;

η —Pressure transmitting coefficient ($\mu m^2 \cdot MPa$);

t —Time (h);

$-$ —Laplace domain;

D —Dimensionless;

w —Well-bore parameter

VI.   CONCLUSIONS

Based on the reasoning mentioned above, the following conclusions can be drawn.

1)   By using Laplace transformation and introducing dimensionless variables, the linear seepage flow model of composite reservoir is transformed into a BVP of composite zero-order modified Bessel equation. Solutions of dimensionless reservoir pressure and dimensionless bottom-hole pressure are obtained and they have similar structures. Base on the continued fractions Equations (8)-(10), we can directly analyze the effect of the well-bore storage and effective radius on the reservoir pressure and bottom hole pressure.

2)   The linear seepage flow model of composite reservoir has the similar structure of solutions under three different outer boundary conditions. The difference is that similar kernel functions are different under three different outer boundary conditions.

ACKNOWLEDGMENT

This work is supported the by the Scientific Research Fund of the Sichuan Provincial Education Department of China (Grant No. 12ZA164).

REFERENCES

[1] D. Peon, AOSTRA, H. Chhina, AEC. Society of Petroleum Engineers. 1989, 5: 28-31.

[2] Fikri J, Kuchuk, Tarek M, Habashy. Transport in Porous Media. 19(1995): 199-232.

[3] P.H Liu , Z.C. Chen, S.C. Li. Journal of Xihua University (Natural Science Edition), 2006, 25(2):23-26.

[4] S.C. Li, T.P. Zheng. Journal of Jilin University (Engineering and Technology Edition), 2004, 34:104-107.

[5] X.R. Deng, S.C. Li. Journal of Xihua University (Natural Science Edition), 2005, 24(2):4-7.

[6] C.X. Xu, S.C. Li, W.B. Zhu. Drilling & Production Technology, 2006, 29(5):39-42.

[7] P.S. Zheng, S.C. Li, W.Z. Xu. Drilling & Production Technology, 2007, 30(3): 49-50(-62).

[8] S.C. Li, P.S. Zheng, Y.F. Zhang. Journal of Mathematics in Practice and Theory, 2008, 38(3):23-28.

[9] Z.G. Yang, Y.X. Chen, X. Wang. Petrochemical Industry Application, 2011,30(3):16-19.

[10] J.X. Luo, L.H. Zhang, Y.L. Zhao, et al. Journal of Yangtze University (Nature Science Edition), 2011,8(2):65-67.

[11] F.X. Zhang, B.C. Wang, Y.T. Fei, et al. Journal of Southwest Petroleum University (Science & Technology Edition), 2010,32(4):99-102.

[12] Liu Shishi, Liu Shida. Special Function. Beijing: China Meteorological Press, 2002.

International Conference on Power Electronics and Energy Engineering (PEEE 2015)

A Fuzzy Trust Evaluation Model in Mobile Commerce

J. Chen, Z.Y. Cai
College of Computer and Information technology
China Three Gorges University
Yichang, China

J. Peng
College of Civil Engineering & Architecture
China Three Gorges University
Yichang, China

Abstract—With the quick development of mobile commerce, lacking of trust and security is becoming a big obstacle to the development of the mobile commerce. In this paper, the reliability and security problem of mobile commerce, especially trust evaluation model in its operation are studied, and a trust evaluation model based on fuzzy mathematics is proposed. Then in order to gain a comprehensive trust value of the whole mobile commerce, the fuzzy recommend and combined algorithm is presented. Lastly the proposed model is verified by an example.

Keywords-mobile commerce; information security; trust evaluation; fuzzy mathematics

I. INTRODUCTION

Since there are great uncertainty and subjective factors in mobile commerce, many early researches have focused on mobile payment trust. Their results have shown that from the point of view of the original trust, the impact of two factors affected the performance and willingness of users [1,5,6]. Trust benefits people who live in a risky and uncertain situation by providing means to decrease complexity. Trust is the key to decision making; thus visualizing trust information could benefit users' behavior and decisions [2]. On one hand, trust research provides empirical evidences to illustrate the coexistence of integrity trust and distrust. Generally, if an individual trusts the integrity of a service provider, the cause is more likely to keep its promise of providing genuine personalized mobile services by a substantial investment in location detection technologies. On the other hand, if the individual distrusts the integrity of a service provider, it is the subsequence of this investment that the service provider has a strong incentive to cheat others in order to justify the investment [3,4].

Recently, theory framework of influence factors of consumers' trust in mobile commerce has gained much attention. Then a questionnaire on trust evaluation of mobile commerce is designed for reliability analysis and validity analysis on the trust evaluation scale respectively[7, 8, 9]. Due to lack of punishment to malicious recommendation from referee (nodes), enormous subjective malicious recommendation nodes are remained in network and they are searched by the next criminal opportunity constantly. This paper proposes a dynamic recommendation trust evaluation model based on mobile e-commerce environment. On the basis of the research questions above, we proposed a Fuzzy Trust Measurement(FTM)model for trust evaluation. Given the fuzzy trust definition of nodes, the model can calculate

local trust with fuzzy inference and obtain recommendation from neighboring nodes. Then the recommendation will be combined with local trust after getting synthesis weight with fuzzy inference to obtain comprehensive trust of nodes finally.

II. FLOW DIAGRAM OF TRUST EVALUATION

The flow chart describes the recommended value, trust degree of recommendation and domain reliability in mobile commerce system, as shown in Figure 1. Trust depends on observation and recommendation of the third party. The design of FTM model is as follows. Firstly, it is to assess the initialized trust value of local node by observing the change in the trust factor and combine it with the fuzzy inference which is to be updated. Secondly, it is to assess the trust value of node by collect in recommendation information from other nodes. Finally, it is to assess recommendation from the local node by using fuzzy reasoning weighted with its comprehensive rights credibility from recommendation trust value, on the basis of which the recommended trust value in the synthesis assessment will form a comprehensive trust value of the system.

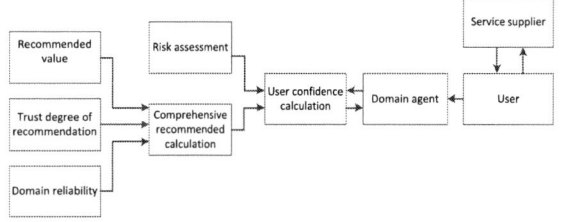

FIGURE I. FLOW DIAGRAM.

Recommendation value of trust values is indicated by vector form $E_{IT} = \left(E_{T1}^{I}, E_{T2}^{I}, E_{T3}^{I}, E_{T4}^{I}, E_{T5}^{I} \right)$. In order to cut energy waste in communication and prevent the recycle loop of the trust, it is limited between neighboring nodes without using pass iteration. The recommended trust value provided by interrelated node is referred to as a recommendation node. As shown in Figure 1, the recommendation nodes would automatically send the trust value of the local nodes that being assessed as comprehensive recommended calculation to evaluate nodes after receiving the evaluation requirement of other nodes. Then its local trust value of the node can be evaluated as a recommendation trust value to the node trust assessment. To sum up, only nodes who are the neighbor with

© 2015. The authors - Published by Atlantis Press

both evaluating nodes and have been evaluated by others can provide recommendation trust value.

Risk was assessed with a distributed algorithm in the mobile commerce, where the two adjacent nodes trust each other, and the user confidence calculation in made inassessment node, while the objects are called evaluated nodes. Therefore, the FTM model is designed as follows. Firstly, it is to assess the local nodes' trust value assessment, thenbe combined with the fuzzy inference to update it by observing the change in the trust factor. Secondly, it is to assess noderecommendation gathered from other nodes to assess the trust value of nodes. Finally, the assessment node is recommended by its local node using fuzzy reasoning with comprehensive right credibility from its recommendation trust value weights, where recommended trust value in the synthesis of the assessment is the comprehensive trust value of the whole system.

Domain agentneedsto collectrecommendation trust values and synthesizethemto an integrated trust value of the local trust value in order to improve the accuracy and robustness of trust assessment for user and service provider. Therefore the comprehensive trust value's vector form is marked as $E_{CT} = \left(E_{T1}^{C}, E_{T2}^{C}, E_{T3}^{C}, E_{T4}^{C}, E_{T5}^{C}\right)$. Node's trust value is decided by two-step process, overall weight computing and synthesis trust value.

III. FUZZY TRUST MEASUREMENT MODEL

In the process of quantizinglocal trust value, the observation of the trust factor is a subjective cognitive process, which is of strong fuzziness, where credibility trust factor with multiple values from different observation factors is a comprehensive assessment of nodes and weights are neededto determine a problem.

Nodes in mobile commerce are usually deployed in densely. However, due to the randomness of the distribution of nodes, there is some special circumstance that nodes might have no recommendations. In this case, the assessment directly from the neighboring node is evaluated as an integrated trustable node.

Evaluated node will be assessed on its own observations by comprehensive trust factorsof multiple nodes to quantify the value of trust result, which iscalled as a local trust value Z_{BC} ,with vector form as $Z_{BC} = \left(Z_{C1}^{B}, Z_{C2}^{B}, Z_{C3}^{B}, Z_{C4}^{B}, Z_{C5}^{B}\right)$.Quantization process of local trust value is divided into two phases, initialization phaseand update phase.

Therefore, node trustis classified as several grades. Let x be the node's credibility, its domain is $T_{X} = \left\{d \mid d \in [0,1]\right\}$.The trust classification nodes are described by five variables: "absolutely incredible", "compared unreliable", "uncertain", "less credible" and "absolutely credible". The corresponding fuzzy subset are $X_{1}, X_{2}, X_{3}, X_{4}$ and X_{5}, the corresponding membership function are $\beta_{X_{1}}, \beta_{X_{2}}, \beta_{X_{3}}, \beta_{X_{4}}, \beta_{X_{5}}$, and according to the trust classification of nodes, the trust value of

the node is represented as a vector

$$T_{D} = \left(r_{D_1}, r_{D_2}, r_{D_3}, r_{D_4}, r_{D_5}\right) \tag{1}$$

where$(k = 1,2, \ldots, 5)$ is the membership degree of the node trust category D_k.

Let the other assignment be 0, that is, Z_{BC}=(0, 0, 1, 0, 0), and the initial value $\lambda_1^{*}, \lambda_2^{*}, \cdots, \lambda_n^{*}$ is 0.In case of lacking prior information, assessing node cannot be trusted to determine the assessmentcondition of nodes and will be evaluated by local trust value node as anuncertain component of the assignment.After the establishment of the neighbor relationship of nodes, all nodes can be reviewed to assess the initial local trust value of the system. To reflect this subjective fuzziness, the node trust value based on fuzzy set theory is determined as shown in Figure 2.

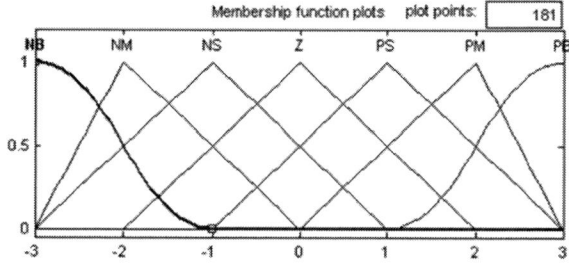

FIGURE II. FUZZY TRUST VALUE.

As we can see from Figure 2, there are 7 fuzzy trust language value, namely {NB,NM,NS,Z,PS,PM, PB}, to define different trust degree.In the mobile commerce networks, subjective fuzziness of trust evaluation nodes mainly appears fuzzy, where trust classification is based onmore value, instead of binary logic. That is to say, we could not simply classify a node into trustable or unreliable, and we should take the middle status into consideration. Trust is not either-or, but both-and, that means a node belongs to not just a category of trust, but more than one trust category in varying degrees.

IV. ALGORITHM OF FUZZY TRUST MEASUREMENT

Suppose that there are n trust factors: $H_1, H_2, \cdots, H_n$ (j = 1, 2, ..., n) can be classified as m grades: $K_{j1}, K_{j2}, \cdots, K_{jn}$. Thus thecorresponding fuzzy subsets of its membership function are $\beta_{kj1}\left(\varepsilon_i\right), \beta_{kj2}\left(\varepsilon_i\right), \cdots, \beta_{kni}\left(\varepsilon_i\right)$,respectively, which represents the I-type trust factor. Thus, the local trust value is calculated as follows.

① Defining fuzzy inference rules. According to rule knowledge and experience of trust reasoning, the fuzzy rules of node trust factor can be gotten by derived category, the total rules isset as W.

Rule 1: if H_1isK$_{j1}$, and H$_2$is K$_{j2}$, and … andH$_n$isK$_{jn}$, thenX is X$_1$;

Rule 2:if $H_1$is K_{j1}, and H_2 is K_{j2}, and … and$H_n$isK_{jn}, thenX is X_1;

Rule W: if H_1isK_{j1}, and H_2 is K_{j2}, and … and$H_n$isK_{jn}, thenX is X_5;

②Building fuzzy implication relationship S_{k-x}. At first, it is to strike implication relations under a single fuzzy rule S_λ = (λ =1,2,…,W) with formula S_λ = $\beta_{K1}^\lambda\left(\varepsilon_1\right) \wedge \beta_{K2}^\lambda\left(\varepsilon_2\right) \wedge \cdots \wedge \beta_{Km}^\lambda\left(\varepsilon_m\right) \wedge \beta_\lambda^x\left(x\right)$, where $\beta_{K1}^\lambda\left(\varepsilon_1\right)$, $\beta_{K2}^\lambda\left(\varepsilon_2\right)$,…, $\beta_{Km}^\lambda\left(\varepsilon_m\right)$ are the first fuzzy rules under the membership functional subset. Secondly all the fuzzy comprehensive implication relations are under the rules.This rule contains multiple relationships as follows.

$$S_{k-x} = \vee_{\theta=1}^W \left(\beta_{K1}^\lambda\left(\varepsilon_1\right) \wedge \beta_{K2}^\lambda\left(\varepsilon_2\right) \wedge \cdots \wedge \beta_{Km}^\lambda\left(\varepsilon_m\right) \wedge \beta_\lambda^x\left(x\right) \right) \quad (2)$$

③Calculating thelocal credibility of every nodes. Using the actual value of the trust factor λ_1^*, λ_2^*, … λ_n^* and synthetic relationship S_{k-x}, fuzzy output of local credibility node can be gotten as formula (3)

$$\beta_x^*\left(x\right) = \vee_{\theta=1}^W \left(\beta_{K1}^\lambda\left(\varepsilon_1\right) \wedge \beta_{K2}^\lambda\left(\varepsilon_2\right) \wedge \cdots \wedge \beta_{Km}^\lambda\left(\varepsilon_m\right) \wedge \beta_\lambda^x\left(x\right) \right) \quad (3)$$

Then using the center of gravity degasification method,the credibility of the local node can be obtained, such as the formula (4)

$$x^* = COG = \int \beta_x^*\left(x\right) \cdot Xdx \;/ \int \beta_x^*\left(x\right) \cdot Xdx \quad (4)$$

④ Integratingthe trust vector of local nodes. Using the node-local credibility and membership function of each node's local trust value, the trust category can be calculated as formula (5).

$$Z_{BC} = \left(\beta_{x1}\left(x^*\right), \beta_{x2}\left(x^*\right), \beta_{x3}\left(x^*\right), \beta_{x4}\left(x^*\right), \beta_{x5}\left(x^*\right)\right) \quad (5)$$

Therefore, FTM model uses fuzzy inference trust value of local node in calculation, not only embodying the trust quantifying subjective process, but also preventing the problem of weight uncertainty of the trust factor in calculating.

V. SIMULATED EXPERIMENT AND ANALYSIS

Experiment is based on MATLAB.The scene is set as follows:assumingthe communication radius of each node is 25 m, and packet loss rate and packet tampering rates are set as 75% to 100%.There are 100 nodes randomly dispenser in a 100 m×100 m detection area, among which 10 nodes are set as malicious nodes.Under the collusion and strategic attack mode,the updating cycle of local trust value is assumed as 0.1s.

The simulation resultsof fuzzy trust measurement compared with traditional crisp trust measurement are shown in Figure 3.

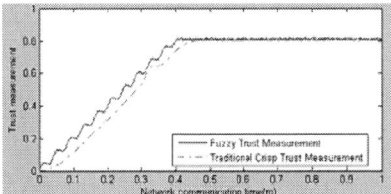

FIGURE III. TRUST MEASUREMENT RESULTS OF FTM MODEL AND TRADITIONAL CRISP MODEL.

The absolutely credible and less credible nodesare designated as malicious nodes in our experiments. We test the validity of FTM model to improve network security by observing the discovered proportion of maliciouslocalnodes, and compare them with the traditionalcrisp model. Figure 2 shows that the two found proportions of malicious nodes are different, and under the model FTM, the proportion of malicious nodes was found rise rapidly in the initialization phase of the network, and then maintained at a high level of 0.8. The reason is that FTM model definite node trust value of fuzzy and quantification fuzzy inference which ensures the accuracy of trust evaluation, and enhances the inclusive analysis of the weights and the robustness of trust assessment.

VI. SUMMARY

Here a trust evaluation model is put forward for mobile commerce based on fuzzy mathematics, and the formal definition of the trust value of nodes is given.Therefore we can quantify the value of a local trust with the method of fuzzy reasoning and obtain recommendations from its' neighbor nodes.Finallycomprehensive trust of thewhole system can be obtainedby trustable nodes.Future research should focused on combinedcomputing with local trust after getting synthesis weight with fuzzy inference

ACKNOWLEDGEMENTS

This research was supported by the National Natural Science Foundation of China (No. 71471102), and Key Projects of Science and Technology Research Program of Hubei Provincial Department of Education in China (Grant No. D20101203).

REFERENCES

[1] Chou, Tao.An Empirical Examination of Initial Trust in Mobile Payment [J]. Wireless Personal Communications, 2014, 77(2): 1519-1531.

[2] Yan, Zheng; Liu, Conghui; Niemi, Valtteri. Exploring The Impact of Trust Information Visualization on Mobile Application Usage [J]. Personal and Ubiquitous Computing, 2013, 17(6): 1295-1313.

[3] Zhang, Ruidong; Chen, Jim Q; Lee, CaJaejung. The Effects of Location Personalization on Integrity Trust and Integrity distrust in Mobile Merchants [J]. Mobile Commerce and Consumer Privacy Concerns, 2013, 53(4): 31-38.

[4] Wang, Nan; Shen, Xiao-Liang; Sun, Yongqiang.Transitionof Electronic Word-of-Mouth Services From Web to Mobile Context: A Trust Transfer Perspective [J]. Decision Support Systems, 2013, 54(3): 1394-1403.

[5] Liu, bailing; Zheng, D; Shi, J; Zhang, L; Cloud-based Trust Establishment Protocol towards Mobile Commerce [C]. International

Conference on Computer, Networks and Communication Engineering (ICCNCE), 2013: 700-702.

[6] Liu,bailing; Wang, Weijun; Wang, Jianzong. A Negotiation Support Framework for Establishing Bilateral Trust in Mobile Commerce [J]. Information-An International Interdisciplinary Journal, 2010, 15(9): 3841-3847.

[7] Piao, Chunhui; Wang, Shuzhen; Wen, Jie. Mobile Commerce Trust Model and Its Application for Third Party Trust Service Platform [C].IEEE 14th International Conference on Commerce and Enterprise Computing (CEC), 2010:120-125.

[8] Piao, Chunhui; Wang, Shuzhen; Yang, Fengtao. Research on Trust Evaluating Model for Mobile Commerce Based on Structural Equation Modeling [C]. 9th IEEE International Conference on e-Business Engineering (ICEBE) / 8th SOAIC / 6th EM2I / 6th SOKMBI / 4th ASOC, 2012: 25-32.

[9] Bhalaji, N; Shanmugam, A. Dynamic Trust Based Method to Mitigate Greyhole Attack in Mobile Adhoc Networks [C]. International Conference on Communication Technology and System Design, 2012, 30: 881-888.

International Conference on Power Electronics and Energy Engineering (PEEE 2015)

A Novel Method for 3D Morphing by Deformation Matrix with Triangle Meshes

C.L. Peng, T.W. Xing, Y. Yu, Y. Zhou, S.D. Du
School of Electronic Science and Engineering
Nanjing University
Nanjing, China

Abstract—**We introduce a 3D morphing method which generates a merged model given a series of triangle meshes. Our morphing, based on a set of parameters between the source and target shapes, can show the process of the transformation from the source to the target smoothly. We choose a model as our reference mesh, and obtain corresponding unified models from other models which may have different number of vertices or facets. Given these unified models, parameters between any two meshes can be computed integrally or separately for each rigid part. Different forms of combination of the parameters can generate different merged models. To address the collapsed situation happened occasionally, shape and pose morphing are separated for some parts in our work. By merging different parts of different models, we can get a merged shape, e.g. an animal with the horse head and the cat body. As an application of our 3D morphing method, quantifying the difference between any two models can be done efficiently, represented by the distance between any two sets of low-dimensional parameters reduced from the initial parameters using Principal Component Analysis (PCA). Character replacement and model driven are another two applications. Characters in 2D images are used to guide our morphing work and depth image sequence is used to drive our merged model to show the same pose as the character in the sequence respectively.**

Keywords-3D morphing; difference quantifying; character replacement; model driven

I. INTRODUCTION

A variety of related work has been done in the morphing of 3D models. One popular approach in [1] is sphere embedding. Athanasiadis *et al.*[2] made some improvements using an optimization technique, a mapping geometrically similar to the original object while preserving connectivity and topology can be obtained. Kraevoy *et al.*[3] propose another method by constructing a common base domain based on his another work, Matchmaker [4]. Given the common base domains, after initial cross-parameterization and compatible remeshing, a final new mesh can be generated by smoothing and refinement. Bronstein*et al.*[5] explore the problem of similarity criteria between nonrigid shapes. Assuming the models have point-to-point correspondences, Allen*et al.*[6] do the morphing between two models by taking linear combination of the vertices.

Different human models have similar scale, while different animals may have a huge difference in shape and size. We take an inspiration in the work of deformation transfer by Sumner and Popović [7]. They use a 3×3 matrix to represent the deformation of each source triangle to the corresponding target. We use this transformation matrix to address the task of 3D morphing.Getting the models with the point-to-point correspondences is just like to warp the template mesh onto the targets. Kaick*et al.*[8] review a series of methods to compute correspondences between geometric shapes. With a model as the deformable template and the other models as the targets, the classical method is an extending Iterative Closest Points(ICP) algorithm for modeling non-rigid objects,such as [9, 10]. However, this method is not stable and good results are yielded only when the inter-frame deformations are small.

In this paper, an efficient 3D morphing method is introduced.With these unified models, we can compute the transformation matrices from the source model to the target,using the method similar to SCAPE[11].Given the transformation and a source model, 3D morphing can be done smoothly.Merging different parts which belong to different models is included in our work,e.g. a shape which consists of a human body and a cat head.The other two applications of our method are character replacement in 2D imageand driving a merged model by a motion sequence.

Our 3D morphing method has two notable contributions:

- The new model generated by using our morphing method maintains the point-to-point correspondences with the template mesh, without increasing the number of mesh vertices and facets.

- Character replacement in an image and model driven by a depth sequence can be done with the help of our 3D morphing method.

II. METHOD

One of the main purposes of our approach is to deform the source shape to the target smoothlyand another is to generate a merged model. By adding sparse markers on any two models, deformation transfer can morph one triangulated mesh to another gradually in the process of an iterated closet point algorithm with regularization. Assuming the meshes have point-to-point correspondences, affine transformation matrices between the paired triangles can be computed like SCAPE. Taken horse and cat as an example, as the the proportion of cat's transformation parameters is increasing, the model generated is more like a cat.

Our database mainly comes from a CG software called Poser. SCAPE datasets are also included.Every model in

© 2015. The authors - Published by Atlantis Press

SCAPE datasets have 12500 vertices and 25000 triangular facets, with point-to-point correspondences.

A. Definitions

The terms used in our method are defined in this way that our datasets consist of a source model S and a set of targets T $=\{T^1, ..., T^N\}$. The source mesh S $= \{V_S, F_S\}$ has a set of vertices $V_S = \{v_1, ..., v_j\}$ and a set of triangular facets $F_S = \{f_1, ..., f_k\}$.

B. Transformation Parameters

Assuming our models have correspondence between each pair of triangular facets, transformation parameters can be computed in the following method. Let the paired triangular facets f_k and $\widehat{f_k}$ have the vertices $v_{k,1}$, $v_{k,2}$, $v_{k,3}$ and $\widehat{v_{k,1}}$, $\widehat{v_{k,2}}, \widehat{v_{k,3}}$. Since we just want to use a affine transformation defined by a 3×3 matrix Q, not using a displacement vector d, deformations are applied in the local coordi- nate system. We represent the deformation as $\widehat{V} = Q \times V$ where $V = [v_2 - v_1, v_3 - v_1]$, $\widehat{V} = [\widehat{v_2} - \widehat{v_1}, \widehat{v_3} - \widehat{v_1}]$. Since it is not enough to determine the affine transformation matrices, we add a regularization to constrain the problem following [6, 7]. With a smoothness constraint which indicates the transforms should be similar in adjacent triangular facets, this problem can be described as the following form:

$$\min_{Q_1,...,Q_N} \sum_{k=1}^{N} \sum_{i=2,3} \left\| Q_k V_{k,i} - \widehat{V_{k,i}} \right\|^2 + w_s \sum_{k=1}^{N} \sum_{j \in adj(k)} \|Q_k - Q_j\|^2 \quad (1)$$

where $V_{k,i} = v_{k,i} - v_{k,1}, i = 2,3$, and N is the number of the triangular facets. This equation can be solved integrally for the entire shape or separately for each rigid part by using the least squares method.

III. SHAPE MORPHING

Given the source model and a set of transformation parameters from the source to the target, 3D morphing can be achieved smoothly. Without loss of generality, each triangular facet f_k can be represented as two vectors $v_{2,1}$, $v_{3,1}$. The generated triangle can be represented as the following formulation:

$$\widehat{V_{k,i}} = (\omega Q_k^m + (1 - \omega)Q_k^n)V_{k,i}, i = 2,3 \quad (2)$$

where Q^m and Q^n belong to any two different models and the weight ω is used to guide the morphing between the two shapes. With the decrement of the proportion one model occupied, the model generated tends to be more like the other. Model size and orientation should be unified firstly. For convenience, we adjust the model so that their boundary is a unit cube and the orientation is along the z-axis. And then we can define a function whose dependent variable is the parameter ω and independent variable is the distance from the current facet to a defined plane, denoted by z-axis coordinate in this case.

As stated above, what we have now are a series of triangular facets. Each one is denoted by two vectors, not three vertices. Generally speaking, the number of the facets in a model is about twice as the number of vertices. So as to resolve the vertices, a

start vertex should be defined firstly. And then this problem can be described as the following form:

$$\min_{v_1,...,v_{N_v}} \sum_{k=1}^{N_f} \left\| (v_{k,i} - v_{k,1}) - V_{k,i} \right\|^2, i = 2,3 \quad (3)$$

where N_v and N_f denote the number of the vertices and facets respectively. This overconstraint problem can also be solved by using the least squares method. During the process of the morphing, interpenetration between different parts sometimes occurs. Figure1(a) (left) shows a situation where the ear goes into the head when a dog model transforms to a cat. Since dog ear is different from cat both in shape and orientation, merging these two parts may cause distortion if the proportion distribution is not reasonable. This problem can be solved by a refinement method introduced in DRAPE [12]. A penalty function is defined in the following form:

$$p(E) = \sum_{v_i \in E} \|\epsilon + n_{v_j}(v_i - v_j)\|^2 \quad (4)$$

where E is the set of vertices belonging to ear, v_j is the nearest vertex in the head to v_i and n_{v_j} is the normal of the vertex v_j. ϵ is a variable which ensures the ear vertices lie outside the head. To regularize the solution, two additional terms are added which named smooth warping and damping:

$$s(E) = \sum_{v_i \in E} \left\| (v_i - \widehat{v_i}) - \frac{1}{|N_i|} \sum_{j \in N_i} (v_i - v_j) \right\|^2 \quad (5)$$

$$d(E) = \sum_{v_i \in E} \|v_i - \widehat{v_i}\|^2 \quad (6)$$

where $\widehat{v_i}$ is the vertices of the ear before transformation and N_i is the set of vertices adjacent to v_i. Then our object function can be defined as:

$$\min_{v_i \in E} p(E) + \omega_s s(E) + \omega_d d(E) \quad (7)$$

This equation can be solved using the least squares method iteratively. Figure 1(a) shows the process where the ear goes out of the head gradually.

The essence of this problem is that ears of dog and horse are not in the same posture. If we treat the ear as a rigid part, a rotation matrix R can be computed. By using this matrix, the dog ear can be rotated to the same orientation as the horse ear. And then shape deformation matrices Q_k can be computed. By this way, pose and shape morphing can be separated. The generated triangle can be represented as the following formulation:

$$\widehat{V_{k,i}} = \text{Rotate}\left((\omega Q_k^m + (1 - \omega)Q_k^n)V_{k,i}, q(\omega) \right), i = 2,3 \quad (8)$$

where $q(\omega)$ is a quaternion used to represent the rotation and Rotate is a function defined to rotate the vector by the quaternion. The rotation matrix R can be converted to a quaternion qi, and $q(\omega)$ is a function of ω, interpolated between qi and $[1,0,0,0]$. Since quaternion is suitable for rotation interpolation, $q(\omega)$ can be computed easily by linear interpolation or spherical linear interpolation. As is illustrated in Figure 1 (b), pose and shape of the ear are merged separately.

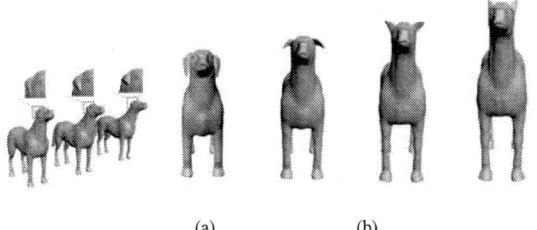

(a)   (b)

FIGURE I. (A): MORPHING PROCESS OF DOG EAR; (B): TRANSFORMING A DOG TO A HORSE.

A. Shape Merging

Our method can merge any number of models, but for convenience we take the merging of two models for example. Figure 2 (a) shows a merged shape whose head is from a horse and other parts are from a cat. To get such a model, we need to determine firstly which model every part should belong to. Then conjunction between different parts should be smoothed. As the new model is generated by using the least squares method, this problem has been eased to a certain extent, but what we want is a smooth model. To address this defect, the mutation at the conjunction can not be allowed. So we apply a method to guide this smooth transition. As is shown in Figure 2(c), we firstly get the user-defined conjunction by labeling an area on the surface mesh manually. Then boundary of this area can be computed easily since these triangular facets are shared by two different parts. After that, the Dijkstra algorithm is applied to compute the geodesic distance from every vertex in the labeled area to the boundary. Figure 2(d) shows the result by labeling gradient colors on the mesh to represent the distances. A linear function is used to determine the proportions the two models account for respectively based on the distances from the current facets to the boundary.

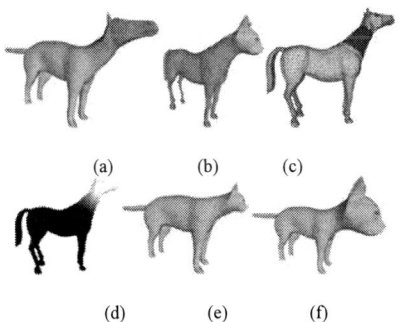

(a)   (b)   (c)

(d)   (e)   (f)

FIGURE II. (A, B): PARTS MERGING; (c, d): CONJUNCTION SMOOTH METHOD; (E, F): SIZE DEFORMATION.

Size deformation is also included in our work. In Figure 2 (e, f), the cat with a small head or a big head is shown.

B. Consensus Correspondence

Our work above is based on the unified models which have point-to-point correspondences. But the models we have in hand do not have this characteristic. So a correspondence system introduced by [6, 7] has been taken. It is just like to wrap a source surface onto the target. We choose the horse model as the source mesh which has N vertices, $|T|$ facets and the other models as targets. To compute the deformed vertices $\widehat{v_1},...,\widehat{v_N}$, a set of affine transformations T_i, $i \in [1...|T|]$ are defined to minimize the object function:

$$E = \omega_S E_S + \omega_I E_I + \omega_C E_C + \omega_M E_M \qquad (9)$$

where $E_S = \sum_{i=1}^{|T|} \sum_{j \in adj(i)} ||T_i - T_j||^2$, $E_I = \sum_{i=1}^{|T|} ||T_i - I||_F^2$, $E_C = \sum_{i=1}^{N} ||v_i - c_i||^2$, $E_M = \sum_{i=1}^{m} ||\widehat{v_{s_k}} - m_k||^2$. E_S is defined to achieve a smooth deformation. This regularization can ensure the transformations for the adjacent facets to be similar when the source mesh transforms to the target. E_I, deformation identity, is defined to make the transformations similar to a identity matrix. This term can prevent the drastic change in the mesh caused by the deformation smoothness term. E_C is defined to represent the distance between each vertex in the source and the closest vertex in the target. To avoid the local minima, a marker error E_M is introduced to guide the deformation. Since corresponding vertices should be found correctly in E_C, we need to make the two models in a similar shape and position firstly. To achieve this purpose, sparse markers need to be added onto the models manually. s_k is the index for marker k on the source mesh and m_k is the corresponding marker on the target mesh.

An iterative approach is used to minimize this energy function. In the first iteration, we ignore the distance error between closest vertices by setting the weights $\omega_S = 1$, $\omega_C =0$, $\omega_M = 10$. The marker error is the dominant constraint to guide the deformation in this phase. After this phase, the markers in the source mesh are transformed to the position of the corresponding markers in the target mesh. And the other vertices move to the target mesh restricted by the smoothness error E_s. Then a set of closest vertices can be computed using this model and the target. In the second phase, the optimization problem is solved by increasing the ω_C from 1 to 50 in four steps preserving $\omega_S= 1$, $\omega_M=1$. After each step, a new model which is more like the target will be generated. And we update the closest points for the deformed source mesh from the target mesh. If the normals of the two corresponding vertices are less than 90° in orientation, this pair is valid.

IV. Application

In this section, we will introduce two applications based on our work above. One is to quantifying the differences between different animals, the other is to replace a character in a 2D image.

A. Differences Quantifying

While people can distinguish different animals easily, it is always difficult for a computer to tell the differences between different animals. Inspired by the work of SCAPE, each model can be represented by a column vector β_i. Since we have obtained the transformation matrices Q_k^i for each instance i and triangular facets k, a simple linear subspace which is used to generate Q^i can be estimated by using PCA:

$$Q^i = \varphi_{U,\mu}(\beta^i) = \overline{U\beta^1 + \mu} \qquad (10)$$

where μ is the mean value of the matrix composed of Q_i and U are the first n eigenvectors computed by using PCA.

Unlike SCAPE, what we want to do is not to learn the shape deformation model, but to get a set of low-dimensional parameters to represent our instances. Given these low-dimensional parameters, differences between any two animals can be quantifying. By computing the Euclidean distance between any two sets of the Ndimensional parameters, the difference can be denoted as:

$$D_{i,j} = \sqrt{\sum_{k=1}^{N}(\beta_k^i - \beta_k^j)^2} \qquad (11)$$

where N is the dimension of β. Figure 3 (a) shows the distances from every instance to the horse model assuming each β have 5 degrees of freedom while (b) is a graph which draw the β of seven models in the 3D space. And these animals can be divided into four groups intuitively: Horse and camel, Cat, lion and wolf, Dog and Bear.

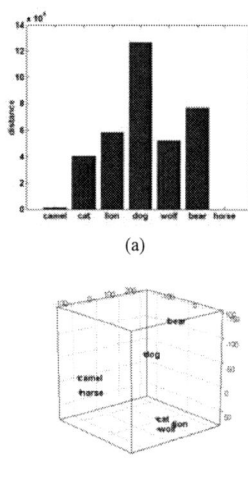

(a)

(b)

FIGURE III. DIFFERENCES QUANTIFYING.

B. Character Replacement

Assuming there is a human in a 2D image and we want to replace it with another character which is composed of a human body and a cat head, what we need to do is to project such a model onto the image. Firstly, we need to get a model whose pose and shape are the same as the character in the image. Pose estimation can be done accurately by locating the projections of joints of the model on the image [13, 14]. We use the method similar to [13]by labeling a series of markers on the image, the positions of which are the joints of the corresponding 3D model, the pose can be obtained by using the scaled orthographic projection method assuming the size of every rigid part is known.

(a)   (b)   (c)   (d)

FIGURE IV. CHARACTER REPLACEMENT.

FIGURE V. MORPHING FROM A HUMAN HEAD TO A CAT HEAD.

Then our 3D morphing method can be used to get a merged shape. Figure 4 shows the result: a shape composed of a human body and a cat head. The human model comes from Poser. The trunk and limbs are completely from the human model, while the head is completely from the cat. The process of morphing is shown in Figure 5.

V. SUMMARY

In this paper, we have introduced a method to do the 3D morphing for any number of unified models by using the transformation parameters between them. Rules for this morphing can be defined by users regardless you want to merge the models integrally or separately. Differences quantifying and character replacement are two major applications of our work.

ACKNOWLEDGMENTS

This work is supported in part by the National Natural Foundation of China under Grant No. 61271231 and Jiangsu Natural Science Foundation under Grant No. BK2011337.

REFERENCES

[1] M. Alexa, Recent advances in mesh morphing. COMPUTER GRAPHICS Forum 200, 3, 1–23. (2002)

[2] T. Athanasiadis, I. Fudos, C. Nikou, and V. Stamati, Feature-based 3D morphing based on geometrically constrained sphere mapping optimization. Proceedings of the 2010 ACM Symposium on Applied Computing, ACM, 1258-1265.

[3] V. Kraevoy, A. Sheffer, Cross-parameterization and compatible remeshing of 3D models. ACM Transations on Graphics 23, 3, 861-869. (2004)

[4] V. Kraevoy, A. Sheffer and C. Gotsman, Matchmaker: constructing constrained texture maps. ACM Transactions on Graphics 22, 3, 326-333. (2003)

[5] A.M. Bronstein, M. M. Bronstein and R. Kimmel, Topology-invariant similarity of nonrigid shapes. International journal of computer vision 81, 3, 281-301. (2009)

[6] B. Allen and B. Curless and Z. Popović, The space of human body shapes: Reconstruction and parameterization from range scans. ACM Transactions on Graphics 21, 3, 587–594.(2003)

[7] R. W. Sumner and J. Popović, Deformation Transfer for Triangle Meshes. Proceedings of ACM SIGGRAPH 2004 23, 3, 399-305.

[8] O. Van Kaick, H. Zhang, G. Hamarneh and D. Cohenor, A survey on shape correspondence. In Computer Graphics Forum, vol. 30, Wiley Online Library, 1681–1707. (2011)

[9] D. Hahnel, S. Thrun and W. Burgard, An extension of the ICP algorithm for modeling nonrigidobjects with mobile robots.In Proc. IJCAI, Acapulco, Mexico (2003)

[10] J. Sümuth, M. Winter and G. Greiner, Reconstructing Animated Meshes from Time-Varying Point Clouds. In Computer Graphics Forum, vol. 27, Wiley Online Library, 1469–1476.

[11] D. Anguilov, P. Srinivasan, D. Koller, S. Thrun and J. Rodgers, SCAPE: Shape completion and animation of people. ACM Transactions on Graphics 24, 4, 408-416. (2005)

[12] P. Guan, L. Reiss, D. A. Hirshberg, A. Weiss and M. J. Black, DRAPE: DRessing Any Person. ACM Transations on Graphics 31, 4 (2012)

[13] C. J. Taylor, Reconstruction of articulated objects from point correspondences in a single uncalibrated image. CVIU 80, 10, 349–363. (2000)

[14] A. Hornung, E. Dekkers and L. Kobbelt, Character animation from 2D pictures and 3D motion data. ACM Transactions on Graphics (TOG) 26, 1, 1 (2007)

International Conference on Power Electronics and Energy Engineering (PEEE 2015)

A Method of Public Policy Refinement Based on OWL and Linear Temporal Logic

D.P. Lang, S.B. Huang, L.S. Shen, T. Zhang, H. Chen

Computer Science and Technology
Harbin Engineering University
Harbin City, Heilongjiang Province, China

Abstract—**It's the consistency, safety and sustainability of public policy that to some extent determines the effectiveness of policy implementation. This paper focuses on whether we can use formal verification methods to verify the policy implementation consistent with their objectives. Based on a number of policy refinement methods, this paper uses ontology language OWL to break the objectives of public policy into several objects which can be represented by linear temporal logic in the second step for their logical reasons. At last model checking technology is applied to an example to verify the entire process. Then we give the performance analysis of the method and prospects.**

Keywords-public policy; refinement; OWL; LTL; model checking

I. INTRODUCTION

A. Public Policy and Policy Analysis

Policy science is an interdisciplinary and comprehensive field of research after World War II in western countries, which was taken as a science revolution in the development of western social science. Late 1970s, along with the historical process of reform and opening up, policy analysis attracted scholars and policy researchers' attention in China.

Policy analysis, also known as policy science [3] is a policy research, formulation, analysis, screening, implementation and evaluation of the whole process of research methods. Back in the 1940s, the American political scientist Harold Lasswell has proposed a *policy science* concept [4]. Then a Israeli scholar Yehezhel Dror wrote the famous trilogy of science policy: The re-examination of public policy, Policy Sciences Vision, Policy Sciences exploration, which was considered the second milestone of policy science.

Policy analysis being developed into an independent social science, deductive reasoning, experience, analysis and other artificial means to determine the principles are still the main methods of policy analysis. As the rapid development of computer and information technology, formalization and formal verification came in to the map.

B. Model Checking Technology

Model checking was proposed by Clarke Emerson [5], Quielle and Sifaki [6] for automatic verification of finite state concurrent systems. Model checking has been successfully applied to computer hardware, communication protocols, control systems, and other aspects of security authentication protocol analysis and verification. Current research shifts to the need for verifying more complex software system. Purposes of software model checking is to extend the field of application of model checking techniques, which applies its reasoning to the program to maximum the automation of verification process.

II. MOTIVATIONS

A. General Idea of Policy Refinement

Policies can be defined on different levels, ranging from high-level business objectives to a single resource policy. Hopefully we can directly or indirectly gain the system policy that is lower policy from the overall goals that are high-level policy. There are also many progresses in this area. With the increasing complexity of the network, the management of related systems becomes increasingly difficult. Managing such a complex and distributed systems faces a series of challenges. Among these challenges, one is how to ensure the operation of the system is consistent to a given overall objectives. In order to solve this problem have been proposed policy-based management system [7].

In a policy-based management system, the expected behaviors are defined in the policy which can be defined as a clear goal or guide of current or future action method [8]. In other words a policy is a set of rules that are used to describe how to achieve an expected behavior. This can be accomplished by policy refinement. The process is to map a higher-level policy to the lower-level policy goals.

B. Use of Model Checking

The essence of policy refinement is to give an accurate format description of the policy through mathematical methods, procedures and methods, formal methods, What we do is to facilitate following verification and implementation. This paper borrows the definitions of policy refinement in the references [9] by Moffett and Sloman. Then we clarify the process of refinement and objectives:

(1) Clearly state all the resources needed in public policy analysis

(2) During the process of transmitting the high-level policy into the low-level policies, the meta policy that composes of low-level policy and can be used to verify properties should be explicitly stated.

© 2015. The authors - Published by Atlantis Press

(3) Verify whether the transformed low-rise policies consistent with high-level policy, in other words we have to verify the properties and specification that the low-level policies satisfy consistent with the ones that high-level policy satisfies.

The concept ontology language OWL mentioned in the second step will be used to describe the meta policy and given in the following section; this paper uses linear temporal logic (LTL, given in Section III) to show the logical relationship among meta policies, namely the lower logical derivation relations in the policies.

III. PRELIMINARIES

A. Policy Refinement

Reference [1] uses formal techniques from KAOS to achieve target refinement, each of which represents a temporal logic rules. And the paper used the refined model to transform the original goals into logical included sub-goals. A set of refined goals is generated after the process. Key step in policy refinement process is to describe the high-level policy objectives using low-level goals which can be achieved by systems. In the reference [5], goal elaboration is achieved by using temporal logic and related evidence to detect the problems that may arise during the process of refinement.

Javier in the paper [9] proposed a policy refinement method based on goal-oriented requirements engineering and module testing technology. The paper derived low-level policy containing logically high-level management objectives through KAOS goal refinement methods.

B. Semnatics of LTL

Linear temporal logic combines propositional logic with timing operations. The analysis method based on LTL is an important formal analysis method that performs validation and description of software architecture characteristics. It takes a path (sequence of states) as an assertion object, whose true value will be confirmed in the state sequence. Linear temporal logic can be easily and accurately used to describe important properties of concurrent systems, such as safety and liveness. Safety is about bad things will never happen; Liveness is about good things will eventually happen. There are syntax and semantics of linear temporal logic from Pnueli's work [8] below.

Definition 1: Grammar of LTL can be recursively defined as follows:

1)Atomic propositional variables $p, q, r, \cdots$ are linear temporal logic formulas.

2)If p, q are linear temporal logic formulas, then the formulas built by Boolean operators such as $\vee$ (or), $\wedge$ (and), $\neg$ (not) or timing operator $\cup$ (until), $\Diamond$ (finally), $\Box$ (always), X (next) are also temporal logic formulas like $p \vee q$, $p \wedge q$, $\neg q$, $p \cup q$, $p \Diamond q$, $p \Box q$ and $p X q$.

IV. AN EXAMPLE OF POLICY REFINEMENT

A. Building the Prototype

Policy refinement has long been considered to be a daunting task. One of the reasons is that the process requires a lot of manual intervention. With the popularity of ontology representation language OWL, people consider the use of the ontology to describe policies and regulations, such as the use OWL to formalize the policy.

In this section we are going to give an example to illustrate the policy refinement method. The example only refine the constraint conditions of the policy and similar to the rest of the policy refinement. We want to define such a policy: When employees retire, if the payment period is longer than 15 years he has the right to enjoy the pension; otherwise he has to fill the difference amounts. After finish the process of policy refinement, the relevant information rises (payment and certain conditions of enjoying the treatment).

There are 10 concepts in this simple ontology and 1 data structure that are shown in the figure 1.

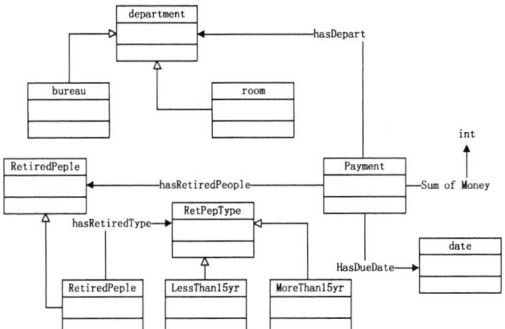

FIGURE I. PAYMENT ONTOLOGY DIAGRAM.

Meanwhile, in this public policy, the following information is also included: the policy subject, the policy object, the policy trigger conditions, the policy constrains and actions. Figure 2 shows the policy ontology diagram.

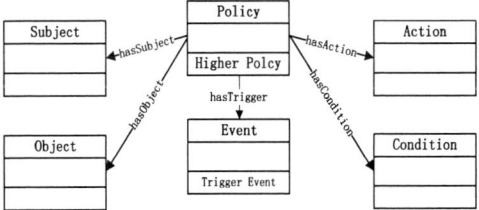

FIGURE II. POLICY ONTOLOGY DIAGRAM.

B. Establishing LTL Formulas

Based on the analysis above, this paper will break down the policy concepts of atomic propositions expressed as follows:

Let P be the meaning that the payment period is greater than or equal to 15 years; then the payment period is less than 15 years would be $\neg P$;

R represents paying the pensions and enjoying pensions is expressed by Q.

So, we want to establish an assertion that if a retired person wants to enjoy the pension, his payment period must be longer than 15 years or he must make a supplementary payment before he enjoys the pension.

$$P\square Q \vee \neg P \wedge R\lozenge Q \vee \neg P \wedge \neg R\square \neg Q \qquad (1)$$

On the last step, we run SPIN as a model checking tool to verify the results in line with expectations. To sum up the policy refinement steps are as follows:

- Goal refinement

The original target should be decomposed into OWL ontology language to describe policy.

- Assign responsibilities to the managed entity

This will assign the refined sub-goals to the appropriate execution module.

- Target implementation

Make sure that the managed entity completes the relevant action.

- Policy encoding

Encode the policy formation expressed in OWL logical language with LTL language

- Model check the LTL formula

Run the SPIN model checker to verify the consistency of formulae with specifications[10].

C. Performance Analysis

Assuming the number of concepts involved in the ontology is m, the concept of number involved in the policy refinement process is n and apparently $m \geq n$. And furthermore the greater m is, the more complex the ontology is, which means we need more time to analyze the ontology. In the policy refinement process, we set up a concept decomposition tree synchronously and the number of nodes of the tree is multiple times of n so the time complexity is $O(n)$.

At the meantime from the aspects of semantic correlation between the concepts, if there are too many levels the semantic relevancy between the low-level constraint set and the original high-level constraint may go relatively lower. This may affect the consistency between refined results and the original target.

V. PROTOTYPE OF THE METHOD

To simulate the whole process of refinement, we need to borrow the definition of agents to represent both sides of policy control. We run the simulation on the platform called JADE which is developed by Java. And based on Protégé OWL API, which we take as semantic reasoning tools to control and analyze the object and subject and the relations between them. The software is demonstrated on Figure 3. And we can insert a new object by insert the necessary agent parameters shown on Figure 4.

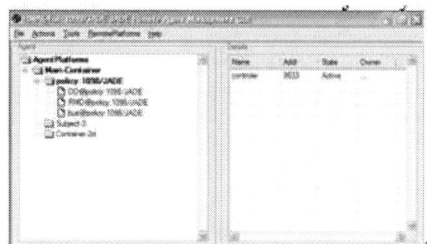

FIGURE III. REMOTE AGENT MANAGEMENT GUI.

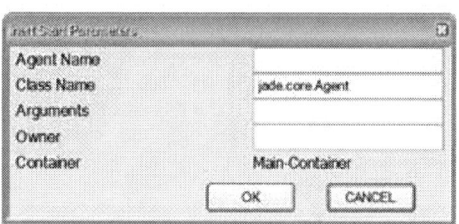

FIGURE IV. INSERT AGENT DIAGRAM.

After we build the policy ontology diagram like what Figure 1 and Figure 2 show, what we have is several group of terms composed of OWL. Next part of this chapter is to demonstrate how to use SPIN verify the model based on OWL.

The basic idea of model checking is to input the specification (1) in the last chapter into the model checker, which is SPIN in this paper.

VI. CONCLUSIONS AND FUTURE WORK

Policy refinement plays a vital role in policy-based management system. This paper borrows the idea of policy refinement and applies the methods to the public policy refinement. Though there exists some policy refinement methods, on the one hand, most of them are confined to a specific application; on the other hand these methods have not been used in the area of public policy analysis. Besides the implementation process is increasingly complex, and the automation is still far from reaching. Whereas we intend to make some progress in the following aspects of work: During the process of policy analysis, we want to set the analysis model and domain knowledge apart, i.e. decoupling. 3. Explore a new language based on OWL-LITE and OWL-DL as a better and more powerful expression tool of public policy.

ACKNOWLEDGMENT

This work is sponsored by National Key Project of Scientific and Technical Supporting Programs under grant number2012BAH08B02, the Fundamental Research Funds for the Central Universities under grant number HEUCFZ1212. Heilongjiang province degree and postgraduate research project of education and teaching reform under grant number JGXM_HLJ_2012036.

REFERENCES

[1] SLOMAN, M.S 1994. Policy Driven Management for Distributed Systems[J]. Journal of Network and Systems Management 2(4):333-360

[2] Arosha K Bandara, A Formal Approach to Analysis and Refinement of Policies. Department of Computing Imperial College London University of London, London, July 2005,pp:14-55

[3] Chen Zhenming. Policy Scienc[M]. Beijing: China Renmin University Press, 1998.

[4] D.Lerner and H.D.LassWell, The Policy Science: Recent Development in Scape and Method, Stanford, CA:Stanford University Press, 1951.

[5] Clarke E, Emerson E. Design and synthesis of synchronization skeletons using branching time temporal logic[J]. Logics of Programs, 1982,pp.52–71.

[6] Queille J, Sifakis J. Specification and verification of concurrent systems in CESAR[A]. International Symposium on Programming[C]. 1982,pp.337–351.

[7] Wies, R., Policies in Network and Systems Management - Formal Definition and Architecture, Journal of Networks and Systems Management, vol. 2, no. 1, pp.63-83, Mar.1994.

[8] Alcantara, O. D. and Sloman, M. QoS Policy Specification A mapping from Ponder to the IETF. 2003.

[9] J. Moffett and M. S. Sloman, Policy Hierarchies for Distributed Systems Management, IEEE JSAC, vol. 11, pp. 1404-14, 1993.

[10] Dapeng Lang, Shaobin Huang, Yuan Cheng, Xinxin Yang and Ya Li, A State Space Abstract Algorithm of Incremental Data Recognition Based on Model Checking, Journal of Computational Information Systems, 10: 4 (2014) 1731-1742

Stochastic Economic Dispatch Using Bacterial Swarm Algorithm

M.S. Li, Y. Hu

School of Electric Power Engineering
South China University of Technology
Guangzhou, China

X. Zhang

The Chinese University of Hong Kong Shenzhen Research
Institute
Shenzhen, China

Abstract—**This paper adopts a stochastic optimization method for solving the Security-Constrained Optimal Power Flow (SCOPF) problem with the consideration of distributed load variations in the grid. The objective function of the dispatch scheme aims to minimize fuel costs of the grid. Compared with conventional dispatch scheme, the computational complexity of proposed method is significantly increased. Therefore, this research adopts an improved Bacterial Swarm Algorithm (BSA) to solve the optimization. Compared with most Evolutionary Algorithms (EAs), BSA is more effective and has better convergence performance. The simulation studies reports the results obtained using an IEEE 30-bus system with uncertain load. A comparison between the results achieved using the proposed method and those obtained from conventional dispatch is given.**

Keywords-economic dispatch; stochastic; bacterial swarm algorithm

I. INTRODUCTION

Economic dispatch has been intensively studied as a network constrained Optimal Power Flow (OPF) problem, since its introduction by Carpenter [1] in 1962. Generally, the economic dispatch problem aims to achieve the minimization of the fuel cost of a model of a power system, by adjusting the control variables, such as power and voltages of each generator, the tap ratios of transformers and the reactive power of volt-ampere reactive of the system, while satisfying a set of operational and physical constraints [2]. As a result, the economic dispatch problem is formulated as a non-linear constrained optimization problem.

Conventional dispatch studies assume that the model of the grid is invariant between the dispatch intervals, which are defined as deterministic dispatch [3]. However, the power system is significantly affected by uncertainty factors, including the renewable energy generators and distributed loads. Environmental uncertainties, such as weather conditions and climate, causes variation of loads. Therefore, it is necessary to develop an economic dispatch scheme for the gird, which is able to deal with the environmental uncertainties [4]. Conventional deterministic dispatch schemes usually consider deterministic objective functions and constraints. In such frames, the load is assumed to be invariant. The power consumed on each bus in deterministic dispatch is considered as a constant value, which is in conflict with actual system [5]. Thus, the control variables obtained is not reliable.

Most recent studies have focused on the uncertainties in the distributed load due to the effects of the unpredictable climate changing. To reduce the influence of such uncertainties, recently proposed stochastic dispatch frameworks minimize the generation cost of the scenario that is most likely to occur the future, and modify the constraints to accommodate other possible scenarios [6]. Some of the existing research has used commitment decisions and implemented multiple stages of preventive and corrective measures to address the uncertainties in the dispatch process. In this paper, we adopt a novel concept of stochastic dispatch, which considers the variations of distributed loads between dispatch actions. Different from deterministic dispatch, stochastic dispatch focuses on simultaneously optimizing the expectation and deviation of the fuel cost of the grid, which avoids the risk of an unpredictable operational status by introducing a mean-variance portfolio [7].

To solve the stochastic dispatch problem, a novel optimization algorithm, BSA, is introduced [8]. The BSA is inspired from the bacterial chemo taxis behavior described in Bacterial Foraging Algorithm (BFA). Moreover, BSA also describes further details of bacterial behaviors, and incorporates the mechanisms of quorum sensing. BSA models two bacterial behaviors: 1) Chemo taxis offer the basic search principle of BSA, which comprises of two basic foraging patterns, tumble and run. The biased random walk performing the local search. In the tumble process, the heading angle of each bacterium is described as a compound angle; 2) Quorum sensing enables BSA to escape from local optima. This is a two-fold operation that can either attract a bacterium to the optimal location or repel it away from the location where bacteria are concentrated. According to previous, BSA demonstrates a superior performance in comparison with other Evolutionary Algorithms (EAs).

II. STOCHASTIC ECONOMIC DISPATCH

The objective function of stochastic dispatch can be formulated as a minimization problem, described as follows:

$$\min F(Y, X) \tag{1}$$

$$\text{s.t. } G(Y, X) = 0 \tag{2}$$

$$H(Y, X) > 0, \tag{3}$$

Where $F(Y, X)$ is the objective function, which is concerned with fuel cost, $G(Y, X)$ is a set of equality constraints, and

© 2015. The authors - Published by Atlantis Press

$H(Y, X)$ is a set of formulated inequality constraints. Y is the vector of dependent variables, which is expressed as:

$$Y^\top = [P_{G_1} V_{L_1} \cdots V_{L_{NG}} Q_{G_1} \cdots Q_{GN_G} S_1 \cdots S_{N_E}], \quad (4)$$

Which includes the slack bus power P_{G_1}, the load bus voltage V_L, generator reactive power outputs Q_G, and the apparent power flow S. X is the set of control variables:

$$X^\top = [P_{G_2} \cdots P_{GN_G} V_{G_1} \cdots V_{GN_G} T_1 \cdots T_{N_T} Q_{C_1} \cdots Q_{CN_C}] \quad (5)$$

Which includes the generator real power output P_G except slack bus P_{G_1}; the generator voltages V_G, the transformer tap setting T, and the reactive power generations of var source Q_C. The detailed notations and formulation for equality constraints and inequality constraints are given in [4].

The objective, F, is formulated to reducing the mathematical expectation and variance of the fuel cost to alleviate the uncertainty in the power system:

$$F = E[f_{cost}] + \lambda_{Var} Var[f_{cost}] \quad (6)$$

Where $E[\cdot]$ is the mathematical expectation of the stochastic function; λ_{Var} is a weight to balance the mathematical expectation and variance; $Var[\cdot]$ is the variance of the stochastic function; and f_{cost} is the fuel cost of the power system. The mathematical expectation of the fuel cost is expressed as:

$$E[f_{cost}] = \frac{1}{N_S} \sum_{i=1}^{N_S} f_{cost_i} \quad (7)$$

where N_S is the number of scenario used in each objective function evaluation, f_{cost_i} is the fuel cost of the power system calculated with the ith scenario. The variance of the fuel cost is expressed as:

$$Var[f_{cost}] = E\left[\left(f_{cost_i} - E[f_{cost}]\right)^2\right] \quad (8)$$

The fuel cost of the ith scenario is calculated as follow:

$$f_{cost_i} = \sum_{j=1}^{N_G} f_{cost_{ij}}, \quad i = 1, 2, \cdots, N_S \quad (9)$$

$$f_{cost_{ij}} = a_j + b_j P_{G_{ij}} + c_j P_{G_{ij}}^2, \quad j = 1, 2, \cdots, N_G \quad (10)$$

In these equations, N_S denotes the number of scenarios, N_G denotes the number of generators, $f_{cost_{ij}}$ is the fuel cost (\$/h) of the jth generator at the ith scenario, a_j, b_j and c_j are fuel cost coefficients, and $P_{G_{ij}}$ is the real power output generated by the jth generator at the ith scenario.

The equality constraints $H(Y, X)$ are the power flow equations:

$$0 = P_{G_i} - P_{D_i} - V_i \sum_{j \in N_i} V_j \left(G_{ij} \cos\theta_{ij} + B_{ij} \sin\theta_{ij}\right) \quad i \in N_0 \quad (11)$$

$$0 = Q_{G_i} - Q_{D_i} - V_i \sum_{j \in N_i} V_j \left(G_{ij} \sin\theta_{ij} + B_{ij} \cos\theta_{ij}\right) \quad i \in N_{PQ}, \quad (12)$$

The inequality constraints $G(Y, X)$ are the limits of the control variables and state variables, which can be formulated as:

$$P_{G_i}^{min} \leq P_{G_i} < P_{G_i}^{max} \quad (13)$$

$$Q_{G_i}^{min} \leq Q_{G_i} < P_{G_i}^{max} \quad (14)$$

$$Q_{C_i}^{min} \leq Q_{C_i} < Q_{C_i}^{max} \quad (15)$$

$$T_k^{min} \leq T_k < T_k^{max} \quad (16)$$

The load variations distributed across the system is described as Gaussian distribution. The real power consumed at the jth load bus is denoted by $\hat{P}_{D_j}$, which is assumed to obey a Gaussian distribution with an expected value of P_{D_j} and a standard deviation of $0.1 P_{D_j}$. The probability density function of the distributed load variations is expressed as follows:

$$P\left(\hat{P}_{D_j}\right) = \frac{1}{\sqrt{2\pi\left(0.1 P_{D_j}\right)^2}} \exp\left(-\frac{\left(\hat{P}_{D_j} - P_{D_j}\right)^2}{2\left(0.1 P_{D_j}\right)^2}\right) \quad (17)$$

III. BACTERIAL SWARM OPTIMIZER

Suppose the pth bacterium, in the tumble-run process of the kth iteration, has a current position X_p^k. The objective of the optimization is to find the minimum of $F\left(X_p^k\right)$. The bacterium also has a rotation angle $\varphi_p^k = \left(\varphi_{p1}^k, \varphi_{p2}^k, \cdots \varphi_{p(n-1)}^k\right)$ and a tumble length $D_p^k\left(\varphi_p^k\right) = \left(d_{p1}^k, d_{p2}^k, \cdots d_{pn}^k\right)$, which can be calculated from φ_p^k via a polar-to-cartesian coordinate transform:

$$d_{p1}^k = \prod_{i=1}^{n-1} \cos\left(\varphi_{pi}^k\right), \quad (18)$$

$$d_{pj}^k = \sin\left(\varphi_{p(j-1)}^k\right) \prod_{i=p}^{n-1} \cos\left(\varphi_{pi}^k\right) j = 2, 3, \cdots, n-1, \quad (19)$$

$$d_{pn}^k = \sin\left(\varphi_{p(n-1)}^k\right) \quad (20)$$

In the tumble-run process of the kth iteration, the pth bacterium generates a random rotation angle, which falls in the range of $[0, \varphi_{max}]$. A tumble action takes place in an angle expressed as:

$$\hat{\varphi}_p^k = \varphi_p^k + \frac{r_1 \varphi_{max}}{2} \quad (21)$$

Where r_1 is a uniform random sequence with a range of $[-1, 1]$. The run action immediately follows the tumble action. Because the run action will be performed more than once, the position X_p^k is recorded as $X_p^{k,0}$, which indicates the position of the pth bacterium at the beginning of the kt iteration.

Once the angle is determined by the tumble step, the bacterium will run for a maximum of N_c run steps. If at the N_f^{th} run step, the bacterium finds a position which has a better fitness value than the current one, the run process also stops. The position of the pthbacterium is updated at the hthrun step in the following way:

$$\hat{X}_p^{k,h} = \hat{X}_p^{k,h-1} + r_2 D_p^k\left(\hat{\varphi}_p^k\right) \quad (21)$$

Where r_2 is a normally distributed random number, and $\hat{X}_p^{k,h}$ is the position of the pth bacterium after the hth run step. For convenience of description, the position of the p thbacterium beginning immediately after the tumble-run process of the kth iteration is denoted by $\hat{X}_p^{k,N_f}$.

Inspired by PSO, the positions of the bacteria moving by attraction are updated as follows:

$$X_p^k = \hat{X}_p^{k,N_f} + r_3\left(X_{best} - \hat{X}_p^{k,N_f}\right) \quad (22)$$

where r_3 is a normally distributed random number with a range of $[-1,1]$, which describes the strength of bacterial attraction, and X_{best} indicates the position of the current best global solution updated after the evaluation of each function.

In BSA, a small number of the bacteria are randomly selected to be repelled. To measure the degree of repelling, a repelling rate is defined by ζ, \emph{i.e.}, in each iteration, 100ζ percent of the bacteria are processed by repelling. Accordingly the attraction rate is $100(1-\zeta)$ percent. The repelling process is based on the random searching principle. If the pth bacterium shifts into the repelling process, a random angle is generated. The bacterium is thereby moved to a random position following this angle in the search space, which can be described as:

$$X_p^k = \hat{X}_p^{k,N_f} + r_4 D_p^k\left(\hat{\varphi}_p^k + \pi/2\right) \quad (23)$$

Where r_4 is a normally distributed random sequence.

IV. SIMULATION STUDIES

The simulation studies are undertaken on the well-studied IEEE 30-bus system. This model represents a portion of the American Electric Power System. The model comprises 30 buses, 6 generators, and 40 branches. The fuel cost coefficients of the generators, given in (10), are listed in Table I. The BSA is evaluated and compared with Genetic Algorithm (GA) [9] and Particle Swarm Optimizer (PSO) [10]. For the PSO parameters, the inertia weight is 0.73, and the acceleration factors are both 2.05, as recommended in [10]. The maximum number of function evaluations for all algorithms is set to 3×10^6. The maximum number of scenario taken in the optimization process, N_S, is set to 200.

TABLE I. FUEL COST COEFFICIENTS IN THE IEEE 30-BUS SYSTEM.

Generator	a	b	c
1	0	2	0.02
2	0	1.75	0.0175
3	0	1	0.0625
4	0	3.25	0.0083
5	0	3	0.025
6	0	3	0.025

Table II lists the mean and variance of the fuel costs obtained by BSA, GA and PSO. The experimental results demonstrated that proposed BSA (807.2853) outperforms GA 809.6017 and PSO 808.6747 on the objective of mean fuel cost. Meanwhile, the fuel cost optimized by BSA also has the smallest standard deviation (10.5451). Thus, the grid operational policy obtained by BSA is robust.

TABLE II. MEAN AND VARIANCE OF THE FUEL COSTS OBTAINED BY STOCHASTIC DISPATCH.

Algorithm	Mean fuel cost ($/h)	Standard deviation of fuel cost ($/h)	Computational time (seconds)
BSA	**807.2853**	**10.5451**	571
GA	809.6017	10.9191	640
PSO	808.6747	11.6947	**544**

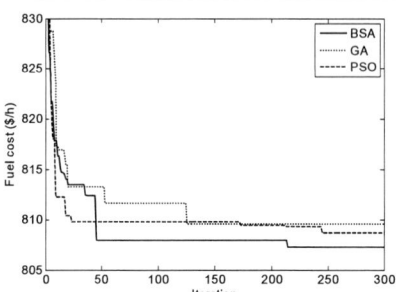

FIGURE I. CONVERGENCE PROGRESS.

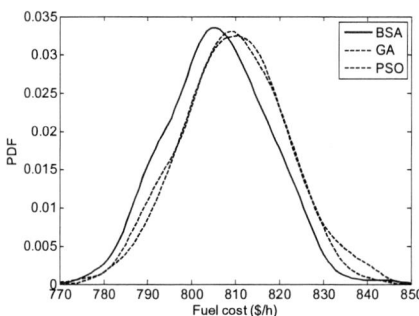

FIGURE II. PDF OF FUEL COST.

The convergence progresses of BSA, GA and PSO are illustrated in Figure 1. It can be found that although BSA converges slowly in the early period of the searching, the quorum sensing prevents the premature result in the optimization, and leads to a better performance in the late stage. Figure 2 shows the distribution of the fuel cost estimated by these three algorithms. The fuel cost estimated by BSA has a small mean value and standard deviation.

V. CONCLUSION

This paper has adopted a novel stochastic model for economic dispatch in an environment that considers distributed load uncertainties. Simulation studies have been conducted on an IEEE 30-bus system with the simultaneous objectives of minimizing the mean value and the standard deviation of fuel cost. The simulation results indicate that BSA provides a more reliable solution set for power system dispatch than GA and PSO due to its excellent convergence performance.

ACKNOWLEDGEMENTS

This research is jointly supported by National Natural Science Foundation of China (51307062) and Guangdong Innovative Research Team Program (No. 201001N0104744201).

REFERENCES

[1] Carpentier, J., Contribution to the economic dispatch problem, Bull. Soc.Franc. Elect, 1962, pp. 431-447.

[2] Yu Xia; Ghiocel, S.G.; Dotta, D.; Shawhan, D.; Kindle, A.; Chow, J.H., "A Simultaneous Perturbation Approach for Solving Economic Dispatch Problems With Emission, Storage, and Network Constraints," Smart Grid, IEEE Transactions on , vol.4, no.4, pp.2356,2363, Dec. 2013

[3] Chamba, M.S.; Ano, O., "Economic Dispatch of Energy and Reserve in Competitive Markets Using Meta-heuristic Algorithms," Latin America Transactions, IEEE (Revista IEEE America Latina) , vol.11, no.1, pp.473,478, Feb. 2013

[4] Li, M.S.; Ji, T.Y.; Wu, Q.H.; Xue, Y.S., "Stochastic Optimal Power Flow using a Paired-Bacteria Optimizer," Power and Energy Society General Meeting, 2010 IEEE , vol., no., pp.1,7, 25-29 July 2010

[5] Chang, C.S.; Fu, W., "Stochastic multiobjective generation dispatch of combined heat and power systems," Generation, Transmission and Distribution, IEE Proceedings- , vol.145, no.5, pp.583,591, Sep 1998

[6] M.S. Li, Q.H. Wu, T.Y. Ji, H. Rao, Stochastic multi-objective optimization for economic-emission dispatch with uncertain wind power and distributed loads, Electric Power Systems Research, Volume 116, November 2014, Pages 367-373

[7] Xiangyu Cui; Xun Li; Duan Li, "Unified Framework of Mean-Field Formulations for Optimal Multi-Period Mean-Variance Portfolio Selection," Automatic Control, IEEE Transactions on , vol.59, no.7, pp.1833,1844, July 2014

[8] M.S. Li, T.Y. Ji, W.J. Tang, Q.H. Wu, J.R. Saunders, Bacterial foraging algorithm with varying population, Biosystems, Volume 100, Issue 3, June 2010, Pages 185-197

[9] Leung, H.C.; Chung, T.S., "Optimal power flow with a versatile FACTS controller by genetic algorithm approach," Advances in Power System Control, Operation and Management, 2000. APSCOM-00. 2000 International Conference on , vol.1, no., pp.178,183 vol.1, 30 Oct.-1 Nov. 2000

[10] Clerc, M.; Kennedy, J., "The particle swarm - explosion, stability, and convergence in a multidimensional complex space," Evolutionary Computation, IEEE Transactions on , vol.6, no.1, pp.58,73, Feb 2002

International Conference on Power Electronics and Energy Engineering (PEEE 2015)

Study on the Comprehension Difference between Managers and Front-Line Employees

W. Jiang, Z.M. Zhu, L.N. Li

Department of Resources and Safety Engineering
China University of Mining & Technology (Beijing)
Beijing, China

Abstract—The safety signs are widely used in safety management and daily life, which is applied to instruct and manage people, safety behavior. In this paper, the author use the questionnaire survey to research the comprehension of safety signs in the workplace among managers and front-line employee in machining workshop from 6 indications namely attention, hazard perception, safety note, safety knowledge, safety awareness, safety habits. Then the author obtains some conclusions: the vast majority (over 60%) of employees consider safety signs in the workplace are useful; over 80% of managers and front-line employees pay attention to the safety signs; over 70% of managers and front-line employees consider the safety signs really work; the front-line employees prefer to care about the effect of the safety signs can improve their hazard perception; over 65% of managers and front-line employees agree that the safety signs in the workplace have improved their safety knowledge; over 75% of managers and front-line employees think that have increase their safety awareness; over 80% of them consider the safety signs help to cultivate safety habits.

Keywords-comprehension difference; managers; front-line employees

I. INTRODUCTION

Safety sign can warn people potential unsafe factors more visually, and guide them to take reaction rapidly to dangerous accidents and environment in order to accidents happening[1,2].So safety signs are widely used in the safety management of the production and daily life, and to guide and control behavior of people[3].In this paper, the author studies the comprehension difference between managers and front-line employees and whether safety signs take effect or not.

II. INTRODUCTION OF THE QUESTIONNAIRE

A. The Questionnaire

The author researches the comprehension difference of safety signs between managers and front-line employee from six indications, namely attention, hazard perception, safety note, safety knowledge, safety awareness, safety habits, and six questions are set for the six indications. Then a questionnaire survey was conducted among employee in a coal machining welding workshop.

TABLE I. THE CONTENT OF QUESTIONS.

No.	Indication	the content
1	attention	attention people paid to the safety signs
2	safety note	the effect of the safety tips
3	hazard perception	the effect of hazard perception
4	safety knowledge	how much the safety signs increase person's safety knowledge
5	safety awareness	how much the safety signs improve safety awareness of human
6	safety habits	how much the safety signs improve people's safety habits

The safety signs used in the survey safety signs in the mine workplace. Each title has three options: the option of A is on the positive attitude towards the relevant evaluation indication; and B is the opinion that evaluation indication of safety signs takes effect but not great; C is on the negative attitude towards the relevant evaluation indication. Taking the evaluation indication of safety knowledge as an example, and the question is as following:

Do you think safety signs increase your safety knowledge?

A: Yes, and the effect is great.

B: Have effect but not great.

C: Do not take effect.

B. The Staff Being Surveyed

The total number of the employee in the repairing factory is 60, and the number of employee in the machining welding workshop is only 30, so send out 60 questionnaires. All the employee in the machining welding workshop took the questionnaires and those left questionnaires were sent out outside the machining welding workshop.

The investigators in this survey are mainly divided into two categories: the front-line employee, who work in the machining welding workshop; managers, the direct leadership of front-line employee or responsible for daily life of the workshop, such as the director, deputy director and engineers. The random sampling has been conducted to determine the sample, and 60 people has been selected: 52 front-line employee and 8 managers, as following:

© 2015. The authors - Published by Atlantis Press

TABLE II. THE STRUCTURE PROPORTION OF THE SAMPLE.

categories	number	Percentage(%)
front-line employee	52	86.7
managers	8	13.3
total	60	100.0

III. ANALYSIS OF SURVEY RESULTS

Results of the survey are drawn in a curves, from which it is convenient to see the whole results of the front-line employee and managers, and the contradistinction of them. And then it can be seen the advantages and disadvantages of safety signs and get targeted suggestions in the further development.

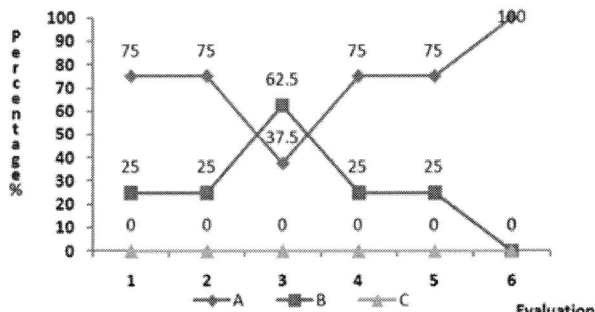

FIGURE I. THE CURVE OF THE RESULTS FROM MANAGERS.

The reason to distinguish different categories is that staffs in different levels hold different comprehension on the concept of safety culture ,safety management systems[4].Managers may pay more attention to safety culture and safety management of the corporate, and what they consider is how to instill the idea of safety to employee in order to increase their safety performance and avoid accidents. However, the front-line employee may pay more attention to the job site, facilities and equipment, and salary. The difference above to safety signs is that managers force on whether safety signs reasonable and appropriate, and whether safety signs can control safety behaviors; while employee force whether safety signs relate to their facilities and equipment and process. The different focus point dues to different survey results.

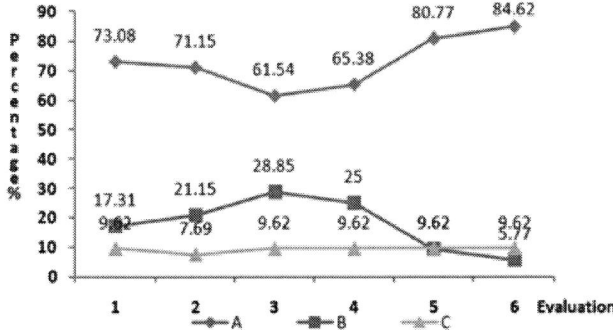

FIGURE II. THE CURVE OF THE RESULTS FROM FRONT-LINE EMPLOYEES.

Through the statistics above, we can see that A is elected more than B and C almost in every question, which means safety signs in the workplace are considered useful from the overall analysis.

From the statistical linear, their is no one in management level to select C, which means the attitude such as "don't know"," don't understand", "don't care" do not exist among managers and they emphasis on safety work and they think whether safety signs reasonable or not can change behaviors of the front-line employees. However ,this precisely reflects the function of safety signs for management-- improving the management's attention and support for safety signs, and strength the communication between managers and front-line employees, and this one-way, unequal communication has more influence on employees[5].However, analyzing the linear trend of option A among the front-line employees, the linear trend of option A of question 3 (hazard perception) fall down, which means people can't exactly know what danger hidden in the operating environment and what illegal behaviors should be avoided. Because some employees did not receive relevant training about safety signs and they did not research safety signs, then they cannot combine safety signs with work environment and operation process appropriately. The influence of management on the front-line employees more reflect on the recessive factors.

However, analyzing the linear trend of option A among the front-line employees, the linear trend of option A of question 3 (hazard perception) fall down, which means people can't exactly know what danger hidden in the operating environment and what illegal behaviors should be avoided. Because some employees did not receive relevant training about safety signs and they did not research safety signs, then they cannot combine safety signs with work environment and operation process appropriately. The influence of management on the front-line employees more reflect on the recessive factors[6]-- the initial psychological state is improved when employees note safety signs, and then they pay more attention to safety signs, and their safety knowledge increases, and their safety awareness enhances, and form the habit of regulation operation.

IV. ANALYSIS OF THE QUESTIONNAIRE SURVEY INDICATIONS

From the concern extent index-personal comparison chart, we can conclude that managerial personnel and front-line staff are basically identical for the indicators of concerning on the safety sign ,the option of A(often focusing on signs)close to 80% demonstrates that safety sign system is concerned by staff ,and the situation that some workers select option C (paying no attention to signs), apart from the reason of working pressure and compliance costs, are also related to enterprise's unsound safety management system, the deletion of the safety training, examination and review. To improve the situation, we should construct the enterprise safety management system at first.

It can be seen from the safety-note indication-personal comparison chart that managerial personnel and front-line staff are basically identical for the indicators of function of the safety sign as a reminder, the number of their choice of option A (safety signs play a great role) and the option B(safety signs play some role as a reminder) is almost same, which indicates that safety signs at workplace is noticed ,translated by personals and do play a role of guiding and alerting employees , meaning that safety sign system is really useful.

From the hazard- perception indication-personal comparison chart ,we can draw a conclusion that managerial personnel and front-line employee have great differences in risk perception perception, personal linear trend is partial normal, option A(one can exactly perceive the security risk in the workplace and illegal behavior) is much higher than option B(one can generally understand) and option C(one can't perceive),The investigation, interview shows that the experience of front-liner employees working in the machine workshop is almost more than 2 years and half of them is more than 10 years, therefore, they have their own perception and understanding for the equipment, operation flow, dangerous and harmful factors. While the phenomenon that the number of option B selected in the management is high is very normal. Because the management concerns the whole workshop from an overall perspective and have a weaker risk perception than front-line employees, the attention and support of safety for managements should be strengthened.

From the safety-knowledge indication-personal comparison chart, it informs that management and front-line staff's view is essentially identical for the issue that whether the safety signs increase the safety knowledge. The conclusion that the number of option A selected (safety signs effectively increase the safety knowledge) is much higher than option B(safety signs increase some safety knowledge) and C(safety signs don't increase the safety knowledge) demonstrates that safety signs system in the workplace is useful and do play a supposed role ,and it extend the safety knowledge of employees and make them adhere to instructions of the safety signs in the daily work voluntarily .

From the safety-awareness indication-personal comparison chart, it informs that management and front-line staff's view is essentially the same for the issue that whether the safety signs increase the safety awareness. The number of option A selected (safety signs have a good effect on strengthening safety awareness) is almost equal and it is much higher than the number of option B(safety signs have some effect) and option C(safety signs have no effect),which manifests that safety signs effectively increase the safety awareness of machining workshop staff , making them take safety seriously in their mind and follow the instructions of safety signs .

From the safety-habit indication-personal Comparison chart ,we can find that management and front-line staff's view is basically the same for the issue that safety signs increase the safety habit, namely, the safety signs system in workplace is effective, and management develop a good safety command habit and operations develop a safety operation habit which indicates safety signs system does have an effect on reminding and instructing staff's behaviors and it improves safety signs system can help to develop safety habit of staff, indicates that safety signs actually take function in safety note, instructions,

guiding behaviors of employees, and improve the safety knowledge, safety awareness and safety habit, and it also improves the habitual behavior of employees, which can greatly reduce disposable unsafe behaviors, to avoid accidents and loss.

V. CONCLUSION

Above all, the paper obtains some conclusions as following:

(1)The vast majority (more than60%) employees think safety signs workplace useful. And managers force on whether safety signs reasonable and appropriate, and whether safety signs can control safety behaviors; while employee force whether safety signs relate to their facilities and equipment and process.

(2)Both managers and front-line employees pay attention to safety signs in the workplace, nearly 80% , which means staff follow with interest in safety signs.

(3)The managers and front-line employees hold the same view on the function of safety signs. Over 70% of managers and front-line employees think that safety signs can prompt danger.

(4)Managers and front-line employees have great difference in hazard perception, and the front-line employees care more about the function that safety signs can improve their hazard perception.

(5)Over 65% of the managers and front-line employees think that safety signs in the workplace can increase their safety knowledge, meaning safety signs in the workplace are useful, and they do take effect and increase safety knowledge of staff.

(6)Over 7 5% of the managers and front-line employees think that safety signs improve their safety awareness. It shows that safety signs improve the safety awareness of employees in the machining welding workshop, making staff value safe in their thoughts, and operate in accordance with safety signs consciously and follow the instructions of safety signs voluntary.

(7)Over 80% of the managers and front-line employees think that safety signs system can help to develop safety habit of staff, indicates that safety signs actually take function in safety note, instructions, guiding behaviors of employees, and improve the safety knowledge, safety awareness and safety habit, and it also improves the habitual behavior of employees, which can greatly reduce disposable unsafe behaviors, to avoid accidents and loss.

ACKNOWLEDGEMENT

The Fundamental Research Funds for the Central Universities(the project number:2013QZ02).

REFERENCES

[1] J.P.Yuan. An empirical study on the factors affecting the effectiveness of safety signs[D]. Zhejiang University,2009.

[2] Alan H S Chan, Annie W Y Ng. Effects of sign characteristics and training methods on safety sign training effectiveness[J]. Ergonomics, 2010, 53(11):1325-1346.

[3] J.Bian. Study on the perception mechanism of safety signs and design factors in safety management--based on neural Industrial Engineering Perspective[D].

[4] Zhejiang University,2014 Paton, Nic. Senior. Managers Fail to Show Competence in Health and Safety[J]. Occupational Health, 2008, 60(3): 6.

[5] Gwang-Hee Kim, Seok-Hoon Nam, S.J.Hwang, Hee-Bok Choi. Evaluation Construction Workers' Understanding of Safety Signs[J]. Applied Mechanics and Materials, 2013:291-294.

[6] Herber, John W..Quality problems as process safety warning signs[J]. Process Safety Progress, 2013(32), 2:175-178.

International Conference on Power Electronics and Energy Engineering (PEEE 2015)

Analysis on Tectonic Style and Forming Mechanism of the Cambrian System in the Central Uplift Belt, Tarim Basin

W. Yin

CNPC Research Institute of Petroleum Exploration & Development, PetroChina
Beijing, China

T. L. Fan

China University of Geosciences
Beijing, China

Abstract—Based on the theory and method of structural geology, through the systemic analysis on seismic data, the author researched on characteristics of structure geometry and kinematics, and its dynamics mechanism: There are a series of half-grabens in the Lower-Middle Cambrian Series of the Central Uplift Belt, and they are located in Bachu uplift and the southwestern area of Tazhong uplift. Based on the special basement, tectonic movements, including extension and compression are the main factors of forming half-grabens. The residual gravity anomaly high is accord with the distribution of half-grabens, the strike of buried fault and half-graben is concordant.

Keywords-half-graben; tectonic style; forming mechanism; cambrian; central uplift belt; tarim basin

I. REGIONAL GEOLOGY

Tarim Basin consists of eight first-order tectonic units and Central Uplift Belt is located in the central part of Tarim Basin. From west to east, it is Bachu Uplift, Tazhong Uplift and Tadong Uplift orderly. Bachu Uplift and Tazhong Uplift are both neighbouring second-order structural units of Central Uplift Belt, but there are obvious differences in tectonic evolution between them. Bachu uplift formed in Hercynian period, and finalized in Himalayan period. Tazhong Uplift formed in the middle Caledonian period, and finalized in the early Hercinian period (Figure 1).

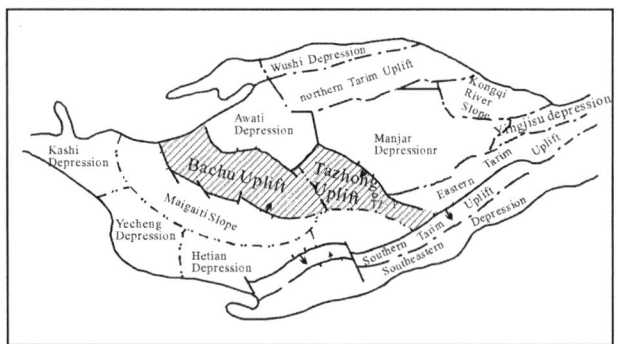

FIGURE I. LOCATION MAP OF WORK AREA.

II. CAMBRIAN TECTONIC STYLE

Tectonic style refers to the total tectonic deformations controlled by the same tectonic movement or the same stress, and it is one of the important contents of the structure research of oil and gas fields. In this paper, target zones is middle and lower Cambrian (the top surface of middle Cambrian is T81, the top surface of lower Cambrian is T82, and the bottom surface of lower Cambrian is T90). The faults of this period have the typical wedge shapes — — half-grabens. In half-grabens, not only formation thickness changes obviously, but also does seismic reflection characteristic. Generally, in the deepest location of half-grabens, the amplitude is weak, and continuity is poor. In the slope and the top of half-grabens, the amplitude is high, and continuity is better (Figure 2).

The scale of single half-graben is small, but so many half-grabens develop in the same belt, and consist of a large scale half-graben, which develops highland, slope and sag at the same time. A large scale half-graben controls regional depositional landform structure and restricts regional strata development and sedimentary facies distribution. In seismic profile, the expression of a large scale half-graben is not obvious, while, the top boundary of T81 was leveled, the tectonic framework of a large scale half-graben is distinct (Figure 3).

Half-grabens experienced 2 periods of tectonic movements: ① in early and middle Cambrian, it mainly experienced extensional movement and block tilt movement, accompanied by a short minor local compressional movement, and local compressional movement did not change the extensional structural framework of lower and middle Cambrain system. ② it mainly experienced compressional and flexing action and formed thrust folds after early and middle Cambrian. Both actions reshaped the half-grabens（Figure 4）[1-2]. In the planar, the trend of half-grabens is mainly northwest-southeast, and strike of half-grabens is mainly northeast-southwest. Half-grabens mainly distribute western Bachu uplift, western Hetianhe block and southern tazhong uplift (Figure 5).

© 2015. The authors - Published by Atlantis Press

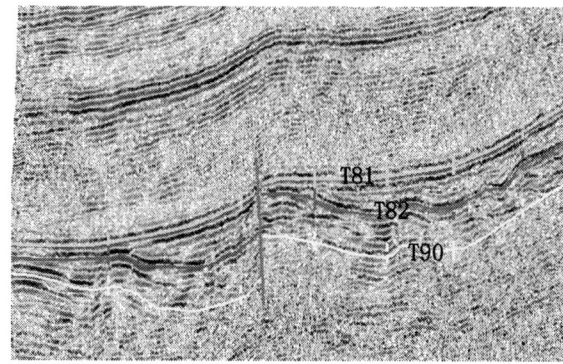

FIGURE II. STRUCTURAL STYLE OF HALF-GRABEN IN HETIANHE BLOCK OF BACHU UPLIFT (HTH-TZ03-244.6SN).

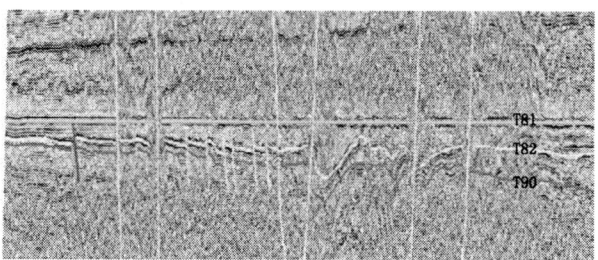

FIGURE III. SECTION STYLE BEFORE THE UPPER CAMBRIAN SEDIMENTATION IN BACHU UPLIFT (BC04-L1).

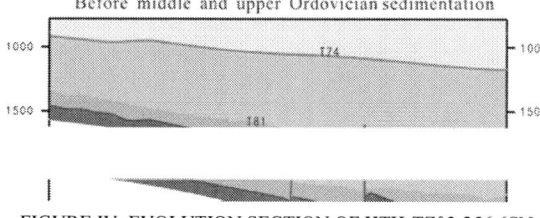

FIGURE IV. EVOLUTION SECTION OF HTH-TZ03-256.6SN.

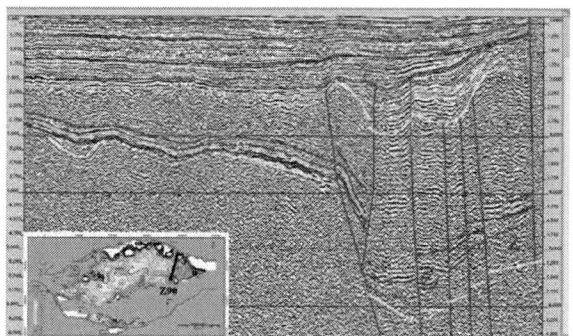

FIGURE V. PLANAR DISTRIBUTION OF HALF-GRABENS ON THE BOUNDARY T82 IN CENTRAL UPLIFT BELT.

III. FORMATION MECHANISM OF HALF-GRABENS

A. Dynamic Mechanism of Half-Grabens[3-4]

A series of important geological events were happened during Sinian-Early Ordovician passive continental margin, where aulacogen was developed in the northeast margin of Tarim Basin during Early Sinian[5]. The development of aulacogen played an important role in the formation of Early-Middle Cambrian half-graben in central uplift belt. Regional seismic line (TLM-Z90) of Tarim Basin, regional seismic line (TZ01-448SN) in Tazhong uplift, and regional seismic line (BC04-L1) of Bachu uplift were selected to analyze the structure features of the depression and its plane changes. Obvious difference of strength were found in the above three lines: TLS-Z90 was closely close to the aulacogen, where the affection by the aulacogen was the greatest, fault-throw can be reached up to 1700ms (Figure 6.); The distance from TZ01-448SN to the aulacogen became larger, where affection became smaller, fault-throw was about 500-800ms (Figure 7.); BC04-L1 in Bachu uplift was the farthest one from the aulacogen, where the affection by the aulacogen was the minimum, leading the smallest fault-throw, about 20-200ms (Figure 8.). According to the changes of fault-throw, it can be concluded that the development of depression was mainly controlled by the aulacogen that was developed in the northeast margin of Tarim Basin, and the affection towards the faulted-sag became weakened from east to west. Therefore, a a reasonable explanation was gained for the development of half-graben under the regional dynamic background.

FIGURE VI. REGIONAL SEISMIC PROFILE TLM-Z90 OF TARIM BASIN.

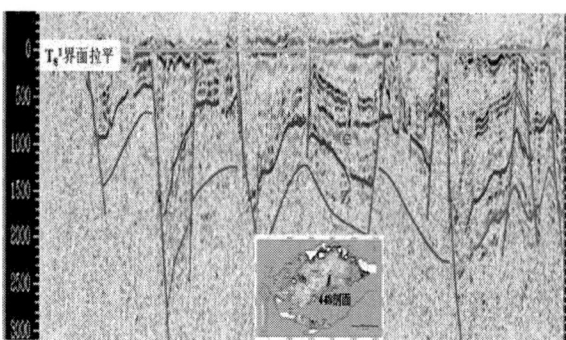

FIGURE VII. REGIONAL SEISMIC PROFILE TZ01-448SN IN TAZHONG UPLIFT OF TARIM BASIN.

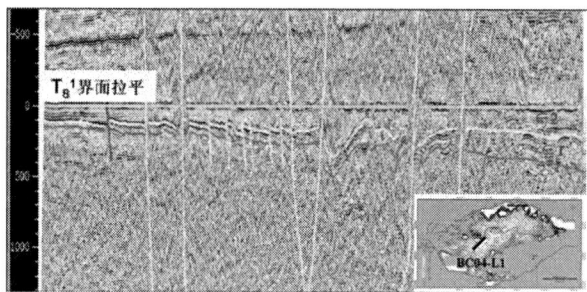

FIGURE VIII. REGIONAL SEISMIC PROFILE BC04-L1 IN BACHU UPLIFT OF TARIM BASIN.

Based on the above analysis, it can be concluded that the development of aulacogen was the dynamic mechanism for the formation of half-graben. However, which factor controlled the distribution and strike of the half-graben? According to the analysis of gravity-magnetic data of the basement in Tarim Basin, the distribution and strike of the half-graben was controlled by the characteristics of the basement of Tarim Basin.

B. Gravity Field Controls Planar Distribution of Half-Grabens

In the residual gravity anomaly map of Tarim Basin, the work area was located in the high-value zone, which was corresponding to upper-mantle uplift of Bachu and upper-mantle uplift of Tazhong. The high-value zones of residual gravity anomaly were easy to form half-graben by the regional horizontal extensional stress in Early-Middle Cambrian [5].

The whole Bachu uplift was located in upper-mantle uplift zone and with large uplift amplitude, so the half-graben was well developed; The south part of Tazhong uplift was located in upper-mantle uplift zone, while the basement of the north part was flat, so the half-graben was well developed in the south part. Based on the above analysis, it can be concluded that the distribution of half-graben was consisted with the high residual gravity anomaly zone, In other words, the high residual gravity anomaly controlled the distribution of half-graben (Figure 9.).

C. Influence of Insidious Basement Faults Over Half-Grabens

Buried basement faults were developed in the basin basement. 41 larger buried basement faults were found in Tarim Basin, some of which promoted the development of Middle-Lower Cambrian half-graben, including ① AF7 and BF8 (buried basement fault) in the central uplift and ② AF1 and AF2 (latitudinal basement fault) in the central (Figure 10.).

The controlling of buried basement faults towards the half-graben was shown in the strike of half-graben. Strike of buried basement fault in Tarim Basin was NW-SE, while half-graben was also NW-SE. Different buried basement faults played the major role in different areas. AF7 and BF8 played the major role in Bachu uplift while AF7 and AF2 played the major role in Tazhong uplift. According to the above analysis, buried basement fault played a relatively strong controlling over the strike of half-graben in overburden [5-6] (Figure 10.).

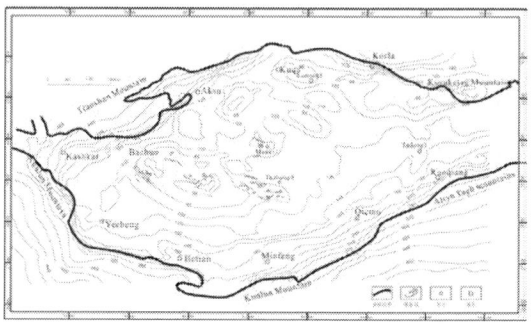

FIGURE IX. CORRESPONDING DIAGRAMS BETWEEN RESIDUAL GRAVITY ANOMALY AND HALF-GRABENS DISTRIBUTION OF TARIM BASIN (ACCORDING TO JIA CHENGZAO, 1997).

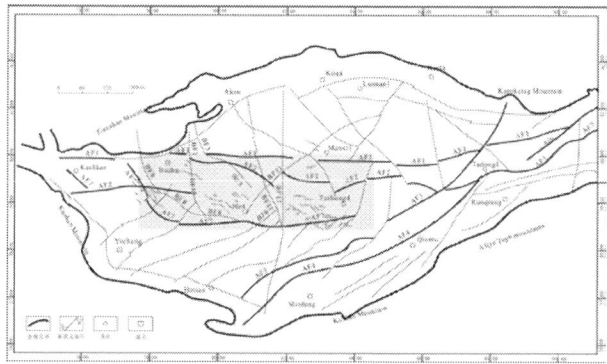

FIGURE X. CORRESPONDING DIAGRAMS BETWEEN INSIDIOUS BASEMENT FAULTS AND HALF- GRABENS DISTRIBUTION OF TARIM BASIN.

IV. CONCLUSION

The extensional background of the Central uplift in Tarim Basin during Early-Middle Cambrian was the major controlling factor for the development of half-graben. The formation of aulacogen in northeast margin of the basin was the dynamic mechanism for the developing of half-graben. The upper-mental uplift zones controlled the distribution of half-graben, and buried basement faults controlled the strike of half-graben. Therefore, the development of half-graben in the

Central uplift was the result of joint-action by the extensional and compression stress in the later stage under the specific basement background.

REFERENCES

[1] Zhensheng Zhang, Mingjie Li & Sheping Liu, Generation and evolution of Tazhong low uplift. Petroleum Exploration and Development, 29(1), pp. 28-31, 2002.

[2] Min Zheng, Gangxin Peng & Ganglin Lei, Structural pattern and its control on hydrocarbon accumulations in WushiSag, Kuche Depression, Tarim Basin. Petroleum Exploration and Development, 35(4), pp. 444-451, 2002.

[3] Fajing Chen, Xinwen Wang & Guangya Zhang, Structure and geodynamic setting of oil and gas basins in China. Geoscience, 6(3), pp. 317-327, 1992.

[4] Hepu Liu, Dynamic classification of sedimentary basins and their structural styles. Earth Science-Journal of China University of Geosciences, 18(6), pp. 699-724, 1993.

[5] Chengzao Jia, Struactural features and oil&gas accumulation in Tarim basin, China[M]. Beijing: Petroleum Industry Press, 1997.

[6] Ying Zhu, Outline of tectonic and deep tectonic features in China and its neighboring area[M]. Beijing: Petroleum Industry Press, pp. 1-32, 2004.

Effect of the Plan for Educating and Training Excellent Engineers on the Experimental Technical Ability of Undergraduate

X.D. Yuan

Materials Science and Engineering Institute
Shandong Jianzhu University
Jinan, Shandong, China

X.J. Yang

Shandong Product Quality Inspection Research Institute
Ji'nan, 250100, China
Shandong Key Laboratory for Testing Technology of Material
Chemical Safety
Ji'nan, 250103, China

G.L. Yuan

Jinan Engineering Quality and Work Safety Supervision Station
Ji'nan, Shandong, 250013, China

Abstract—The training system of 1234 engineering and technical personnel was implemented in requirements of the plan for educating and training excellent engineers by School of materials science and engineering, Shandong Jianzhu University. The construction of three-dimensional platform between the laboratory, the enterprise and the industry was strengthen and the experiment technical ability of undergraduate was improved by using of teaching tutorial and ideological mentor of advantage resources. The influence of 1234 training system of engineering and technical personnel on the experiment technical ability of undergraduates was studied taking the study on friction and wear properties of PTFE coatings as an example.

Keywords-experiment teaching; outstanding engineers; PTFE; wear

I. INTRODUCTION

The plan for educating and training excellent engineers (referred to as "excellence program"), is a major reform project implementing the "national medium and long-term educational reform and development plan (2010-2020)" and "national long-term talent development planning outline (2010-2020)". The pilot colleges and universities in the country draw up training objective of the engineering and technical personnel of the school combined with the school characteristic and the talent training standards under the guidance of general standards, since the beginning of June 23, 2010. Studies, including the teaching

reform of school and professional, personnel training mode, school enterprise cooperation and the reform of teaching practice were carried out by some colleges and universities in the "excellence program" requirements [1-8]. Material forming and control engineering in school of materials science and engineering is approved by the 2013 third batch of "excellence program" organized by the ministry of education. In requirement the training program of professional, studies about the

demand for technical personnel materials engineering in the new period was carried out combing the characteristics of the material science itself by Shandong Jianzhu University. 1234 Training system of engineering and technical personnel was implemented And the experiment technical ability of undergraduates was improved. The so called talent training system of 1234 engineering technology includes the following four aspects: improving engineering practice ability of college students, the introduction of Double Tutorial System, three platforms of the industry, enterprise and laboratory, and four stages referring to tamp foundation, internship, painstaking research and independent innovation.

II. TRAINING SYSTEM

The influence of 1234 training system of engineering and technical personnel on the experiment technical ability of undergraduates was studied in this paper taking the study on the friction and wear properties of PTFE coatings as an example.

Training system of the engineering and technical personnel is training system by which engineering practical ability of students can be improved

The freshman makes efforts to learn theoretical knowledge, and consolidates the foundation. The sophomore experiences "Internship", understands the needs of enterprises. The junior participates or declares laboratory research projects, participates in the international conference and all kinds of associations of industry, innovates in thinking, and researches with great concentration. The senior carries out the work of graduation design combining engineering technical problems of enterprises, solves problems independently, and realizes the independent innovation. The implementation mode was shown in Figure 1.

© 2015. The authors - Published by Atlantis Press

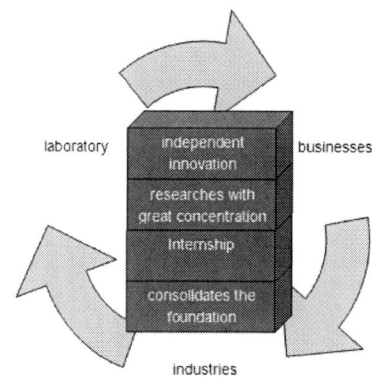

FIGURE I. TRAINING SYSTEM OF THE ENGINEERING AND
TECHNICAL PERSONNEL.

III. EXPERIMENTS

A. The Design of Friction and Wear Experiment of PTFE Coatings

Polytetrafluoroethylene (PTFE) is a popular polymer solid lubricant due to its resistance to chemical attack in a wide variety of solvents and solutions, high melting point, low coefficient of friction, and biocompatibility. It is widely used in bearing and seals applications [9]. The machine parts experience early failure and leakage problem because of poor wear and abrasion resistance of PTFE [10-11]. To minimize the problem, various suitable fillers were added to PTFE.

The friction and wear properties of PTFE coatings were investigated in this paper. And the experiment parameters are as follows:

The friction and wear behavior of PTFE coatings was carried out on a ball-on-disk wear tester under vacuum conditions (10^{-4}Pa). All PTFE coatings were slid against a GCr15-bearing steel ball. All sliding tests were carried out at a linear velocity from 0.2 to 2.4m/s, at the load of 6N, and the distance of 1000m. In this work, three samples were tested at each condition. The friction coefficient and wear were the average values of these tests for each condition.

B. Characterization of Friction and Wear Performance of PTFE Coatings

The friction coefficient and wear of PTFE coatings under different sliding velocity are shown in figures 2 and 3, respectively. It can be seen from figure 2 that the friction coefficients of PTFE coatings under vacuum and air conditions are similar to each other and have similar trend curves. It also can be observed that the friction coefficients of the PTFE coatings first increase with the increase of sliding velocity and then decrease with the increase of sliding velocity when the sliding velocity is higher than 1.2m/s. The variation of wear with sliding speed under the vacuum condition is shown in Figure3. The result shows that the wear of the PTFE coating first decreases with the increase of sliding speed and then increases as the sliding speed increases.

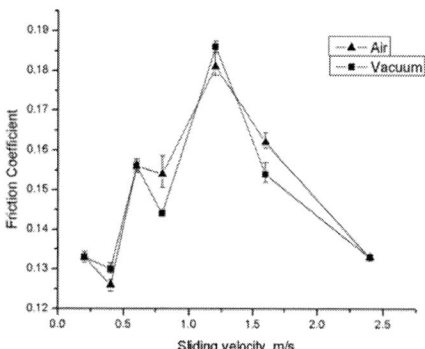

FIGURE II. VARIATIONS OF FRICTION COEFFICIENT WITH SLIDING
VELOCITY (LOAD, 6N).

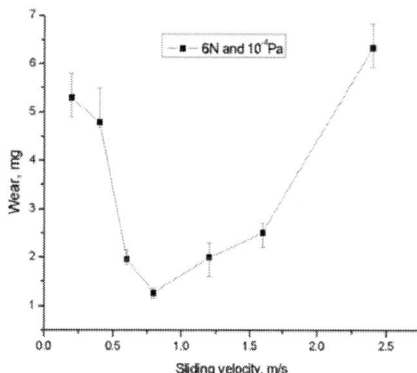

FIGURE III. VARIATIONS OF WEAR WITH SLIDING VELOCITY
(LOAD, 6N AND PRESSURE, 10-4PA).

IV. CONCLUSIONS

The 1234 training system of engineering and technical personnel was implemented by School of materials science and engineering of Shandong Jianzhu University. And the experiment technical ability of the undergraduate was improved

ACKNOWLEDGEMENT

Sponsored by the Research Fund for the College science and technology plan of Shandong Province (Project No. J12LA11).

REFERENCE

[1] Huijuan Fu, Yong Xin. Pursuing "Engineer Excellence Program" to cultivate practical talents of technical engineering. *Experimental technology and management*, 2011, 28(11): 155-158.

[2] Dong Chen. Research on teaching reform of steel structure course based on outstanding engineers plans [J]. Journal of Chongqing university of science and technology (social sciences edition), 2011(6): 174-176.

[3] Lili Huang, Haisheng Zhang, Xiaohu Huang. Study on Educational Administration based on "Plan for Educating and Training Outstanding Engineers"[J].The 3rd International Annual Conference on Teaching Management and Curriculum Construction, 2012(5): 178-180.

[4] Yuhua Dai, Kai Gu, Jianping Huang. Development of an Enterprise Training Program for A Plan for Educating and Training Outstanding

Engineers [J].Research and exploration in laboratory, 2012,31(12): 159-162.

[5] Wensong Lin, Liang He, Yanhui Liu. Experimental teaching reform of engineering materials course based on "plan for educating and training outstanding engineers" [J].Experimental technology and management, 2012,29(12): 20-22.

[6] Qisheng Wu, Changsen Zhang, Baoxiang Jiao et al..The research of training mode for excellent engineer plan [J]. Chemical engineering education,2012(3):14-18.

[7] Shuangqing Lv, Yu Zhang. Research of software professional training plan Based on the "outstanding engineers plans" [J]. Vocational technology, 2011(7):132.

[8] Zhenzhong Sun, Shenggui Chen, Shouyan Zhong. Training mode of enterprise engineering practice for outstanding engineers[J].Research and exploration in laboratory, 2012,31(8): 285-289.

[9] W. Gregory Sawyer, Kevin D. Freudenberg, Praveen Bhimaraj, Linda S. Schadler, A study on the friction and wear behavior of PTFE filled with alumina nanoparticles[J]. Wear, 2003 (254) 573-580.

[10] D. Gong, Q. Xue, H. Wang, Study of the wear of filled polytetrafluoroethylene[J]. Wear, 1989 (134) 283-295.

[11] T.A. Blanchet, F.E. Kennedy, Sliding wear mechanism of (PTFE) and PTFE composites[J]. Wear, 1992 (152) 229-243.

International Conference on Power Electronics and Energy Engineering (PEEE 2015)

Study on the Experimental Teaching for Undergraduates in Requirements of the Plan for Educating and Training Excellent Engineers

X.D. Yuan

Materials Science and Engineering Institute
Shandong Jianzhu University
Jinan, China

X.J. Yang

Shandong Product Quality Inspection Research Institute
Ji'nan, 250100, China
Shandong Key Laboratory for Testing Technology of Material
Chemical Safety
Ji'nan, 250103, China

G.L. Yuan

Jinan Engineering Quality and Work Safety Supervision Station
Ji'nan, 250013, China

Abstract—In requirements of the plan for educating and training outstanding engineers, the experimental resources were integrated by School of materials science and engineering, Shandong Jianzhu University, the opening experimental teaching system was founded, and the experiment technical skills of students were improved, especially metallographic skills. The influence of the opening experimental teaching system on the experiment technical ability of undergraduates was studied taking the study on the microstructure of Cr-Rare-boronized layer under low temperatures as an example, and reasons were also analyzed.

Keywords-experiment teaching; outstanding engineers; Cr-rare earth-boronizing; steel 45

I. INTRODUCTION

The plan for educating and training outstanding engineers (referred to as "outstanding program"), is a major reform project implementing the "national medium and long-term educational reform and development plan (2010-2020)" and "national long-term talent development planning outline (2010-2020)". The universities and colleges formulate training goals of the engineering and technical personnel of the school in accordance with the industry standard requirements and combined with the school characteristic and the talent training standards under the guidance of general standards, since the beginning of June 23, 2010. In the "outstanding program" requirements, studies, which include the teaching reform of school and professional, personnel training mode, school enterprise cooperation and the reform of teaching practice were carried out by some universities and colleges[1-7].

Shandong Jianzhu University is a provincial university giving priority to engineering, with balanced development of engineering, science, management, economy, arts and law. Over the past years, applied talents with the ability of engineering practice were cultivated. Material forming and control engineering in school of materials science and engineering is approved as the 2013 third batch of "outstanding program", which was organized by the ministry of education. In the requirements of "outstanding program", studies about the demand for technical personnel materials engineering in the new period was carried out combing the characteristics of the material science itself. The experimental resources were integrated, the opening experimental teaching system was founded, and the experiment technical skills of students were improved, especially metallographic skills.

In this paper, the influence of the opening experimental teaching system on the experiment technical ability of undergraduates was studied taking the study on the microstructure of Cr-Rare-boronized layer under low temperatures as an example, especially metallographic skills, and reasons were also analyzed.

II. IMPLEMENTATION OF THE OPENING EXPERIMENT TEACHING SYSTEM AND IMPROVEMENT OF THE ABILITY TO SOLVE PROBLEMS

The opening experiment teaching system was implemented and the duration was one month. Students design the experiment independently, operate the equipment independently, and deal with the experimental data independently.

Students design the experiment scheme according to practical problems encountered in the enterprise, apply for the new experiment project by themselves (they may receive the money, 2500 Yuan, from the school). Practical engineering problems can be solved through the experimental study. Experimental results can also be used to participate in various national university science and technology competition, from which the students' employment competitiveness can be improved. Each year the University subsidized student autonomous application topic more than 30 items, a total

© 2015. The authors - Published by Atlantis Press

313

funding of research funds was more than 1 00000 Yuan. The ability of analyzing the question, solving the problem, and hands-on activities was enhanced through the implementation of opening experiment teaching system. The specific experimental teaching mode was shown in figure 1.

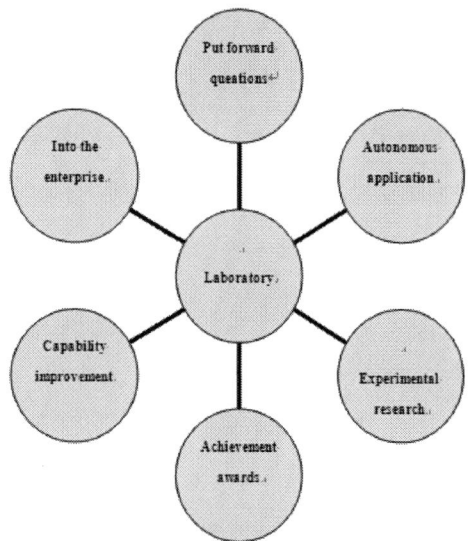

FIGURE I. MODE OF OPENING EXPERIMENT TEACHING SYSTEM.

III. IMPROVEMENT OF THE ABILITY OF METALLOGRAPHIC SAMPLE REPARATION

A. Experiment Process

The main material for the experiment is steel 45. The samples were first quenched in water at 860℃ for 30min. The Cr-Rare earth-boronizing process was performed under 600℃ ×6 h on the surface of the quenched samples. The boronized samples were then set on an inlaying machine, polished using electronic grinder until the specimen surface was smooth. Then samples were grinded by using 200# sandpaper gently, and the grinding direction can not be changed for 3 minutes so that the specimen surface further can be bright. The samples were grinded on sandpaper immediately, which was coated with white cat detergent on the surface. The general milling time was 5 minutes. Then samples were polished, cleaned, and dried. Metallographic observation was performed on an optical microscopy.

B. Microstructure of the Cr-Rare-boronized Layer

The microstructure of the boride layer was shown in Figure 2. It can be seen that the boride layer is integrity, continuous, and uniform.

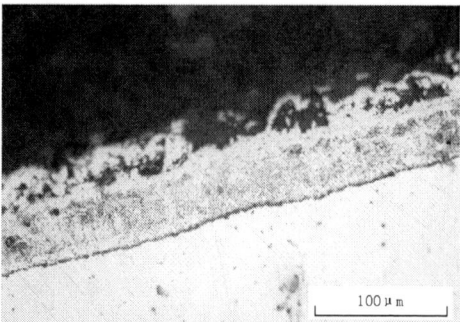

FIGURE II. MICROSTRUCTURE OF THE BORIDE LAYER.

Parts of boride layer shedded during the process of grinder grinding due to its higher hardness and brittleness. Black holes and micro cracks were observed. The shedding phenomenon continues and the microstructure was damaged during the sandpaper grinding. Those problems can not be solved by traditional grinding methods. The agent which contains a surface active agent was used to solve the above problems. The surface active agents play a major role in all kinds of detergent materials, which are sometimes referred to as the "active ingredients" or "effective". It is capable of reducing solvent surface tension of material at low concentrations. The dirt was washed off due to the existence of surfactants in detergent, which lead to the wetting emulsification, dispersion and produce, foaming, and solubilization. The hard particles produced in the process of sandpaper grinding will be removed under the action of surfactant, and does not exist between the surface of the sp0ecimen and sandpaper acting as a "second phase particles". The boride layer will not fall off.

The presence of sodium silicate in detergents, which was known as the "water glass" and "sodium silicate" plays an important role in regulating the PH of the detergent and maintaining the detergent solution to certain alkalinity. The existence of sodium carbonate in detergent can improve the ability of removing oil. The detergent itself is alkaline due to the above substances, which will corroded the materials and reduce the later corrosion time. Furthermore, the corroded materials will be removed effectively and the damage degree of boride layer will be reduced.

IV. CONCLUSIONS

Opening experiment teaching system was implemented by school of materials science and engineering of Shandong Jianzhu University after the "outstanding program" was approved. The three-dimensional platform between the laboratory, businesses and industries was used, and the ability of metallographic sample reparation was improved comprehensively.

ACKNOWLEDGEMENT

Sponsored by the Research Fund for the College science and technology plan of Shandong Province (Project No. J12LA11).

REFERENCE

[1] Sun Z Z, Chen S G, Zhong S Y. Training mode of enterprise engineering

practice for outstanding engineers [J].*Research and exploration in laboratory*, 2012, 31(8):285-289.

[2] Chen S. Reform of practice teaching according to mode of applied cultivation [J]. *Research and exploration in laboratory*, 2012, 31(8):1-4.

[3] Zhao J H. Review of studies on the effectiveness of undergraduate experimental teaching in the west [J]. *Research and exploration in laboratory*, 2012, 31(10):91-94.

[4] Zhang X M. Innovative research and practice on "project supermarket" mode of opening experiment. [J]. *Experimental technology and management*, 2012, 29(10):33-36.

[5] Chen N N. Opening laboratory in engineering colleges and universities and reform of experiment teaching. [J]. *Experimental technology and management*, 2006, 23(4):110-112.

[6] Lai Z H. On experimental teaching reform and practice of civil engineering major [J]. *Research and exploration in laboratory*, 2010, 29(11):319-321.

[7] Wang H Q, Wu Q F, Liang X J. Discussion on reform of university physics experiments by taking application as guidance. [J]. *Experimental technology and management*, 2012, 29(9):148-150.

Study on the Opening Experiment Teaching for Undergraduates in Requirements of the Credit Management System

X.D. Yuan

Materials Science and Engineering Institute
Shandong Jianzhu University
Jinan, Shandong, China

G.L. Yuan

Jinan Engineering Quality and Work Safety Supervision
Station
Jinan, Shandong, China

X.J. Yang

Shandong Product Quality Inspection Research Institute
Ji'nan, China
Shandong Key Laboratory for Testing Technology of Material Chemical Safety
Ji'nan, China

Abstract-In requirements of the credit management system, the experimental resources were integrated by School of materials science and engineering, Shandong Jianzhu University, the opening experimental teaching system was established, and the experiment technical skills of students were improved. The influence of the opening experimental teaching system on the experiment technical ability of undergraduates was studied taking the failure analysis of ammonia-flume as an example, and reasons were also analyzed.

Keywords-experiment teaching; credit system; failure analysis; metallurgical structure inspection

I. INTRODUCTION

The credit system is a kind of the elective system. The study quantity of students was calculated by credits, and students can apply for graduate until they obtain certain credits.

The learning of students under credit system is elastic. They learn different courses and receive course credits. Students can apply for graduate until they obtain the lowest credits which satisfied the requirements of the school. Researches, including the personnel training mode, the teaching reform of school and professional, school enterprise cooperation and the reform of teaching practice were carried out by some colleges and universities in the "credit system" requirements [1-7].

Shandong Jianzhu University is a provincial university, in which engineering subjects are dominant, with balanced development of other subjects. The university began to try the student credit system management, and draw up educational objectives since 2012. Studies were carried out about the demand for technical personnel materials engineering in the new period combing the characteristics of the material science itself. The opening experiment teaching system was implemented relying on foundry cleaner production engineering technology research center of Shandong province and experimental teaching demonstration center, by which the experiment technical ability of undergraduates was improved.

In this paper, the influence of the opening experimental teaching system on the experiment technical ability of undergraduates was studied taking the failure analysis of ammonia-flume as an example, and reasons were also analyzed.

II. IMPLEMENTATION OF THE OPENING EXPERIMENT TEACHING SYSTEM

The opening experiment teaching system was implemented for one month. In this month, the experiment was designed by students independently, the equipment was operated by students independently, and the experimental data was deal with by students independently. The experiment scheme was designed by students according to practical problems encountered in the enterprise, and the new experiment project was applied for by themselves (they may receive the money, 2500 Yuan, from the school). Practical engineering problems can be solved through the experiment. Results can also be used to participate in a variety of national university science and technology competition, which leads to the improvement of students' employment competitiveness. The University subsidized student autonomous application topic more than 30 items each year, a total funding of research funds was more than 1 00000 Yuan. The ability to analyze the question, solve the problem, and hands-on activities was improved through the implementation of opening experiment teaching system. The specific experimental teaching mode was shown in Figure 1.

© 2015. The authors - Published by Atlantis Press

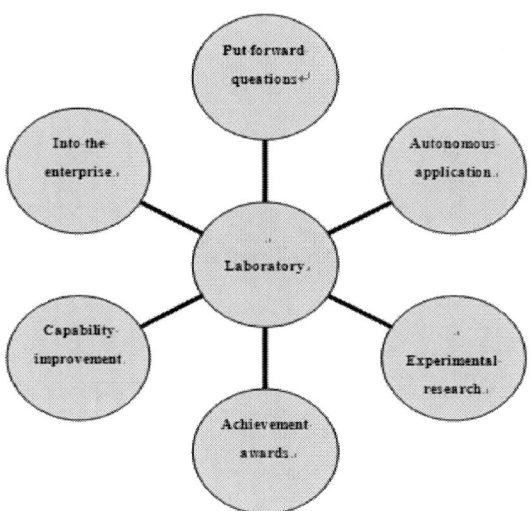

FIGURE I. MODE OF OPENING EXPERIMENT TEACHING SYSTEM.

III. IMPROVEMENT OF EXPERIMENT TECHNICAL ABILITY OF UNDERGRADUATES

A. Objective and Methodology

An ammonia water tank of the factory leaked in the process of using. The tank was made of steel Q235B plate. It was welded by seven steel plates. The temperature of ammonia water was 68 to 73 ℃.

The failure analysis of ammonia-flume was conducted by macro appearance observation, chemical composition analysis and metallurgical structure inspection.

B. Macro Appearance Observation and Composition Analysis

The illustration of the tank was shown in Figure 2. The leakage occurred in the parent metals. Macro-cracks can be observed. The composition was in accordance with the requirement of GB/T700-2006.

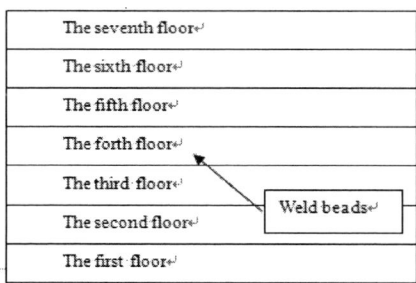

FIGURE II. THE ILLUSTRATION OF THE TANK.

C. Metallurgical Structure Inspection

Samples were cut from two directions which were parallel to and ventricular to the weld seam. After mosaic, grinding and polishing, samples were observed by the optical microscope. It can be seen that there are some dendritic cracks as shown in Figure 3. After corrosion with 4% nitric acid alcohol, samples were observed by the optical microscope, the microstructure was shown in Figure 4. It can be seen from Figure 4 that there are some intergranular cracks and the cracks are close to the weld beads.

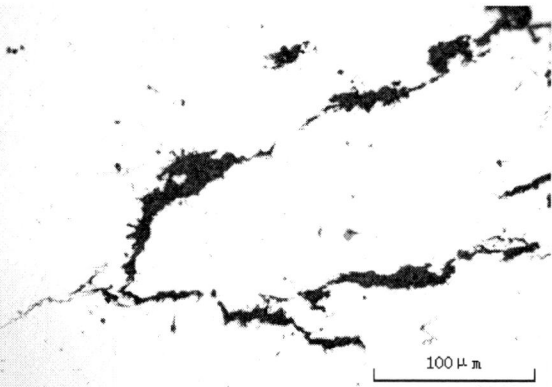

FIGURE III. DENDRITIC MORPHOLOGY OF THE CRACK.

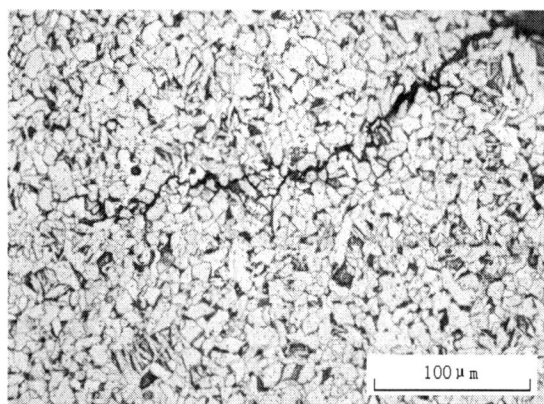

FIGURE IV. NTERGRANULAR CRACKS.

D. Failure Analysis

There are some dentric cracks in the leakage section of the parent metal, and cracks spread along the grains. It can be preliminarily concluded from those experimental data that stress corrosion occurred in the parent metal, which lead to the leakage of the tank.

It is well known that material factors, environmental factors and stress factors are the basic conditions of stress corrosion. In stress corrosion system, stress and corrosion are not the simple superposition, but promote each other. It is reported that carbon steels commonly suffer from stress corrosion in H2S solution, NaOH solution, nitric acid and nitrate solution [8]. A small amount of researches on the corrosion of carbon in ammonia were reported [9-10].

Tank bears only the action of gravity at work. It is expected that the stress, resulting in the leakage of the tank, comes from the pipe manufacturing and the welding stress.

It is reported that a protective film will be formed on the grain boundaries. However it is not stable and dissolves easily.

Cracks will be generated along grain boundaries under stress, where the selective dissolving will occur and then stress corrosion happen leading to the leakage of the tank.

IV. Conclusions

The opening experiment teaching system was implemented by the school of materials science and engineering of Shandong Jianzhu University since the implement of the credit system. The advantages of three-dimensional platform between the laboratory, the enterprise and the industry were fully developed. The experiment technical ability of undergraduate was improved, especially the ability of failure analysis.

Acknowledgement

Sponsored by the Research Fund for the College science and technology plan of Shandong Province (Project No. J12LA11).

Reference

[1] Chen S. Reform of practice teaching according to mode of applied cultivation [J]. *Research and exploration in laboratory*, 2012, 31(8):1-4.

[2] Zhao J H. Review of studies on the effectiveness of undergraduate experimental teaching in the west [J]. *Research and exploration in laboratory*, 2012, 31(10):91-94.

[3] Zhang X M. Innovative research and practice on "project supermarket" mode of opening experiment [J]. *Experimental technology and managemen*, 2012, 29(10):33-36.

[4] Chen N N. Opening laboratory in engineering colleges and universities and reform of experiment teaching [J]. *Experimental technology and management*, 2006, 23(4):110-112.

[5] Lai Z H. On experimental teaching reform and practice of civil engineering major [J]. *Research and exploration in laboratory*, 2010, 29(11):319-321.

[6] Wang H Q, Wu Q F, Liang X J. Discussion on reform of university physics experiments by taking application as guidance. [J]. *Experimental technology and management*, 2012, 29(9):148-150.

[7] Liang L Y, Hu B L, Zhu L. Reform of experimental teaching for environmental engineering. [J]. *Experimental technology and management*, 2012, 29(9):126-129.

[8] Zhang Z. Study on Carbon Steel Sulfide Stress Corrosion Cracking Suspectibility in Refinery [D]. *Dalian University of Technology*: 2009. 3-4.

[9] Zhou Y L, Liu Q G. 3P_1402B cracking analysis of ammonia flume [J]. *Physical Testing and Chemical Analysis Parta Physical Testing*, 2002, 38(1): 565-567.

[10] Wang Y L, Yu Q. Analysis of leakage causes in circulating ammonia pipes and countermeasures. *Anhui Metallurgy*, 2010(2): 39-42.

Study on the Talent Cultivation Mode in Requirements of the Credit Management System

X.D. Yuan
Materials Science and Engineering Institute
Shandong Jianzhu University
Jinan, Shandong, China

G.L. Yuan
Jinan Engineering Quality and Work Safety Supervision Station
Shandong Ji'nan, China

X.J. Yang
Shandong Product Quality Inspection Research Institute
Ji'nan, China
Shandong Key Laboratory for Testing Technology of Material Chemical Safety
Ji'nan, China

Abstract-1234 Training system of the engineering and technical personnel was implemented by School of materials science and engineering, Shandong Jianzhu University in requirements of the credit management system, and the cultivation quality for undergraduates were improved. The influence of the1234 training system of the engineering and technical personnel on the cultivation quality for undergraduates was studied taking the study on the friction and wear properties of Cr-Rare earth-Boronized layer as an example, and reasons were also analyzed.

Keywords-credit system; 1234 training system; Cr-rare earth-boronizing; steel 45

I. INTRODUCTION

The credit system is a kind of the elective system, in which the study quantity of students was calculated by credits, and students can apply for graduate until they obtain certain credits. The learning of students under credit system is elastic. They learn various courses and receive course credits. Students can apply for graduate when they obtain the lowest credits which satisfied the requirements of the school. Studies, including the teaching reform of school and professional, the personnel training mode, school enterprise cooperation and the reform of teaching practice were carried out by some colleges and universities in the "credit system" requirements [1-6].The university tried he student credit system management and formulated educational objectives since 2012. Researches on the demand for materials engineering technical personnel were carried out combing the characteristics of the material science itself by Shandong Jianzhu University. 1234 Training system of the engineering and technical personnel was implemented and the experiment technical ability of students was improved.

II. TRAINING SYSTEM

1234 Training system of the engineering and technical personnel is a training system which focuses on integrating resource superiority of teaching tutorial (young teachers) and ideological mentor (counselor), making full use of advantages of three-dimensional platform among the laboratory,

businesses and industries, and improving engineering practical ability of students gradually. In the first year, students make efforts to learn theoretical knowledge and consolidate the foundation. In the second year students experience "Internship", understand the needs of enterprises. In the third year students participate or declare laboratory research projects, participate in the international conference and all kinds of associations of industry, innovate in thinking, and researches with great concentration. In the forth year, students carry out the work of graduation design combining engineering technical problems of enterprises, solve problems independently, and realize the independent innovation. Implementation mode is shown in Figure 1.

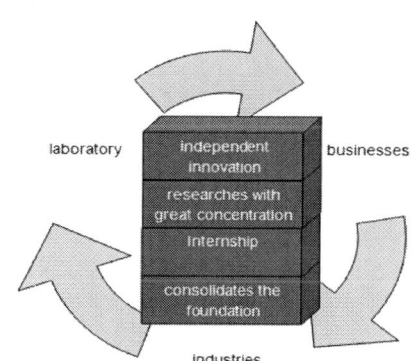

FIGURE I. 1234 TRAINING SYSTEM OF THE ENGINEERING AND TECHNICAL PERSONNEL.

III. EXPERIMENTS

A. Objective

Boronizing is the technical process to add boron atoms into the surface of materials and hence to form borides eventually. This process happens due to the chemical or electrochemical reactions by heating the target materials within boracic environment. This advanced technique can make boron atoms disperse into the target materials via chemical reaction to form

© 2015. The authors - Published by Atlantis Press

the boride layer with extreme hardness, excellent wear resistance and stronger corrosion resistance [7-9].

B. Experimental Details

1) Materials

Materials mainly used in this study were steel 45. Parts of samples were quenched in water at 850℃ and tempered at 200℃. Parts of samples were Cr-Rare earth-boronized in a penetrating tank at 650 for 6h.

2) Experimental Approach

The friction and wear behavior of samples was performed on a wear tester. All samples were slid against a steel T12. All sliding tests were carried out at an angular velocity of 200 r/min, loads of 1500N, and the duration was 5 min. five samples were tested at each condition. The friction coefficient and weight loss were the average values of these tests. The friction coefficient of samples was determined by measuring the friction torque, while the friction torque was detected by a torque measuring system. The sliding distance was calculated by diameter and angular velocity. The weight loss was calculated by the mass of samples before and after each test to an accuracy of 0.01 mg.

C. Results and Discussion

Variations of weight loss with the sliding distance were shown in Figure 2. It can be seen from Figure 2 that the weight loss of quenched samples was lower than that of Cr-Rare earth-Boronized samples during the early period of friction. It was because of that the surface of Cr-Rare earth-Boronized samples was looser with a lot of holes. And the weight loss of quenched samples was higher than that of Cr-Rare earth-Boronized samples with a further increase of sliding distance.

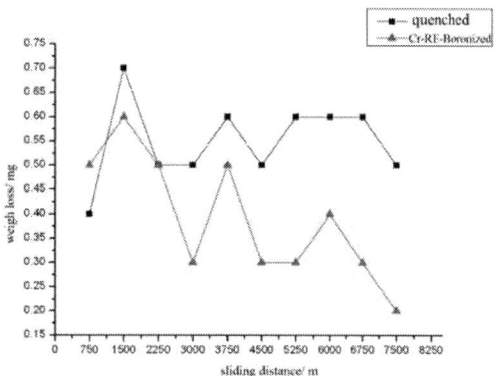

FIGURE II. VARIATIONS OF WEIGHT LOSS WITH THE SLIDING DISTANCE.

Variations of friction coefficient with the sliding distance were shown in Figure 3. It can be seen from Figure 3 that the friction coefficient of quenched samples was higher than that of Cr-Rare earth-Boronized samples. It was in accordance with the results of Figure 2. It was probably because of that the vacancies in phase Fe_2B were occupied by chrome and rare earth atoms, leading to the increase of density. Meanwhile, the

compounds formed by the rare earth atoms and the impurities in grain boundaries resulting in the decrease of brittleness of boride layer. And then the origination and propagation of cracks were inhibited.

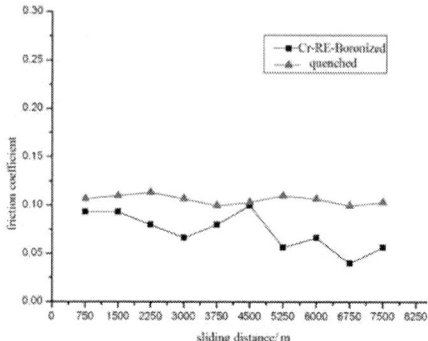

FIGURE III. VARIATIONS OF FRICTION COEFFICIENT WITH SLIDING DISTANCE.

The wear resistance can be represented by the reciprocal of weightlessness. The weight loss of Cr-Rare earth-Boronized samples and quenched samples are 3.9mg, and 5.5mg. And the wear resistance of Cr-Rare earth-Boronized samples is 0.256 which is superior to that of quenched samples 0.182.

The relative wear resistance is represented by the weight loss of quenched samples divided by the weight loss of Cr-Rare earth-Boronized samples, which is 1.41. It can be concluded that the wear resistance of Cr-Rare earth-Boronized samples is 1.4 times higher than that of quenched samples.

IV. CONCLUSIONS

1234 Training system of the engineering and technical personnel was implemented by the school of materials science and engineering of Shandong Jianzhu University since the implement of the credit system. The cultivation quality of undergraduates was improved.

ACKNOWLEDGEMENT

Sponsored by the Research Fund for the College science and technology plan of Shandong Province. (Project No. J12LA11).

REFERENCE

[1] Xu Z B, Xu J H. on the teaching management in the full credit system [J]. *journal of hebel normal univer sity/educational science edition*, 2009,11(2):90-93.

[2] Yan G F, Zai X J. The credit system of colleges and universities charge [J]. *china academic journal electronic publishing house*, 2007(8):58-60.

[3] Feng X D. to practice the credit system: a profound transform in teaching and opinion [J]. *systemjournal of higher education*, 2003,24(6):59-63.

[4] Lv X L, Dai J F, Hu M G. open experimental teaching based on credit system [J]. *research and exploration in laboratory*, 2012,31(8):121-124.

[5] Zhou Q M. analysis on supervision of teaching quality under credit system [J]. *systemjournal of higher education*, 2003,24(5):80-82.

[6] Luo Q L, Chen Y L. The measure of accumulated-points in the credit system [J]. *Application of statistics and management*, 2004,24(6):48-53.

[7] P.X.Yan,Y.C.Su.Metal surface modification by B-C-nitriding in a

two-temperature-stage process. *Materials Chemistry and Physics* 1995, 39: 304-308.

[8] M. Keddam. Computer simulation of monolayer growth kinetics of Fe_2B phase during the paste-boriding process: Influence of the paste thickness. *Applied Surface Science* 2006, 253:757-761.

[9] M.A. Béjar , R. Henríquez. Surface hardening of steel by plasma-electrolysis boronizing. *Materials and Design* 2009, 30: 1726-1728.

International Conference on Power Electronics and Energy Engineering (PEEE 2015)

Analysis on the Characteristics of Particles Spectral of Ice caused by Freezing Rain in Nan-yue Mountain of Hunan Province

L.C. Zhang, J.L. Yang, B. Liu, K.J. Zhu
China Electric Power Research Institute
Beijing 100055, China

X.Y. Li
Institute of Atmospheric Physics
Chinese Academy Sciences Beijing 100029
Beijing, China

Abstract—In order to know the formation and development micro-physical mechanism of iced-conductor in freezing rain, using the instrument of Parsivel laser droplet distribution to carry out the field observations of icing process under freezing rain in 2013 to 2014 winter in Nan-yue mountain of Hunan province.

Keywords-freezing rain; transmission line; droplet distribution; power system

I. INTRODUCTION

Severe icing can cause mechanical and electrical faults grid, which cause enormous economic losses. For example, the sustained bad weather (low temperature, freezing rain and snow) oc-curred in a large area of South China in 2008 caused serious influence to the safety of the grid, the direct property losses was about 10,450,000,000 RMB, and the reconstruction cost up to 39,000,000,000RMB. In order to reduce the influence of icing on power system, many domestic and foreign scholars started research work in the field of overhead transmission line icing. Jones studied the ice thickness in different return periods with freezing rain in America and Canada, providing a basis for the design of overhead transmission line [1]. Makkonen explored the rela-tionship between ice thickness of wire and the meteorological factors, and established several numerical models for icing calculation [2-3]. In China, the study on transmission line icing mainly focused on the effects of various parameters for transmission line icing, and on ice melting for AC/DC transmission lines [4-10]. However, analyses on icing particle spectrum feature are less, mainly concentrated in Guizhou and parts of Hubei Provinces [11, 12].

In this paper, the data of particle spectrum was obtained through field observation of freezing rain and ice in Hunan Nanyue Mountain in winter of 2013-2014, and relevant parameters (freezing rain intensity, the total rainfall, radar reflectivity, number and diameter of particles) were analyzed. The results have important significance to study the Microphysical Characteristics of icing in freezing rain in the mountain area.

II. PLACE AND DEVICE FOR THE OBSERVATION

The place of field Meteorological observation was selected in NanYue mountain in the city of Hengyang, Hunan province (112.6893 degrees E, 27.2974 degrees N, respectively 1266m above sea level), in the winter of 2013-2014. The type of disdrometer is OTT Parsivel2, made in Germany (shown in Figure 1). Using this device, we can observe various parameters we want, such as freezing rain drop spectrum, freezing rain intensity, the number of particles, the raindrops falling speed etc.. Droplet spectrum observation data including 32 speed grades and 32 particle scale grades, the particle number for each level will be recorded and outputted every 15 seconds.

III. ANALYSIS OF OBSERVATION RESULTS

The freezing rain icing process occurred eight times in Nanyue Mountain in the winter of 2013 to 2014. One of the eight freezing rain processes was occurred in December 15th 2013, 08:47-18:18, which was called process A. This paper analyzed the particle spectrum characteristics in process A.

Figure 2 to Figure 5 shows the observed time series of freezing rain intensity, accumulative rainfall, radar reflectivity and particle number for process A respectively. It can be seen from FIGURE I that, the maximum freezing rain intensity of process A was about 4-4.5mm/h, according to the three minutes of moving average curve, the main peak of freezing rain intensity was about 2.5mm/h, it was below 2mm/h in most of the time section.

© 2015. The authors - Published by Atlantis Press

FIGURE I. PARSIVEL LASER RAINDROP SPECTRUM INSTRUMENT.

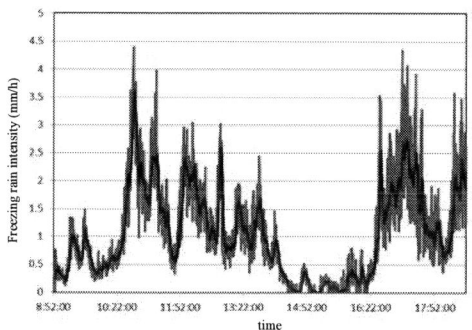

FIGURE II. TIME SERIES OF FREEZING RAIN INTENSITY (RED LINE) AND THE AVERAGE MOVING TIME SERIES (BLACK LINE) FOR PROCESS A.

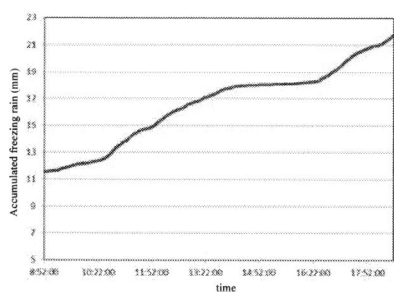

FIGURE III. ACCUMULATED RAINFALL TIME SERIES OF PROCESS A.

It can be seen from Figure 3 that, the accumulated rainfall was about 10mm in the time period of 08:47 to 18:18 during the freezing rain process. However, the accumulated rainfall in the time period of 08:00 to 20:00 was about 14.1mm by artificial observation, which was higher than the former. It is due to the observation error between different instruments at different observation periods.

It can be seen from Figure 4 that, in the freezing process of December 15th 2013, the basic reflectivity in Nanyue Mountain changed from weak to strong. The radar echo was bellow 20dBZ at 09:00 AM, and it changed between 20 to 25dBZ at 12:00 AM.

In the afternoon, the new cloud formatted, and moved north to Nanyue Mountain area, then the maximum radar reflectivity reached 30 to 35dBZ. In the picture of radar reflectivity calculated by raindrop spectrum, the radar reflectivity was below 20dBZ at 09:00 AM, and it changed between 20 to 25dBZ at 12:00 AM. At 18:00, the radar reflectivity was about 30dBZ, which was in good agreement with the result of Doppler radar observations. The main radar echo was below 30dBZ in process A, 20-25dBZ in the main precipitation period.

It can be seen from Figure 5 that, the maximum quantity of freezing rain particles in 15 seconds of the observation period achieved 450, the three minutes sliding average particle number had reached 350, the particle number distribution and the change of freezing rain intensity had the same trend. However, the precipitation particle quantity was less at 16:00, but the freezing rain intensity was still as much as the strength of 11:00, which means the freezing rain particle size having larger increase in the afternoon.

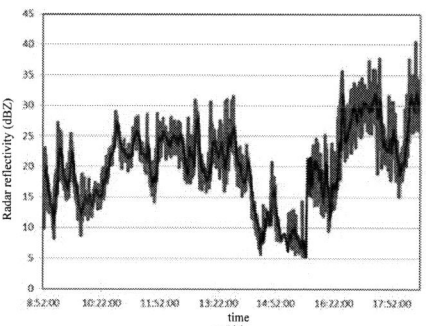

FIGURE IV. RADAR REFLECTIVITY TIME SERIES (RED DASHED LINE) AND RADAR REFLECTIVITY MOVING AVERAGE TIME SERIES (BLACK LINE) OF PROCESS A.

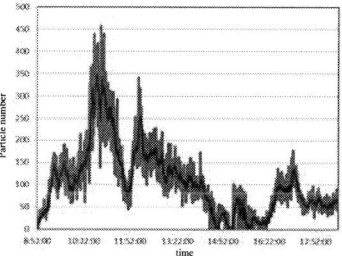

FIGURE V. PARTICLE NUMBER TIME SERIES (RED DASHED LINE) AND PARTICLE NUMBER MOVING AVERAGE TIME SERIES (BLACK LINE) OF PROCESS A.

In the different freezing processes, the relationship between diameter of freezing rain drop spectrum and the particle number was closer. It can be seen from Figure 6 that, during process A, the corresponding particle diameters were 0.5mm and 0.625mm in the high value area of particles number (more than 60000). The other freezing rain processes had a similar phenomenon. Therefore, in the high altitude mountain, freezing rain is composed of rain and drizzle, and the particle number of rain was more than that of drizzle.

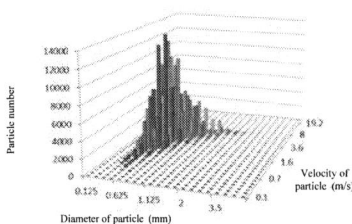

FIGURE VI. (A) DROPLET SPECTRUM DIAMETER, VELOCITY AND PARTICLE NUMBER DISTRIBUTION MAP OF PROCESS A.

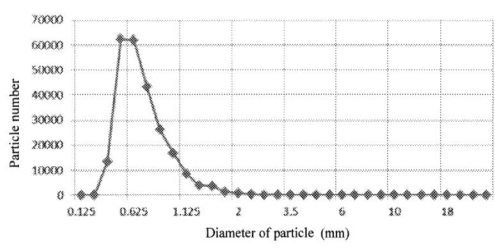

FIGURE VI. (B)RELATIONSHIP BETWEEN DROPLET SPECTRA DIAMETER AND PARTICLE NUMBER.

IV. SUMMARY

(1) The maximum range of radar reflectivity for freezing rain was 30 to 35dBZ, the radar echo in most of the freezing rain process was bellow 25dBZ, and the radar reflectivity calculated by spec-trometer was good agree with the radar observations.

(2) During freezing rain processes, the raindrop diameters were mainly concentrated in the drizzle level (<0.5mm) or above, most of them were about 0.625mm.The number of particles (whose diameter were larger than 0.5mm) is more than the number of drizzle particles. In the ic-ing process of high altitude mountain, the contribution for ice growth making by cooled raindrops (whose diameter is greater than or equal to 0.625mm) is greater than that of the drizzle (< 0.5mm).

REFERENCES

[1] Jones, K. F, Neal L. Ronald T. Extreme ice from freezing rain. New Hampshire: American Lifelines Alliance, 2004.

[2] Sundin E, Makkonen L. Ice loads on a lattice tower estimated by weather station data. Journal of applied meteorology, 1998, 37(5): 523-529.

[3] Makkonen l, Kari A. Climatic mapping of ice loads based on airport weather observatioins. Atmospheric Research, 1995, 36(3-4): 185-193

[4] Xingliang Jiang, Jun Ma, Shaohua Wang, et al. Transmission lines' ice accidents and analysis of the formative factors. Electric Power, 2005,38(11): 27-30.

[5] JiahaoXu, Wei Zhang, Xiaoning Huang, et al. Transmission line icing prediction model under micro-meterological conditions. Electric Power, 2014, 47(2): 58-63.

[5] Jiazheng Lu, Hongxian Zhang, JiwenPeng, et al. Calculation of Hunan power grid icing recurrence interval based on extreme-value type I probability distribution model. High Voltage Engineering, 2012, 38(2): 464-468.

[6] GuoyiNie. Discussion about the icing load on conductors of overhead transmission lines. Electric power construction, 1992, 13(7): 46-54.

[7] Guanri Tan, Bing Wang, Jian Qin, et al. Calculation of the extreme ice thickness for conductors. Electric Technique, 1983(2):8-14.

[8] Jiazhen Lu, Hongxian Zhang, JiwenPeng, et al. Calculation of Hunan power grid icing recurrence interval based on Pearson III type probability distribution. Transactions of China electrotechnical society, 2013, 28(1):80-86.

[9] ZhichongGuo, Jiazheng Lu, Zhangqing Yu, et al. Analysis on induced voltage of deicer for UHV double-circuit transmission line on the same tower. Power system technology, 2013,37(11):3015-3021.

[10] Ran Jia, ShengjieNiu, Rui Li. Observational study on microphysical characteristics of wire icing in west Hubei, ScientiaMeteorologicaSnica, 2012, 30(4): 481-486.

[11] ShengjieNiu, Yue Zhou, Ran Jia, et al. Preliminary study of the microphysics of ice accretion on wires: Observations and simulations. Science China Earth Science, 2011, 41(12): 1812-1821.

International Conference on Power Electronics and Energy Engineering (PEEE 2015)

Changes in Content and Component of Purple Corn (Zea Mays L.) Anthocyanin during the Extraction and Preparation

D. Wang, Y. Ma, P.P. Liu, C. Zhang, X.Y. Zhao

Beijing Vegetable Research Center, Beijing Academy of Agriculture and Forestry Science
Ministry of Agriculture
Key Laboratory of Biology and Genetic Improvement of Horticultural Crops (North China)
Beijing, P.R. China
Key Laboratory of Urban Agriculture (North)
Ministry of Agriculture
Beijing Key Laboratory of agricultural products of fruits and vegetables preservation and processing
Beijing, P.R. China

Abstract—**The contents of the total anthocyanin and color were detected in the purple corn at different process. The anthocyanin content at concentrating and spry drying process was decreased by 10.93%, 21.46% respectively compared to the extraction. The redness (a*) decreased significantly during the preparation process. Furthermore, the components of acylated group will decrease, whereas the unacylated group will increase during the extraction and preparation process.**

Keywords- purple corn; anthocyanin; color

I. INTRODUCTION

Anthocyanin is a novel natural pigment that is responsible for the blue, purple, orange and red color in plant. In recent years, many studies have reported that anthocyanins have strong antioxidants, which could lower the size of adipocytes and anticarcinogenic properties [1,2,3,4]. Meanwhile, it can protect the DNA from being damaged by free radicals [5].

Purple corn is an important source of anthocyanin. The cob, bract and seed of it were generally used to extract the anthocyanins. The major anthocyanins have been characterized previously. Pelargonidin-3-glucoside, peonidin-3-glucoside, cyaniding-3-glucoside, cyaniding-3-(6'-malonylglucoside) were found to be the major compositions [5]. However, the change of the individual composition during the preparation process has not been researched in detail. In this study, the HPLC-ESI-MS spectrometry was used to identify the change of these compositions of purple corn anthocyanin at extracting, concentrating and spray drying process. Meanwhile, the color and the total anthocyanin were also studied.

II. MATERIALS AND METHODS

A. Materials

Purple corn bract was supplied by Beijing Vegetable Research Center, Beijing Academy of Agriculture and Forestry Science.

B. Extraction of Anthocyanin

Bract was macerated with 60% ethanol. The pH was adjusted to 3 by adding 1 mmol/L HCL. The samples were extracted at temperature 50℃ for 4 h. The product to solution was kept at 1:65 (W/V). The samples were obtained by filtering through the filtered paper. The extraction solvent were then concentrated at 8.5 kPa, 35 r/min, 50℃ for 3 h. The remaining aqueous extract was brought to a spray drying equipment (BÜCHI Mini Spray Dryer B-290), which the drying temperature was 160 ℃.

C. Determination of Anthocyanin

The total anthocyanin content in purple corn was determined using the pH differential method. The samples were diluted in buffers at pH 1.0 (KCL-HCL) and pH 4.5 (CH₃COONa-HCL), respectively. The resulting mixture was allowed to stand for 10 min at room temperature, followed measuring on the UV-Vis spectrophotometer at 512 nm and 700 nm. Absorbance (A) was calculated as follows:

$$A = (A_{\lambda_{max}} - A_{700})_{pH1.0} - (A_{\lambda_{max}} - A_{700})_{pH4.5} \qquad (1)$$

where $(A_{\lambda_{max}} - A_{700})_{pH1.0}$: the absorbance of the sample at pH1.0;

The anthocyanin concentration was calculated using the following equation:

$$C = (A \times MW \times DF \times 1000)/(\varepsilon \times L) \qquad (2)$$

where C is the mass concentration (g/L), MW is the molecular weight of cyanidin-3-O-glucoside (449.2 g/mol), DF is the dilution factor, ε is the molar extinction coefficient of cyanidin-3-glucoside (26900 L cm⁻¹mol⁻¹), *and L* is length of the optical path (1cm).

© 2015. The authors - Published by Atlantis Press

D. Color Analysis

Color values of the extracted anthocyanin were measured using a Hunter colorimeter (Spectrophoto METER CM-3700d). Colors of the samples were indicated by CIELAB. L^*(brightness), a^*(redness), b^*(yellowness) were determined using an illumination (illuminant D65, 10°observer). C^* and $h°$ were calculated from L^*, a^* and b^* value.

$$C^* = (a^{*2} + b^{*2})^{1/2} \qquad (3)$$

$$h° = \arctan(b^*/a^*) \qquad (4)$$

E. Chromatographic Conditions

The HPLC analysis was performed on an Agilent 1200 series. The separation was achieved on a Waters XBridge C_{18} column (4.6×250 mm, 5 μm) using formic acid/water (5%, v/v) and acetonitrile at thermostat 30 °C. An aliquot of 20 μL was injected onto the program and the flow rate is 0.8 mL/min. The gradient program was set as follows: 0-15 min, 5%-10% B; 15-30 min, 10%-13% B; 30-34 min, 13%-20% B; 34-40 min, 20%-25% B; 40-43 min, 25%-100% B. A diode-array detector was monitored at 520 nm.

The HPLC was coupled to a 6000 ion trap with an ESI source in a positive ionization mode. The data was dealt with an Agilent Chemstation Rev.A.09.01 software (Agilent, Palo Alto, CA). The MS parameters were as follows: MS scanning rage from 100-1500; drying temperature 350 °C; nitrogen flow rate 12 L/min; capillary current 34 nA; nebulizer pressure 0.31 MPa.

F. Statistical Analysis

The figure was treated by the software origin 8.0. And all the experiments were performed three times.

III. RESULTS AND DISCUSSION

A. Changes in the Content of Anthocyanins in Purple Corn during Processing

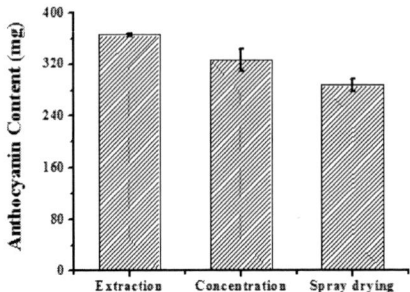

FIGURE I. THE CHANGE OF THE ANTHOCYANIN CONTENT DURING THE PROCESSING.

Purple corn is a rich source of anthocyanins. In this study, the impact of process on the changes of the total anthocyanins was evaluated by comparing the purple corn with extraction, concentration and spry drying. As shown in Figure 1, the anthocyanin content decreased with the preparation process. The anthocyanin content at concentration and spry drying was decreased 10.93%, 21.46% respectively compared to the extraction.

B. Changes in Component of Purple Corn

Table 1 shows the anthocyanin composition of the purple corn and the changes of their relative amounts that calculated by their peak area during the process. As Table 1 displayed, the major anthocyanin identified were cyaniding-type, pelargonidin-type and peonidin-type. What is more, all of them were acylated in the glucose moiety. Furthermore, it can be seen that the acylated group were decreased, whereas the unacylated group increased during the extraction and preparation processing.

TABLE I . CHANGES IN COMPONENT OF PURPLE CORN DURING THE PROCESSING.

Compound identity	Percentage of anthocyanin calculated from the peak area (%)		
	Extraction	Concentration	Spry drying
Cyanidin-3-O-glucoside	10.289	11.211	15.011
Pelargonidin-3-O-glucoside	1.964	1.317	2.705
Peonidin-3-O-glucoside	6.184	6.55	7.828
Isocyanidin-3-(6"-malonylglucoside)	25.182	26.763	34.898
Pelargonidin-3-(6"-malonylglucoside)	7.903	7.607	8.181
Cyanidin-3-2malonlyglucoside	40.061	38.549	26.234
Pelargonidin-3-2malonlyglucoside	4.984	4.568	3.131
Peonidin-3-2malonlyglucoside	3.433	3.435	2.012
Total (%)	100	100	100

C. Color Analysis

The effect of extraction, concentration and spry drying on the visual appearance of athocyanin was shown in Figure 2. As can be seen, redness (a*) decreased significantly with the preparation process. The a* value denotes the red color of the product. The color of the athocyanin showed the similar trend with athocyanin content, which indicated that the changes of the red color is greatly related to the athocyanin content [6]. Furthermore, some studies had reported that the acylated group attributed to the color intensity as well [7]. The more glycoside substituents, the redder of the anthocyanins.

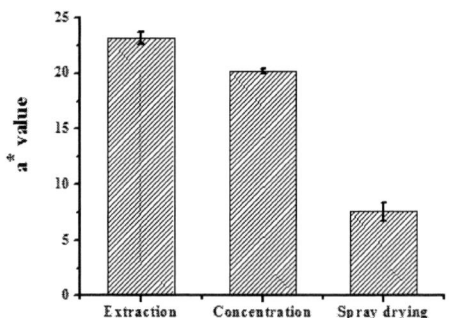

FIGURE II. THE COLOR CHANGE OF THE ANTHOCYANIN DURING THE PROCESSING.

IV. CONCLUSIONS

The results suggested that the athocyanin was not stable. The total athocyanin content decreased with the extraction and preparation process. The athocyanin contributed to the color of the product. In addition, the anthocyanin composition of the purple corn varied with the process.

ACKNOWLEDGEMENTS

This work was supported by Beijing Nova Program (Z131105000413023) and Scientific and Technological Innovation Ability Foundation of Beijing Academy of Agriculture and Forestry Science (KJCX20140204).

REFERENCES

[1] U. Szymanowska,U. Złotek, M. Karaś and B. Baraniak, Anti-inflammatory and antioxidative activity of anthocyanins from purple basil leaves induced by selected abiotic elicitors, Food Chemistry. 172 (2015) 71–77.

[2] Juadjur, C. Mohn, M. Schantz, M. Baum, P. Winterhalter, E. Richling, Fractionation of an anthocyanin-rich bilberry extract and in vitro antioxidative activity testing, Food Chemistry 167 (2015) 418–424.

[3] M.K. Jin, S.C. Lian, K.G. Ngoh, F.C. Tet, R. Brouillard, Analysis and biological activities of anthocyanins, Phytochemistry 64 (2003) 923–933.

[4] M.B. Spela, Z. Lovro, P. Tomaz, V. Andreja, P.U. Natasa, A. Veronika, T. Federica, P. Sabina, Bilberry and blueberry anthocyanins act as powerful intracellular antioxidants in mammalian cells, Food Chemistry 134 (2012) 1878–1884.

[5] D. Wang, X.R. Wang, C. Zhang, Y. Ma, X.Y. Zhao, Calf Thymus DNA-Binding Ability Study of Anthocyanins from Purple Sweet Potatoes (Ipomoea batatas L.), J. Agric. Food Chem, 59 (2011) 7405–7409.

[6] K. Şelale, E.A. Erçelebi, Thermal degradation kinetics of anthocyanins and visual colour of Urmu mulberry (Morus nigra L.), Journal of Food Engineering 116 (2013) 541–547.

[7] Y. Zhang, E. Butelli, C. Martin, Yang Zhang, Eugenio Butelli and Cathie Martin, Current Opinion in Plant Biology, 19 (2014) 81–90.

International Conference on Power Electronics and Energy Engineering (PEEE 2015)

Study and Practice on Teaching System Reform of Engineering Graphics Based on Training Engineering Consciousness

D.T. Xu
School of Mechanical Engineering & Automation
University of Science and Technology
Anshan, Liaoning, China

Y.J. Feng
School of Mechanical Engineering & Automation
University of Science and Technology
Anshan, Liaoning, China

J.L. Shi
School of Mechanical Engineering & Automation
University of Science and Technology
Liaoning, Anshan, China

Abstract— It is very important and meaningful to improve college students' engineering consciousness and project practice ability in engineering graphics teaching, so that excellent engineers, who adapt to the development of economy and society, can be cultivated. Based on the research and analysis situation of the home and abroad in engineering graphics teaching, many methods to cultivate students' engineering consciousness are proposed, concrete measures include: constructing the teaching method system of case-based teaching technique; building the content system which begins with the three-dimensional configuration design as a starting point, lays equal stress on a variety of expression modes, emphasizes the cultivation of ability to freehand drawing and teaching of engineering application, and pays attention to practical teaching.

Keywords-engineering graphics; case teaching; teaching reform; engineering consciousness

I. INTRODUCTION

Engineering graphics is technology basic course which has strong engineering consciousness and the practical requirements, and it is the first important compulsory course for engineering students [1]. The problems which the engineering graphics teacher must be meet, think about and solve are: how to improve the students' engineering consciousness, how to bring up students' engineering ability in order to develop students to adapt to social and economic development, with the development of manufacturing industry and science and technology [2].

Based on the analysis of American, Japanese engineering graphics teaching contents, teaching methods, and combined with Chinese teaching practice of engineering graphics, this article structure engineering graphics' curriculum system including to teaching content system, teaching method system

based on training university students' engineering consciousness.

II. THE SURVEY OF THE TEACHING ACTUALITIES OF ENGINEERING GRAPHICS

Some American engineering graphics teaching revolves around product design to bring up students' engineering expressive ability, and combining with engineering practice and the requirements of enterprises in teaching. Their teaching are not only pay attention to the cultivation of students' researching thinking, but also emphasize to develop students' practical ability and expression skills through combining with the practice teaching and classroom teaching [4]. For example, California University pay attention to the cultivation of students' ability to solve practical engineering problems and manipulative ability, combined with practical teaching and classroom teaching [5].

In Japan, graphics curriculum emphases are different for different professional students, in order to combine with the subsequent course. In addition, Japanese graphics teaching pays more attention to the combination of enterprise requirement. Teaching contents are integrated with product manufacturing and designing, and emphasize to cultivate students' creative thinking and practical ability.

In China, owing to long-term influence of examination oriented education, it is difficult to realize the transformation from theory education to engineering education when students entered the university stage. The students' engineering consciousness is eroded so that it is very difficult to combine knowledge and engineering practice, so our students' design is unrealistic.

In recent years, some domestic universities carried out some exploration and research of engineering graphics teaching reform. The Supervision Board for teaching engineering

© 2015. The authors - Published by Atlantis Press

graphics education of Ministry of education revised "Basic requirements for Engineering Graphics Teaching in ordinary colleges and universities", it introduced the ideas about industrial product design, it proposed engineering graphics teaching should begin with training of engineering culture quality and setting up innovation design thinking, should base on carrying out the new standard of engineering drawing, configuration expression design, and configuration design, should take basic theory, basic knowledge, and basic skills of graphical representation as the basis to cultivate three kinds of ability to the drafting, gauge drawing and computer drawing.

III. CONSTRUCTION OF ENGINEERING GRAPHICS TEACHING SYSTEM BASED ON CULTIVATING STUDENTS' THE ENGINEERING CONSCIOUSNESS

A. Construct "Case-Based" Classroom Teaching Method System

The "case" teaching method can help students to understand the relevant knowledge and principle about teaching object by case analysis, so as to set up the engineering concept and enhance the engineering practice consciousness.

First, teaching case selection: In the whole course of teaching engineering graphics, choose cases following three aspects: First, cases should serve for theoretical knowledge. The case selected should have effect of guidance and inspiration. Trigger students interest easily, hence stimulate students' enthusiasm in exploration and innovation; second, the case must be typical and representative. Usually display the most common characteristics, meaning in the same kind of things, through the study of one case, we can explore the inherent law of the same kind cases; third, case should be practically in engineering and close to the production practices. Case selection is better from the project site, it can enhance students engineering consciousness and make up for students' lack of engineering background, greatly reduce the gap between teaching situations and practical engineering situations.

Second, design the teaching process: "Case-based" classroom teaching is usually describe the subject and teaching goal of this knowledge at start, and leads to the basic concepts, characters of teaching content and explore problem-solving ideas by using teaching cases, and finally solve the problem raised by the case, as shown Figure 1.

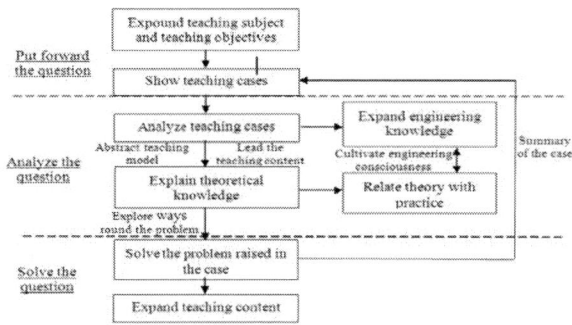

FIGURE I. TEACHING PROCESS DESIGN

The whole teaching process is based on "case" as the clue, using the process of cognize things that people are most familiar which is "ask questions, analyze and solve problems" to complete the teaching process of the whole teaching knowledge points. First, raise questions through teaching cases. Cases can take full advantage of illustrated multimedia, present engineering practice scenes with a rich three-dimensional multimedia, help students understand some principles of the machine, analyze background of teaching cases, expand students fundamental knowledge of engineering. Next, analyzing the basic concepts, nature of the teaching content which be led by the case, and exploring the solutions of raised questions. Guide students to think actively about how to solve the problems raised in the teaching case, so that students know where to use what they have learned; and finally, use of theoretical knowledge to solve practical problems raised in the teaching case, improve students' ability to solve practical problems. At the same time, it requires the ability to draw inferences, and to expand the teaching content.

The last, an example of teaching process: The teaching process about "intersection of a plane and a cylinder" can begin with the "ball valves" teaching case. When teaching case is showed, all the parts of ball valve can be dismantled, and its internal structure can be show through the teaching video. By the analysis of working principle, the structure and function of the valve stem can be clearly grasp, then the stem can be simplified to a cylindrical tenon model. At last, the projection of intersection of a plane and a cylinder can easily be draw. The teaching process is as shown in   Figure 2.

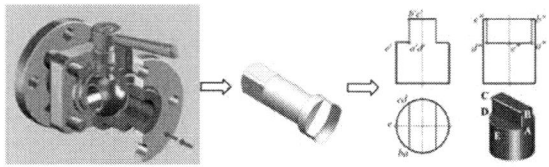

FIGURE II. TEACHING CASE——"BALL VALVES"

329

B. The Construction of Teaching Content System Based on Cultivating Students' the Engineering Consciousness

First, build teaching content system along the main line of designing three-dimensional configuration: The product design process is often around the subject of product design and manufacturing, start from the three-dimensional configuration and expression to a two-dimensional view, and highlight a variety of visual expressions in the integrated application in the process of design and manufacture. Based on studying object configuration method, students are required to master a variety of expression methods of three-dimensional shape, to meet the requirements of product design. Combined with requirements of designing and processing product and based on three-dimensional expression, teaching a two-dimensional view is started, and from an engineering point, the basic view, cross-sectional view and direction view is introduced.

Second, integration of a variety of drawing methods, emphasizing cultivating students' ability to freehand drawing and the part of teaching engineering application: The three parts of descriptive geometry, engineering drawing, and computer drawing are integrated into the whole teaching process in order to achieve comprehensive training three graphics capabilities of manual sketches, instruments drawing, and computer drawing. In the early stage of product configuration design, freehand drawing three-dimensional and two-dimensional sketch could express the design intent in the most directive and convenient way, more importantly, which can play a key role in quickly capturing design ideas, inspiring design idea in the design process. Therefore, the freehand drawing ability is an important capability which students should have. Comprehensive applications of expression method, technical requirements, standard parts and commonly-used parts, detail drawings and assembly drawings are applied in actual production and reflect the engineering awareness, so teaching about these parts should be strengthened. Cut off part of the contents of descriptive geometry, as to those problems which can be solved easily by graphic and the difficult and abstract part that automatically generated intersection line when the three-dimensional is modeled could be deleted, and recommends drawing intersecting lines in a simplified way. It doesn't affect manufacturing and could save drawing time; reduce the difficulty of drawing so that students accepted easier.

At last, strengthening practice: Practical part of drawing integrates the drawing, mechanical practices with survey and draw together. In teaching organization, understanding modern mechanical design and manufacturing technology, applying basics mechanical drawing knowledge, learning parts survey and draw, and taking part in redesigning typical products are taken as important teaching part, built practical skills and project's awareness of students. In practically training of survey and draw, from measuring the simplest assemble model to shaft-sleeve group, to the parts of the case-housing group and fork-frame group, and learn the component structure, process requirements and the functions of parts in the machine. Through drawing sketches, then understanding the effects and requirements of the process structure, and lay the foundation for subsequent courses of study, make students to possess basic capability of engineering application. Strengthen the practice content of computer graphics. Use computer graphics software to complete the entity modeling, three view drawing, section view drawing, detail drawing, assembly drawing and other special training.

IV. SUMMARY

Nowadays, with the rapid development of the manufacturing industry and science and technology, and training of excellent engineers, means not only updating educational ideas and concepts, but also the need to reform the teaching content system, teaching methods system. How to combine the specific circumstances of different universities for different engineering professional features, further deepen classroom teaching reform of engineering graphics, it needs for further research and try in the future.

ACKNOWLEDGEMENTS

This work is financially supported by the young teacher' teaching reform special project of University of Science and Technology Liaoning (qnjj-2013-09); Teaching reform and research project of Liaoning province the common college undergraduate (UPRP20140291).

REFERENCES

[1] YANG Wen-tong, LI Yang et al. Study on the Whole Procedure Education System of Engineering Graphics for the Practice and Innovation [J]. Journal of Engineering Graphics, 2006, 27(5):137-141.

[2] JIAO Yong-he, ZHANG Tong, CHEN Jun, et al. The Investigation and Research on the Current Education Situation in Universities in China [J]. Journal of Engineering Graphics, 2004, 25 (4):125-130.

[3] FENG Juan. Analysis on Engineering Graphics Education in U.S. Universities [J]. Journal of Engineering Graphics, 2008, 29 (3):139-144.

[4] Graham M, Slocum A, Moreno S R. Teaching high school students and college freshmen product development by deterministic design with PREP [J]. Journal of Mechanical Design. 2007, 129(7): 677-681.

[5] MAO Wen-wu. Studying on the Teaching of Engineering Graphics in University of California, San Diego [J]. Journal of Engineering Graphics, 2007, 28(5): 172-177.

AUTHOR INDEX

Bai, W. L. 160
Cai, Y. H. 215
Cai, Y. Z. 27
Cai, Z. Y. 27, 252, 285
Cao, H. C. 36
Cao, J. D. 118
Chen, C. W. 243
Chen, H. 294
Chen, J. 285
Chen, P. Y. 240
Chen, S. 14
Chen, S. J. 211
Chen, S. M. 115
Chen, T. 206
Chen, W. D. 245
Chen, X. 122
Chen, X. M. 157
Chen, Y. Q. 86
Chen, Z. G. 79
Cheng, X. M. 86
Chu, C. 225
Cui, W. H. 163
Deng, X. Z. 30
Deng, Y. 215
Ding, J. S. 222
Ding, S. F. 109, 109
Ding, S. X. 167
Ding, X. M. 143
Dong, L. J. 266
Dong, N. 74
Dong, X. X. 277, 281
Du, F. 89
Du, S. D. 289
Fan, H. C. 86
Fan, P. 93
Fan, T. L. 306
Feng, F. 147
Feng, L. 61
Feng, Y. J. 328
Fu, C. 163
Fu, X. C. 206
Gan, H. G. 167
Gao, C. 134
Gao, G. W. 12
Gao, Q. 93
Gong, K. Q. 163
Gu, F. 177
Guan, Z. W. 89
Gui, D. D. 277, 281
Guo, H. 69, 129
Guo, J. M. 19
Guo, J. Q. 126
Guo, Q. 12
Guo, X. B. 112
Guo, Y. H. 185

Guo, Z. M. 231
Hall, P. 177
Han, Q. 228
Hou, Z. 122
Hu, C. 206
Hu, H. L. 240
Hu, K. T. 51
Hu, R. F. 93, 243
Hu, Y. 298
Hua, J. S. 198
Huang, C. X. 19
Huang, F. 19
Huang, H. T. 61, 134
Huang, J. J. 157, 167
Huang, L. 101
Huang, L. J. 83
Huang, Q. Q. 109, 273
Huang, S. B. 219, 294
Huang, X. B. 74
Huang, X. X. 195
Huo, B. R. 174
Ji, S. C. 122
Jia, J. G. 126
Jiang, W. 302
Jin, P. 228
Jin, X. 97
Jing, L. S. 206
Kou, B. Q. 36
Kou, M. 83
Lai, C. 170
Lang, D. P. 294
Leng, J. 112
Li, B. 55
Li, C. M. 101
Li, C. Z. 86
Li, D. J. 237
Li, F. 225
Li, F. T. 41
Li, G. H. 51
Li, H. E. 277, 281
Li, H. J. 19, 122
Li, J. 122, 255
Li, J. D. 30
Li, J. J. 51
Li, L. N. 302
Li, M. S. 298
Li, Q. X. 109
Li, S. C. 277, 281
Li, W. L. 36
Li, X. 273
Li, X. B. 273
Li, X. L. 23
Li, X. Y. 322
Li, Y. 147, 160

AUTHOR INDEX

Li, Y. P. 1
Li, Y. R. 27
Li, Z. X. 3
Liang, B. 191
Liang, C. H. 61
Liang, S. 1
Liao, Z. L. 93
Lin, T. 231, 234
Liu, B. 322
Liu, C. H. 89
Liu, C. L. 157
Liu, D. J. 55
Liu, D. W. 126
Liu, H. 30
Liu, H. M. 261
Liu, H. W. 71
Liu, J. 245
Liu, L. W. 252
Liu, P. P. 325
Liu, Q. 74
Liu, Q. D. 147
Liu, Q. S. 23
Liu, T. 12
Liu, X. 258
Liu, X. F. 160
Liu, X. G. 58
Liu, Y. 202
Liu, Y. M. 7
Liu, Y. Y. 198
Liu, Z. M. 58
Liu, Z. Y. 3
Lu, S. R. 195
Luo, D. M. 65
Luo, J. 225
Luo, M. H. 118
Lv, Y. M. 58
Ma, W. Q. 7
Ma, Y. 143, 325
Meng, H. 69
Meng, T. 219
Miles, N. J. 177
Min, Y. 185
Mo, H. D. 195
Pan, J. 134
Pan, R. 202
Peng, C. L. 289
Peng, J. 285
Qin, Y. 237
Quan, H. 48
Ran, Y. 65
Rao, L. 139
Ren, B. K. 51
Ren, L. 30
Shao, H. P. 231, 234
Shen, L. S. 294

Shen, X. Z. 211
Shen, Y. 118
Sheng, L. Y. 170
Shi, G. F. 222
Shi, G. Q. 222
Shi, J. 30
Shi, J. L. 328
Shi, T. Z. 115
Sun, H. J. 19
Sun, L. L. 185, 188, 215
Tai, L. J. 243
Tang, J. 147
Tang, L. 163
Tang, M. Z. 261
Tang, X. 14
Tang, Y. J. 30
Tian, H. Y. 41
Tian, X. T. 83
Wan, G. F. 48
Wan, L. 48
Wan, Y. S. 89
Wang, C. L. 240
Wang, D. 325
Wang, D. D. 219
Wang, D. F. 115
Wang, G. Q. 55
Wang, K. 198
Wang, L. L. 202
Wang, P. 245
Wang, Q. D. 23
Wang, Q. J. 191
Wang, S. L. 147
Wang, S. Y. 225, 270
Wang, X. J. 101
Wang, Y. 157, 277, 281
Wang, Y. C. 23, 55
Wang, Y. Q. 74
Wang, Z. Q. 195
Wang, Z. Y. 83
Wei, C. 195
Wen, C. X. 3
Wu, A. B. 51
Wu, J. D. 71
Wu, J. J. 134
Wu, M. 58
Wu, S. J. 231
Wu, Z. X. 225
Xi, T. F. 170
Xia, J. J. 206
Xiao, T. 225
Xiao, W. 215
Xin, J. B. 1
Xin, N. 69
Xing, T. W. 289
Xiong, Y. 157

AUTHOR INDEX

Xu, D. T. 328
Xu, W. 1
Xu, Y. 12, 30
Yan, H. 245
Yang, C. 195
Yang, C. B. 19, 122
Yang, G. N. 174
Yang, J. 225
Yang, J. L. 322
Yang, J. T. 237
Yang, X. J. 310, 313, 316, 319
Yao, S. Y. 219
Yao, W. 198
Yin, W. 306
You, X. L. 198
Yu, F. 134
Yu, W. 65
Yu, X. T. 112
Yu, Y. 289
Yu, Z. L. 71
Yuan, D. 261
Yuan, G. L. 255, 258, 310, 313, 316, 319
Yuan, X. D. 255, 258, 310, 313, 316, 319
Zang, L. M. 195
Zeng, O. 225
Zhan, T. F. 252
Zhang, B. R. 41
Zhang, C. 325
Zhang, C. M. 219
Zhang, F. A. 195
Zhang, G. B. 65
Zhang, H. M. 106
Zhang, J. 58, 153
Zhang, L. 69, 143, 185, 188, 234
Zhang, L. C. 322
Zhang, L. X. 106
Zhang, P. N. 86
Zhang, Q. 79
Zhang, S. F. 167
Zhang, T. 294
Zhang, W. 7, 79
Zhang, W. Q. 14
Zhang, X. 249, 298
Zhang, X. F. 118
Zhang, X. W. 191
Zhang, Y. H. 215
Zhang, Z. J. 97
Zhang, Z. Q. 97
Zhao, J. J. 153
Zhao, X. Y. 325
Zhao, Z. 86
Zheng, H. 234
Zheng, M. X. 174
Zhong, X. P. 228
Zhou, M. R. 151

Zhou, T. M. 79
Zhou, X. 191
Zhou, Y. 58, 289
Zhu, H. J. 139
Zhu, K. J. 322
Zhu, X. X. 270
Zhu, Z. M. 302
Zuo, R. L. 153